T0213403

HARVARD SEMITIC SERIES
VOLUME VI

CRESCAS' CRITIQUE OF ARISTOTLE

CRESCAS' CRITIQUE OF ARISTOTLE

PROBLEMS OF ARISTOTLE'S *PHYSICS* IN JEWISH AND ARABIC PHILOSOPHY

BY

HARRY AUSTRYN WOLFSON

NATHAN LITTAUER PROFESSOR OF
HEBREW LITERATURE AND PHILOSOPHY, EMERITUS,
HARVARD UNIVERSITY

CAMBRIDGE

HARVARD UNIVERSITY PRESS

PRINTED IN THE UNITED STATES OF AMERICA

PREFACE

MEDIAEVAL philosophy is no longer considered as a barren interval between ancient and modern philosophy. Nor is it any longer identified with works written solely in Latin. Scholarship recognizes it more and more as a formative period in the history of philosophy the records of which are to be found in a threefold literature—Arabic, Hebrew and Latin. In certain respects, the delineation and treatment of the history of philosophy should follow the same lines as the delineation and treatment of the political and social history of Europe. The closing of the philosophic schools at Athens early in the sixth century is analogous in its effect to the fall of Rome toward the end of the fifth century. Like the latter, it brought a dying past to its end, and prepared the way for a shifting of scene in a phase of history. The successive translations of Greek treatises into Syriac, Arabic, Hebrew and Latin correspond, in philosophy, to the spread of the diverse elements of Roman civilization with the successions of tribal wanderings, of invasions, and of conversions. Both accomplished similar results, transforming something antiquated and moribund into something new, with life in it. By the same token, just as one cannot treat of the new life that appeared in Europe during the Middle Ages as merely the result of the individual exploits of heroes, or of the eloquence of preachers, or of the inventive fancy of courtiers, so one cannot treat of the development of mediaeval philosophic thought as a mere interplay of abstract concepts. There is an earthly basis to the development of philosophic problems in the Middle Ages—and that is language and text. The present work is an attempt to trace the history of certain problems of philosophy by means of philological and textual studies.

In form this work is a study of certain portions of Ḥasdai Crescas' *Or Adonai* ("The Light of the Lord"). In substance it is a historical and critical investigation of the main problems of Aristotle's *Physics* and *De Caelo*. Its material, largely unpublished, is drawn from the general field of Jewish philosophy and from related works in Arabic philosophy, such as the writings of Avicenna and Algazali, and particularly the commentaries of Averroes on Aristotle. The scope of this work, confined as it is to a closely interdependent group of writings, did not call for citations from works outside the field of Greek, Arabic and Jewish philosophy. Yet the material is such that the discussion of the history of the various problems will furnish a background for corresponding discussions of the same problems in scholastic philosophy. The notes, which form the greater part of the work, are detachable from the text and can be used in connection with similar texts in other works. Many of the notes exceed the bounds of mere explanatory comments, being in fact extended investigations of the development of certain philosophic concepts by means of a study of the interpretation and criticism to which Aristotle's writings were subjected in two forms of mediaeval philosophic literature—the Arabic and the Hebrew.

Ḥasdai Crescas, whose work is the subject of the special investigation, was a true representative of the interpenetration of the Arabic and Hebrew philosophic traditions. Born in Barcelona in 1340, he died in Saragossa in 1410. He flourished, it will be seen, two centuries after Maimonides (1135-1204), who was the last of that line of Jewish philosophers, beginning with Saadia (882-942), whose works were written in Arabic for Arabic speaking Jews. During these two intervening centuries the centre of Jewish philosophic activity had shifted to non-Arabic speaking countries—to Christian Spain, to Southern France and to Italy—where the sole literary language of the Jews was Hebrew. In these new centres, the entire philosophic literature written in Arabic by Jews as well as almost everything

of general philosophic interest written by Moslems was translated into Hebrew, and thereby Hebrew literature became also
the repository of the whole Aristotelian heritage of Greek
philosophy. Acquaintance with the sources of philosophy
acquired by means of these translations stimulated the production of an original philosophic literature in Hebrew, rich
both in content and in volume. It also gave rise to a new
attitude toward philosophy, an attitude of independence, of
research and of criticism, which, among those who continued to
be opposed to philosophy, manifested itself in a change in the
temper of their opposition, while among those who were aligned
on the side of philosophy, it took the form of incisive, searching
studies of older texts and problems. Of the vast learning so
attained by fourteenth century Jewish scholars and also of the
critical attitude which inspired their studies Crescas is the
fruition. In his work are mirrored the achievements of five
centuries of philosophic activity among Moslems and Jews,
and in his method of inquiry is reflected the originality and
the independence of mind which characterize the Jewish philosophic writings of his time—an originality and independence
which is yet to be recognized. Crescas' method has been
described elsewhere in this work (pp. 24-29) as the hypothetico-
deductive method of Talmudic reasoning, usually called pilpul,
which is in reality the application of the scientific procedure to
the study of texts. Applied by Crescas to the study of the texts
of others, this method is here applied to the text of his *Or
Adonai.*

The *Or Adonai* is divided into four Books (*ma'amarim*), the
first three of which are subdivided into Parts (*kelalim*), or, as
the Latin translators from the Hebrew would more accurately
call them, *summulae*, and these are again subdivided into
Chapters (*peraḳim*). The first twenty-five chapters of Part I
of Book I are written in the form of proofs of the twenty-five
propositions in which Maimonides summed up the main prin-

ciples of Aristotle's philosophy. The first twenty chapters of
Part II of Book I are written in the form of a criticism of twenty
out of the twenty-five propositions. The present work deals
with these two sets of chapters, with the proofs and the criticisms.
Together they compose about one sixth of the entire work.
A separate study of Part III of Book I and of the remaining
chapters of Parts I and II will be published shortly under the
title *Crescas on the Existence and Attributes of God.* In reprinting
the text I have changed somewhat its original order by placing
the criticism of each proposition immediately after its respective
proof. The text is edited on the basis of the first edition and of
eleven manuscripts; it is accompanied by an English translation
and is followed by a commentary in the form of notes on the
translation. There is also an Introduction, which is divided
into six chapters. Chapter I discusses literary and historical
problems. Chapters II to V contain a systematic presentation
of the main problems dealt with in the text and the notes.
Chapter VI interprets some of the larger aspects of Crescas'
philosophy and endeavors to appraise him as one of the first
to forecast that which ever since the sixteenth century has been
known as the new conception of the universe. Translation,
commentary and introduction are interdependent and mutually
complementary.

The study of a text is always an adventure, the adventure of
prying into the unknown recesses of the mind of another. There
is sleuthing in scholarship as there is in crime, and it is as full
of mystery, danger, intrigue, suspense and thrills—if only the
story were told. In a work of this kind, however, the story is
not the thing. What one is after is the information it uncovers.
Accordingly, no attempt has been made to recount the processes of the search. Only the results arrived at are set down,
and the corroborative data are so marshalled as to let them
speak for themselves and convince the reader by the obviousness
of the contention.

A considerable part of this work—the study of the first proposition dealing with infinity, including text, translation, notes and introduction—was completed in 1915. Three years later, in 1918, the entire work was brought to a conclusion and the part on infinity thoroughly revised. When in the fall of 1927, through the liberality of Mr. Lucius N. Littauer, means were provided for the publication of the work, the manuscript was again gone over, to prepare it finally for the press. In addition, English translations were made of all the Hebrew passages quoted in the notes, and, wherever necessary, references to Aristotle were filled out with passages quoted from available English translations of his works. This, it is hoped, will open up the notes to a wider circle of readers.

The work could not have been complete without good will and cooperation from many quarters. In the years 1912-14, while I was in Europe in search for manuscript material, I enjoyed the privileges of the libraries of Paris, Munich, Vienna, Parma, the Vatican, the British Museum, Jews' College, Oxford and Cambridge. The library resources and facilities of Harvard University have made it possible to correlate the special studies of Hebrew texts with the larger field of philosophic literature. In the collection of Hebrew manuscripts in Columbia University, through the kindness of Professor Richard Gottheil and the librarians, I was able to find several Hebrew manuscripts which, during the final stages of the printing of the book, it became necessary for me to consult. Mr. Adolph S. Oko, of the Hebrew Union College Library, generously supplied me with many books which I had to use constantly. Dr. Joshua Bloch, Chief of the Jewish Division of the New York Public Library, always responded to my distant requests for bibliographical data. Professor Alexander Marx, of the Jewish Theological Seminary, not only opened to me the great treasures of the library of which he is the head, but also directed my attention to rare books and manuscripts in its possession. Professor Julius

Guttmann, of the Hochschule für die Wissenschaft des Judentums, Berlin, was kind enough to bring to my knowledge the existence of the Bloch manuscript of the *Or Adonai*, now in the possession of the Akademie für die Wissenschaft des Judentums, and to procure a photostatic copy of it for my use. For help in securing a greater degree of textual impeccability I am indebted to Professor Isaac Husik, of the University of Pennsylvania, Professor William Thomson, of Harvard, and Professor Ralph Marcus, of the Jewish Institute of Religion, who have read in proof considerable portions of the work. Dr. George Sarton, of Harvard, was kind enough to read Chapter VI of the Introduction and to reassure me when I entered on uncertain ground. Of inestimable aid in the final clarification of some of the views presented in this work was the opportunity I had for several successive years to ventilate them in the Seminar on Aristotle in which I was associated with Professor James H. Woods, and in the frequent discussions with Professor Horace M. Kallen, of the New School for Social Research, and Professor Henry M. Sheffer, of Harvard. To all these my grateful acknowledgments.

And finally I wish to record my gratitude to two men under whose guidance I entered upon this work and whose encouragement has sustained me throughout its progress. In Professor David Gordon Lyon I have found an ideal exemplar of teacher and friend, through whose broad conception of the fields of Semitic learning opportunities were created for this undertaking. To the teaching and friendship of Professor George Foot Moore I shall always feel myself profoundly indebted. During my labors on this work, whenever I was confronted with a perplexing problem, I found in his wide learning and sage counsel the illumination I needed.

H. A. WOLFSON

CONTENTS

CHAPTER III

MOTION

CHAPTER IV

TIME

CHAPTER V

MATTER AND FORM

CHAPTER VI

FORESHADOWING A NEW CONCEPTION OF THE UNIVERSE.

universe, 118.—Crescas' establishment of a complete homogeneity and continuity of nature, 119.—The characteristic feature of atomism in Arabic and Jewish philosophy, 120.—Crescas as tending toward a revival of atomism, 121.—Crescas' unification of the forces of nature by bringing magnetic attraction under the general laws of natural motion, 121.— The contrast between Aristotle's and Spinoza's conceptions as to the relation of God to the universe, 122.—Crescas' stopping short of Spinoza's conception, 123.—Implications of the expression "God is the place of the world", 123.—The three stages in the history of the opposition to Aristotle as reflected in Crescas, 124.—Crescas, Newton and Galileo, 126.

INTRODUCTION

CHAPTER I

SOURCES, METHOD, OPPOSITION AND INFLUENCE

I

THE power of generalization which is so remarkably displayed by Maimonides in all his writings, whether philosophic or Talmudic, is nowhere employed by him to greater advantage than in his introduction to the second part of the *Guide of the Perplexed*. Within the limited range of twenty-five propositions he contrived to summarize in compact and pithy form the main doctrines of Aristotle, which, supplemented by some from Avicenna, form the premises upon which are built his proofs for the existence, unity and incorporeality of God. Of these propositions Maimonides says that "some may be verified by means of a little reflection," while "others require many arguments and propositions, all of which, however, have been established by conclusive proofs in the *Physics* and its commentaries and partly in the *Metaphysics* and its commentaries."[1] But Maimonides himself did not consider it as part of his task to reproduce those proofs, for, as he again and again declares, "in this work it is not my intention to copy the books of the philosophers."[2] To the students of the *Guide*, however, the explanation and proofs of these propositions offered a wide field of research, and among the numerous commentaries which in the course of time have clustered around the *Guide* quite a few dealt exclusively with the propositions Four commentaries of this latter kind were written during the thirteenth and fourteenth centuries, by Altabrizi, Hillel of Verona,

[1] *Moreh Nebukim* II, Introduction, Prop. XXV: מהם מה שהוא מבואר במעט התבוננות ... ומהם מה שיצטרך למופתים והקדמות רבות, אלא שכבר התבארו כולם במופת אין ספק בו, קצתם בספר השמע ופירושיו וקצתם בספר מה שאחר הטבע ופירושיו.

[2] *Ibid.* שאין כונת המאמר הזה להעתיק ספרי הפיל סופים בו.

1

Zeraḥia Gracian, and Jedaiah Bedersi.[3]　It is to this class of literature that Crescas' treatment of the twenty-five propositions in his *Or Adonai*, completed in the early years of the fifteenth century, should be assigned.

There is, however, a difference between Crescas and his predecessors. None of his predecessors has acted upon Maimonides' suggestion of going directly to the works of Aristotle and his commentators for the proofs of the propositions. What the nature of Bedersi's commentary was there is no way of determining, as the work is no longer extant. Zeraḥiah Gracian admits that for a complete explanation of the propositions one would have to resort to the sources out of which they sprang, but evidently awed by the enormity of the labor that such a task would involve he decided to restrict himself to brief explanatory notes in which, he says, he would especially endeavor to explain the order and sequence of the propositions.[4]　Hillel of Verona, too, realized the need of a complete and comprehensive commentary upon the propositions and expressed the hope that some day either he himself or some one else would undertake to write it, but for the present, he said, he would give only a brief discussion of certain general topics.[5]　Nor does the commentary of Altabrizi do more justice to the subject. Though

---

[3] Friedländer, *The Guide of the Perplexed*, Vol. III, Preface, pp. xix–xxiii; Steinschneider, "Die hebräischen Commentare zum 'Führer' des Maimonides" in *Festschrift zum siebzigsten Geburtstage A. Berliner's*, pp. 345–363.

[4] MS. Paris, Bibliothèque Nationale, Cod. Heb. 985: אמר המפרש, אחי בן אוני,
בעלות על רעיוני עניי אלו ההקדמות, כי הן צריכות בידיעות מופתיהן אל חכמות רבות נשגבות
ורמות, היו בעיני כל מעיין בהם נ)ה(נעלמות, לפי שצריך לכל משכיל לדעת הסקומות, אשר לקחו
מהם, ובאיזה ספר מן החכמות הן רשומות . . . על כן אני כונתי לבאר אלו ההקדמות,
לא באריכות אך בקצרה . . . ואולי אחדש בהם דבר להודיע למה זו קדמה לחברתה, אחת מן
ההקדמות לזולתה אחריה.

[5] Introduction to Hillel of Verona's commentary on the Twenty-five Propositions: האחד, פירוש כל
ודע, אחי, כי צריך לך ולכל מבין בביאור אלו ההקדמות שני ענינים.
ההקדמה בעצמה, ר'ל פירוש נוסחא ההקדמה, והשני, כונת כל הקדמה, ר'ל על איזה תכלית
כיוון בה הרב ז'ל והעיון המבוקש הנוכח ממנה. וחלק הביאור אפרש לך בקצרה . . . ולכן
שמעתי בקולך וכתבתי אליך בפירושם מה שיכלתי ובקצרה . . . ואולי לימים עוד אועד אליך,
או אני או חכם אחר גדול ממני, שיעמידך על תכלית אמתתם ותגל נפשך מאד.

his discussions of the propositions are full and elaborate, they reflect only faintly the original works of Aristotle; his material is drawn mainly from the works of Arabic authors. In the first proposition, for instance, Altabrizi cites none of the arguments given by Aristotle; the three arguments he advances are taken from later sources. The statement made by Narboni in connection with the propositions may be quoted here as expressing the general attitude of all those who undertook to comment upon them. "My object has been to discuss the meaning of the Master's propositions and not to give you the proofs by which they may be demonstrated. Their proofs are to be found in the works from which the propositions are taken, and were I to reproduce them the result of my effort would be a book instead of a commentary."[6] It was left for Crescas to undertake the task from which his predecessors had steered clear and to compile a commentary on the propositions, or rather a book, as Narboni would call it, along the lines indicated by Maimonides himself.

Crescas, however, did not start out to write a mere commentary. He was primarily a critic of philosophy. His main object was to show that the Aristotelian explanation of the universe as outlined by Maimonides in his propositions was false and that the proofs of the existence of God which they were supposed to establish were groundless. But not wishing to appear as if he were arguing in the absence of his opponent, he felt it was necessary for him to present Aristotle's case before trying to demolish it. He therefore divides his treatment of the propositions into two parts, the proofs and his criticism of the proofs. In the proofs, as he himself avers, he intended to do nothing but to collect the arguments he had found in various sources and to present them in orderly and logical form according to a scheme of his own design. No such statement is made by him

---

[6] Narboni on Prop. XXV: ואני אמנם כונתי להבינך מאמרי הרב לא לאמתחם במוחלט, והמה מאומתים במקומותם, ולא יהיה זה פירוש כי אם חבור.

with regard to his criticism. But we shall see that his criticism is likewise made up of material drawn from other sources, its originality—and there is a considerable amount of originality in it—consisting merely in the use made of this material and in the particular purpose it was made to serve, for Crescas uses his sources as the poet uses his words and the artist his paints. In fact, the history of the criticism of Aristotle is inseparable from the history of the interpretation of his works. His commentators were not mere expositors. They were investigators, constantly looking for new problems, discovering difficulties, raising objections, setting up alternative hypotheses and solutions, testing them, and pitting them against each other. What was therefore meant by them primarily to be an interpretation inevitably became a criticism, albeit a friendly criticism, carried on by indulgent disciples in the spirit of a search for the true understanding of the Master who had to be justified at all costs. It was only necessary for one like Crescas to free himself from the bondage of discipleship in order to convert these special pleadings into hostile criticisms.

Nowhere, however, does Crescas give a complete account of his sources. In his prefatory statement to the first book, to be sure, he speaks of "Aristotle in his works the *Physics* and the *Metaphysics*; then his commentators, such as Themistius and Alexander, and the later commentators, such as Alfarabi and Averroes; then the authors after Aristotle, such as Avicenna, Algazali and Abraham ibn Daud."[7] But this list was not intended by Crescas as a catalogue of his own sources. It is rather a statement of the main authorities who prior to Maimonides had applied philosophical reasoning to the problem of the existence of God. Within the body of the commentary itself Crescas mentions the "Ancients"[8] (i. e., the pre-Aristotelian philoso-

---

[7] See below p. 131.

[8] הקדמונים Prop. X, Part I; Prop. XV, Part I.

phers), Aristotle,[9] Alexander,[10] Themistius,[11] Avicenna,[12] Algazali,[13] Avempace,[14] Averroes,[15] Altabrizi,[16] and Narboni.[17] Vague references are also made by him to "authors other than Aristotle,"[18] "commentators of [Aristotle],"[19] "the multitude of philosophisers,"[20] "they,"[21] "one of the later,"[22] "one of the commentators [of the *Guide*],"[23] and "followers [of Avicenna and Algazali]."[24] He names also several books by their titles: *Physics*,[25] *Metaphysics*,[26] *De Caelo et Mundo*,[27] Averroes' commentary on the *Physics*,[28] and the *Conic Sections* [of Apollonius].[29] All these names and titles, however, give us neither a complete nor an accurate idea as to the sources actually used by Crescas in the composition of his study of the twenty-five propositions. On the one hand, the extent of Crescas' indebtedness to other authors, named or unnamed by him, is much larger than one

[9] והנה ההקדמה הזאת חקר עליה ארסטו Prop. I, Part I (p. 134) et passim.

[10] Prop. VII, Part I.

[11] *Ibid.*

[12] Prop. II, Part II; Prop. III, Part I; Prop. X, Part II.

[13] *Ibid.*

[14] אבובכר, i. e., Abu Bekr Mohammed ibn Yaḥya ibn al-Saig ibn Badja: Prop. I, Part II, (p. 184); Prop. VII, Parts I and II.

[15] Prop. I, Parts I (p. 144) and II (p. 184); Prop. II, Part II; Prop. III, Part I; Prop. VII, Part II; Prop. X, Part II; Prop. XII, Part II.

[16] Prop. I, Parts I (p. 148) and II (p. 188); Prop. III, Part II; Prop. IV; Prop. VII, Part II; Prop. VIII, Part II. Prop. XXIII.

[17] Prop. VIII, Part II; Prop. XXIII.

[18] וזולתו מהמחברים Prop. I, Part I (p. 176).

[19] ומפרשי ספריו *ibid.*; Prop. X, Part I; המפרשים Prop. VII, Part I.

[20] המון המתפלספים Prop. V.

[21] הקשו . . . השיבו Prop. IX, Part I; זכרו Prop. IX, Part II.

[22] מהאחרונים Prop. I, Part I (p. 170) and Part II (p. 184).

[23] קצת המפרשים Prop. III, Part II.

[24] הנמשכים אחריהם Prop. X, Part II.

[25] Prop. I, Part I (p. 134); Prop. III, Part II; Prop. VIII, Part II; Prop. XII, Part I.

[26] Prop. I, Part I (p. 134); Prop. III, Part II.

[27] Prop. I, Part I (p. 134); Prop. XII, Part II.

[28] אבן רשד בבאורו לספר השמע Prop. II, Part II.

[29] ספר החרוטים Prop. I, Part II (p. 206).

would be led to believe from his own acknowledgments and, on the other hand, many of the names and titles he mentions do not at all indicate sources which he had directly consulted; they are rather names quoted by him from other works.

The failure on the part of Crescas to mention his sources, which is to be observed also in other places of his work, has been noted by one of his critics.[30] Still there is no question of bad faith involved in it, for in omitting to give more specific information as to his immediate sources, Crescas was simply following the accepted literary practice of his time—a practice especially in vogue in philosophic writings. The scope and contents of philosophic writings at the time of Crescas, especially those which revolved around the works of Aristotle, were limited to certain sets of problems which by constant repetition became philosophic commonplaces and a sort of stock-in-trade. The existence of a large number of philosophic treatises of compendious and encyclopedic nature in which each author tried to present a complete catalogue of opinions on any given question and all the pros and cons of any given argument resulted in stripping philosophic discussions of their individual authorship and to invest them with a kind of anonymity. Crescas no more felt the need of mentioning authorities than do we when we deal with generally accepted views found in school text-books.

The information which we fail to find in Crescas himself we have been able to obtain by a close comparison of his work with the entire field of philosophic literature which was available to Crescas and with which we have reason to believe he was acquainted. By means of such a comparison we have been able to identify the immediate sources used by Crescas and to trace the history of almost every argument employed by him. His sources, on the whole, fall within his own classification of the philosophic literature prior to Maimonides, namely, Aristotle,

---

[30] *Neveh Shalom* VIII, 9, p. 144b: ‏ואם שזה החכם לא הזכיר הדברים בשם אומרם‎.

his various commentators, and those who expounded Aristotle in independent works.

Aristotle was unknown to Crescas in the original Greek. He was also unknown to him in the Arabic translations. He was known to him only through the Hebrew translations which were made from the Arabic. It would be, however, rash to conclude on the basis of this fact that his knowledge of Aristotle was hazy and vague and inaccurate, for, contrary to the prevalent opinion among students of the history of philosophy, the translations of Aristotle both in Arabic and in Hebrew have preserved to a remarkable degree not only clear-cut analyses of the text of Aristotle's works but also the exact meaning of his terminology and forms of expression. The literalness and faithfulness with which the successive translators from one language into another performed their task, coupled with a living tradition of Aristotelian scholarship, which can be shown to have continued uninterruptedly from the days of the Lyceum through the Syriac, Arabic and Hebrew schools of philosophy, enabled Crescas to obtain a pretty accurate knowledge of Aristotle's writings. That knowledge, to be sure, was traditional and one-sided, but the tradition upon which it was based, like the various traditional interpretations of the Bible text before the rise of independent critical scholarship, was clear and definite and suffered comparatively little corruption. In the present work we have shown how often terms and expressions used even in indirect paraphrases of Aristotle reflect the original Greek.[31] We have also shown how commentators, who knew no Greek, speculated as to what was the original statement in Aristotle—and often guessed right.[32] In one place we have shown, how the Hebrew word for "limit" has preserved the different shades of meaning it had acquired through its being indirectly a translation of several

[31] Cf. n. 16 (p. 337) on Prop. I, Part I; n. 3 (p. 398) on Prop. I, Part II; n. 8 (p. 700) on Prop. XXV.

[32] Cf. n. 54 (p. 410) on Prop. I, Part II.

different Greek words.[33] Crescas' knowledge of Aristotle, furthermore, was extensive. He seems to have had the works of Aristotle on the tip of his tongue, and was always ready to use them at a moment's notice. He knew his Aristotle as he knew his Bible and Talmud. With an apparent ease and freedom he draws upon him whenever he is in need of some apt expression or statement for the purpose of illustrating a point or clinching an argument.[34] He never had to hunt Diogenes-like after a needed quotation nor had he ever to pray for a windfall.

The immediate source of Crescas' knowledge of Aristotle was the series of works by Averroes known as the Intermediate Commentaries as distinguished from his Long Commentaries and Epitomes. In these commentaries, the text of Aristotle, sometimes translated and sometimes paraphrased, was interspersed with Averroes' own comments and discussion. To a reader unacquainted with the text of Aristotle's own works it would often be difficult to distinguish within those Intermediate Commentaries between Aristotle's original statements and Averroes' elaborations. Crescas, however, seems to have been able to distinguish between them. In one place, for instance, he reproduces what is supposed to be Aristotle's argument against the existence of an infinite number. The argument, however, though given in the Intermediate Commentary on the *Physics*, is not to be found in Aristotle's *Physics*. Subsequently, when Crescas takes up that argument for criticism, he significantly remarks that the argument "has indeed been advanced by Averroes in his commentary on the *Physics*."[35] This is the only time that he directly refers to the "commentary" of Averroes as the source from which he has reproduced Aristotle's arguments and it would have been entirely uncalled for unless he meant to indicate thereby that

[33] Cf. n. 84 (p. 358) on Prop. I, Part I.
[34] Cf. notes 3 (p. 398), 79 (p. 456), 96, (p. 462) 104 (p. 464) and 126 (p. 472) on Prop. I, Part II.
[35] Prop. II, Part II, and n. 5 (p. 477).

the particular argument under discussion was not found in the original work of Aristotle. We have therefore reason to conclude that Crescas had another source of knowledge of Aristotle's writings. As there were no independent Hebrew translations of Aristotle's *Physics*, it must have been Averroes' Long Commentary which furnished him with a direct knowledge of the genuine text of Aristotle, for in that commentary the text of Aristotle was reproduced in such a way as to be distinguishable from the commentator's explanatory remarks. The same conclusion is to be drawn also from other instances where Crescas makes use of certain phrases and expressions which are to be found only in the Long Commentary.[36] In a few instances direct borrowing from the Long Commentary on the Physics can be discovered, though it is possible that the borrowing was made through some intermediary source.[37] As for the Epitome, which is a free and independent paraphrase of the problems dealt with in Aristotle's works, there is no positive evidence that Crescas has made use of it.[38]

Two Hebrew translations of the *Intermediate Physics* are known, one made by Zeraḥiah Gracian and the other by Kalonymus ben Kalonymus. Of these, Crescas seems to have used the latter.

Though Crescas frequently refers to Alexander, Themistius and Avempace in connection with the interpretation of certain passages in the *Physics*,[39] there is no evidence that he had a direct knowledge of their commentaries on the *Physics* which, as far as known, were never translated into Hebrew. His references to them are all taken from Averroes. On the other hand, extensive use was made by him of Gersonides' supercommentary on Averroes' Intermediate Commentary on the *Physics*, and

[36] Cf. notes 5, 7 and 8 (p. 541) on Prop. VII.
[37] Cf. n. 54 (p. 437) on Prop. I, Part II.
[38] Cf. list of quotations from the *Epitome of the Physics* in the "Index of Passages".
[39] Cf. above p. 5, notes 10, 11, 14.

perhaps also of his supercommentary on *De Caelo*, though no reference is ever made to either of them. In many places, in fact, both Aristotle and Averroes are reproduced through Gersonides. For this there is abundant evidence of a literary nature.[40] On the basis of many similarities, though not on direct literary evidence, it may also be inferred that Crescas has made use of Narboni's supercommentary on the *Intermediate Physics*.[41] This work, too, is never mentioned by Crescas.

As for the original works of Arabic authors he mentions, there is no evidence that he made use of Avicenna's writings. All the references to Avicenna can be traced to intermediary sources. Of Averroes' original works, Crescas may have used the Hebrew text of the *Sermo De Substantia Orbis*, for an important point in his criticism of Aristotle is based upon a distinction made by Averroes in that work.[42] However, the same distinction occurs also in the *Intermediate De Caelo* which we know to have been used by him.[43] It is certain, however, that he has made use of Algazali's *Maḳaṣid al-Falasifah* (*Kawwanot ha-Pilosofim*), though the work is never mentioned by title and no direct quotation from it can be discerned. This work, translated into Hebrew many times[44] and commented upon by Narboni and Albalag, was a popular source book of philosophic information and was used as a text book in the instruction of philosophy to the young until late in the sixteenth century.[45] It must have

[40] Cf. notes 91,97,99, 100 and 103 (p. 365 f.) on Prop. I, Part I; notes 13, 16, 17 (p. 403) and 40 (p. 424) on Prop. I, Part II; n. 8 (p. 556) on Prop. VIII.
[41] Cf. notes 40, 44 and 48 (p. 424) on Prop. I, Part II; n. 8 (p. 478) on Prop. II.
[42] Prop. XII, Part II and n. 7 (p. 612).
[43] *Ibid.*
[44] Steinschneider mentions three translations (*Die hebraeischen Uebersetzungen des Mittelalters*, p. 309, §174). But a comparison of the different MSS. would seem to point to an intermingling of these translations.
[45] Cf. Alexander Marx, "Glimpses of the Life of an Italian Rabbi of the First Half of the Sixteenth Century", *Hebrew Union College Annual* I (1924), pp. 613, 617.

been this work, too, that furnished him with information about Avicenna, for the work is nothing but a summary of Avicenna's philosophy. He may have also made use of Narboni's commentary on that work.[46]

The question as to whether Crescas was acquainted with Algazali's *Tahafut al-Falasifah* (*Happalat ha-Pilosofim*) and to what extent it had influenced his own critical attitude toward philosophy requires special consideration.

A tradition has already grown up among modern students of Jewish philosophy that Crescas' criticism of Aristotle was inspired by Algazali's *Tahafut al-Falasifah*.[47] The source of this tradition would seem to be nothing but a vague surmise based on a general impression and on a haphazard combination of irrelevant facts. Algazali, it must have been reasoned, is known as an opponent of philosophy, and also to have influenced Jewish philosophers. Crescas is a Jewish philosopher and an opponent of philosophy. Furthermore, Crescas happens to mention Algazali. Hence, it was concluded, it must have been Algazali who inspired Crescas in his criticism of philosophy.

In order to prove the influence of the *Tahafut al-Falasifah* on the *Or Adonai* it is necessary first to determine whether it was possible for Crescas, who derived his knowledge of Arabic philosophy from Hebrew translations, to have used the *Tahafut*, for there is no direct reference in the *Or Adonai* to the *Tahafut* and whenever the name of Algazali is mentioned the reference is always traceable to the *Maḳaṣid al-Falasifah*.[48] Such a possi-

---

[46] Cf. n. 54 (p. 437) on Prop. I, Part II. Cf. Index of Passages: Narboni.

[47] Cf. Joël, *Don Chasdai Creskas' religionsphilosophische Lehren*, p. 3; Kaufmann, *Geschichte der Attributenlehre*, p. 134; Broyde, "Ghazali", *Jewish Encyclopedia*, V, 649; Husik, *Hist. of Med. Jewish Phil.*, p. 392.

[48] Joël seems to have based his conclusion as to Algazali's influence upon Crescas upon the vague references to Algazali which are to be found in the *Or Adonai*, without realizing that none of them is to the *Tahafut*. He also speaks of Abravanel as one who had noticed a resemblance between Crescas and Algazali (*op. cit.*, p. 80, Note III). Abravanel's reference (בספרו אבוחמד שכתב מה

bility, it must be admitted, existed. While the *Tahafut* itself was probably not translated into Hebrew until after the completion of the *Or Adonai*,[49] there had existed a Hebrew translation of Averroes' *Tahafut al-Tahafut* (*Happalat ha-Happalah*) ever since the early part of the fourteenth century [50] and this work incorporated the work of Algazali. The *Tahafut* was thus available to Crescas, but was it ever used by him in the composition of his *Or Adonai?*

An answer to this question was undertaken by Julius Wolfsohn in a treatise devoted especially to the evidence of Algazali's influence upon Crescas.[51] He deals with the subject under four headings. First he discusses the influence of Algazali on Crescas as to the general tendency of his philosophy (pp. 8–33). Then he takes up in succession the following special topics: Attributes (pp. 34–46), Unity of God (pp. 47–55), and Free Will (pp. 55–72). We shall examine his arguments one by one.

Under the first heading the author tries to prove the dependence of Crescas upon Algazali by showing certain similarities in their general attitude toward philosophy: that both come out for the liberation of religion from philosophy (pp. 8–11), that both undertake to refute philosophy by the reasoning of philosophy itself (pp. 15–18), and that both refute philosophy not only when it is opposed to tradition but also when it is in agreement with it (pp. 23–28). That such similarities exist between them cannot be denied, but general similarities of this kind, even when not offset by a more impressive list of differences that

---

באלהיות) is likewise to the *Makasid.* Abravanel, as we shall see later, did not believe that Crescas had any knowledge of the *Tahafut* at the time of his writing of the *Or Adonai.*

[49] The *Or Adonai* was completed in 1410. Don Benvenisti, for whom Zerahiah ha-Levi ben Isaac Saladin translated the *Tahafut al-Falasifah*, died in 1411. See Steinschneider, *Die hebraeischen Uebersetzungen des Mittelalters*, p. 328.

[50] Translated by Kalonymus ben David ben Todros shortly before 1328. See Steinschneider, *op. cit.* p. 332.

[51] *Der Einfluss Gazali's auf Chisdai Crescas* 1905.

can easily be drawn up, do not in themselves establish a literary relationship. Crescas had no need for an inspiration from without to take up the cudgels in behalf of tradition as over against speculation. The rise of philosophy to a dominant position in any religion inevitably brings its own reaction, and as far as Judaism is concerned the native opposition to philosophy which had appeared simultaneously with the rise of the philosophic movement itself, is sufficient to account for the particular position taken by him. Still less convincing is the author's attempt to establish a literary influence by the fact that both Algazali and Crescas argue for the creation of the world, for God's knowledge of particulars, and for bodily resurrection and reward and punishment (pp. 18–23). These are common problems to be found in almost any work on theology of that period, and Crescas' attitude on all these problems reflects the traditional Jewish view, and there is no need for assuming a foreign influence.

In his chapter on attributes the author again shows a similarity in the general attitudes of Algazali and Crescas without establishing a literary relationship between their works. It is indeed true that both Algazali and Crescas raise objections to the theory of negative attributes, but Algazali's objections as reproduced by the author are unlike those reproduced by him in the name of Crescas (pp. 35–40). It is also true that both Algazali and Crescas try to justify the admissibility of positive attributes, but beyond the fact that both believed that positive attributes are not incompatible with the simplicity of the divine essence, the author establishes no similarity in their arguments. That Crescas' attempt to justify positive attributes would have to contend that they do not contradict the simplicity of the divine nature was only to be expected—that much Crescas could have gathered from Maimonides' polemic against the upholders of positive attributes. But what was it that made Crescas override Maimonides' objections and assert with certainty that there was no contradiction? Were his reasons the same as Algazali's? I believe it can be shown that Algazali and Crescas justify the admissibility of positive essential attributes on entirely different grounds. To Algazali the justification is to be found principally in his contention that the concept of necessary existence does not preclude an inner plurality; to Crescas it is to be found in a moderately nominalist conception of universals.[52]

In his discussion of the unity of God the author adduces only one argument from Crescas which bears some relation to a similar argument by Algazali. Both argue against the philosophic contention that two deities could not adequately divide their fields of activity within the world and try to show that some adequate division of labor could exist between them. In Algazali the contention is that such a division of labor can be found in the fact that one deity may be the cause of the celestial sphere and the other of the sublunar elements, or that one may be the cause of the immaterial beings and the other of the material beings (p. 51). Crescas argues somewhat similarly that, while within this universe there could not be any adequate division of labor between two deities in view of the fact that the universe is an organic unit in which all parts are interconnected, there is still the possibility of a division of labor on the assumption of the existence of more than one universe, in which case one deity may be the cause of one universe and the other of another. That there is some relation between these two arguments may be granted. Still it does not follow that Crescas had knowledge of the Tahafut, for Algazali's argument is reproduced, without the mention of the name of Algazali, in Narboni's commentary on the Moreh Nebukim, and we know that Crescas had made use of that commentary.[53]

Similarly unconvincing is the author's discussion of the prob-

[52] See H. A. Wolfson, Crescas on the Existence and Attributes of God.
[53] Ibid.

lem of the freedom of the will wherein, again, the reasoning is based upon vague and general similarities.

If general similarities of this kind are to be the basis of establishing the influence of Algazali on Crescas, a more imposing number of them might have been gathered. In the commentary on the text I have called attention to all such instances. Two of these are of particular importance as they contain arguments which are individual to Algazali and which form some of the crucial points in Crescas' criticism. First, Algazali contends that the concept of necessary existence precludes only external causation and is not incompatible with an inner composition of the essence. Crescas repeats a similar contention several times in his criticism of the proofs of the existence of God.[54] Second, Algazali argues that the motion of the celestial sphere should be regarded as natural instead of voluntary, as was the general assumption. Crescas has a similar contention which he repeats several times referring to it as "our own view" in contradistinction to the commonly accepted view of the philosophers.[55] In both these instances, however, as well as in other similar instances, we have shown that there are other sources, with which Crescas is known to have been acquainted and from which he could have taken these views.[56]

Not only are all these evidences inconclusive, but there is evidence which shows quite the contrary, that Crescas could not have known the *Tahafut*. In one place Crescas lines up two groups of philosophers as to the question of the possibility of an infinite number of disembodied souls. Algazali is placed by him among those who admit that possibility. This is quite in agreement with Algazali's view as given in the *Maḳaṣid* where he only restates the views of Avicenna, without necessarily committing himself to them. In the *Tahafut*, however, Algazali

[54] *Ibid.*
[55] Cf. n. 11 (p. 535) on Prop. VI.
[56] Cf. *ibid*

explicitly rejects the possibility of an infinite number of disembodied souls.[57] Had Crescas known the *Tahafut* he certainly would not have allowed that fact to pass unnoticed.

The question as to whether Crescas had knowledge of Algazali's *Tahafut al-Falasifah* or of Averroes' *Tahafut al-Tahafut* at the time of writing the *Or Adonai* has already been raised by a mediaeval Jewish author. The question comes up in the following connection.

In the chapters on the problem of creation in the *Or Adonai* Crescas refutes a certain argument which he quotes in the name of Gersonides. The same argument is also found in Algazali's *Tahafut*. In another work, the *Bittul 'Ikkere ha-Nozerim*, Crescas makes use of the very same argument which has been rejected by him in the *Or Adonai*.

Joseph ben Shem-tob, the Hebrew translator of the latter work of Crescas, after calling attention to the origin of Crescas' argument in Gersonides and Algazali and to Crescas' own refutation of the argument in the *Or Adonai*, suggests that Crescas' *Bittul 'Ikkere ha-Nozerim* must have been written after his *Or Adonai* and that after he had written the latter work he must have changed his mind with regard to the validity of the argument under consideration.[58] Isaac Abravanel accepts this suggestion of Joseph ben Shem-tob, adding that Crescas' change of view must have resulted from his reading of Algazali's *Tahafut al-Falasifah* or of Averroes' *Tahafut al-Tahafut* after he had written the *Or Adonai*.[59] Furthermore, on the basis of other evidence, Abravanel tries to show that Crescas could not have

----

[57] Cf. n. 6 (p. 485) on Prop. III.

[58] *Bittul 'Ikkere ha-Nozerim*, ch. III, p. 30: והרב הזה העתיקו הנה נגד הנוצרים
בהולד הבן ואצילות הרוח, ואח'כ שב מדרכיו, שהוא כבר טען ע'ז מופת בספרו אור ה'. ואדמה
שהוא חבר המאמר הזה אחר חברו אותו הספר.

[59] *Shamayim Hadoshim* III, p. 28: ואחשוב אני שאחרי שעשה הרב חסדאי ספרו ראה
דברי אבוהמד ואבן רשד וחזר להחזיק בטופת הר'ל אשר נער בו. ולכן במאמר אשר עשה בלשון
ארצו בספקות אמונת האומה הנצרית, בפ'ג ממנו, הקשה כנגדם בההולדה התמידית אשר שמו
בתאר הבן, ועשה עליו המופת הזה שעשה הר'ל כנגד הקדמות וחייב להם כל הבטולים האלה.

known of these two works at the time of the writing of the
Or Adonai.[60]

As for the accuracy of the conclusion that the *Biṭṭul Iḳḳere
ha-Noẓerim* was composed after the *Or Adonai*, it is open to
grave doubt. The Fourth Book of the *Or Adonai*, according to a
colophon which occurs in most of the manuscripts, was completed
in 1410,[61] which is probably also the year of the author's death,
whereas the *Biṭṭul Iḳḳere ha-Noẓerim* would seem to have been
written in 1398, for it refers to the Great Schism (1378) as having
occurred twenty years previously.[62]

In mitigation of this doubt, however, the following two
considerations may be urged:

First, the composition of the *Or Adonai* must have extended
over many years, for the discussion of the Messiah (III, viii),
which occurs not far from the end of the book, was written five
years before the completion of the entire work.[63] It is not impossible, therefore, that the problem of creation (III, i) was
written before 1398.[63]

Second, it would also seem that the *Or Adonai* was not
written in the order in which it is now arranged. Certain chap-

---

ושבח הטופה ההוא והענדהו עטרות לו, וכמו שהעיר עליו החכם ר' יוסף אבן שם טוב שהעתיק
אותו המאמר ללשון הקודש.

[60] *Ibid.* pp. 27–28: או לא ראה שלא לפי חסדאי להר' קרה הזה הטעות שכל לך דע ואתה
עיין בדברי אבוחמד בספר ה פ ל ת ה פ י ל ו ס פ י ם ובדברי אבן רשד בן מחלוקתו בספר
ה פ ל ת ה ה פ ל ה . . . ואחשוב אני שאחרי שעשה הר' חסדאי ספרו ראה דברי אבוחמד ואבן
רשד וחזר להחזיק במופה הר"ל אשר נער בו. Cf. *Mif'alot Elohim* IX, 7, p. 67vb.

[61] והיתה ההשלמה בחדש זיו שנת מאה ושבעים לפרט האלף הששי ליצירה בסרקוסטה אשר
במלכות ארגון. This colophon evidently does not come from the hand of the
author. It does not occur in the *editio princeps* nor in the Paris manuscript.
The Parma manuscript, which seems to have been written by a student of
Crescas, reads here as follows: המחבר ז'ל השלימו בעיר סרקסתה במלכות ארגון שנת
ק'ע לפרט האלף הששׁ. The same reading occurs also in the Jews' College manuscript. Cf. also colophon of Turin MS. quoted at the end of Bibliography I.

[62] Chapter 8: כי עוד היום בעבור שבכל יותר הדברים בין האמונה הנוצרית קרוב לכ' שנה
יש להם שנים ראשונה (שני ראשים) אפיפיורים, וכל אחד מהם ומהנמשכים אחריהם חושב לזולתו
למנודה ונעשנ לשמים. Cf. Graetz, *Geschichte der Juden*, Vol. VIII, Note 2.

[63] *Or Adonai* III, viii, 2: עתה שהיא שנת אלף ושלש מאות שלשים ושבעה לחרבן הבית.
This is the correct reading according to the Munich, Paris, Vienna and New

ters in Book IV bear the unmistakable internal evidence of having been written originally as a sort of preliminary studies to problems dealt with in earlier parts of the work. Thus the discussion as to "whether there is only one world or whether there are many worlds at the same time" in IV, 2, seems to have been written as precursory to the same problem dealt with at the end of Prop. I, Part II, and similarly the discussion as to "whether the celestial spheres are animate and intelligent beings" in IV, 3, seems to have been written as precursory to the same problem discussed in Prop. VI. In both these instances, the problems are treated in greater detail and in a spirit of greater impartiality in Book IV than in the earlier parts of the work. It is thus not impossible that the problem of creation was among the first to have been taken up by Crescas and to have been written by him long before 1398.

But whatever value one may attach to the conclusions of Joseph ben Shem-ṭob and Abravanel, there is no positive evidence of Crescas' acquaintance with the *Tahafut al-Falasifah*. Even if we assume his acquaintance with that work and recognize it as the source of all those arguments for which we find parallels in it, it is far from being the predominant influence upon the *Or Adonai*. The most that can be said is that it is one of the many works from which Crescas has borrowed certain arguments which he has incorporated in his own work. It is not impossible that his knowledge of the *Tahafut*, assuming that he had any knowledge of it, he obtained not from a study of the book itself but from his pupil Zeraḥiah Saladin who was versed in Arabic and later translated the *Tahafut* into Hebrew.[64]

Another class of sources of the *Or Adonai* are the commentaries on the *Moreh*. Of these the most widely used by Crescas is Altabrizi's commentary on the twenty-five propositions.

York manuscripts. The editions and some of the other manuscripts have here corrupt readings.

[64] See above p. 11, n. 48.

The commentary of Altabrizi was originally written in Arabic. Its author was a Persian Mohammedan, who flourished probably in the thirteenth century. From a remark in his introduction it may be inferred that the author had intended to interpret the entire work of the *Moreh*,[65] but whether he really did so or not there is no way of determining. Two Hebrew translations of this commentary are extant, one of which, done by Isaac ben Nathan of Cordova or Xativa, was published in Venice, 1574, and the other, anonymous, is found only in manuscript form.[66] The fact that this anonymous commentary is a translation of Altabrizi was first noticed by Steinschneider.[67] There is, however, this to be added to the description of this work. While indeed it is nothing but a translation of Altabrizi, there is sufficient evidence to show that the translator, whoever he was, wished to have that fact unknown and to have his work passed off as an original composition or, at least, as a compilation made by himself out of different Arabic sources. The deliberate purpose of the translator to mislead his readers is evident at the very outset of the work. In Isaac ben Nathan's translation, Altabrizi begins with that inevitable jingle of glorifications, exaltations and elevation to the Creator, Causator, and Originator of this our universe, from which he passes to a second topic wherein he gives an account of himself and of his genealogy and concludes with a eulogy of Maimonides and his works. All these are omitted by the anonymous translator in the three out of the

---

[65] Cf. Altabrizi's Introduction in the Vienna manuscript of Isaac ben Nathans translation: אמר עבד האלה מחמד אבובכר בן מחמד אלתבריזי. זה החלק אשר כתבו [בנדרפס: אמר עבד האלה מחמד אבובכר בן מחמד אלתבריזי.
סדר בו] הנכבד השר [בנדרפס נוסף: הראש] משה עבד האלהים הישראלי הקרטבי מהספר [בנדרפס:
מהספרים] אשר נחשוב לבארו ולנלותו, והוא הספר הרשום בהוראת [בנדרפס: להוראת] הנבוכים.
My inference as to the author's intention of writing a commentary on the entire *Moreh* is based upon the expression מהספר אשר נחשוב לבארו. It is quite possible, however, that the clause אשר נחשוב לבארו refers to החלק.

[66] Six MSS. are recorded by Steinschneider in *Die hebraeischen Uebersetzungen*, p. 362.

[67] See *Catalogus Librorum Hebraeorum in Bibliotheca Bodeliana*, p. 1143.

six extant manuscripts which I have examined in Paris, Vienna, and London. But beginning with the third topic of Altabrizi's Introduction which contains a brief description of the twenty-five propositions, the translator adds a long statement of his own, the evident purpose of which is to create the impression that his work is a compilation of various Arabic commentaries supplemented by numerous remarks of his own, which, however, he modestly says, are not differentiated by him from the unoriginal portions of the work, as his main object, he concludes, is to impart information.[68] Upon examination, however, his claim seems to be rather exaggerated. The commentary faithfully follows the single work of Altabrizi with a few exceptions where the translator either omits some passage found in the original, or, acting upon a suggestion of Altabrizi himself, expands certain brief statements of the author. The following examples will illustrate the nature of what the translator has claimed as his own original contributions.

(1) In Proposition I, after the third argument against the existence of an infinite magnitude, the translator remarks that his restatement of the arguments is the fine flour of the lengthy discussions of the numerous commentators.[69] As a matter of fact, his text is a faithful translation of Altabrizi except for the omission of a few digressions found in the original.

(2) In Proposition IV, Altabrizi has a brief illustration of the phenomenon of expansion, which is included among the subdivisions of quantitative change. That illustration is more

---

[68] הפרוש בעזרי על קצת מאנשי החכמה, וראיתי באורם וסופתיהם באלה ההקדמות, ושענין הבנתן עמוק, ושהרב ס"כ לא זכרם אלא בזכרון פרוץ, כי הוא אומר שאריסטו הביא מופת על כל אחת מהן, והענין אשר עמדתי עליו ממבארי ההקדמות האלה היה בלשון הערב, אמרתי גם אני אכחבנו בלשון העברי, כדי שיקבלו תועלת ממנו מחכמי אומתנו אשר אין להם דרך בלשון הערב, ואהיה מזכה וזכה, ומה שיתחדש לי גם אני בביאור הזה אכללנו עם באור זולתו, כי אין הכונה אלא להוסיף התועלת.

[69] וההו המופת התבאר ובוו זאת ההקדמה, ולא יצא זה אלא אחרי בלבולים רבים וקושיות, והוא סולת דבריהם.

elaborately restated by the anonymous translator. In substance, however, the two illustrations are identical.

(3) In Proposition VI, after discussing various classes of motion, Altabrizi remarks: "The tabulation of the motions under this class can be done by yourself.":[70] In the translation a complete list is given introduced by the words: "I shall now draw up the classification myself."[71]

(4) In Proposition XVII, the translator says: "As for the meaning of motion according to essence, many have been confused concerning it and have advanced a variety of explanations, but we shall restate here the fine flour of their views."[72] Here, too, excepting his omissions of several alternative views stated by Altabrizi, the translator closely follows the original text.

These two translations of Altabrizi represent the two different styles of philosophic Hebrew, the Arabicized and the native, which were used in the translations from the Arabic and the classic examples of which are to be found in the two translations of Maimonides' *Moreh*, the one by Samuel ibn Tibbon and the other by Judah al-Ḥarizi. Isaac ben Nathan uses the Arabicized form of expression; the anonymous translation is written in the native form of rabbinic Hebrew. Of these, Crescas has used Isaac ben Nathan's translation.

Next in importance as a source used by Crescas is Narboni's commentary on the *Moreh*. Crescas mentions this commentary in several places,[73] but his indebtedness to it is evident in many other places where no mention of it is made.[74] As Norboni often

---

[70] ויצא מהם מהזדווג אלה החלקים קצתם עם קצת שמנה חלקים לתנועה ההכרחית, ועליך בחינת הפרדתם בהמשלם.

[71] והנה יעלה בידינו מזיווג אלה החלקים קצתם עם קצתם שמנה חלקים לתנועה הכרחית ואני אסדר לך חלקיהם ומשליהם עליך.

[72] ואמנם פירוש המתנועע מעצמו נתבלבלו בי רבים בפירושים משתנים, וסולת הפירוש בו נזכיר.

[73] Cf. above p. 5, n. 17.

[74] Cf. n. 16 (p. 492) on Prop. III; notes 8 (p. 507), 9, 11 and 16 on Prop. IV; n. 8 (p. 534) on Prop. VI; notes 4 and 10 (p. 551) on Prop. VIII; n. 5 (p. 605) on Prop. XI; n. 2 (p. 682) on Prop. XIX; n. 5 (p. 697) on Prop. XXIV; n. 6 (p. 700) on Prop. XXV.

follows Altabrizi's method in expounding the proposition, it is sometimes not clear as to which of these sources he directly follows.[75] Besides Altabrizi and Narboni, no other commentary on the *Moreh* is mentioned by Crescas, but it is not impossible that he made use of the *Moreh ha-Moreh* and also of Hillel of Verona's commentary on the twenty-five propositions.[76] It is certain, however, that Crescas had no knowledge of Maimonides' own comments on Propositions IV, XXIII and XXIV, contained in his letter to Samuel ibn Tibbon, for Crescas gives entirely different interpretations of those propositions.[77]

In addition to these works there is the entire body of philosophic Hebrew literature extant at the time of Crescas. Whether any of these Hebrew works is mentioned by him or not and whether it is directly used by him in the *Or Adonai* or not, we have reason to assume that he was acquainted with it and we are therefore justified in drawing upon it for the reconstruction of the historical background of his ideas. One can speak, however, with greater certainty as to Crescas' direct indebtedness to the *Emunah Ramah*. Not only is its author Abraham ibn Daud mentioned by him in the general list of Maimonides' philosophic predecessors,[78] but one can discover in several places not merely parallels to some of Crescas' arguments but concrete literary relationships.[79]

Close observation of Crescas' proofs of the propositions reveals the fact that with the exception of propositions I, VIII, XII, XIV, XXIV, XXV, all of them start out with an opening based on Altabrizi and that even of those which do not start with such an opening all, with the exception of XXIV and XXV, contain

[75] Cf. n. 8 (p. 534) on Prop. VI; n. 3 (p. 540) on Prop. VII; n. 4 (p. 551) on Prop. VIII.

[76] See "Index of Passages" under these names.

[77] Cf. n. 3 (p. 502) on Prop. IV; n. 2 (p. 690) on Prop. XXIII.

[78] Cf. above p. 4, n. 7.

[79] Cf. n. 73 (p. 354) on Prop. I, Part I; notes 7, 8, 9, 13, 16 (pp. 571–579), 26 and 27 (p. 598) on Prop. X; notes 6 and 7 (p. 670) on Prop. XVII.

some elements which can be traced to Altabrizi. Then also the Hebrew text of seventeen propositions (II, III, IV, VI, VII, VIII, XII, XIII, XIV, XVII, XVIII, XIX, XX, XXI, XXII, XXIII, XXV) is taken from Isaac ben Nathan's translation of Altabrizi, the text of five propositions (I, IX, XI, XV, XVI) is taken from Ibn Tibbon's translation of the *Moreh*, two of these (XI, XV), however, containing some phrases from Altabrizi. Propositions V and XIV read alike in both translations, and Proposition X is composed of parts taken from both translations. The inference to be drawn from this is that Crescas has taken Isaac ben Nathan's translation of Altabrizi as the basis of his own commentary on the propositions, departing from it only when he finds it unsatisfactory or insufficient for his purpose. In most cases his departure from Altabrizi consists merely in amplifying the former's discussion by the introduction of material drawn from other sources. But sometimes he departs from Altabrizi completely and follows entirely new sources. An example of this is the first proposition, where the entire structure of the proof is independent of that of Altabrizi, though within it are incorporated also the arguments of Altabrizi. It is not impossible that the collection of material and especially the abstracts of literature used in the composition of the work were prepared by students, for Crescas informs us that in preparing the work he is to avail himself of the assistance of a selected group of associates[80]—"associates" being a polite Talmudic term applied by teachers to their advanced students. This may explain the inadequacy of some of these abstracts, the unevenness of their style and their occasional misplacement in the text.[81]

[80] Cf. *Or Adonai, Hakdamah*, p. 2a: ובהסכמת החברים ובעזרתם, and p. 2b: עם חשובי החברים.

[81] See, for instance, notes 104 (p. 374) and 107 on Prop. I, Part I; n. 6 (p. 611) on Prop. XI; n. 6 (p. 699) on Prop. XXV.

II

The research into the literary sources of Crescas undertaken in the present study was not a matter of mere idle play or even of intellectual curiosity. It was essentially necessary for the understanding of the text. Crescas like all mediaeval philosophers operates on the whole with conventional concepts of his time which to a large extent are foreign to our way of thinking and to understand which we must acquaint ourselves with their origin and background. But there is even something more than this in Crescas' method of literary composition. He not only re-echoes the ideas of his predecessors but he collocates torn bits of their texts. The expository part of his work is a variegated texture into which are woven many different strands. Mosaic in its structure, it is studded with garbled phrases and expressions torn out of their context and strung together in what would seem to be a haphazard fashion. At times the text is entirely unintelligible and at times it is still worse—misleading. We read it, and think we understand it. If we do happen to come across some ambiguity, some abrupt transition, some change of point of view, or some unevenness of style, we are apt to attribute it to an inadequacy of expression on the part of the author and try our best, by whatever general information we may happen to possess or may be able to gather, to force some meaning upon it—and trying, we think we succeed. But sometimes by a stroke of good luck we may happen to stumble upon the immediate source of Crescas' utterances and at once our eyes are opened wide with surprize and astonishment, ambiguities are cleared up, certainties call for revision and what has previously seemed to us meaningless or insignificant assumes an importance undreamed of.

The critical part of Crescas' works offers still greater difficulties to the modern reader on account of its adherence to what may be called the Talmudic method of text study. In this method the starting point is the principle that any text that is deemed worthy of serious study must be assumed to have been written with such care and precision that every term, expression, generalization or exception is significant not so much for what it states as for what it implies. The contents of ideas as well as the diction and phraseology in which they are clothed are to enter into the reasoning. This method is characteristic of the Tannaitic interpretation of the Bible from the earliest times; the belief in the divine origin of the Bible was sufficient justification for attaching importance to its external forms of expression. The same method was followed later by the Amoraim in their interpretation of the Mishnah and by their successors in the interpretation of the Talmud, and it continued to be applied to the later forms of rabbinic literature. Serious students themselves, accustomed to a rigid form of logical reasoning and to the usage of precise forms of expression, the Talmudic trained scholars attributed the same quality of precision and exactness to any authoritative work, be it of divine origin or the product of the human mind. Their attitude toward the written word of any kind is like that of the jurist toward the external phrasing of statutes and laws, and perhaps also, in some respect, like that of the latest kind of historical and literary criticism which applies the method of psycho-analysis to the study of texts.

This attitude toward texts had its necessary concomitant in what may again be called the Talmudic hypothetico-deductive method of text interpretation. Confronted with a statement on any subject, the Talmudic student will proceed to raise a series of questions before he satisfies himself of having understood its full meaning. If the statement is not clear enough, he will ask, 'What does the author intend to say here?' If it is too obvious, he will again ask, 'It is too plain, why then expressly say it?' If it is a statement of fact or of a concrete instance, he will then ask, 'What underlying principle does it involve?' If it is a broad generalization, he will want to know exactly how much it is to include; and if it is an exception to a general rule, he will want to know how much it is to exclude. He will furthermore want to know all the circumstances under which a certain statement is true, and what qualifications are permissible. Statements apparently contradictory to each other will be reconciled by the discovery of some subtle distinction, and statements apparently irrelevant to each other will be subtly analyzed into their ultimate elements and shown to contain some common underlying principle. The harmonization of apparent contradictions and the inter-linking of apparent irrelevancies are two characteristic features of the Talmudic method of text study. And similarly every other phenomenon about the text becomes a matter of investigation. Why does the author use one word rather than another? What need was there for the mentioning of a specific instance as an illustration? Do certain authorities differ or not? If they do, why do they differ? All these are legitimate questions for the Talmudic student of texts. And any attempt to answer these questions calls for ingenuity and skill, the power of analysis and association, and the ability to set up hypotheses—and all these must be bolstered up by a wealth of accurate information and the use of good judgment. No limitation is set upon any subject; problems run into one another; they become intricate and interwoven, one throwing light upon the other. And there is a logic underlying this method of reasoning. It is the very same kind of logic which underlies any sort of scientific research, and by which one is enabled to form hypotheses, to test them and to formulate general laws. The Talmudic student approaches the study of texts in the same manner as the scientist approaches the study of nature. Just as the scientist proceeds on the assumption that there is a uniformity and continuity in nature so the Talmudic student proceeds on the assumption that there is a uniformity and continuity in human reasoning. Now, this method of text interpretation is sometimes derogatorily referred to as Talmudic quibbling or pilpul. In truth it is nothing but the application of the scientific method to the study of texts.

A similar attitude toward texts and a similar method of interpretation was introduced by Jewish thinkers into the study of philosophy. One need only look into some of the commentaries upon Averroes, or upon Maimonides, especially the commentary of Abravanel upon the *Moreh*, to become convinced of the truth of this observation. It is well-nigh impossible to understand their writings and to appreciate the mode of their reasoning unless we view them from this particular angle. It is still less possible to give an accurate account of their philosophy without applying to them the same method that they applied to their predecessors. The mere paraphrasing of the obscurities of their texts is not sufficient. Still less sufficient is the impressionistic modernization of their thought. We must think out their philosophy for them in all its implications and rewrite it for them in their own terms. We must constantly ask ourselves, concerning every statement they make, what is the reason? What does it intend to let us hear? What is the authority for this statement? Does it reproduce its authority correctly or not? If not, why does it depart from its authority? What is the difference between certain statements, and can such differences be reduced to other differences, so as to discover in them a common underlying principle? We must assume that their reasoning was sound, their method of expression precise and well-chosen, and we must present them as they would have presented them had they not reasoned in symbols after the manner of their schools. In the case of Maimonides we have his own statement as to the care he exercised in the choice of terms, and in the arrangement of his problems, declaring that what he has written in his work "was not the suggestion of the

moment; it is the result of deep study and great application."[82]
Similarly Crescas declares that everything in his work, though
briefly stated, was carefully thought out and is based upon long
research.[83]

Now this Talmudic method of reasoning is intelligible enough
when it is fully expressed, when its underlying assumptions are
clearly stated and every step in the argument distinctly marked
out. But in the literature in which this method is followed, owing to the intimacy of the circle to which it was addressed, the
arguments are often given in an abbreviated form in which the
essential assumptions are entirely omitted or only alluded to,
the intermediary steps suppressed or only hinted at, and what
we get is merely a resultant conclusion. This abbreviated form
of argumentation is characteristic of the recorded minutes of
the school-room discussions which make up the text of the
Talmud. It was continued in the rabbinic novellae upon the
Talmud, reaching its highest point of development in the
French school of the Tosafists which began to flourish in the
twelfth century. Shortly after, it was introduced into the philosophic literature in the form of novellae upon standard texts,
resembling the Talmudic novellae in their external literary form
even to the extent of using the same conventional phrases by
which questions and answers are introduced.[84] Crescas' work
belongs to that type of novellae literature, conforming to the
Talmudic novellae literature in all its main characteristics, its
attitude toward texts, its method of text interpretation, its
abbreviated form of argumentation. Again and again Crescas
declares in his *Or Adonai* as well as in his *Biṭṭul 'Iḳḳere ha-
Noẓerim* that whatever he has to say will be expressed by him

---

[82] *Moreh Nebukim*, Introduction: כי המאמר הזה לא נפלו בו הדברים כאשר נזדמן,
אלא בדקדוק גדול ובשקידה רבה.

[83] *Or Adonai, Haḳdamah*, p. 2b: וזה אמנם בעיון גדול ושקידה רבה.

[84] E. g., such expressions as יש להקשות, ואם תאמר, etc.

with the utmost brevity,[85] and to this declaration of his he has lived up faithfully.

But it seems that Crescas' vaunted brevity was too much even for those who had been used to that form of expression. It often bordered upon obscurity. Joseph ben Shem-ṭob, the Hebrew translator of his *Biṭṭul 'Iḳḳere ha-Noẓerim* was in one place compelled to give a free paraphrase of a certain passage in order to make it intelligible, justifying himself for so doing in the following declaration: "This is how the words of the Master, of blessed memory, are to be understood here. In translating them I have expanded their meaning, for his original words in this passage are all too brief and all too abstruse, so that I have not met anybody who was able to understand them. Hence, in this passage, more than in any of the other passages of his book, I have allowed myself to overstep the bounds of what is proper in a translation."[86] A student of Crescas, in a marginal note on his copy of the *Or Adonai* preserved at the Biblioteca Palatina at Parma, has the following characterization of his master as lecturer and writer: "When I studied under my Master I could not fathom the full meaning of his view on this subject . . . The Master, of blessed memory, was accustomed to express himself with the utmost brevity both in speaking and in writing."[87] This statement would also lead us to believe that the *Or Adonai* had its origin in class-room lectures and discussions. We know of other instances where Hebrew philosophic works were the result of class-room lectures. It was while thus addressing himself to a group of initiated students, expecting to be interrupted with questions whenever he failed

---

[85] Cf. Prop. I, Part I, p. 178: בקצור מופלג; *Biṭṭul 'Iḳḳere ha-Noẓerim*, p. 11: .וזה יהיה בתכלית הכללית והקצור, כל אריכות דברים נעזוב

[86] *Biṭṭul 'Iḳḳere ha-Noẓerim*, Ch. III, pp. 27–28: הנה על זה האופן ראוי שיובנו דברי הרב ז״ל במקום הזה. ואני הרחבתי הביאור בהעתקתי אותם, כי לשונו קצר ועמוק במקום .הזה, לא ראיתי מי שיבינהו, ולכן עברתי חק ההעתקה בזה יותר משאר מאמריו

[87] .והנה הרב ז״ל היה מדבר גם כותב בקצור מופלג The same note occurs also on the margin of the Jews' College manuscript.

to make himself clear, as is evidenced from his former student's remarks, that his style assumed that allusive and elliptical form by which it is characterized. In order, therefore, to understand Crescas in full and to understand him well, we must familiarize ourselves with his entire literary background. We must place ourselves in the position of students, who, having done the reading assigned in advance, come to sit at his feet and listen to his comments thereon. Every nod and wink and allusion of his will then become intelligible. Words previously quite unimportant will become pregnant with meaning. Abrupt transitions will receive an adequate explanation; repetitions will be accounted for. We shall know more of Crescas' thought than what is actually expressed in his utterances. We shall know what he wished to say and what he would have said had we been able to question him and elicit further information.

A faint echo of the class room discussion of Crescas' lectures on philosophy has reached us indirectly in the work of his student Joseph Albo. In several instances, and as far as the scope of this chapter is concerned we may mention only the discussion of place and of time, he makes use of several specific arguments which are found in the *Or Adonai*. He does not mention the *Or Adonai* in any of these instances. Nor does his restatement of the arguments bear any specific, verbal resemblance to the corresponding originals in the *Or Adonai*. Sometimes the arguments are considerably modified and are made to prove different conclusions.[88] Sometimes also a well developed and clearly expressed argument in Albo's *'Ikkarim* has as its counterpart in the *Or Adonai* only a meaningless ejaculation.[89] All this would seem to point to the fact that what we get in the *'Ikkarim*, at least in these instances and in a few others like them, is not direct borrowings from the *Or Adonai* but rather material of

---

[88] Cf. notes 66 (p. 448) and 78 (p. 456) on Prop. I, Part II; n. 23 (p. 556, 558) and 33 (p. 663) on Prop. XV.

[89] Cf. n. 80 (p. 457) on Prop. I. Part II.

those class room discussions out of which the *Or Adonai* was composed.

The period which witnessed the rise of opposition to philosophy among Jews was also the period of the greatest philosophic activity among them. The knowledge of Aristotle which became widespread through the Hebrew translations of Averroes created a genuine interest in the study of philosophy as an independent discipline, irrespective of its bearing upon problems of religion. The works of Aristotle were included as a subject in the school curriculum. Expositions and studies of Aristotle became a popular form of literature. In certain families specialization in the works of Aristotle or Averroes became a tradition. Especially notable for this was the Shem-ṭob family, the two brothers, Joseph and Isaac (fifteenth century) and the son of the former, Shem-ṭob. Sons and grandson of Shem-ṭob Ibn Shem-ṭob, who was active as an opponent of philosophy, they became champions of philosophy and strict partisans of Averroes—not to be confused, however, with the hybrid Averroism of the Scholastics. It was therefore quite natural for them to come out in the defense of Aristotle as against Crescas. All these three authors appear as critics of Crescas. For our present purpose only two are important, Isaac ben Shem-ṭob and his nephew Shem-ṭob ben Joseph ben Shem-ṭob.

Isaac ben Shem-ṭob was more prolific a writer than he is generally considered. He was the author of at least fourteen works, of which eight are still extant.[90] Among these are four commentaries on Averroes' *Intermediate Physics*, evidently successive revisions of lectures delivered before students. We shall designate them as *first, second, third, fourth* successively. The *first, third,* and *fourth* are preserved in the library of Trinity College, Cambridge, bearing no name of author, but his authorship of

[90] See H. A. Wolfson, "Isaac Ibn Shem-tob's Unknown Commentaries on the *Physics* and His Other Unknown Works" in *Freidus Memorial Volume.*

these works has been established by the present writer.[91] Of the *second*, there are two copies, one in Munich, wrongly ascribed to Isaac Albalag, and the other in the University Library, Cambridge. In all but the *fourth* there are refutations of Crescas. In the *second*, the name of Crescas is mentioned in two places, where he is referred to as Ibn Ḥasdai.[92] In three other places references to "one may say," "one may raise a doubt" and "a certain one of the philosophers" can be traced to Crescas.[93] In his *first* commentary references to Crescas can be discerned under the guise of such expressions as "one may ask," "one may object," "some one has asked," "some one has objected'[94] or in the commentator's excessive zeal to justify a certain statement of Aristotle which, upon examination, is found to have been assailed by Crescas.[95] In the *third* commentary there is one discussion introduced by "some one asks," which probably has reference to Crescas.[96]

His nephew Shem-ṭob ben Joseph ben Shem-ṭob is best known for his commentary on the *Guide*, which is printed together with the text in almost every edition of the work. He is also the author of a supercommentary on Averroes' *Intermediate Physics* of which only one copy is extant in the Bibliothèque Nationale in Paris. In both of these works he takes occasion to criticise Crescas' commentary on the twenty-five propositions, referring to him either as Rabbi Ḥasdai or as Rabbi Ibn Ḥasdai.[97] But more than his criticism is of interest to us his personal estimate

[91] *Ibid.*

[92] חסדאי 'ן, see n. 40 (p. 424) on Prop. I, Part II; n. 8 (p. 479) on Prop. II.

[93] See n. 1 (p. 395) on Prop. I, Part II (יש לאומר שיאמר); n. 44 (p. 428) on Prop. I, Part II (ויש למספק שיספק); n. 22 (p. 650) on Prop. XV, Part II (ועוד יש כמו שחשב חכם אחד מן החוקרים) and למספק שיספק.

[94] See notes 1 (p. 396, ויש לשאול), 4 (p. 398, ויש מי שישאל), 40 (p. 425 ויש מי שהקשה) and 48 (p. 431 ויש להקשות) on Prop. I, Part II.

[95] See n. 44 (p. 428) on Prop. I, Part II.

[96] See n. 4 (p. 398) on Prop. I, Part II (ויש מי שישאל).

[97] See notes 1 (p. 394, הרב חסדai), 44 (p. 427, הרב ן' חסדאי) and 57 (p. 441, הרב ן' חסדאי) on Prop. I, Part II; n. 23 (p. 549, הרב ן' חסדאי) on Prop. VII.

of Crescas. In his commentary on Maimonides he concludes his proof of the first proposition with the following words: "When you have grasped the meaning of these two arguments you will be able to answer all the objections against the Master raised by Rabbi Ḥasdai in his commentary on this proposition, for against these two arguments no doubt and objection can be raised except by a perverse fool who is incapable of understanding. Similarly all the objections and criticisms levelled by Rabbi Ḥasdai against the Aristotelian proofs of this proposition are mere figments of the imagination, for the truth of these proofs can be understood by anyone whom God has endowed with reason and understanding to be able to distinguish between truth and falsehood."[98] In his commentary on Averroes he also uses words to the same effect: "To this we answer that his [Rabbi Ḥasdai's] contention is quite right, but Aristotle is addressing himself here to men of intelligence and understanding . . . inasmuch as thou, who art of sound mind, already knowest . . ."[99] Again, "Aristotle is addressing himself here to a man of good sense."[100] The implication of these passages is quite clear, Crescas is a "perverse fool" and is lacking in good sense and understanding. There is the note of an *odium philosophicum* here which has in it more odium than the proverbial *odium theologicum*. To a confirmed Aristotelian like Shem-ṭob, evidently, any attempt to question the veracity of his master's teachings could not be explained except on the ground of a perversity of judgment. Or, perhaps, Shem-ṭob was merely re-echoing a prevalent contemporary opinion about Crescas.

---

98 וכשתבין אלה הדרכים יבוטל מעליו כל הטענות שעשה הרב ר' חסדאי על הרב על ביאור זאת ההקדמה, כי באלו השני מופתים אין ספק ולא דחיה אל מסכל מתעקש ובלתי מבין הדברים. ואף גם כן כל הקושיות וההדחיות שעשה הרב ר' חסדאי על הביאורים שעשה אריסטו על זאת ההקדמה הם הזיות, יבינם מי שנתן לו השם שכל ודעת להבין האמת והשקר.

99 Cf. n. 1 (p. 394) on Prop. I, Part II: . . . . אבל אריסטו ידבר עם אנשי השכל והתבונה

אבל אחר שאתה, הבריא השכל, כבר ידעתה.

100 Cf. n. 44 (p. 427) on Prop. I, Part II: שאריסטו ידבר עם בעל שכל.

The approval which Crescas failed to receive from the Jewish Aristotelians was granted to him in generous measure by the non-Jewish opponents of Aristotle. With the setting in of the reaction against Aristotle, which is marked, if indeed not brought about, by a revival of the views of the early Greek philosophers, Crescas came into his own. The exponents of that movement saw in Crescas a kindred spirit, for he, too, fought against Aristotle by setting up in opposition to him the views of pre-Aristotelian or post-Aristotelian philosophers. One of these, Giovanni Francesco Pico della Mirandola, in his work *Examen Doctrinae Vanitatis Gentium*, draws frequently upon Crescas for the confirmation of his own views in the discussion of such problems as vacuum, place, motion and time.[101] Sometimes the name of Crescas is mentioned, and in such instances he is referred to as Hebraeus R. Hasdai, or Hebraeus Hasdai or R. Hasdai. The passages from the *Or Adonai* are sometimes translated but more often paraphrased. The accuracy of these translations or paraphrases of Crescas would indicate that he must have received his knowledge of Crescas from some learned Jew, for even if he himself had been a student of Hebrew as his more celebrated uncle Giovanni Pico della Mirandola he could hardly have known enough of the language to read and understand Crescas' work.[102] This confirms us in the belief that a great deal of Jewish philosophy was transmitted orally to non-Jews through the medium of Jewish assistants and that one must not confine the study of Jewish influence upon mediaeval philosophy to Hebrew works which happened to have been translated into Latin. Ever since the time of Emperor Frederick II, Jewish scholars had been used

---

[101] Cf. notes 4 (p. 398) 10, 12 (pp. 402–3), 22, 24, 26, 29, (pp. 412–17) 33, 34, 36 (pp. 41 –22), 66, 68 (p. 449) and 78 (p. 456) on Prop. I, Part II; n. 14 (p. 560) on Prop. VIII; n. 5 (p. 564) on Prop. IX; notes 20 and 22 (p. 625) on Prop. XIII; notes 22 (p. 650), 23 (p. 658), 27 (p. 661), 30 (p. 662) and 31 (p. 663) on Prop. XV.

[102] Cf. Joël, *Don Chasdai Crescas' religionsphilosophische Lehren*, pp. 9 and 83.

in Europe as intermediaries. Of some the names are known; but there must have been others whose names are unknown to us.

If it was possible for Giovanni Francesco Pico della Mirandola to become acquainted with some of Crescas' criticisms of Aristotle through some unknown Jewish scholar, we have reason to believe that it is not a mere fortuitous coincidence that many of Giordano Bruno's strictures on Aristotle have a reminiscent ring of similar strictures by Crescas. The name of Crescas is not mentioned by Bruno, but still one cannot help feeling that there must be some connection between them. While any single one of his arguments might have occurred to any one who set out to study Aristotle critically, the accumulation of all of those arguments creates the impression that there must have been some connecting link between Crescas and Bruno. Like Crescas, Bruno argues that Aristotle's definition of place does not apply to the place of the uttermost sphere.[103] Again, like Crescas, Bruno tries to prove the existence of a vacuum by arguing that according to Aristotle himself the nothingness outside the finite world must be a vacuum and that since that vacuum cannot be limited by a body it must be infinite.[104] Like Crescas, he argues against Aristotle's denial of the existence of an infinite force in a finite body by drawing a distinction between infinite in extension and infinite in intensity.[105] Both of them argue against Aristotle's theory of the lightness of air by the use of the same illustration, the descent of air into a ditch.[106] But more important than these individual arguments is Bruno's refutation of Aristotle's arguments in De Caelo against the possibility of circular motion in an infinite body, which bear a striking resemblance to the criticism levelled against them by Crescas. Both of them dismiss all these arguments by declaring that those who believe

[103] Cf. n. 58 (p. 443) on Prop. I, Part II.
[104] Cf. n. 36 (p. 422) on Prop. I, Part II.
[105] Cf. n. 7 (p. 613) on Prop. XII.
[106] Cf. n. 23 (p. 414) on Prop. I, Part II.

the universe to be infinite claim also that it is immovable.[107]
Both of them argue that the infinite would be figureless,[108] that
it would have no weight and lightness,[109] that it would have
neither end nor middle,[110] and that when an infinite acts upon
a finite or upon another infinite the action would be finite.[111]
Both of them at the conclusion of their refutation of the arguments against infinity take up Aristotle's discussion of the impossibility of many worlds and refute it by the same argument.[112]
That two men separated by time and space and language, but
studying the same problems with the intention of refuting Aristotle, should happen to hit upon the same arguments is not
intrinsically impossible, for all these arguments are based upon
inherent weaknesses in the Aristotelian system. But knowing
as we do that a countryman of Bruno, Giovanni Francesco Pico
della Mirandola, similarly separated from Crescas in time and
space and language, obtained a knowledge of Crescas through
some unknown Jewish intermediary, the possibility of a similar
intermediary in the case of Bruno is not to be excluded.[113]

There was no need for some unknown intermediary to furnish
Spinoza with his undoubted knowledge of Crescas' work. Crescas' revised form of the cosmological proof of the existence of
God is reproduced by Spinoza with the acknowledgment that he
has found it "apud Judaeum quendam Rab Ghasdai vocatum."[114]

---

[107] Cf. n. 102 (p. 664) on Prop. I, Part II.
[108] Cf. n. 122 (p. 470) on Prop. I, Part II.
[109] Cf. n. 49 (p. 431) on Prop. I, Part II.
[110] Cf. n. 125 (p. 472) on Prop. I, Part II.
[111] Cf. n. 111 (p. 466) on Prop. I, Part II.
[112] Cf. notes 126 (p. 472) and 130 (p. 476) on Prop. I, Part II.
[113] General suggestions as to a similarity between Crescas and Bruno
have been made by the following authors: Joël, *Don Chasdai Crescas' religionsphilosophische Lehren*, p. 8; Julius Guttman, "Chasdai Crescas als
Kritiker der aristotelischen Physik" in *Festschrift zum siebzigsten Geburtstage
Jakob Guttmanns*, p. 45, n. 3; Waxman, *The Philosophy of Don Hasdai Crescas*, p. 45.
[114] Cf. Epistola XII olim XXIX.

But more than this. His entire discussion of the infinite, both the restatement of the arguments against its existence and his refutation of these arguments, are directly based upon Crescas. This conclusion does not rest upon similarities between restatements of individual arguments or between individual refutations, for each of these individually could be accounted for by some other source. But there are certain intrinsic difficulties in Spinoza's presentation of the views of his "opponents" which could not be cleared up unless we assumed that he had drawn his information from Crescas. Furthermore, there is something in the literary form in which the problem is treated by him in two independent sources, in the *Ethics* and in his correspondence, which seem to suggest Crescas as his immediate source. In the *Ethics* Spinoza enumerates three "examples" by which the philosophers have tried to prove the impossibility of an infinite. In his letter to Ludovicus Meyer he declares that the problem of the infinite is considered "most difficult, if not insoluble," owing to a failure to make three "distinctions." Now, it happens that these three "distinctions" are suggestive of three refutations advanced by Crescas against three of Aristotle's arguments which correspond to Spinoza's three "examples."[115]

Perhaps one should be careful not to overestimate the importance of Crescas' influence upon these men in evaluating their philosophy. One cannot, however, altogether overlook the importance of the striking resemblances between them if one wishes to evaluate the place of Crescas in the general history of philosophy. He anticipated these men in his criticism of Aristotle; his criticism, like theirs, took the form of a revival of the views of pre-Aristotelian Greek philosophers; and what is of still greater importance, he opened for us the vistas of a new conception of the universe.

[115] See H. A. Wolfson "Spinoza on the Infinity of Corporeal Substance" in *Chronicon Spinozanum* IV (1924–26), pp. 79–103; cf. notes 1 (p. 394), 37 (p. 423) and 112 (p. 466) on Prop. I, Part II.

CHAPTER II

Infinity, Space and Vacuum

Towards the end of his proof of the first proposition denying the possibility of an infinite magnitude—a proof made up of material drawn from other sources—Crescas sums up his own contribution to the subject. In the first place, he says, he "has recast those arguments in their logical form." Then, he has "restated them in exceeding brief language." Thirdly, he has strengthened "some of them by introducing points not mentioned by any of the other authors." Finally, he has arranged the arguments according to some logical plan, for in their original form, he claims, they lacked any orderly arrangement. These claims of Crescas are only partly true. It is true indeed that he "has recast those arguments in their logical form," if by this he means to refer to his method of presenting every argument in the form of a syllogism. It is also true that he "has restated them in exceeding brief language," if by this he means that he did not reproduce his authorities verbatim. But his statement that he has strengthened some of the arguments "by introducing points not mentioned by any of the other authors" is not altogether true, unless he means by it that he has strengthened some of the arguments advanced by one author by points taken from the arguments of another author. As a matter of fact, Crescas did not introduce new arguments of his own; what he did was simply to introduce into the Aristotelian arguments taken from Averroes the arguments advanced by Altabrizi or to incorporate within them some remarks by Gersonides. Nor is it altogether true that the arguments in their original form were lacking any orderly arrangement. As a matter of fact, the argu-

[1] This chapter is based upon Propositions I, II and III.

38

ments are presented in a well-ordered fashion by both Aristotle and Averroes, and that order of arrangement has been retained by Crescas practically intact. What he has done is simply to have modified somewhat the original plan of classification.[2]

[2] The following analysis will bring out the relation between Averroes' arrangement of the arguments and that of Crescas.

Averroes

I. Argument against the existence of an incorporeal infinite magnitude arranged in the order of (a), (b), (c), (d).

II. Arguments against an infinite existing as an accident in sensible bodies, divided and subdivided as follows:

A. General or logical argument.

B. Four physical arguments: 1, 2, 3, 4 (a), 4 (b).

(These two classes of arguments are to be found in the *Intermediate Physics*).

III. Arguments from motion, divided and subdivided as follows:

A. Six arguments to prove that an infinite could not have circular motion: 1, 2(a), 2(b), 3, 4, 5, 6(a), 6(b).

B. Two arguments to prove that an infinite could not have rectilinear motion: 1(a), 1(b), 2.

IV. Four general arguments: 1, 2, 3, 4.

(These two classes of arguments are to be found in the *Intermediate De Caelo*).

Crescas

His "First Class of Arguments" corresponds to Averroes' I, but parts (a) and (d) are merged together and parts (b) and (c) are given in reversed order. See n. 7 (p. 332) on Prop. I, Part I.

This class of arguments includes also the following additions:

1. Arguments against the existence of a vacuum, taken from Averroes. See Prop. I, Part I, p. 139.

2. Two reinforcing arguments, taken from Averroes, but given in reversed order. See n. 49 (p. 344) on Prop. I, Part I.

3. One of the three arguments of Altabrizi. See Prop. I, Part I, p. 149.

His "Second Class of Arguments" corresponds to Averroes' II, but with the following variations:

1. Averroes' II B 2 is omitted. See n. 65 (p. 351) on Prop. I, Part I.

2. Crescas' *second physical* argument corresponds to Averroes' II B 3. See *ibid.*

3. Crescas' *third physical* argument corresponds to Averroes' II B 4 (a). See n. 68 (p. 352), *ibid.*

4. Crescas' *fourth physical* argument corresponds to Averroes' II B 4 (b) into which is incorporated a restatement of Aristotle' discussion about place also taken from Averroes. See n. 73 ff. (p. 354f.), *ibid.*

His "Third Class of Arguments" corresponds to Averroes' III, but with the following variations:

In order to enable ourselves to recapitulate Crescas' critique of Aristotle's rejection of infinity without having to restate Aristotle's own arguments, we shall first briefly outline the main drift of Aristotle's discussion.

The infinite, according to Aristotle, may mean two things. It may mean that which is limitless because it is excluded from the universe of discourse of limitation just as a voice is said to be invisible because it is excluded from the universe of discourse of visibility. Or it may mean that a thing which is capable of being limited is limitless. Dismissing the term infinite in the first sense as something outside the scope of his discussion, he confines himself to the discussion of infinity as applied to some kind of extension or magnitude which, though capable of being finite, is infinite. He shows that there can be no infinite incorporeal extension on the ground that no incorporeal extension exists. He then shows by five arguments that no corporeal extension can be infinite. All these are discussed in the *Physics* and in the *Metaphysics*. He further proves the impossibility of an infinite extended body by showing that none of the sublunar

1. The order of A and B are reversed in Crescas. See n. 90 (p. 365), *ibid.*

2. Under *rectilinear* motion Crescas gives *three* arguments. The *first* does not correspond to Averroes' arguments from *rectilinear* motion but rather to his II B 2 (see notes 106, p. 375, and 116, p. 376, *ibid.*), incorporating within it, however, certain other elements (see n. 91, p. 365, *ibid.*). The *second* corresponds to Averroes' III B 1 (b), incorporating within it, however, a passage from Averroes' III B 1 (a). (But see notes 104, p. 364, and 107, p. 375, *ibid.*). The *third* corresponds to Averroes' III B 2.

3. Under *circular* motion Crescas follows Averroes' enumeration of six arguments, but with the following variations:

At the end of the *first* argument he adds an argument from Altabrizi. See n. 133 (p. 381) *ibid.*

The *second* argument reproduces only Averroes' III A 2 (a). See n. 136 (p. 382) *ibid.*

The *third* argument is composed of Averroes' III A 2 (b), III A 3, and another one of Altabrizi's arguments. See n. 141 (p. 383) *ibid.*

The *sixth* argument reproduces only Averroes' III A 6 (a).

His "Fourth Class of Arguments" reproduces only Averroes' IV 1 and IV 2. See n. 157 (p. 390) *ibid.*

elements could be infinite, for the sublunar elements are endowed with rectilinear motion and no infinite can have rectilinear motion, and also by shewing that neither could the translunar element be infinite, for the translunar element is endowed with circular motion and no infinite can have circular motion. These last two classes of arguments are discussed in *De Caelo*. Though Crescas in his critique tries to refute all these arguments, it is not his intention to establish the existence of an infinite extended body. His main purpose is to establish the existence of an incorporeal extension and to show that that incorporeal extension can be infinite. We shall therefore reverse the order of his argument and leave the discussion of an incorporeal extension to the end.

There is a common fallacy, contends Crescas, running through five of Aristotle's arguments. In all of these, Aristotle argues against the existence of an infinite from the analogy of a finite. Conceived in terms of a finite magnitude, the infinite, according to Aristotle, cannot have existence because as a magnitude it must be contained by boundaries,[3] it must have gravity or levity,[4] it must have a spherical figure,[5] it must revolve round a centre,[6] and finally, it must be surrounded by external perceptible objects.[7] All of these assumptions, argues Crescas, however true with regard to finite magnitudes, are ill-conceived with regard to an infinite. The infinite, if it exists, will not be contained by boundaries,[8] will be devoid of both gravity and levity,[9] will be shapeless with regard to figure,[10] moving circularly but

[3] Cf. Prop. I, Part I (p. 151), n. 57.

[4] *Ibid.* (p. 161), n. 106.

[5] *Ibid.* (p. 173) n. 144.

[6] *Ibid.* (p. 175) n. 158.

[7] *Ibid.* (p. 177), n. 160.

[8] Cf. Prop. I, Part II (p. 191), n. 40.

[9] *Ibid.* (p. 195), n. 49.

[10] *Ibid.* (p. 213), n. 122.

not round a centre,[11] and, finally, though moving by volition, will not require external objects to act upon it as stimuli.[12] In fine, if an infinite exists, it must not be conceived in any of the terms by which a finite object is described.

Nor would it follow that the infinite can be neither composite nor simple.[13] Quite the contrary it can be either composite or simple.

In the first place, the infinite may well be a composite body, consisting of an infinite number of elements. To be sure, Aristotle has rejected the possibility of an infinite number of elements. But his rejection is based upon an assumption that the elements must be known whereas an infinite number cannot be known. But why, asks Crescas, should the elements have to be known?[14]

In the second place, the infinite may be conceived to be either a composite body consisting of a finite number of elements one of which is infinite in magnitude, or a simple body consisting of one infinite element. Both of these possibilities have been rejected by Aristotle on the ground that no infinite element could exist among finite elements, for whatever that infinite element may be, whether one of the four known elements or some other element outside the four, it would have to possess characteristic properties of its own, radically distinct from those of the other elements, but, being infinite, it would in course of time overwhelm and destroy the other finite elements.[15] Crescas, however, contends that an infinite element outside the four elements is not impossible. That element, while it would indeed be distinct from the four other elements, would not have to possess positive qualities of its own. It could be conceived as

[11] *Ibid.* (p. 215), n. 125.
[12] *Ibid.*
[13] Prop. I, Part I (p. 151), n. 60.
[14] Cf. Prop. I, Part II (p. 193) n. 44. See also refutations of this argument quoted in the note (p. 426).
[15] Cf. Prop. I, Part I (p. 151), n. 63.

being without any form and quality but only capable of assuming all kinds of possible forms and qualities. It could furthermore be conceived in its relation to the other four elements as matter to form or subject to quality. Consequently though infinite, it would never cause the corruption of the other finite elements, for its relation to them would not be as one element to another but rather as matter to form.[16] Crescas cites the case of the celestial element, which, according to Aristotle, though distinct from the four sublunar elements, is devoid of any positive qualities whatsoever.[17]

Again, Aristotle enforces his preceding argument by a statement that if one of the elements were infinite, it would have to be so in all its dimensions, and so there would remain no room in the universe for the other elements.[18] This does not follow, according to Crescas, for it is quite possible to conceive of an infinite element that is infinite in only one dimension. Infinity, in the present argument, is not assumed by Aristotle to be something essential to the element; it is only accidental to it, as any other accidental quality. As such, the assumption that one of the dimensions is infinite would not necessarily lead to the assumption that the other dimensions would likewise be infinite.[19]

Another argument against a corporeal infinite magnitude advanced by Aristotle is based upon his conception of place.[20] Aristotle himself divides this argument into two parts. First, from the fact that place has only a finite number of directions, namely, up and down, right and left, before and behind, he infers that everything that exists in place must be finite. Second, from the fact that each of these six directions is finite, he infers that

---

[16] Cf. Prop. I, Part II (p. 193), n. 45. This would seem to be the point of Crescas' argument in that passage.

[17] *Ibid.* (p. 193), n. 46.

[18] Prop. I, Part I (p. 151), n. 64.

[19] Prop. I, Part II (p. 195), n. 48.

[20] Prop. I, Part I (p. 153), n. 68.

the object existing in place must be finite. In restating the second part of Aristotle's argument, Averroes introduces Aristotle's formal definition of place and makes the entire argument hinge upon that definition. Similarly Abraham ibn Daud advances an argument against the existence of an infinite based upon Aristotle's formal definition of place. Probably following these precedents Crescas likewise makes of the second part of Aristotle's argument from place an independent argument in which he reproduces a complete summary of Aristotle's discussion leading up to his definition of place.[21]

Place is defined by Aristotle as the limit of the surrounding body. This definition is the result of a discussion of the nature of place in which Aristotle lays down three conditions. First, place must surround that of which it is the place. Second, it must be equal to the thing surrounded by it; it can be neither smaller nor greater than the thing surrounded. Third, it must not be a part of the thing surrounded by it but something separate from that thing.[22] In some of the works of Arabic and Jewish philosophers a brief summary of these three conditions is sometimes ascribed to Aristotle as the definition of place. Following these precedents, therefore, Crescas restates Aristotle's definition of place as the surrounding, equal and separate limit, that is to say, the limit of the surrounding body, equal to the body surrounded, but separate from it.[23]

The implication of Aristotle's definition is that there can be no place unless one body is contained by another body, for it is only then that there is a surrounding, equal and separate limit. Inasmuch as everything within the universe is surrounded by something else and all things are ultimately surrounded by the all-surrounding outermost sphere, everything within the

---

[21] *Ibid.* (p. 153), n. 71 (p. 352) and n. 73 (p. 354).

[22] *Ibid.* (p. 153), n. 75.

[23] The relation of this phrasing of the definition of place to Aristotle's phrasing is fully discussed in n. 89 (p. 362) on Prop. I, Part I.

universe is in place. Thus, for instance, in the case of the four
sublunar elements, earth is surrounded by water, water by air,
air by fire, and fire by the lunar sphere, and similarly in the
case of the celestial spheres, each sphere is surrounded by another sphere until we come to the outermost sphere. But how
about that outermost sphere which is not surrounded by anything on the outside, is it in place or not? To this question the
following answer is given by Aristotle: "But heaven is not, as
we have said, anywhere totally, nor in one certain place, since
no body surrounds it; but so far as it is moved, so far its parts
are in place, for one part adheres to another. But other things
are in place accidentally, as, for instance, soul and the heaven,
for all the parts are in a certain respect in place, since in a circle
one part surrounds another."[24] To the commentators of Aristotle
this passage seemed to bristle with all kinds of difficulties. The
question was raised as to what did Aristotle mean by the term
"heaven." Did he mean by it the universe as a whole, or only
the outermost sphere, or every one of the spheres? Again, what
did he mean by the term "accidentally" which lends itself to
several interpretations? No less than six interpretations have
been advanced.[25] But for our present purpose only two of these
interpretations are necessary.

According to Themistius the term "heaven" refers only to the
outermost sphere. That outermost sphere, not having anything
surrounding it, has as its place the limit of the body surrounded
by it, that is, the convex surface of the sphere immediately surrounded by it. Thus the place of the outermost sphere is an
equal and separate limit but not a surrounding limit; it is rather
a surrounded limit. The outermost sphere, furthermore, is said
to be in place only accidentally. All the other spheres, however,
have as their place the limit of the body surrounding them, that

[24] *Physics* IV, 5 212b, 8–13.
[25] See discussion on this point in n. 54 (p. 432) on Prop. I, Part II.

is, the concave surface of the spheres which respectively surround them. Thus, in contradistinction to the place of the outermost sphere, the place of all the other spheres is a surrounding, equal and separate limit, and it is what is called an essential place.[26]

According to Avempace and Averroes not only the outermost sphere but also all the other spheres have as their place the convex surfaces of the spheres that are respectively surrounded by them. They maintain that Aristotle's definition of place as the surrounding limit refers only to the sublunar elements. In the case of the celestial spheres, however, place is the surrounded limit. But there is the following difference between Avempace and Averroes. According to the former, all the spheres are in place essentially; according to the latter, all the spheres are in place accidentally.[27]

With these preliminary remarks, we may now turn to Crescas' criticism. His discussion may be arranged under three headings: First, his refutation of Aristotle's argument from the definition of place against the existence of an infinite. Second, his criticism of that definition. Third, his own definition of place.

The infinite, argues Aristotle, could not exist in place since place is the limit of a surrounding body and the infinite cannot be surrounded by anything. The argument is inconclusive. True, the infinite cannot have a surrounding limit, but still it can have a surrounded limit, namely, the convexity of the sphere which it surrounds, for in this manner is the place of the outermost sphere conceived by Aristotle according to most of his interpreters.[28]

Aristotle's definition of place furthermore will give rise to many difficulties and absurdities:

First, if we accept Themistius' interpretation of Aristotle's view as to the place of the "heaven," the term place when ap-

[26] Ibid.
[27] Ibid.
[28] Prop. I, Part II (p. 195), notes 50–54.

plied to the outermost sphere and the other spheres will have to be understood in different senses, for in the case of the former it will mean the surrounded limit whereas in the case of the latter it will mean the surrounding limit.[29]

Second, if we accept the interpretation of Avempace and Averroes, a still greater absurdity will follow. According to both of them, the place of the celestial spheres is the centre round which they rotate. Now, according to Aristotle, bodies are naturally adapted to be in their place, and toward their place they tend. Consequently, according to Avempace's and Averroes' interpretation, the celestial bodies must be assumed to be naturally adapted to abide in something beneath them. But that is absurd. For not even fire is adapted to anything beneath it.[30]

Third, Avempace's and Averroes' views as to the place of the celestial spheres rests upon the Aristotelian assumption that the rotation of a sphere implies the existence of a fixed, round magnitude, distinct from the sphere itself, upon which the sphere rotates as its centre. This is an impossible absurdity. There is nothing but the mathematical point at the centre, and this cannot be the place of the sphere.[31]

Fourth, if as Aristotle claims the proper place of the elements is that to which they naturally tend, then the centre of the universe should be the proper place of earth.[32] But the centre is a point, and cannot be place.[33]

Fifth, there is the following difficulty. According to Aristotle, place must satisfy three conditions: it must surround the body, it must be something distinct from it, and it must be equal to

[29] *Ibid.* (p. 197) notes 58–59.
[30] *Ibid.* (p. 197) notes 67–69.
[31] *Ibid.* (p. 199) notes 70–73.
[32] As for the differences of opinion with regard to the place of earth, see n. 64 (p. 445) on Prop. I, Part II.
[33] Prop. I, Part II (p. 199), n. 78.

it. Again, according to Aristotle, the parts of a continuous body have no independent motion *in* the whole but move together *with* the whole, and that motion of theirs is to be described as essential. Furthermore, the parts of a continuous body are said to exist in that body as parts in a whole and not as things in a place. The question may therefore be raised, what is the place of the parts of a continuous body? Will their place satisfy the three conditions mentioned? To take a concrete example: Air is a continuous body. The proper place of air as a whole is the concavity of fire. But what will be the proper place of any part of air taken from the middle? That it must be in its proper place is clear enough, since no part of air is moved independently without the whole and no element is without motion when out of its proper place. Two alternatives are possible. First, that the place of the part of air is identical with that of the whole. But then, the place will not be equal to the object occupying it. Second, that the place of the parts of air will be the other parts of air surrounding it. But then, the place will not be distinct from its occupant. Furthermore, the place of the whole of the . air and of any part thereof will not be the same.[34]

Sixth, if we accept Aristotle's definition of place, that it is the limit of the surrounding body, the place of the same cubic block, for instance, will be smaller when existing as a whole than when broken into parts. But it is absurd to think that the place of the same object as a whole would be smaller than the sum of the places of its parts.[35]

Crescas has thus shown that Aristotle's definition of place as the surrounding, equal and separate limit of the contained object is erroneous, and furthermore that "proper place" cannot be described as that toward which the elements are naturally moved. But before adopting his final definition of place, Aris-

[34] See notes 60–66 (pp. 443–449) on Prop. I, Part II.
[35] See p. 199, and n. 80 (p. 457), *ibid.*

totle has tentatively discussed three other provisional definitions, one of which asserted that the place of a thing is the interval or the vacuum or the distance which is occupied by the thing.[36] This definition, which has been rejected by Aristotle, is now adopted by Crescas.[37] Place is thus according to him the interval or the vacuum or the distance of a thing. Not that there is no distinction between vacuum and place, but the distinction is not in their essential character. What is called vacuum when it contains no body, becomes place when it contains a body.[38] This, of course, would imply the existence of a vacuum, but its existence, as we shall see, is maintained by Crescas on independent grounds. According to this definition of place, the Aristotelian proper places are dispensed with, for wherever an object happens to be, that is its proper place. Furthermore, the part is as much in its own place as is the whole. Finally natural motion is not to be explained by any tendency toward a proper place, which, according to this new definition of place, does not exist. Natural motion, as we shall see later on, is explained by Crescas in another way.[39]

In rejecting the existence of an infinite sublunar element, Aristotle employs the following argument. The infinite could not be a simple element of infinite magnitude, because it would then be unable to perform rectilinear motion. Nor could it be a composite element consisting of an infinite number of heterogeneous parts, for as every part requires a proper place, it would follow that there would be an infinite number of proper places. But an infinite number of proper places is impossible, for the very idea of proper places is derived from natural motion, and natural motion is finite in kind. Now, that natural motion is finite in kind is an empirical fact. Motion is either from the

[36] Prop. I, Part I (p. 155) notes 79–80.
[37] Prop. I, Part II, notes 55 (p. 441), 75 (p. 455) and 82 (p. 458).
[38] See n. 31 (p. 417) on Prop. I, Part II.
[39] See n. 76 (p. 456) on Prop. I, Part II.

centre of the universe, or towards it, or round it; that is to say, upward, downward, or circular. Motion being thus finite in kind, it is argued, the proper places of elements endowed with motion must likewise be finite.[40]

It is the conclusion that is found fault with by Crescas. Assuming the existence of an infinite element composed of an infinite number of heterogeneous parts, Crescas endeavors to show that an infinite number of proper places is not impossible. While it is true, he argues, that the proper places must be finite in kind, they can still be infinite in number. Suppose then we say that the universe consists of an infinite number of concentric spheres. The motions would then be still finite in kind, centrifugal or centripetal, determined by their direction with regard to a common centre, but the centrifugal or upward motion would be infinite in number since there will be an infinite number of circumferences. Take, for instance, the motion upward, from the centre of the universe to the circumferences of the infinite number of spheres: all such motions from the centre to the infinite circumferences are one in kind, the sphere being concentric, but they will be infinite in number since they are individually different, each having a proper place of its own at the concavity of an individually different sphere. Thus since the number of these proper places are infinite, the number of the elements may be infinite.[41]

To be sure, such a conception of the universe may be objected to on the ground that in an infinite number of concentric spheres there could be no absolute upper place to correspond to its absolute lower place, which is the centre; but the very distinction of upward and downward, it may be replied, is based upon the conception of a finite universe. If you admit its infinity, as do the Atomists, no such distinction must needs be assumed.[42]

[40] Prop. I, Part I (p. 157), n. 91 ff.
[41] Prop. I, Part II (p. 203), notes 97–98.
[42] See n. 98 (p. 463) on Prop. I, Part II.

It may indeed also be argued that if the infinite consists of an infinite number of heterogeneous elements, those elements would have to be not only infinite in number but also infinite in kind, and consequently the infinite number of corresponding places would have to be not only infinite in number but also infinite in kind. But this argument, too, is inconclusive, for according to Aristotle himself, while the number of places must correspond to the number of elements, those places, unlike the elements, must not necessarily be all different in kind. Take, for instance, the sublunar elements, which are four in number and differ from each other in kind. Their corresponding places are likewise four in number; but as to kind, they are less than four, for the only generic distinction between them is that of above and below. Hence there is no reason why there should not exist an infinite composite element, consisting of an infinite number of heterogeneous parts, each of which would have its proper place in one of the infinite number of circumferences.[43]

Thus disposing of Aristotle's argument against the existence of an infinite rectilinearly moving sublunar element, Crescas then examines Aristotle's arguments against the existence of an infinite circularly moving translunar element. Starting with the proposition that the distance between the radii at the circumferences of an infinite sphere would have to be infinite, Aristotle proceeds to show by two arguments that the infinite sphere could not complete a revolution, inasmuch as no infinite distance is traversible.[44] It is the initial proposition that Crescas endeavors to disprove.

In the first place, he tries to show that to assume that the distance between two infinite radii at the circumference of the infinite sphere is infinite is intrinsically absurd. For if this assumption were true, it would have to apply to any pair of radii,

[43] See n. 103 (p. 373) on Prop. I, Part I.
[44] Prop. I, Part I (p. 169), n. 126 ff.

forming any angle at the centre. Suppose then that we take any point in the alleged infinite distance between any pair of infinite radii and through it draw a new radius. This new radius will form an angle at the centre with either of the other two radii, and still the distance between them will be finite, contrary to the assumption.[45]

In the second place, he tries to show that though the radii of an infinite sphere are infinite, the distance between them is always finite, for distance must be measured between two points by which it is bounded. Again, these points in the radii are at a finite distance from the centre, and, therefore, the distance between them must be finite. The distance is said to be infinite only in the sense of indefinite, that is to say, whatever distance you assume you may always assume one greater than it, since the radii are infinite. The distances are, therefore, infinite only in capacity, that is, they are always capable of increase, but not in energy. This distinction between potential and actual infinity is applied by Aristotle to number. To corroborate his view about the finitude of the distance, Crescas refers to Apollonius' discussion of the asymptote and quoting Aristotle's dictum that "every pair of contraries falls to be examined by one and the same science"[46] he concludes with a favorite type of Talmudic reasoning, the argument a minori ad majus. If in the case of infinitely approaching limits the distance always remains finite; a fortiori must the same hold true in the case of infinitely parting limits.[47]

Finally, he concludes that since the distance between any two points in the infinite radii is finite, the infinite sphere will be capable of completing a revolution, for at any given point the sphere, though infinite, will revolve on a finite axis. Though it

[45] Prop. I, Part II (p. 209), notes 108–110.
[46] *Metaphysics* XI, 3, 1061a, 19. Cf. n. 104 (p. 464) on Prop. I, Part II.
[47] Prop. I, Part II (p. 207), notes 103–107.

is impossible to perceive by the imagination how this could be done, still reason proves it to be so. For we can conceive by reason many things which we cannot perceive by the imagination.[48]

The underlying assumption in three other arguments[49] advanced by Aristotle against the existence of an infinite revolving sphere is that an infinite has no first point and that an infinite distance cannot be traversed in finite time. With this as a starting point it is argued that if an infinite revolving sphere existed, two infinite lines moving on a centre in contrary directions, or one moving and the other fixed, would have to meet at some first point and would have to be passed through in finite time. To this Crescas' reply may be restated as follows: Motion has no absolute beginning, for there can be no first part of motion, since motion is infinitely divisible. By the same token, the time of motion has no absolute beginning. When, therefore, two infinite lines meet, they do not meet at any absolute first point, nor is there any absolute beginning in the time when they first meet. Consequently, you cannot speak of two infinite lines meeting at a first point, or of an infinite distance being passed through in finite time. But, as said above, a revolving infinite sphere will revolve on a finite axis. Any distance, therefore, traversed by it in finite time will be finite.[50]

Having shown that Aristotle's arguments against a corporeal infinite magnitude are all inconsequent, Crescas proceeds to disprove also his arguments against an incorporeal infinite magnitude. The main objection against an incorporeal infinite magnitude is that no magnitude can be incorporeal. Every magnitude, by its nature, contends Aristotle, implies the existence of body. That is not true, says Crescas. It is a corollary of Aristotle's own proposition that there is no vacuum within

[48] *Ibid.* (p. 211), n. 112.
[49] *Second, third* and *sixth.* Prop. I, Part I (pp. 171–175).
[50] Prop. I, Part II (p. 211), notes 114–120.

or outside the world. But if we assume the existence of a vacuum, there exists also an incorporeal magnitude,[51] for a vacuum is nothing but extension devoid of body.[52] And thus Crescas enters into a minute discussion of Aristotle's arguments against the existence of a vacuum.

In his *Physics* Aristotle enumerates two theories which were held by early philosophers with regard to a vacuum. First, the vacuum is inseparable from the corporeal objects of the world, it is everywhere dispersed throughout the pores of the bodies, thus breaking up the continuity of the world. Second, there is no vacuum within the world, the world itself being continuous, but there is a vacuum beyond the world. The first of these views is ascribed to the Atomists, the second to the Pythagoreans.[53] Allusions to these two views occur also in Maimonides.[54] Five arguments in support of the existence of a vacuum are reproduced by Aristotle in the name of those philosophers.[55] One is based upon the assumption that without a vacuum motion would be impossible; or, in other words, the vacuum is the cause of motion. This assumption, however, is shown by Aristotle to be untenable, for the vacuum, he argues, could not be the cause of motion in any of the four possible senses of the term cause.[56] It is against this argument that Crescas now endeavors to uphold the existence of a vacuum.

Aristotle's refutation, contends Crescas, is based upon a misunderstanding of the Atomists' statement that the vacuum is the cause of motion. They had never considered the vacuum as the sole producing cause of motion. The vacuum to them was only an accidental cause, or rather a condition of motion,

[51] Prop. I, Part I (p. 139), n. 14 f.
[52] Prop. I, Part II (p. 189).
[53] See n. 7 (p. 400) on Prop. I, Part II.
[54] *Ibid.*
[55] These five arguments are divided by Crescas into two groups, one argument being negative and four being positive. See Prop. I, Part I (p. 139), n. 18.
[56] Prop. I, Part I (p 139), n. 19.

without which the latter, though its producing causes were present, could not take place. For they contend, and support their contention by various natural phenomena, that had there been no vacuum, bodies could not perform their motion on account of their impenetrability. Being thus only a condition of motion, and not its cause, the vacuum may exist even if it cannot be any of the four causes enumerated by Aristotle.[57]

Nor is Aristotle's next argument, namely, that the existence of a vacuum would make motion impossible,[58] more conclusive than the preceding one.[59] Having already explained that to the Atomists the vacuum is only an accidental cause, or rather a condition, of motion, removing as it does the possible obstruction that motion would encounter in a plenum, Crescas now inquires as to what would be the producing cause of motion if a vacuum existed. The producing cause of motion within a vacuum, says he, could be the same as is now assumed by Aristotle in a plenum, namely, the natural tendency of the sublunar elements towards their respective proper places, which is, for instance, the concavity of the lunar sphere with respect to fire and the centre of the universe with respect to earth.[60] It is with reference to those proper places that the motion of each element would be designated as being either natural or violent. It is natural when the element tries to escape from a foreign place and seeks to reach its own natural place; it is violent, when the element is forced away from its own natural place. But, argues Aristotle, in a vacuum the elements would have no reason for trying to escape one part in order to reach another, inasmuch as a vacuum is devoid of any definite character and all parts thereof are alike.[61] True enough, says Crescas.

[57] Prop. I, Part II (p. 181), n. 4.
[58] Prop. I, Part I (p. 141), n. 25.
[59] Prop. I, Part II (p. 183), notes 7–12.
[60] As for differences of opinion with regard to the place of earth, see n. 64 (p. 445) on Prop. I, Part II.
[61] Prop. I, Part I (p. 143).

The vacuum, throughout its entire extent from the earth to the lunar sphere, is the same in one part as in another, in so far as its own nature, or lack of nature, is concerned. But with reference to the earth and·the lunar sphere some parts of the vacuum may be called nearer while others may be called farther—an entirely external relation which is compatible with the neutral character of the vacuum itself. This difference in distance it will be which will make the elements within the vacuum try to escape one part in order to reach another. They will always tend to draw nearer to their proper places.[62] This explanation of motion within a vacuum, it should be noted, is advanced by Crescas only to show that Aristotle's theory of natural motion and proper places could be maintained even if a vacuum is assumed to exist. His own theory of motion is explained later.[63]

The argument from motion is still less applicable to the Pythagorean theory of the existence of a vacuum beyond the world. For if such a vacuum is conceived, the object within it would not move rectilinearly but rather circularly. Now circular motion, according to Aristotle, does not imply the existence of opposite termini and places. It is motion within one place, and is possible even within a homogeneous vacuum wherein there is no distinction of a *terminus a quo* and a *terminus ad quem*.[64]

Another argument against the existence of both a vacuum and an infinite is based upon what may be called Aristotle's laws of motion. According to Aristotle's laws of motion, the times of two motions, all things being equal, are proportional to the tenuity of the media in which the motion is performed, or to the weight of the moving objects, or to the motive forces of these objects. From these he infers that should the medium be a vacuum, or should the weight of the moving object or its motive

[62] Prop. I, Part II (p. 183), n. 10.
[63] See below p. 79.
[64] Prop. I, Part II (p. 183), notes 11–12.

force be infinite, the time would equal zero; that is to say, motion would be performed in no-time, which to him is impossible. Hence Aristotle concludes that neither a vacuum nor an infinite has actual existence.[65]

This view, however, was opposed by Avempace. The time of motion, according to him, is not due to the medium. Motion must be performed in a certain time, even if that motion were to take place within a vacuum. That time, in which motion is performed independently of its medium, is called by him the original time of motion, which remains constant and never disappears. The medium to him is not the cause of motion but rather a resistance to it. Aristotle's law that the time of two motions is proportional to their respective media is, therefore, erroneous. It is only true to say that the excess in the time of two motions over their original time is proportional to the resistance offered by their media.[66]

In opposition to Avempace and in defence of Aristotle, Averroes argues that the media are not mere resistances of motion; they rather determine the nature of the motion. The velocity of an object in air is greater than that of the same object in water not because air offers less resistance than water, but because motion in air is of an entirely different nature than motion in water. "For the motion in air is faster than that in water in the same way as the edge of an iron blade is keener than that of a bronze blade." Motion without a medium would be impossible, and the medium which causes its existence likewise determines its nature and velocity.[67]

In order to prove that both a vacuum and an infinite are possible, Crescas adopts Avempace's theory of an original time of motion, and proceeds to defend it in a rather indirect manner.

[65] Prop. I, Part I (p. 143), n. 31 f.
[66] See n. 13 (p. 403) on Prop. I, Part II.
[67] *Ibid.*

If Averroes' contention that the medium is a necessary condition of motion be accepted, it would likewise have to be true that the medium is a necessary condition in the existence of weight and lightness.[68] For weight and lightness are defined by Aristotle in terms of motion. "I call that simply light which is always naturally adapted to tend upward, and that simply heavy which is always naturally adapted to tend downward."[69] If Crescas, therefore, could prove that weight and lightness are independent of a medium he would thus indirectly establish that motion is likewise independent of a medium. This is exactly the line of attack he follows. He first tries to show how weight and lightness could be explained in such a way as would completely dispense with the requisite of a medium. The explanation which he offers is not original with Crescas; it is taken from the works of Aristotle, where it is attributed to the Atomists and Plato. According to this new explanation, the difference in the weight of the elements is explained as being due to a difference in their internal structure, which Crescas characterizes by saying "that both weight and lightness belong to the movable elements by nature." Or, in other words, there exists no absolute lightness, as is assumed by Aristotle, but all bodies possess some amount of weight.[70]

Since weight and lightness are not conditioned by the medium, it is not necessary to assume that the medium is essential to the existence of motion. In fact all natural elements tend toward the centre by reason of their weight. Thus it is only downward motion that may be called natural. Upward motion, on the other hand, is not natural; it must be explained by some mechanical principle. The cause of upward motion, says Crescas, and is in effect quoting the view of Democritus and Plato, is due to the

---

[68] See n. 20 (p. 410) on Prop. I, Part II.

[69] *De Caelo* IV, 4, 311b, 14–15.

[70] See notes 20–21 (p. 410) on Prop. I, Part II.

pressure of the more heavy elements upon the less heavy. All the elements being heavy, naturally tend toward the centre; but the heavier reach there sooner and thus compell the less heavy to move upward.[71]

Thus far Crescas has argued for Avempace's theory of an original time of motion and in opposition to Aristotle and Averroes, in order to show the possibility of temporal motion in a vacuum. But suppose we follow the view of Aristotle and Averroes that the medium is a prerequisite of motion and that within a vacuum motion would have to be in an instant, even then, Crescas contends, the theory of an original time may still be maintained. We may say, that since every motion requires a medium, there is an original medium of motion and hence an original time. That original time is constant, and remains the same even when the magnitude of the moving object is infinitely increased or decreased. It is only the excess over the original time that varies in proportion to the increase in the resistance of the medium and to the decrease in the magnitude of the object. Aristotle's laws of motion, namely, that the whole time of motion is proportional to its medium and to the magnitude, is, therefore, erroneous. It is only the time of the motion additional to the original time that is so proportional. Hence, if we admit the existence of an infinite body, it would not have to perform motion without time, for the original time would still remain.[72]

Another argument against the existence of a vacuum advanced by Aristotle is based upon the impenetrability of bodies. A vacuum by definition is tridimensionality devoid of body. Now, if a vacuum existed and could despite its tridimensionality be penetrated by a body, why could not bodies penetrate into each other.[73] The assumption underlying this argument is that the

[71] Prop. I, Part II (p. 185), n. 22.
[72] Prop. I, Part II (p. 183), notes 13–16.
[73] Prop. I, Part I (p. 147), n. 44.

impenetrability of bodies is due solely to their tridimensionality. In attacking this argument Crescas, therefore, tries to show that tridimensionality is not the sole cause of impenetrability of bodies, but tridimensionality in so far as it is also corporeal. The vacuum, to be sure, is tridimensional like bodies, but it differs from bodies in that its tridimensionality is incorporeal, whereas that of bodies is corporeal. This difference between a vacuum and bodies is that which makes a vacuum penetrable and a body impenetrable, for the impenetrability of bodies is not due to their tridimensionality, which they share in common with the vacuum, but to their corporeality, in which bodies differ from a vacuum. Now, that there is a difference between the corporeal dimensions of bodies and the incorporeal dimensions of a vacuum is admitted by Aristotle's commentators, but they argue that the mere difference as to corporeality could not result in a difference as to impenetrability, and that corporeality could not be the sole cause of impenetrability but that its sole cause must be found in tridimensionality, which both bodies and a vacuum share in common. But as for this, argues Crescas, granted that corporeality alone could not explain the impenetrability of bodies, neither could tridimensionality alone explain it.[74]

With the refutation of Aristotle's arguments against a vacuum Crescas now undertakes to show that according to Aristotle himself there must exist a vacuum, at least the Pythagorean conception of a vacuum beyond the world. He furthermore shows that a vacuum may be classified as an incorporeal continuous magnitude. And finally he shows that this incorporeal magnitude must be infinite.

According to Aristotle the world is finite, and beyond the outermost sphere there is no body. The absence of a body beyond the universe naturally means the absence of a plenum. The absence of a plenum must inevitably imply the presence of

[74] Prop. I, Part II (p. 187), notes 26–28.

a non-plenum. Now, a non-plenum necessarily means some kind of potential space, actually devoid of any bulk, which, however, it is capable of receiving. Such a potential space is what is called a vacuum, for by definition a vacuum is nothing but incorporeal intervals or extensions. Thus, beyond the universe there must be a vacuum.[75]

The terms generally used in describing the quantity of a vacuum are not "much" and "few" but "great" and "small." Furthermore, a vacuum is measured by a part of itself.[76] All these tend to show that a vacuum is not a discrete but rather a continuous quantity. Now, of continuous quantities there are five: line, superficies, body, place, and time, of which the first four are called magnitudes. As a vacuum is obviously not time, it must necessarily be a magnitude.[77] Hence, the vacuum is an incorporeal, continuous magnitude.[78]

If we now raise the question as to the finitude or infinity of that incorporeal continuous magnitude, we must necessarily arrive at the conclusion that it is infinite. For were it finite we may ask again, what is beyond its limits, and as there can be no plenum there, we will have to assume that beyond them there is another vacuum and beyond that still another and so on to infinity, which really means the existence of an infinite vacuum, or incorporeal extension, beyond the universe.[79]

Thus Crescas has shown that according to Aristotle himself there must exist a vacuum outside the world, and that that vacuum must be infinite. With this he now comes back to Aristotle's original investigation as to whether an infinite incorporeal

---

[75] *Ibid.* (p. 187), notes 30–32 and 36.

[76] As for the meaning and history of this statement, see n. 34 (p. 418) on Prop. I, Part II.

[77] A discussion of the various classifications of quantity is to be found in n. 35 (p. 419) on Prop. I, Part II.

[78] Prop. I, Part II (p. 189).

[79] *Ibid.* (p. 189).

magnitude has existence or not. Aristotle has rejected it because, by his denial of the existence of a vacuum, he could not conceive of the existence of an incorporeal magnitude. Crescas, however, accepts it because a vacuum to him has existence, and a vacuum is an incorporeal extension or magnitude.

But how is this infinite extension or magnitude to be conceived? To begin with, the infinite incorporeal extension is to be infinite by its nature and definition, for the incorporeal can have no accidents. Furthermore, being incorporeal, it is simple and homogeneous. But here a difficulty would seem to arise. Infinity, as we have seen, is used by Aristotle in the sense of that which, though capable of being finite, is infinite. This implies that the infinite must be divisible. But if the incorporeal extension which is infinite by its nature and definition is divisible, then its parts would have to be infinite, which would imply that an infinite is composed of infinites—a difficulty encountered by Aristotle himself in the course of his tentative discussion of the possibility of different conceptions of infinity. In order to remove this difficulty Crescas alludes, rather cryptically, to the analogous case of a mathematical line. He does not, however, explain how the analogy of a mathematical line would remove the difficulty. But evidently what he means to say is this. A distinction is to be made between two kinds of divisibility, one of which implies composition and the other of which does not imply composition. Take, for instance, a syllable. It is divisible into letters, and is also composed of letters. Here indeed divisibility implies composition. But, on the other hand, take a mathematical line. It is said to be divisible, and is infinitely divisible, into parts which are linear. Still it is not composed of those parts into which it is divisible, for the linear parts into which it is divisible, by definition, are bounded by points, and consequently if it were composed of these linear parts it would also be composed of points, but a line is not composed of points. Or in other words, when a thing is discrete and heterogeneous, it is divisible into its component parts and is also said to be composed of those parts, its parts being co-existent with the whole. When a thing is, however, continuous and homogeneous, it is only divisible into its parts but is not composed of them, for it is divisible only in capacity, and the parts into which it is divisible are not actually co-existent with the whole. By the same token, the infinite, simple, homogeneous, incorporeal extension can be divisible despite its being simple; and though divisible into parts each of which is infinite, it will not be composed of those parts. It is simple in the same sense as a mathematical line is simple; that is to say, it is not composed of heterogeneous parts. It is, again, divisible like a mathematical line into parts of its own self. The parts of the infinite, to be sure, will be infinite, just as the parts of the line are lines, but the infinite will no more be composed of infinites than a line is composed of lines, for those infinite parts never actually co-exist with the infinite whole, just as the linear parts never actually co-exist with the linear whole.[80]

Against an infinite incorporeal extension there is now only one argument, that of Altabrizi, which awaits an answer. The gist of the argument is this. If an infinite extension exists, by assuming two lines which are finite on one side and infinite on the other, one may arrive at the absurdity of having one infinite greater than another.[81]

The argument, says Crescas, is based upon a misunderstanding of the meaning of the term infinite as used in the statement that one infinite cannot be greater than another. The term infinite has two meanings. In the first place, it means to have no limits. In the second place, it means to be incapable of measurement. Now, it is possible to have an infinite in the sense

[80] For a full discussion of this interpretation of Crescas' brief statement, see n. 1 (p. 391) on Prop. I, Part II.

[81] Prop. I, Part I (p. 149). For the history of this argument, see n. 54 (p. 346).

of not being capable of measurement which may not be infinite in the sense of having no limits. Such is the case of the two lines in Altabrizi's proof. In so far as the lines are immeasurable neither of them can be greater than the other, for things immeasurable are incomparable. But in so far as both the lines have limits on one side, one of them may be said to be greater than the other in the sense of its extending beyond the other at their finite end.[82] That this is a true distinction may be shown by the fact that in the problem of the creation of the universe, both those who believe in eternity and their opponents will have to resort to it in order to get out of a common difficulty.[83]

The discussion so far has dealt with the impossibility of an infinite magnitude, which is the subject of Maimonides' first proposition. The impossibility of an infinite number is the subject of the second and third propositions. Inasmuch as it is characteristic of number that it involves the idea of both unity and plurality, applying as it does to a group within which the individuals are distinguishable from one another by some kind of difference, it is clear that only such things can be numbered as possess certain individual distinguishing marks. Such individual distinguishing marks which make number possible are, according to the sixteenth proposition of Maimonides, of two kinds. First, in the case of corporeal objects, they are to be found in the relative positions the objects occupy in space or in the accidental qualities which they all possess. Second, in the case of incorporeal beings, like the Intelligences, which do not exist in space and have no accidental qualities, number is possible only in so far as they are differentiated from each other by some external relation, such as the relation of cause and effect, for the Intelligences, according to Maimonides and Avicenna, are related to each other as causes and effects.[84] It is because

[82] Prop. I, Part II (p. 191), n. 37 (p. 423).
[83] Ibid. (p. 191), notes 38–39.
[84] Prop. XVI.

number may be understood in these two different senses that Maimonides has treated the problem of infinite number in two different propositions. The second proposition denies the possibility of an infinite number of corporeal objects, whereas the third proposition denies the infinite number of incorporeal beings, or as he puts it, the infinite number of causes and effects.[85]

That an infinite number of corporeal magnitudes is impossible is demonstrated by a simple argument. It follows as a corollary from the first proposition, for an infinite number of finite magnitudes will make one infinite integral magnitude.[86] To prove, however, the impossibility of an infinite series of cause and effect, more complicated arguments were required.

There is, to begin with, the argument given by Aristotle himself which is intended to show the impossibility of a series which has no beginning as well as that which, having a beginning, has no end, or in other words, the impossibility of an infinite series in the upward direction as well as in the downward direction. This argument of Aristotle has been freely restated by Avicenna, from whom it was taken over by Altabrizi. Crescas reproduces it, with some slight modifications, from Altabrizi and alludes to its origin in Aristotle.[87]

Then, in a comment upon a passage in the *Physics* Averroes disproves the possibility of infinite number on the ground that number must be divisible into odd and even, which an infinite could not be. This argument, though not original with Averroes, for we find it in the writings of Algazali,[88] is quoted by Crescas in the name of the former, and is taken by him to apply with

[85] See n. 2 (p. 480) on Prop. III.

[86] Prop. II, Part I.   This is Altabrizi's proof.   Aristotle's own proof is reproduced in n. 2 (p. 476).

[87] The various restatements of Aristotle's proof are given in n. 4 (p. 482) on Prop. III.

[88] See n. 3 (p. 477) on Prop. II.

equal force to infinite material magnitudes as well as to infinite immaterial beings.[89]

Finally, the first part of Aristotle's argument, the argument against the possibility of an infinite series in the upward direction, is reproduced by Narboni in a statement to the effect that had the universe had no first cause at the beginning nothing could have come into actual existence. This argument occurs repeatedly in various works in connection with the problem of creation, but Crescas quotes it directly from Narboni's commentary on the *Moreh*, introducing it in the name of "one of the commentators."[90]

All these arguments are subjected by Crescas to a searching analysis. He refutes Averroes' argument by pointing out that it is only finite number, because of its being actual and limited, that must be subject to the division into odd and even; infinite number, were it admitted to be possible, would not have to be subject to that division.[91]

Narboni's argument is likewise subtly analyzed and rejected. Causes, contends Crescas, may either precede their effects in nature and co-exist with them in time, or they may precede them both in nature and in time. While Narboni's argument, continues he, may reasonably prove the impossibility of an infinite series of causes and effects when temporally preceding one another, it is insufficient to prove the impossibility of such a series when there is only a natural, without any temporal, precedence, such as is assumed in Maimonides' third proposition. Furthermore, he argues, even in the case of temporal precedence, Narboni's argument is unconvincing. For those who believe in the eternity of the universe draw a distinction in the case of temporally successive causes and effects between essential and

[89] See n. 8 (p. 488) on Prop. III.
[90] See n. 16 (p. 492) on Prop. III.
[91] Prop. II, Part II (p. 219). For sources of this refutation, see n. 9 (p. 488) on Prop. III.

accidental causes, and while they deny the possibility of an infinite series of the former they admit it in the case of the latter. And so, concludes Crescas, since such a distinction is made, and since also an infinite series of temporally successive, accidental causes is admitted to be possible, there is no convincing reason why we should deny the possibility of an infinite series of essential causes of the same description. To say that essential causes are in this respect less possible than accidental causes is a purely arbitrary assertion.[92]

Finally, he refutes the first part of Aristotle's argument which tries to show the impossibility of an infinite series in the downward direction though finite in the upward direction. But in order to show the refutability of this argument, he had to establish first the possibility of an infinite number of incorporeal beings.

As we have seen, under the guise of the denial of an infinite series of causes and effects, Maimonides really aims to deny the possibility of an infinite number of incorporeal beings which have neither accidental qualities nor spatial relations and cannot consequently be numbered except as causes and effects. The question therefore arises: Suppose we find some incorporeal beings which, though without spatial, accidental or causal relations, are still capable of being numbered by some kind of individual distinction in their respective degrees of perfection, could these be infinite in number? Now, such numerable incorporeal beings are found, if we believe in individual immortality, in the case of the human souls which survive after death, for these human souls, if we assume their immortality to be consequent upon certain individual perfections acquired during lifetime, retain their individual distinction even after death. Concretely stated, the question is this: Can the immortal souls after their separation from their bodies be infinite in number?[93] It is Altabrizi who

[92] Prop. III (p. 227) and notes 17–20 (pp. 293–496).
[93] For the history of this problem, see n. 6 (p. 484) on Prop. III.

raises this question, but leaves its solution to God whose knowledge is limitless. Crescas, however, enters into a full discussion of the subject.[94] He finds that authorities differ on that point. Avicenna, he says, followed by Algazali and Maimonides, admits the existence of an infinite number of immortal souls, whereas Averroes denies it. That such a controversy existed is true enough. But Crescas does not seem to be aware that the view he ascribes to Algazali is one which the latter held to be the view of the philosophers, Avicenna and perhaps also Aristotle, with which, however, he himself did not necessarily agree; nor does he seem to reproduce quite accurately the reason for Averroes' denial of an infinite number of disembodied souls.[95]

By refuting the alleged argument of Averroes against the infinity of immortal souls, Crescas, of course, espouses the view of the opposing school, namely, that the infinite number of immortal souls is possible. As a consequence, it would no longer be true to lay it down as a general rule that incorporeal beings can never be infinite in number; it would only be true to say, as Maimonides indeed did say, that they cannot be infinite in number when they are numbered on account of their mutual relation as causes and effects. When incorporeal beings are capable of being numbered on account of some other individual distinction, as, e. g., the immortal souls of the dead, they can be infinite in number. Suppose, now, these infinite immaterial beings be all effects, arising simultaneously from a given un-caused cause, as are, for instance, the Intelligences in the view of Averroes. We would then have an infinite number of pure effects, and there is no reason why that should be impossible. It is thus quite conceivable to have an infinite number of incorporeal beings standing in the relation of effects to one uncaused cause. With this established, Crescas then proceeds to ask,

[94] Prop. III, Part I, notes 5–8.
[95] See notes 6 (p. 484) and 8 (p. 488) on Prop. III.

why should it not be equally possible, with that uncaused cause as a starting point, to have all its infinite effects proceed from one another as causes and effects among themselves and so continue infinitely downward? What should render it less possible when they all proceed from the first cause as a series of causes and effects than when they proceed from it simultaneously? If it is possible for them to be infinite in the latter case, why not also in the former?[96] Still more significant is Crescas' conclusion. Maimonides' Proposition, he says in effect, does not follow Aristotle in denying the possibility of a series of causes and effects which are infinite in the downward direction. It only aims to deny the possibility of the series when it is infinite in the upper direction, for Maimonides is only interested in showing that at the beginning of any series, be the series infinite or finite, there must be an uncaused cause.[97]

[96] Prop. III, Part II, notes 10–13.
[97] *Ibid.* n. 21.

CHAPTER III

Motion[1]

THE terms "change" and "motion," according to Aristotle, are not synonymous. Change is the more comprehensive term, including as it does any kind of transition, whether from non-being into being, or from being into non-being, or from one state of being into another. Motion, more restricted in its meaning than change, applies only to a transition within being itself between one state or condition and another. In Aristotle's own language motion is said to be the change from a certain subject to a certain subject whereas change may be from a subject to a non-subject or from a non-subject to a subject. Accordingly, there is no motion in the category of substance, inasmuch as generation and corruption, which constitute the two opposite changes in the category of substance, are changes from a non-subject to a subject and from a subject to a non-subject. In strict conformity with this distinction, Aristotle is always careful to enumerate under the term change four categories, namely, substance, quantity, quality and place, and under the term motion only three categories, namely, quantity, quality and place. To this generalization there are only a few exceptions, the most notable of which is a passage in the *Categories* wherein he uses the term motion as the subject of his classification but includes under it the category of substance. In that passage he also resolves substance into generation and corruption and quantity into growth and diminution and uses for quality the term alteration, and thus instead of speaking of the four cate-

[1] This chapter is based upon Propositions IV, V, VI, VII, VIII, XIII, XIV, XXV, XVII, XVIII and IX in the order given.

70

gories of motion he speaks of six species of motion, namely, generation, corruption, growth, diminution, alteration, and locomotion.[2]

The distinction between change and motion is generally observed by Arabic and Jewish authors. Formally the distinction is stated by them to be as follows: Change is timeless, motion is in time.[3] Like Aristotle, they insist that if the term motion is used as the subject of the classification the category of substance is to be omitted, and if the term change is used the category of substance is to be included. But again like Aristotle they sometimes deviate from that rule. On the whole we find three types of classifications in the literature of the period. First, there are works which follow Aristotle's *Categories* and enumerate six species of motion reducible to the four categories of substance, quantity, quality and place. Second, there is an Avicennean classification which, using the term motion and hence, in conformity with Aristotle, excluding substance, adds the category of position and thus continues to speak of four categories of motion, namely, quantity, quality, place and position. Third, there is the classification adopted by Maimonides which, using the term change, enumerates the four categories of substance, quantity, quality and place.[4]

But here a question arises with regard to Maimonides' fourfold classification of the categories of change. Why should some of the other categories be excluded from the classification? It is true, Aristotle has stated that there is no *motion* in the categories of relation, action, and passion, but he did not explicitly say that there is no *change* in those categories. Furthermore,

---

[2] A discussion of the different classifications of the categories of change μεταβολή and motion κίνησις as given by Aristotle is to be found in n. 3 (p. 498) on Prop. IV.

[3] See n. 4 (p. 503) on Prop. IV. See contradictory statements in Index: Motion.

[4] A discussion of the different classifications of the categories of change and motion in Arabic and Jewish philosophy is to be found in n. 3 (p. 500) on Prop. IV.

in one place at least, Aristotle has stated quite the contrary, namely, that there is motion in the categories of action and passion. Knowing, as we do, the loose sense in which Aristotle sometimes uses the term motion, why not try to reconcile these two contradictory statements by taking the term motion in the last passage to mean change, and thus there would be more than four categories of change? Indeed, Aristotle never enumerates more than four categories of change, but we have no evidence that he ever meant to give an exhaustive list of the categories of change. In fact, the Stoics have included the categories of action and passion under motion. And the Avicenneans, too, mention the category of position among the categories of motion.[5]

Considerations like these, if not actually these very considerations, must have formed the background of Crescas' question why Maimonides has restricted the categories of change to four—a question already raised by Altabrizi.[6]

In answer to this difficulty Crescas draws upon a distinction between two subjects of change which has been only slightly suggested by Aristotle but fully developed by his commentators.[7] If any concrete perceptible object, call it A, is undergoing a change in any of its accidents, say color, or size or place, passing from one opposite to another, call those opposites B and C, two subjects may be considered in the process of the change. First, A may be considered as the subject of the change, inasmuch as A is that which underlies the opposites B and C and is that in which the change takes place and which sustains the change. A may be therefore called the *sustaining subject*. This sustaining subject exists only in the categories of quantity, quality and place, for it is only in these categories that the subject is some-

---

[5] See notes 6–7 (pp. 504–507) on Prop. IV.
[6] See n. 5 (p. 504) on Prop. IV.
[7] For a full discussion as to the meaning, origin and history of this distinction between the two 'subjects' of change see n. 8 (p. 507 f.) on Prop. IV.

thing concrete and perceptible. In the category of substance there is no such perceptible sustaining subject, though the matter underlying the processes of generation and corruption may be called an imperceptible sustaining subject.[8] Second, the accident which is being changed from one opposite to another, say from whiteness to blackness, may be considered as .the subject of the change, inasmuch as it is that accident, say color, which has these two opposites, whiteness and blackness. This accident may be called the *material subject* or rather the subject-matter of the change.

Now, if you consider change with reference to the sustaining subject, it may be found also in some of the other categories, say the category of action, for in action, too, there is always a sustaining subject which undergoes the change, for now that subject acts and now it does not act. But if you take it with reference to the material subject, it is to be found only in such categories where the two opposites may be each designated by some positive and concrete term. There are only three such categories: quantity, which has the opposites of increase and diminution; quality, which has, for instance, the opposites black and white; place, which has the distinction of up and down and other similar distinctions. In none of the other categories are there such opposites as may be designated by positive opposite terms, an *a quo* and an *ad quem*, between which the change is to take place, and consequently there can be no change between them. Take, for instance, the category of relation. Whatever the relation may be, whether that of reciprocity, as father and son, or whether that of comparison, as greater and smaller, the relation as such cannot suffer any change. It always remains the same relation. If a change takes place at all, the change is always in the objects reciprocally related to each other or compared with each other but not in the relation itself. Similarly in the categories of posses-

[8] *Ibid.* p. 512 f.

sion, action and passion, possession as such, action as such and passion as such cannot change from one opposite to another. In the category of time, indeed, there is the opposite of past and future, and consequently there should be change or motion in the category of time. But the reason why time is not mentioned as one of the categories of motion is that time, according to Aristotle, is itself defined in terms of motion and would be entirely inconceivable without motion. When therefore Maimonides speaks of change, he uses the term with reference to the material subject, and is thus compelled to confine himself only to these three categories of quantity, quality and place, where the material subject undergoes a change between two opposite accidents within one perceptible sustaining subject. Substance was not to be mentioned by him, inasmuch as change in the category of substance is something unique in that its sustaining subject is imperceptible and its opposites generation and corruption are not the opposites of an accident residing within a perceptible sustaining subject. Still Maimonides mentions also change of substance because it is involved in the other three categories of change.[9]

We thus have change and motion. Of change, again, we have two kinds, one considered with reference to its material subject and the other with reference to its sustaining subject. The former kind of change is found only in the four categories of substance, quantity, quality and place. The latter kind of change is found in some of the other categories.

The term motion is to be particularly used with reference to the category of place.[10] Motion is thus primarily locomotion. Indeed, in quantitative changes, such as growth and diminution, there is some sort of locomotion, but that locomotion is hardly perceptible enough to justify the proper application of the term motion to the category of quantity.[11] Still in a general sense the

[9] Prop. IV, notes 9–15.
[10] Maimonides in Prop. IV.
[11] Prop. IV, notes 17–19.

changes of quality and quantity may be called motion. Change in the category of substance, however, and any other change that is timeless, cannot be called motion. Thus while every motion is change, it is not every change that is motion.[12]

There are three formulations of the definition of motion, two given by Aristotle and one by Maimonides. Aristotle's first definition reads: 'Motion is the actuality of that which is in potentiality in so far as it is in potentiality'. His second definition is somewhat differently phrased: 'Motion is the actuality of that which is movable in so far as it is movable'. Maimonides' definition is phrased as follows: 'Motion is a change and transition from potentiality to actuality'. The relative merits of these three definitions as well as the relation of Maimonides' definition to those of Aristotle have been a matter of discussion.[13] Crescas himself finds that Maimonides' definition is only a restatement of Aristotle's first definition. The object of both these definitions is to establish the nature of motion as something which is neither a pure potentiality nor a complete actuality but a potentiality in the process of realization. He finds fault, however, with these definitions on the score of their use of the term potentiality, which might lead to a difficulty. For if every transition from potentiality to actuality is motion, then the transition of a motive agent from the state of a potential motive agent to that of an actual motive agent will be motion. Every motivity then will be motion. As every motion requires a motive agent, every motivity will also require a motive agent. But this is contradictory to Aristotle's view as to the existence of a prime immovable mover.[14] He therefore considers Aristotle's second definition as an improvement upon the first and concludes that while in a general way motion is the process of the actualization of that which is in potentiality, the term potentiality is to be under-

12 Prop. V, n. 2.
13 See notes 5 (p. 523) and 11 (p. 529) on Prop. V.
14 See note 10 (p. 526) on Prop. V.

stood as referring only to a potentiality for receiving motion and not to a potentiality for causing motion.[15]

Besides the classification of motion according to the categories, Aristotle has another scheme of classification. Motion may be essential, that is, the translation of a body as a whole from one place to another, and it may be accidental, by which are meant two things, first, the motion of some accident of a body by reason of the motion of the body itself, and, second, the motion of part of the body by reason of the motion of the whole body. This second kind of accidental motion is sometimes called by him "motion according to part" or "motion according to something else," as contrasted with essential motion which is "motion according to itself." Then motion may again be divided into that which has the principle of motion within itself and that which has the principle of motion outside itself, designated respectively as natural and counternatural or violent. These classifications of motion are scattered in different parts of Aristotle's work and the scheme we have presented is made up of several different classifications by Aristotle.[16] Now, Maimonides, evidently in an attempt to summarize the various classifications of Aristotle, gives a fourfold classification—essential, accidental, partial, and violent.[17] Crescas, having before him the various classifications of Aristotle as well as an elaborately detailed classification by Altabrizi, which is based upon Aristotle, takes Maimonides' classification merely as a general statement to the effect that motion is classifiable and proceeds to work out on the basis of it a more detailed scheme of classification, in accordance with Aristotle and Altabrizi.[18] Motion, according to his revised plan, is divided into the following divisions and subdivisions: A. Essen-

[15] See note 11 (p. 529) on Prop. V.
[16] See n. 3 (p. 531) on Prop. VI for a discussion of the various classifications of motion in Aristotle and in Arabic and Jewish philosophers.
[17] Prop. VI.
[18] See n. 3 (p. 533) on Prop. VI.

tial, subdivided into (a) natural, (b) violent, and (c) voluntary. B. Accidental. C. Violent, subdivided into (a) essential, and (b) accidental. D. Partial, subdivided into (a) violent and (b) natural.[19]

Essential motion is defined by Maimonides as the translation of a thing from one place to another. Now, the celestial spheres in their rotation are not translated from one place to another, their motion being within one place. Indeed, it is on this account that Avicenna does not include the circular motion of the spheres in the category of motion in place. He calls it rather motion in the category of position.[20] It would thus seem that, according to Maimonides' definition of essential motion, the motion of the celestial sphere is not essential.

In his endeavor to prove that the motion of the sphere is essential, Crescas enters upon a discussion of the nature and cause of the motion of the sphere.

The spheres, according to the dominant view, are animate beings. Like all animate beings their soul is the principle of their motion. Their motion is therefore called voluntary and is said to differ from the motion of the sublunar elements which is called natural. The proof of this view rests upon the assumption that matter is inert and that the four sublunary elements have each a proper place in which it is their nature to remain at rest. But as they are occasionally expelled from their respective proper places by some external force, they are then set in motion by a natural reflux to their proper abodes. It is this reflux to their proper resting places that is called natural motion, and the proper places are said to act upon the elements as final causes. This natural motion, therefore, cannot be continuous, for it must come to a stop as soon as each element arrives at its proper destination. Now, since the spheres never leave their

[19] Prop. VI, notes 4–8.
[20] See n. 10 (p. 535) on Prop. VI.

proper places, they would be expected to remain permanently at rest. Still the spheres are continuously in motion, rotating as they do on a centre in their own place. What therefore is the cause of their continuous circular motion? The only answer that could be given was that they are moved by an internal principle called soul.[21] Consequently the motion of the spheres is called voluntary in contradistinction to the motion of the sublunar elements which is called natural.

In opposition to this there was another view which maintained that the motion of the spheres, like that of the sublunar elements, is natural.[22] Crescas adopts this view and argues that there is no need of explaining the circular motion of the spheres by a psychic principle or soul any more than there is need for such an explanation in the case of the motion of the sublunar elements. For matter is not inert; it is naturally endowed with motion. To be always in motion is the essential nature of all the elements, sublunar as well as translunar. But this motion with which all the simple elements are endowed by nature differs with respect to direction in accordance with the inner structure and constitution of each particular element. The celestial element is so constituted as to move in a circular direction whereas the other elements are so constituted as to move either in an upward or in a downward direction. Thus the celestial spheres may be said to be naturally endowed with circular motion just as the sublunar elements are said to be naturally endowed with either upward or downward motion.

Crescas' rejection of the Aristotelian explanation of the circular motion of the sphere is followed by his rejection of Aristotle's theory of absolute lightness. The contrast between lightness and weight, according to Aristotle, corresponds respectively to the

[21] *Moreh Nebukim* II, 4.

[22] See n. 11 (p. 535) on Prop. VI for the history of the view that the motion of the spheres is natural.

contrast between upward and downward motion. Fire is said to be light and earth heavy in the sense that the former has a natural tendency upward whereas the latter has a natural tendency downward. These natural tendencies in opposite directions on the part of the elements is furthermore explained, as we have seen, as a reflux toward proper places which are supposed to exist above and below. Against these views Crescas inveighs on several occasions. To begin with, he denies the existence of proper places.[23] Then he also denies that natural motion is due to the alleged reflux toward those proper places the existence of which he denies; motion is explained by him as being due to the inner structure of the elements themselves. Finally, all the elements are endowed with a natural motion downward, and every apparent motion upward, such as that of fire, is to be explained on the ground of a mechanical cause, namely, on the ground of pressure exerted from below. Consequently, if by weight and lightness is to be understood a natural downward and upward motion there is no such a thing as absolute lightness, for all the elements have only a natural downward motion and are therefore to be described as heavy, though some may be heavier than others.[24]

With this new theory of motion Aristotle's division of motion into natural and violent becomes erroneous. The upward motion of fire can never be called natural, and its downward motion is in no sense unnatural. But, remarks Crescas, while this may be urged as a criticism against Aristotle, it cannot be urged as a criticism against Maimonides' proposition, for in his illustration of violent motion Maimonides does not mention the motion of fire downward. He only mentions the motion of a stone upward, which is indeed violent, being due to an external force.[25]

[23] See n. 76 (p. 456) on Prop. I, Part II.
[24] Prop. VI, notes 14–19.
[25] Prop. VI end.

So much for Maimonides' definition of essential and violent motion. His definition of accidental motion is likewise criticized by Crescas. Accidental motion, according to Maimonides, is to be found only in the motion of accidental qualities which are moved together with the essential motion of the bodies in which they inhere. This, he says, is not altogether accurate. It may be also found, according to Aristotle, in the motion of something which is not an accidental quality, as, for instance, the extreme point of a line. That the motion of the extreme point of a line is to be considered as accidental rather than as essential or partial has been shown by Averroes.[26]

Change and motion, according to Aristotle, imply corporeality and divisibility, and therefore objects capable of change and motion must be corporeal and divisible. That they must be corporeal is self-evident. Change in the category of place, or, what is called motion proper, cannot exist without a body, for place, by definition, is peculiar to body. Change in the other categories, namely, substance, quality and quantity, must likewise imply corporeality. For quality and quantity are accidents which must inhere in a body; and similarly change between being and non-being in the category of substance must imply the existence of matter. That change and motion likewise imply divisibility is demonstrated by Aristotle by the fact that both of these, by definition, are partly potential and partly actual. This demonstration proves that all the four categories of change, including the timeless change of substance, imply divisibility.[27]

To this general proposition, however, two exceptions may be pointed out. First, the mathematical point at the extremity of a line in a body, though it may be moved accidentally with the body,[28] is not divisible nor is it corporeal. Second, both the

[26] Prop. VI, notes 12–13.
[27] Prop. VII, Part I.
[28] Prop. VII, Part I, end.

rational and the sensitive faculties of the soul undergo change, the former undergoing a timeless change in passing from ignorance to knowledge and the latter undergoing a change in time in passing through the emotions of pleasure and pain and their like. Still the soul is incorporeal and indivisible. These exceptions, however, argues Crescas, do not invalidate the proposition, for upon examination it will be found that both these exceptions involve changes which are only accidental, and so all that is necessary in order to justify the proposition is to restrict its application only to such changes and motions that are essential.[29]

In order to prove that there is an immovable mover, that is to say, a mover which moves unlike any other mover in the universe, Aristotle had to prove first that motion is eternal and second that no motion can be eternal unless it is "according to its essence" καθ' αὐτό and "by its essence" ὑφ' αὐτοῦ. The expressions "according to its essence" and "by its essence" mean two different things. The first expression means that the object moved must be moved essentially as a whole and not accidentally as a quality of something else or as a part of something else. The second expression means that the object moved must have the principle of its motion within itself and not outside itself, the latter being known as violent motion. According to Aristotle, for motion to be eternal it must be neither accidental nor violent. In Arabic versions of Aristotle, it would seem, the term violent used in the original text was replaced by the term accidental. Maimonides, therefore, in restating Aristotle's principle, simply says that everything that is moved accidentally must of necessity come to rest, meaning by the term "accidentally" both what is generally known as accidental motion and what is more specifically called violent motion.[30]

[29] Prop. VII, Part II.
[30] See n. 4 (p. 551) on Prop. VIII for a full discussion as to the history of the interpretation of this Proposition.

This Aristotelian proposition, however, is qualified by Crescas. It is true only, he says, if it means to affirm that no accidental motion can of itself be eternal. It is not true if it means to affirm that no accidental motion can under any circumstances be eternal, for it can be shown that accidental motion can be eternal if it is inseparable from some eternal essential motion.[31]

The reason why no accidental motion can of itself be eternal is to be found in the nature of the accidental. Anything accidental, depending as it always must upon some cause, is by its own nature only possible. Its existence, while it endures, is thus always subject to the alternatives of continuing to be or of ceasing to be. At any given time, to be sure, only one of the alternatives can be in a state of actuality, the other alternative, however, must always be regarded as held in reserve, capable of springing into realization at the proper opportunity. Thus while it cannot be said singly of either one of the possible alternatives that it must become realized, it can be said of both the alternatives that within an infinite time they will both have to have been realized. In other words, it is inconceivable that any one of the possible alternatives should remain forever in a state of actuality to the exclusion of the other, inasmuch as possibility is not only the opposite of necessity but is also the opposite of impossibility.[32] Consequently, accidental motion cannot of its own nature continue for an infinite time.[33]

Motion is said to be one in the three senses, generically, specifically, and individually. Upward and downward motions, for instance, may be called one in the sense that they belong to the same category or genus of place, but specifically they constitute two different motions. The upward motion of two different objects, on the other hand, are called one specifically, seeing that

[31] Prop. VIII, Part II.
[32] See n. 2 (p. 693) on Prop. XXIII.
[33] Prop. VIII, Part I, notes 2–3.

they belong to the same species of upward motion under the genus place, but individually they constitute two different motions. The upward motion of one object, taking place during one continuous time, however, is called one in an individual and numerical sense.[34] Again, the term continuous as applied to motion may have two meanings, one in the sense of everlasting motion and the other in the sense of unbroken and coherent motion.[35] Of all the categories of motion only circular locomotion may be said to be continuous in the sense of both everlasting and unbroken. All the other motions, qualitative, quantitative, spatial and substantial, are never continuous in the sense of everlasting. They may, however, be continuous in the sense of unbroken, provided that they are individually one. Motions which are specifically different, still less motions which are generically or numerically different, can never be continuous in either of the senses.[36]

That the specifically different motions of one object, though taking place in a time which is apparently one, cannot be continuous is shown by Aristotle by the following argument. Motions which are specifically different are invariably in opposite directions, and between motions in opposite directions there must always be an instant of rest. This Aristotle proves by induction to be true in the case of the specifically different motions of all the categories—generation and corruption in substance, whitening and blackening in quality, and upward and downward in locomotion.[37]

The case of locomotion is furthermore proved by an additional argument. When a motion returns upon itself, says Aristotle, it must mark an actual point at its turning point. In other

[34] See n. 2 (p. 615) on Prop. XIII.
[35] See n. 6 (p. 617) on Prop. XIII for an Aristotelian basis for these two usages of the term "continuous".
[36] Prop. XIII, Part I, notes 3–6,
[37] Ibid. notes 7–12,

words, when two motions run in opposite directions with reference to a given point, that point must be actual. But having an actual point in motion always implies a pause. Consequently there must be a pause when a rectilinear motion returns upon itself. Since there is a pause between them, the two opposite motions cannot have a common limit at their meeting point. The end of the first motion must be actually different from the beginning of the second motion. And so the two motions cannot be considered as one, for if it were so, the time during which the motions took place would likewise have to be one, but this is impossible, for inasmuch as there is an actual point between the two opposite motions there must be a corresponding actual instant in the two times of two motions. Now, if these two motions were one motion, the two times would likewise have to be one time, despite their being divided by an actual instant. But this is impossible, for time is a continuous quantity and cannot have an actual instant in the middle.[38]

In his criticism of this view Crescas tries to show that motions or changes in opposite directions may be one and continuous. In the first place, argues Crescas, it is not true that there must be a period of rest between two opposite qualitative changes. Two such opposite changes may be continuous, that is to say, the juncture at which the change of direction takes place may be like all the other instants in time which have no separate, actual existence, but constitute the end of the past and the beginning of the future. If an object that has been blackening begins to whiten, the blackening and whitening processes may be considered as constituting one continuous motion taking place in one continuous time. Still it could not be contended, as is done by Aristotle, that at the instant during which the change in direction takes place the motion would have to be at once both blackening and whitening. By no means. As a

[38] *Ibid.* notes 13–16.

point in time, to be sure, that instant is the common boundary of both the past and the future; as a point in the process of motion, however, it is only the boundary of the past motion. And this is a good Aristotelian distinction. For according to Aristotle, in every continuous motion you may take any instant, which as an instant in time will belong both to the past and the future but as a point in motion will belong only to the past. Take, for instance, the qualitative motion of blackening and represent it as moving from A to B. The time AB as well as the motion AB is continuous. Now, take any point C in AB. As an instant in time, says Aristotle, it belongs to both AC and CB. As a point in motion it marks only the end of AC. Still Aristotle calls the motion AB continuous. Why not say the same of the two opposite motions AB and BA. B as an instant of time will belong to both AB and BA, thus preserving the continuity of time. B as a point in the motion will only mark the end of AB. Still the opposite motions AB and BA could be continuous, no less so than the motions AC and CA, and you could not say that at B the motion would run at once in both the opposite directions.[39]

Furthermore, the assumption that between two opposite motions there must always be a pause is absurd. Suppose body A in its motion upward strikes body B, which is in its downward motion, and thereupon A changes its direction and begins to come down. If you say that A must come to rest before it changes its direction, B, too, would have to come to rest. But this is impossible, for the downward motion of B is admittedly continuous.[40]

Finally, Crescas refutes the argument which Aristotle has advanced in the case of locomotion. He denies the initial assumption of that argument. It is not true at all, when two motions

[39] Prop. XIII, Part II, n. 20.
[40] *Ibid.* n. 21.

run in opposite directions with reference to a given point, that the point must be actual. He proves this from the analogy of substantial and qualitative change. The change between generation and corruption or between one generation and another is a substantial, continuous, and timeless change. Now, every substantial change involves a corresponding qualitative change. And so any change from one generation to another will simultaneously register a change from one quality to another. These two qualitative changes will be in opposite directions, inasmuch as, by taking the common limit between the two generations as the point of departure, the one will move towards it and the other will move away from it. And still these two qualitative changes, though in opposite directions, are one and continuous as are their concomitant substantial changes.[41]

Consequently, if it is not necessary to assume an actual instant of rest between two opposite changes of quality and of substance, why should it be necessary to have one between two opposite motions in place?

Let us return to Aristotle. No opposite motions, according to him, can be one and continuous, be they motions in substance, quantity, quality, or place. Now, since the world is finite in magnitude, in quality and in place, there cannot be an infinite spatial, quantitative or qualitative change in one direction. Consequently, if these changes were to continue infinitely, they would have to change their direction. But as soon as they change their direction they must come to a pause; and upon resuming their motion, it will no longer be their old motion that they will resume, but rather entirely a new one. Consequently, none of these changes can be infinite. There is one kind of motion, however, that does not come to a stop even though it changes its direction. That is circular motion. The reason for this exception is that in circular motion there are no absolutely

<hr />

[41] *Ibid.* n, 22,

opposite directions, for at the same time the motion is from
and toward the same given point. No point in it is therefore
assumed to be actual, and it must not necessarily come to a
rest. Consequently, circular motion may be continuous and
eternal.[42]

If we assume the world to have existed from eternity, as Aristotle in fact does, which of the four kinds of motion was first
to appear? It is locomotion; for the locomotion of the spheres
have co-existed from eternity with the prime mover. Then, the
changes of generation, growth, quality, diminution and corruption follow in order of succession. Thus locomotion is prior in
time to all the other motions. But it is also prior in nature to
all the other motions, for all the other motions in a way involve
locomotion, they never occur without the occurrence of some
degree of locomotion, whereas locomotion may take place singly
and independently. Finally, circular motion is prior in essence
or reason to all the other motions, for it is the most perfect, and
the perfect, according to Aristotle, logically precedes the imperfect. The perfect nature of circular motion is attested by its
continuity, by its uniform velocity, and by the excellency of its
subject, namely, the fifth, celestial substance. Unlike all other
motions, the circular is not an incomplete energy; it is an energy
complete and perfect.[43]

The order of temporal priority, however, is to be reversed if
we assume the world to have been created *ex nihilo* in time.
For then assuredly generation was the first of motions. By the
same token, assuming even the universe as a whole to be uncre-
ated, the individual generated beings within the universe, have
generation as the first of their motions. Motion of absolute
quantity, in the shape of corporeal form, is the next motion.
Qualitative motion and afterwards the motion of accidental

42 Prop. XIV, Part I.
43 Prop. XIV, Part I, n. 3; Part II, n. 9,

quantity follow when the elements become possessed of their four natural forms. It is only then that locomotion appears.[44]

Motion is not a self-contained process. Its inception as well as its continuation must be due to some cause. This is true of all the categories of motion, including motion in the category of substance, i. e., the assumption and the casting off of forms, for matter cannot be the cause of its own motion.[45]

The cause of motion, while it must always be distinct from the object in motion, may either be physically external to it or reside internally within it. Thus, for instance, in the case of the violent motion of an inanimate object in a direction contrary to its nature, as that of a stone upward, it is clear that the motive cause is an external force applied from without. And so it is also generally agreed that in the case of the voluntary motion of animate beings the cause is a vital principle, a soul, operating from within. The case of the so-called natural motion of the elements in their appropriate directions, however, is doubtful.[46] That the motive cause of the elements is something distinct is sure enough; but is it also external to them or does it reside within them? On this point we have two conflicting views, the Avicennian and the Averroian.[47]

To Avicenna, the natural motion of the elements, like the voluntary motion of animate beings, may be called motion by an internal cause. The elements move in their respective natural directions by themselves, because, like animate beings, they contain within themselves their principle of motion. To be sure, there is a difference in the action of the internal motive principle of the natural elements and in that of animate beings. In the case of the former, the action is mechanical and is restricted to

[44] Prop. XIV, Part II, notes 10–13.
[45] Prop. XXV.
[46] Prop. XVII.
[47] See n. 7 (p. 672) on Prop. XVII for a discussion of the views of Avicenna and Averroes.

one definite direction, whereas in the case of the latter, the action is voluntary and is operated at large in all directions. Still they both belong to the same order of nature—the motive principle in either case may be identified with some form of the object. In animate beings, that form is the soul, for soul is the form of the body. In the inanimate natural elements, that form is corporeality, or corporeal form, which is the first form that matter assumes.[48] As the form of an object constitutes its nature, nature is thus said to be the principle of motion.[49]

Against this conception of motion, which may be called dynamic, Averroes maintains a view which may be called static. According to him, who indeed only interprets Aristotle, there is only one kind of motion which may be said to contain its motive principle within itself, and that is the voluntary motion of animal beings. All the other motions, including that of the elements, have their motive cause outside themselves. The elements, he maintains, are by their own nature endowed only with a potentiality for motion, which passes into actuality by the action of a series of external causes which ultimately end in the prime mover. Those external causes, indeed, act upon the elements through their specific forms, and thus their forms may in a certain sense be called the cause of their motion. The proper cause of their motion, however, is something external.[50]

As to which of these views was held by Maimonides it is a matter of controversy among his commentators. Crescas is silent on this point.[51]

Motion, properly speaking, is change in place, and, as we have seen, it is not a self-contained activity. It always implies the existence of a motive agent. By the same token, any other kind of change or transition from potentiality into actuality requires an

[48] See n. 18 (p. 579) on Prop. X.
[49] *Ibid.*
[50] *Ibid.*
[51] *Ibid.*

agent or cause to bring about that transition. The proximate cause of motion, as we have seen, is distinct from the object moved but not neccesarily external to it. Its remote or ultimate cause, however, is both distinct and external. Thus in every form of transition from potentiality to actuality the ultimate cause is not only distinct from the object but also outside of it. This view is not the result of a *priori* reasoning; it is rather based upon inductions from actual observations. Whatever form cf change we take, we shall find that the cause is always distinct from the object as well as external to it.[52]

Though action is change and change is a transition from potentiality into actuality, it is not always that a change of action implies a change in the nature of the agent producing the action. Action means the operation of an agent upon an object under given conditions. Any change in action may be therefore due to a change in any of these three causes: the agent, the condition or the object. It is therefore quite possible to have a change within the action or from non-action into action without implying a change in the nature of the agent, as when, for instance, the change or transition can be traced to the nature of the object only. Thus, if you conceive God to have created the world in time, the transition from non-action into action does not mean a change in the divine nature.[53]

A motive agent may act upon its object either as a final cause or as an efficient cause, in the latter case its action is performed in one of the following four ways: drawing, impelling, carrying, and rolling. As a final cause the motive agent may produce motion without itself being moved. As an efficient cause, however, it cannot produce motion without itself being moved at the same time.[54] The case of a magnet, which seems to produce

[52] Prop. XVIII, notes 1-9.
[53] *Ibid.* n. 9.
[54] Prop. IX, Part I, n. 2.

motion in an object as an efficient cause by means of drawing without itself being moved, was advanced as an apparent contradiction to the general rule and called forth various explanations. On the whole, four explanations are discussed in various works in Jewish literature.[55]

First, the magnet does not act as a motive agent in its attraction of iron. It is the iron itself which is moved toward the magnet by reason of a certain disposition it acquires when it comes within the vicinity of the magnet. This explanation is quoted by Averroes in the name of Alexander.

Second, the motion of the iron toward the magnet is brought about by means of certain corpuscles which issue forth from the magnet and come in contact with the iron and draw it toward the magnet. This explanation is attributed to the Stoics. It is also described by Lucretius. It is quoted by Averroes in the name of Alexander and is found in Maimonides.

Third, the magnet possesses a certain force which attracts the iron. Thales calls this force a soul. Plato and, according to Gershon ben Solomon, also Galen deny that this force is a soul but designate it simply by the term power. It is similarly called peculiar power by Joseph Zabara and peculiar property by Altabrizi.

Fourth, magnetic attraction is explained by the same principle as the natural motion of the elements. There is a certain affinity between the iron and the magnet analogous to the affinity which exists between the elements and their respective proper places. The magnet therefore does not act as the efficient cause of the motion of the iron but rather as its final cause. This explanation is advanced by Averroes and is also discussed by Gershon ben Solomon and his son Gersonides.

[55] See notes 5 (p. 563) and 10 (p. 565) on Prop. IX for a history of the various theories of magnetic attraction as are to be found in Jewish philosophical literature.

Crescas adopts the last explanation but modifies it somewhat in accordance with his own explanation of the natural motion of the elements. As we have already seen, Crescas does not attribute the natural motion of the elements to the alleged action of proper places upon the elements as final causes. According to him all the elements are moved downward by their own nature due to some peculiarity in their own physical structure and composition. Similarly in the case of magnetic attraction, he argues, the motion of the iron may be due to some peculiarity in its own physical structure and composition.

CHAPTER IV

TIME[1]

THE relation between time and motion is one of the pivotal points in Crescas' criticism of Aristotle. Aristotle defines time as the number of motion according to the prior and posterior.[2] As against this Crescas defines time as the measure of the duration[3] of motion or of rest between two instants. By this definition Crescas means to disestablish the connection between time and motion which Aristotle's definition has established. But how this end is achieved by Crescas' new definition is not quite clear. The substitution of the term 'measure' for 'number' certainly does not bring about that result, for, besides the irrelevancy of this change of terms to the question in hand, Aristotle himself interchanges these terms in his definition of time.[4] Nor does the addition of the term "rest" make time independent of motion, for Aristotle himself admits that rest, too, is measured by time, but argues that since rest is only the privation of motion, it is measured by time only accidentally.[5] Finally, the substitution of the phrase "between two instants" for Aristotle's "according to prior and posterior" is of no real significance, for Aristotle, too, by his statement that time is the number of motion according to prior and posterior means that motion is numbered or measured by time when it traverses a certain distance between two instants.

[1] This chapter is based upon Prop. XV.
[2] The variety of versions of Aristotle's definition of time in Arabic and Jewish philosophy is discussed in n. 9 (p. 636).
[3] A justification for translating the underlying Hebrew term by 'duration' is to be found in n. 23 (p. 654).
[4] See n. 24 (p. 658).
[5] See n. 22 (p. 646).

The real difference between these two definitions, therefore, cannot be obtained by the mere counting of the words and phrases in which they are couched and by abstracting them from one another. We must first find out what these definitions exactly mean. Now, as for the exact meaning of Aristotle's definition, it can be easily gathered from his own discussion of time.[6] But as for the exact meaning of Crescas' definition, his own discussion on the subject does not lend us any help. We must therefore resort to other discussions which may be found in the philosophic literature spanning the centuries between Aristotle and Crescas and out of these try to get whatever help we can in constructing Crescas' own view.

Aristotle does not approach the problem of time with that feeling of awe with which some later philosophers begin their discussion of the same problem. The term 'time' had not as yet become obscured by the incrustation of layers upon layers of metaphysical speculation. As used by Aristotle, it was still the word of the common speech of the ordinary man. When Aristotle asks himself what time is, he is really asking himself what people mean when they speak of time, and it is from his observations of what people usually mean by time in their every day speech that he arrives at a definition of the nature of time. There is no use of speculating as to the existence of time, he begins his discussion, and there is still less use in attempting to deny the existence of time, when in the daily speech of every man time is treated as something existent. Assuming then that time does exist, Aristotle proceeds with the question, what time is.[7]

In order to know what a thing is, it is first necessary to know to what class of beings it belongs. Now, all beings, according to Aristotle, fall into two classes, substances and accidents. The question is therefore whether time is a substance or an accident.

[6] *Physics* IV, 10 ff.

[7] See n. 7 (p. 634), where also a discussion is to be found as to the different restatements of the pre-Aristotelian definitions of time.

It was very easy for Aristotle to show that it was not a substance, for a substance is something which exists in itself, whereas time is something fleeting, consisting of past and future, neither of which has any actual existence. It must therefore be an accident, existing in something else, just as color and shape and size exist in something else.[8]

But what is that something else in which time exists? Aristotle's answer is that it is motion, for psychologically, he argues, we have no perception of time unless we have a perception of motion. The manner in which our perception of time is formed is shown by an analysis of motion. Motion is a transition from one point to another over a certain magnitude. In the magnitude itself, these points are co-existent, but in motion they are successive, some of them being prior and others posterior. These prior and posterior points in motion are transformed by our mind into past and future, and the past and future when combined furnish us with what we usually call time. Furthermore, motion is numbered, and this is done in two ways, first, according to distance, as when we describe motion by the distance traversed, and, second, according to speed, as when we describe motion as swift or slow. But the swift and the slow are in common speech measured by time, "since that is swift which is much moved in a short time, and that is slow which is but a little moved in a long time."[9] Consequently, Aristotle arrives at the definition of time as being the number of motion according to the prior and posterior.[10]

The implications of this definition are many and far-reaching. Time, according to this definition, while not identical with motion, is still inconceivable without motion.[11] Time thus always implies the existence of some corporeal object in motion; and

<hr/>

[8] See notes 2 (p. 633), 10, 11 and 12 (pp. 640 f.).
[9] *Physics* IV,10, 218b, 15–17; Cf. n. 12 (p. 641).
[10] See notes 13, 14, 15 and 16 (pp. 642 f.).
[11] Prop. XV, Part II, n. 4.

while indeed the object need not be actually in motion, it must be capable of motion.[12] Furthermore, time as now defined has a certain kind of reality and actual existence outside the mind, due to the reality of the moving object to which it is joined, though this reality is to be understood only in a limited sense, for since time is not motion itself but only the number of motion, to that extent, like number, it must be conceptual.[13] Moreover, eternal beings that are incorporeal and immovable, like God and the Intelligences, cannot have the attribute of time, inasmuch as the attribution of time would imply corporeality and movability.[14] Finally, if we accept Aristotle's definition of time but reject his view as to the eternity of the universe, as does Maimonides, we will have to assume the creation of time as well as the creation of matter, inasmuch as time, under this definition, could not have existed prior to the existence of matter and motion.[15]

In order now to understand how Crescas' counter-definition divorces the idea of time from that of motion, we must first call attention to another definition of time, opposed to that of Aristotle, which had been current in Greek, Arabic and Jewish philosophy down to the time of Crescas and which continued to be discussed by philosophers after his time. In the light of this new definition we shall be able to get the full significance of Crescas' definition.[16]

According to this new definition the essence of time is not motion but duration. Unlike motion, duration does not depend upon external objects for its existence, and it does not arise in

[12] See notes 19 (p. 645) and 22 (p. 646).
[13] See n. 28 (p. 661).
[14] See notes 21 (p. 646) and 31 (p. 662).
[15] See n. 33 (p. 663).
[16] A full documented discussion of this definition of time, its rise in Plotinus and its history in Arabic and Jewish philosophy, will be found in n. 23 (pp. 654–658).

our mind out of the motion of things outside ourselves. It is rather the continuity and flow of the activity of the thinking mind. This thinking mind may be God, or the universal soul, in such philosophies as assume the existence of a universal soul, or even our own mind, if our mind is assumed to have an activity and life of its own. Given therefore a thinking mind, even were there no external reality, there would be such duration. But this duration itself would be indefinite and indeterminate. It would have no end and no parts. In order that it might become determinate, there must be some external standard of determination. Such a standard is motion. When duration is determined and measured by motion, the measured part of duration becomes time. Still, while we cannot get time, or that measured-off part of duration, without motion, time is essentially as independent of motion as is the pure, undiluted duration itself, for time is only measured by motion, but is not generated by motion. Unlike Aristotle, then, this definition maintains that it is not time that measures motion but it is rather motion that measures time.[17] This definition may be hewn out of the lengthy discussions of Plotinus, and traces of it may be found in the writings of the Iḥwan al-Safa, Saadia and Altabrizi. In the work of Joseph Albo, a pupil of Crescas, there is a clear-cut statement of it. It can also be traced throughout the writings of Bonaventura, Duns Scotus ,Occam, Suarez, Descartes, Spinoza and Locke.[18] Students of Bergson, too, may perhaps find in it some suggestion of his distinction between "pure duration" and "mixed time."

This is exactly what is meant here by Crescas' definition. In its essence time is duration, and duration is in the mind and is independent of motion. Motion comes in only as a measure by

[17] *Ibid.* p. 655. But see n. 22 (p. 646).

[18] Cf. H. A. Wolfson, "Solomon Pappenheim on Time and Space and his Relation to Locke and Kant", in *Israel Abrahams Memorial Volume*, 1927, pp. 426–440.

which a definite portion of duration is set off. Time is thus formally defined by Crescas as "the duration of motion . . . between two instants." But in order to get that definite portion of the duration, or the time, of a thing it is not necessary for the thing itself to be in motion. It is not even necessary for it to be capable of motion. The measure can be supplied by our mind by its merely conceiving of motion, for, as Crescas says, time may be measured "by the supposition of motion without its actual existence." Now, the thing whose duration is measured by the "supposition of motion" and is itself neither in motion nor capable of motion is described by Crescas as being at rest, using the term 'rest,' unlike Aristotle, not in the sense of the privation of motion in things capable of it but in the sense of absolute immovability.[19] He thus introduces into his definition the additional expression "and of rest."

The implications of this new definition are quite the opposite of those which follow from the definition of Aristotle. Since in its essence time is duration, it implies no external existence, still less the existence of something movable. For a thing to be in time, therefore, it need not be either actually in motion or capable of being in motion. Furthermore, time has no reality whatsoever,[20] inasmuch as it exists in the mind of a knower and could have existed there even were there nothing outside the mind of the knower in existence. Consequently, beings that are incorporeal and immovable, like God and the Intelligences, may be described by attributes of time without implying that they are corporeal and movable.[21] Finally, if the world is assumed to have been created, prior to creation there had existed duration which is the essence of time.[22]

---

[19] On Crescas' use of 'rest' in the sense of 'immovability', see n. 22 (p. 646 f.).
[20] See n. 28 (p. 661).
[21] Prop. XV, Part II (p. 291) and notes 31 and 32; cf. *Or Adonai* I, iii, 3, and H. A. Wolfson, *Crescas on the Existence and Attributes of God.*
[22] See Prop. XV, Part II (p. 291) and n. 33 (p. 663).

CHAPTER V

MATTER AND FORM[1]

IN MEDIAEVAL philosophy it was customary to divide 'being' into that which exists in itself and that which exists in another. To the latter the name accident is given. Accident is then subdivided into that which not only exists in another but exists through the other, and that which, while existing in another, is the cause of the existence of the other. The former is again called accident, the latter is called form. Thus in the accepted terminology of the time, the term accident had two meanings, a general and a specific, the one used to include substance, for form is a substance,[2] and the other used as the opposite of substance. It must have been in order to avoid this confusion of terms that Maimonides introduces the term "force" to take the place of the term "accident" in its general sense. "Force," therefore, designates existence in something else, and it is used by Maimonides in Propositions X, XI, XII, and XVI, to include accidents, forms, the lower faculties of the rational soul, the internal principle of motion, and the universals, all of which require something else in which to exist.[3]

The distinction of matter and form is deduced, after Aristotle, from the phenomenon of the reciprocal transformation of the elements. Water, for instance, becomes air and air becomes water. This process of transmutation, it is argued, cannot be merely the alteration of one thing into another, for the elements represent opposites, and nothing can become its opposite unless

[1] This chapter is based upon Propositions X, XI, XII, XVI, XIX, XX, XXI, XXII, XXIII and XXIV.
[2] See n. 9 (p. 573) on Prop. X.
[3] See n. 15 (p. 577) on Prop. X.

it is first completely destroyed. The transmutation of the elements therefore implies the destruction of one thing and the generation of another. But when one thing is destroyed, it can no longer give rise to another thing, for from nothing, nothing can be generated. It is therefore necessary to assume the existence of a certain substratum common to all the four elements within which the transmutation takes place. That substratum is matter, and the four elements are the four different forms which the matter assumes. Thus every one of the four natural elements is composed of matter and form.[4]

The matter underlying the four elements is known in Jewish philosophy as 'absolute body' and the four forms which it assumes are variously known as the 'elementary,' 'natural,' 'proper,' 'specific' or 'essential' forms[5]. This common, underlying, proximate matter of the four elements, however, was not considered to be completely formless. It was supposed to be composed of another matter, known as 'prime' or 'intelligible' matter, and another form known by various names. Simplicius calls it 'corporeal form,' by which name it is commonly known in Arabic, Jewish and scholastic philosophy. In Plotinus it is also designated by the term 'quantity,' which term is also used in the Arabic philosophic encyclopedia of the Iḥwan al-Safa. The terms 'corporeity' and 'first form' are also applied to it.[6]

There is no reference to 'corporeal form' in Aristotle. It was introduced into his system by his followers in order, probably, to account for the difference in the nature of his prime matter and his common matter of the four elements. The prime matter of Aristotle was generally understood to be incorporeal and inextended. The common matter of the four elements, however, it was argued, had to be something extended. It was therefore

---

[4] See notes 3–7 (pp. 569–572) on Prop. X.
[5] See the list of terms in n. 16 (p. 577) on Prop. X.
[6] *Ibid.*; cf. n. 18 (p. 579) on Prop. X.

inferred that the prime inextended matter is not identical with the common extended matter of the elements, and that between these two matters there must be an intermediate form which endows the prime matter with extension. That form is the first or corporeal form which prime matter assumes.[7]

Once this form was introduced, speculation became rife as to its nature. Three views are recorded in Arabic and Jewish literature, which we shall restate here under the names of their chief exponents, Avicenna, Algazali and Averroes.

According to Avicenna the corporeal form is a certain predisposition in prime matter for the assumption of tridimensionality. As for tridimensionality itself, he considers it as an accident under the category of quantity which accrues to the elements subsequently. Algazali agrees with Avicenna that tridimensionality is only an accident. But he disagrees with him as to the nature of the corporeal form. The latter, according to him, is not a predisposition in matter for tridimensionality but rather the cohesiveness or massiveness of matter in which tridimensionality may be posited. In opposition to both of them, Averroes identifies the corporeal form with tridimensionality itself but he distinguishes between indeterminate and determinate tridimensionality. The former, he says, constitutes the corporeal form, the latter are only accidents. A similar difference of opinion existed among Jewish philosophers. Crescas, in his restatement of the definition of corporeal form, however, uses vague language which lends itself to any of these three interpretations.[8]

The proof for the existence of matter and form from the transmutation of the elements, as we have seen, establishes only the existence of the common matter of the elements and the elementary forms. It has no application at all to the 'prime matter' and

[7] See n. 18 (p. 579 ff.) on Prop. X for a discussion of the origin, history and meaning of the "corporeal form".
[8] *Ibid.* p. 588.

the 'corporeal form.' In order to prove the existence of the latter a new argument had to be devised. This new argument is in its main outline analogous to the argument from the transmutation of the elements, but instead of reasoning from the destruction and generation of elements it reasons from the continuity and division of matter. It runs as follows: Matter which is continuous loses its continuity and becomes divided. Continuity and division are opposites, and opposites cannot be the recipients of each other. Hence, they imply the existence of a substratum capable of assuming both these opposites. This substratum is the prime matter.[9]

It has thus been shown that in the successive stages of matter and form the lowest is the opposition of 'prime matter' and the 'corporeal form.' The combination of these two constitutes the 'common matter' of the four elements. The corresponding form of the latter is the four 'proper' or 'natural' forms of the elements, and so the stages of matter and form go on until the highest pure form is attained. Neither matter nor form can have actual existence by itself—not even the common matter of the four elements, though it is already composed of matter and form. The first actually existent sublunar substances, according to Maimonides, are the four elements.[10] Though form only is to be considered as the cause of the existence of an object, still both matter and form are essential factors in the process of becoming, and consequently both of them are substances.[11] So is also the concrete individual object, composed of matter and form, a substance. For, substance, as defined by Aristotle, has four characteristics: (a) It is that which does not exist in a subject, or, if it does exist in a subject, (b) it is the cause of the existence of that subject, (c) it also constitutes the limits which define the

[9] Evidence for the view expressed in this paragraph as to the existence of such a new proof is to be found in n. 22 (p. 591) on Prop. X.

[10] Maimonides in Prop. X and Crescas in Prop. X, Part I, n. 16.

[11] Prop. X, Part I, notes 8–9

individuality of the subject, and (d) it is its essence.[12] Matter and the concrete thing are substances in the first sense of the term, form is a substance according to the other three senses. Accidents, however, differ from form by the fact that they not only cannot exist without a subject but their existence is not at all essential to the existence of their subject.[13] All the accidents may be classified under nine categories. These, again, may be subdivided into separable and inseparable accidents. The inseparable are quantity, figure, which is a subdivision of quality, and position; the separable are all the other accidents.[14]

The chief points in this theory of matter and form are two. In the first place, the 'common matter' of the four elements is itself a composite, consisting as it does of two elements, the prime matter' and the 'corporeal form.' In the second place, this common, composite matter of the four elements has no actual existence by itself. Actual existence accrues to it by virtue of its 'specific' or 'elementary' form. Against this conception of matter and form Crescas raises no objection as long as its proponents maintain it consistently, as do in fact Avicenna and Maimonides. To both of them the distinction of matter and form is to be found in all material substances, translunar as well as sublunar. The celestial substance, known as the fifth element, is, according to their view, composed of matter and form as are the four sublunar elements. In opposition to Avicenna, however, Averroes draws a distinction between the sublunar and translunar elements. The sublunar elements, he agrees with Avicenna, consist of (a) the 'prime matter,' (b) the 'corporeal form' and (c) the 'specific' or 'elementary' form. The translunar element, that is, the substance of the spheres, however, consists only of

---

[12] For the definition of substance and the enumeration of substances, see notes 8 and 9 (pp. 573–576) on Prop. X.

[13] Prop. X, Part I, notes 13–14.

[14] For the classification of accidents, see notes 4–8 (pp. 686–690) on Prop. XXII.

(a) the 'corporeal form' and (b) the 'specific form' which each sphere possesses, the former being related to the latter as matter to form. Furthermore, the 'corporeal form' of the celestial spheres, unlike the combination of 'prime matter' and 'corporeal form' of the sublunar elements, has actual existence without its 'specific' form.[15]

It is this distinction made by Averroes between the sublunar and the translunar elements that Crescas takes as the point of departure in his criticism of the accepted theory of matter and form. He argues for the elimination of the 'prime matter' in the sublunar elements just as it has been eliminated by Averroes in the translunar element. The 'common matter' of the four elements will thus be something simple, not composed of matter and form, and will also be extended. Furthermore, it will be something actual and will not depend for its existence upon its form.[16] Consequently, Aristotle's definition of form will also have to be modified. It is no longer to be considered as the cause of the existence of a thing. In that respect form is an accident like all the other accidents. It is to be considered a substance only in so far as it constitutes the limits which define the individuality of the subject and is its essence. In these two respects only does form differ from accident.[17]

"Forces" residing in a corporeal object, as we have seen, either exist through the object or are the cause of the existence of the object. To the former class belong the manifold accidents; to the latter class, according to Aristotle, belong the various forms and in a certain sense also the prime inextended matter, inasmuch as like form it is one of the constituents of body without which no body can be conceived. Now, the material object in which these

[15] The history of the question as to whether the celestial spheres are composed of matter and form is discussed in n. 24 (p. 594) on Prop. X.

[16] Prop. X, Part II, notes 25–28.

[17] *Ibid.* notes 29–32.

forces exist is capable of division and disintegration. How that division and disintegration affect the "forces' residing in the material object is the subject of Maimonides eleventh proposition. On the whole, he lays down no hard and fast rule of distinction between these two classes of "forces" with regard to divisibility. In both cases some are divisible with the division of the body and some are not. Of accidents, some secondary qualities, like color and size, participate in the division of the body in which they inhere, while others, like its figure, do not participate in its division. Likewise in the case of substantial "forces," the prime inextended matter is subject to division, whereas the corporeal form is indivisible in the physical sense of the term, though it is capable of some kind of conceptual division.[18] Again, in the case of the soul, which is the form of the body and a substance, the vegetative and animal faculties are divisible, whereas the rational faculty, even the lowest stage thereof, namely, the hylic intellect, is indivisible. Though Maimonides considers the hylic faculty to be a "force" within the body, and is accidentally moved with the body, still he admits it to be not co-divisible with the body, inasmuch as it is not a force distributed throughout the body.[19]

The motive faculty of the soul, like the hylic faculty, is also a "force" residing in a body. Consequently the soul of the sphere which constitutes its principle of motion is a "force" residing in the sphere and must therefore be finite, inasmuch as every body must be finite and no infinite force can reside in a finite body. This is a good Aristotelian proposition. In proof of this proposition, it is first recalled that an infinite body is impossible. Then it is shown that should an infinite force reside in a finite body it

[18] Prop. XI, notes 1–3.

[19] *Ibid.* notes 4–5. See n. 5 (p. 605) for a discussion as to the analogy between the relation of soul to body and the Intelligences to the spheres and as to the difference of opinion between Averroes and Maimonides.

would ensue either that motion could take place in no-time or that a finite and an infinite force could move in equal time.[20]

As over against this, it is Crescas' contention that an infinite motive force is possible. In the first place, Crescas refers to his own refutations of the arguments against the possibility of an infinite body.[21] Then, referring to Avempace's theory of an original time of motion, he argues that assuming the existence of such an original time of motion we may have an infinite force within a finite body without being driven to the absurdity of non-temporal motion or to the equal absurdity of the absence of any temporal distinction between the motion produced by a finite force and that produced by an infinite force. Indeed, argues Crescas, even if you discover a single instance where the finite and the infinite force would produce motion in equal time it is not a sufficient argument to disprove the existence of an infinite motive force.[22] Finally, drawing upon an old distinction between infinite in time and infinite in intensity,[23] which Crescas makes much use of on several occasions, he argues that Aristotle's proof has only established the impossibility of a force of infinite intensity existing in a finite body. It does not prove, however, that a force of finite intensity could not continue its activity in a finite body for an infinite time.

If, therefore, an infinite force within a body is possible, infinite though only in time, there is no need for the assumption of a prime cause, which, according to Maimonides, must be separate from the sphere and exist in addition to the prime mover which is within the sphere.[24] The eternal motion of the sphere might as well be explained as being due to the action of a force, finite in

[20] Prop. XII, Part I.
[21] Prop. XII, Part II, n. 4.
[22] *Ibid.* notes 5–6.
[23] For the origin of this distinction, see n. 7 (p. 612) on Prop. XII, Part II.
[24] See n. 5 (p. 606) on Prop. XI, and H. A. Wolfson, *Crescas on the Existence and Attributes of God.*

intensity, to be sure, but infinite in time, residing within the sphere itself. That such a force should act infinitely, indeed, it would be necessary to find a certain kind of motion and a certain kind of substance which by their nature could continue forever, inasmuch as not every kind of motion and not every kind of substance is capable of continual existence. But such a kind of motion and such a kind of substance are known to exist. Circular motion, according to Aristotle, may be continual, and the celestial substance, again according to him, is eternal. And so the eternal circular motion of the sphere may be due to the action of a certain force residing within it, there being no need for the assumption of a prime cause separate from it.[25]

Furthermore, the eternal circular motion of the sphere may be explained without the postulate of an internal resident force no less than without the postulate of an external separate force. The circularity of the sphere's motion, as has already been shown above,[26] is not due at all to any soul within it but rather to the very nature of the substance of the sphere itself. By the same token, it may be argued, that the eternity of the sphere's motion is not due to any resident force within it but rather to the constituent nature of the sphere itself.[27]

Like accidents, forms and some of the faculties of the soul, the universals may be also called "forces." For universals, in the Aristotelian sense, have no real existence; they are said to exist only in the mind. However that phrase may be interpreted, and whatever the relation of universals to the individuals may be, the universals of Aristotle may be described as "forces" in a body, in the sense that they can have no actual existence apart from individuals. It is only through the material objects in which they exist that universals become individualized and

[25] Prop. XII, Part II, notes 8–11.
[26] See above p. 78.
[27] Prop. XII, Part II, n. 12.

distinguishable, for material objects inevitably have the distinction of time and space and accidental qualities, and it is through such differences that material objects become numerable even when they are one in their universal character.[28] Consequently no incorporeal beings can be subject to number unless they are incarnate in bodies. Without bodily existence there is no distinction of few and many. Number implies the idea of plurality as well as that of unity, and there can be no plurality unless there are material objects which exist in time and space, and are endowed with accidental qualities.[29]

But still there are immaterial beings which are generally admitted to be numerable. The Intelligences of the spheres, for instance, are pure, immaterial spirits, and still they possess individuality and number, the latter being determined by the number of the spheres. What is it then that differentiates the individual Intelligences from one another, notwithstanding the fact that they do not possess the ordinary differentiae of time and space and of accidental qualities?

Two views are recorded, the Avicennian, which is also that of Maimonides, and the Averroian. The Avicennian view considers the Intelligences as evolving from one another by a process of emanation. They are mutually interrelated as causes and effects. There is thus a distinction of cause and effect between them, and it is this distinction that furnishes the basis for their numerability and individuality. The Averroian view denies the existence of any causal interrelation between the Intelligences. It considers them all as co-ordinate beings, proceeding directly and simultaneously from God. But it admits the existence of a difference of value between the Intelligences. Some of them are more simple in their nature and more perfect

[28] See n. 2 (p. 664) on Prop. XVI, where it is shown that Crescas takes the first part of Maimonides' Proposition to be a restatement of Aristotle's theory of universals.

[29] Prop. XVI, Part I.

in their conception of the divine essence than others. It is this difference in the degree of their perfection that accounts, according to this view, for the individuality, and, hence, the numerability, of the immaterial Intelligences.[30]

Another class of immaterial beings which are numerable, and one in which there is no interrelation of cause and effect, is found by Crescas in the case of the departed, immortal souls. If immortality is individual, the immortal part is either the substance of the rational soul itself, which is Crescas' own view, or what is known as the acquired intellect, which is the view of some other philosophers. In either case there are individual distinctions between disembodied souls, distinctions due to the respective perfection attained by individual human beings during their lifetime either in their union with God, as is the view of Crescas, or in their intellectual endowments, as is the view of other philosophers. But, says Crescas, this class of immaterial beings are distinguished from those about which Maimonides generalizes in his proposition in that their individuality has been acquired during a previous existence in material bodies.[31]

Existences are divided according to Aristotle into three classes—the eternally immovable, the eternally movable, and temporarily movable.[32] God, the celestial spheres, and the sublunar beings respectively correspond to these three classes. Again, Aristotle defines the term "necessity", when not taken in its ordinary sense of "compulsion," to mean the eternal continuation of a thing in the same state, or, to use his own words, "that which cannot be otherwise."[33] He also defines the term "possibility," in one of its several senses, as the possibility of a thing to be otherwise, or, again, to use his own words, "a principle

[30] See n. 7 (p. 666) on Prop. XVI.
[31] Prop. XVI, Part II.
[32] This and also the next few paragraphs are based upon n. 1 (p. 680) to Prop. XIX.
[33] *Metaphysics* V, 5, 1015a, 33–34.

of change in another thing or in the same thing *qua* another."[34] From these definitions it is clear that God, who is eternal and immutable, must be called necessary, and that, on the other hand, the sublunar elements, which by their own nature are transitory and changeable, must be called possible *per se*. A question, however, arises with respect to the celestial spheres. These are imperishable and have an eternal, uniform motion. They should on that account be called necessary. But the question is, are they imperishable and eternal on account of their own nature or on account of something else? Avicenna, influenced by Alexander, maintains that the spheres by their own nature could not have eternal motion. For to have eternal motion by one's own nature implies the possession of an infinite motive force. The celestial spheres, however, are finite magnitudes, and, according to Aristotle, no finite magnitude can possess an infinite force. The eternal motion of the spheres must, therefore, be due to an external cause, the prime mover, which, in passing, we may note, according to Avicenna, is not identical with God.[35] Consequently, the spheres are necessary only by virtue of the necessity of their cause; in themselves they are only possible.[36]

With the introduction of that new distinction, we thus have according to Avicenna the following threefold classification of Being—God who is necessary *per se;* the transitory, sublunar beings which are possible *per se;* and the celestial spheres which are possible *per se* but necessary by their cause. Consequently, Aristotle's definition of necessity can no longer stand, since, as has been shown, a thing may continue eternally in the same state without being necessary *per se*. In order therefore to differentiate between necessary *per se* and necessary by a cause, or absolute and relative necessity, absolute necessity is defined by Avicenna in terms of self-sufficiency or the absence of

[34] *Ibid.* V, 12, 1020a, 5–6.
[35] See below p. 606.
[36] See n. 1 (p. 680) on Prop. XIX.

causation. God alone has absolute necessity in that sense. Nothing which has been brought about by a cause can be called necessary.[37]

Averroes disagrees with this view. To him the spheres have eternal motion by their own nature, due to an infinite motive force inherent within them. That an infinite force cannot exist in a finite body is true enough, but that only applies to an infinite in intensity. A motive force, however, may be finite in intensity and still be infinite in the time of its operation. The eternity of the spheres' motion may therefore be due to their own nature, and it is by their own nature that the spheres may be called necessary. Necessity thus retains its original Aristotelian meaning, the eternal continuation of a given state. And so a thing may have a cause and still be necessary.[38]

Necessity thus in the Avicennean sense came to mean causelessness. But it does not merely mean the absence of external efficient causation. It implies as well the absence of any other kind of causation.[39] Consequently, no composite object, be its composition actual or potential, physical or conceptual, real or formal, can be called absolutely necessary. For any composition is conceived to consist of parts, the aggregation of which is not identical with the whole, and so the whole may be said to depend upon its parts as its cause.[40]

Since no composite object can be necessary, no corporeal object can be necessary, whether it be eternal or not. For every corporeal object inevitably contains the conceptual distinction of matter and form and must also possess certain inseparable qualities.[41] Being composite, it cannot be necessary, even though it be eternal. Possibility, as we have seen, means the "may-be-

[37] Prop. XIX.
[38] See n. 1 (p. 680) on Prop. XIX.
[39] Prop. XX.
[40] Prop. XXI.
[41] Prop. XXII.

come" of an object, designating its contingent, inconstant, and transient nature. It implies changeability in an absolute sense and is opposed to impossibility and necessity both of which imply constancy and immutability. Potentiality, on the other hand, is to be taken only in relation to some definite state or quality to which a possible object may change, but prior to its change thereinto. If, for instance, an object may change from A to B, that object is said to be possible in a general sense, but it is said to be potential only in relation to B as long as it has not become B. On its becoming B, it ceases to be potential with respect to B. It is now B in actuality, though the object may still be described as possible, inasmuch as the change from A to B was not impossible nor was it effected by necessity. Potentiality is thus the opposite of actuality. In Greek the term δύναμις is used by Aristotle to designate both possibility and potentiality. In Arabic and in Hebrew one term is used for the former, and another term for the latter.[42]

Possibility, change, or becoming always implies the transition from the state of potentiality to that of actuality. By the phenomenon of becoming, too, as we have seen, Aristotle proves the existence of matter and form. Now, the distinction of matter and form is not simply one of non-being and being; it is rather a distinction between potential being and actual being. Matter is thus the potential, form is the actual. Every object therefore which is composed of matter and form, has a certain actual existence in so far as it possesses form; it has a certain potentiality in so far as it possesses matter. In the many successive stages of existent beings, however, if one goes down the scale, one comes to prime inextended matter, which is absolutely formless, devoid of any actuality and of purely potential existence. On the other hand, if one goes up the scale of existence, one arrives at God

---

[42] For the difference between "potentiality" and "possibility", see n. 2 (p. 690) on Prop. XXIII.

who is pure form and complete actuality. Hence the two propositions of Maimonides in Crescas' interpretation: "Whatsoever is in potentiality, and in whose essence there is a certain possibility, may at some time not exist in actuality," as, e. g., the prime matter.[43] Again, "whatsoever is potentially a certain thing is necessarily material, for possibility is always in matter."[44] In criticism of these propositions, Crescas refers to his own view that prime matter has an actual existence of its own.[45] He also points out that there is a certain possibility which is not in matter, as, e. g., the possibility of a form to alight on matter.[46]

[43] Prop. XXIII.
[44] Prop. XXIV.
[45] Prop. XXIII, Part II.
[46] Prop. XXIV.

CHAPTER VI

FORESHADOWING A NEW CONCEPTION
OF THE UNIVERSE

IN PLOUGHING through the heavy pages of Crescas' critique of
Aristotle one gets the impression, and a true impression it is, that
his discussion has no central point from which it proceeds and
no definite direction in which it is aimed. He seems to pass me-
chanically from argument to argument, scoring a point here and a
point there, setting up counter-theories only as a matter of contention, without trying, after his case has been stated and his
points scored, to set forth what he himself believes to be the right
view, as he invariably does in his discussion of purely theological problems in other parts of his work. This failure to set forth
positive views of his own is not unpremeditated and undesigned.
Crescas, in fact, did not mean to be anything but negative and
destructive in his treatment of the physical problems of Aristotle.
All he wished to accomplish was to undermine the principles
upon which were based the Aristotelian proofs for the existence
of God. As he himself declares at the outset of his discussion,
his arguments are to be *ad hominem*,[1] not to attain to the truth
of the matter but rather to confound his opponent.

Still, within this destructive criticism and within these arguments which are only *ad hominem*, we may discern certain
positive tendencies in the direction of the early Greek philosophers the revival of whose views is the common characteristic
of all those who long after Crescas struggled to emancipate themselves from the thralldom of Aristotle. These stray positive
tendencies we shall now try to gather together and to mould

[1] See n. 14 (p. 326) on Introduction to Book I.

114

into some systematic unity, showing their adumbration of some of those views which form what is called our new conception of the universe.

If we were to give an orderly and systematic presentation of Aristotle's philosophy of nature, we would logically have to start with his view as to the limited extent of the universe. Aristotle's universe, conceived as a system of concentric spheres, of necessity had to have a limit at which to terminate. While the number of the concentric spheres was not fixed by him, still he considered it to be finite, so that there had to be a last outermost sphere which formed, as it were, the top of the universe, and were it only possible for a human being to get up to that top, he would have been able to jump off from it.

But where would he have jumped? He would have had to jump 'somewhere,' but 'somewhere' implies place, and place, according to Aristotle, exists only where bodies exist; and as outside the universe, again according to Aristotle, there were no bodies, there could be no place there. Nor could he have jumped into a vacuum, for Aristotle's, if not nature's, abhorrence of a vacuum made its existence impossible not only within the universe but also outside the universe.

It was this lack of explanation as to what existed outside the universe that proved to be the vulnerable spot in Aristotle's conception of a finite universe. The difficulty is raised again and again by his own followers. Some of them, like Averroes, Gersonides and Albo, tried to solve it by maintaining that outside the universe there was neither a vacuum nor a plenum. What there was there was simply 'nothing'.[2] But Crescas, as later Bruno,[3] was reluctant to accept this explanation. 'Nothing' is not a middle term between plenum and vacuum, and therefore by the law of excluded middle, that which is outside the finite

[2] See n. 36 (p. 421) on Prop. I, Part II.
[3] Ibid.

universe must be either the one or the other. By the force of such reasoning Crescas found himself compelled to conclude that beyond the outermost sphere there must be a vacuum. As the vacuum could not be limited by anything else, he was further compelled to conclude that the vacuum must be infinite.[4] The bounds of the universe were thus extended by Crescas to infinity. The universe is not that finite system of concentric spheres of Aristotle's conception but rather the infinite vacuum within which Aristotle's finite universe is contained as in a receptacle

But what is that infinite, all-containing vacuum which is not simply 'nothing'? Several expressions are used by Crescas in describing it. "It is an extension (or distance or interval or dimension) separated from physical objects."[5] It is "extensions existing apart from matter"[6] or "incorporeal extensions," and "incorporeal extensions" are defined by him as "empty space capable of receiving corporeal extensions".[7] In order to understand the full significance of all these expressions it is necessary to recall that Crescas is trying to establish by them, as over against Aristotle, the distinction between space and place. Aristotle himself makes no such distinction. Space to him is only the remote place of a thing,[8] and neither space nor place has existence except when there is a body or rather when one body is contained by another body, for place is defined by Aristotle as the circumambient limit of a body.[9] But Crescas defines space as extension or distance which may be occupied by a body or may remain free of the occupancy of a body. When it is occupied by a body, then the space becomes the particular place of that body; when it remains unoccupied, then the space is called vacuum or in-

[4] Prop. I, Part II (p. 189).
[5] Prop. I, Part I (p. 147).
[6] Prop. I, Part II (p. 187).
[7] Prop. I, Part II (p. 189).
[8] See n. 69 (p. 352) on Prop. I, Part I.
[9] For the various Arabic and Hebrew versions of Aristotle's definition of place, see n. 89 (p. 362) on Prop. I, Part I.

corporeal extension.[10] Now, this space or vacuum or incorporeal extension, being, on the one hand, not a plenum, and, on the other hand, not simply 'nothing', must of necessity be conceived as a 'something' which differs, either in kind or degree, from that 'something' which constitutes a plenum. Logically, therefore, Crescas' vacuum is to be regarded in its relation to the plenum as the universal ether is regarded in its relation to the plenum by those modern physicists who postulate its existence. It is not an absolute void, but rather matter of a different order. And so, when Crescas argues for the existence of an infinite vacuum, he is arguing for the existence of an infinite extension or space, which is really matter of a different order, and which is to serve as a medium within which this material world of ours is contained.

But this material world of ours, Crescas further argues, is not the only world in existence. Here, again, he comes out in direct opposition to Aristotle, for Aristotle rejects the possibility of many worlds, that is, of many independent systems of concentric spheres, and he does this by an array of arguments which seem to be quite impressive.[11] Crescas, however, dismisses these arguments as inconclusive. On the ground of mere reasoning, he maintains, the possibility of many worlds is not to be excluded.[12] He does not, however, definitely say how many worlds may exist. He only contends for the existence of "many worlds". But knowing of his rejection of Aristotle's denial of an infinite number of magnitudes and of his contention as to the existence of an infinite space, we may reasonably infer that the number of Crescas' many worlds may rise to infinity.[13]

[10] See n. 31 (p. 417) on Prop. I, Part II.
[11] De Caelo I, 8; cf. n. 128 (p. 474) on Prop. I, Part II.
[12] Prop. I, Part II (p. 217) and see n. 130 (p. 474).
[13] Though in one place he describes the Talmudic reference to 18,000 worlds as hyperbolical (Book I, iii, 4; but cf. Book IV, 2).

We thus now get a clear view of Crescas' conception of the universe—an infinite space within which are floating an infinite number of worlds. It is perhaps not altogether a new conception. It had been adumbrated by certain Greek philosophers such as the Atomists, and before them by many others up to Anaximander, all of whom believed in the existence of innumerable worlds in an infinite void. But it is exactly these views of ancient Greek philosophers which about two centuries after Crescas were revived by Bruno and through him were introduced into modern thought. There is, however, the following difference between Bruno and Crescas. Bruno's worlds are Copernican worlds, whereas the worlds of Crescas, for the lack of any statement by him to the contrary, are still Ptolemaic worlds, with stationary earths at the centre, enclosed by a number of concentric spheres.

Another important point on which Crescas differs from Aristotle is what may be described as the principle of the continuity and homogeneity of nature. In Aristotle's conception of the universe, despite his assumption of an interconnection between the various parts of the universe and a continuity of motion running throughout its parts, there was still a certain break and discontinuity and heterogeneity in nature. This break occurs at the juncture of the translunar and the sublunar parts of the universe, and as a result of it nature becomes divided into two distinct realms. The break is of a twofold kind. In the first place, there is a difference in the nature of the motions which respectively characterize the sublunar and the translunar bodies. The rectilinear motion of the sublunar elements is described as natural, being brought about by certain centrifugal and centripetal forces which act upon the four elements and bring about their refluxes to their natural places. In the translunar elements, however, the motion, which is circular, is described as voluntary and appetitive, being brought about by a principle of motion inherent within the celestial bodies, acting upon them from within after the manner of a soul.[14] In the second place, there is a difference in what may be called the ultimate constitution of the sublunar and translunar elements. The four elements out of which the sublunar bodies are constituted are fundamentally different, according to Aristotle, from the ether which constitutes the heavenly bodies. While there may be some question as to whether Aristotle regarded the ether as a fifth element, it is certain that he regarded it as totally different from the sublunar elements. The former is constant, incorruptible and eternal; the latter are changeable, corruptible and transient. Among Arabic and Jewish Aristotelians the distinction between them is sometimes expressed in a different way. In the sublunar bodies, it is said, there is an inextended matter which is pure potentiality and to which tridimensionality is added as what is called corporeal form.[15] In the translunar bodies, there is no inextended, purely potential matter.[16] Logically, the break which these two differences between the sublunar and translunar bodies have produced within Aristotle's universe is analogous to the break which would have been produced in our conception of the universe, if we had assumed that the law of gravitation operates in one part of the universe but not in another and that the ultimate constitution of the matter of the terrestrial bodies is intrinsically different from that of the celestial bodies.

Now, this discontinuity and heterogeneity in nature is eliminated by Crescas. As over against Aristotle's distinction between the nature of the circular motion of the heavens and the rectilinear motion of the sublunar bodies, Crescas argues that such a distinction does not exist but that the motion of both

[14] See n. 11 (p. 535) on Prop. VI.
[15] For the origin, history and meaning of "corporeal form", see n. 18 (p. 579) on Prop. X.
[16] See n. 24 (p. 594) on Prop. X.

celestial and terrestrial bodies is what may be described as natural.[17] While this view, as we have shown, is not altogether original with Crescas,[18] still his repeated emphasis of it is of the utmost importance, for it was not until astronomers had rid themselves, as did Crescas, of the Aristotelian principle that the motion of celestial bodies was unlike that which prevails on earth that any real progress could be made in the proper understanding of celestial mechanics.[19] Then he also denies that there is any distinction between the matter of the celestial spheres and the matter of the sublunar elements, insisting that they are both alike, that in both cases matter is tridimensionality and has actual existence without having its actuality conferred upon it by form.[20] By this Crescas does away with what is the essential characteristic of Aristotle's theory of matter and form, though he retains Aristotle's vocabulary. Furthermore, in his discussion of this question we get a glimpse of the historical development of the view which ultimately resulted in the identification of matter with extension in the philosophy of Spinoza.

Historically, in Greek philosophy, the rival of Aristotle's theory of matter and form was Atomism. In modern philosophy, too, the emancipation from Aristotle's theory of matter and form was a gradual movement in the direction of atomism which was ultimately established experimentally by Dalton. Crescas' criticism of Aristotle, on the face of it, would seem to be outside this movement. He does not directly espouse the atomistic theory, although this theory was known in philosophic Hebrew literature through the Moslem Kalam and an allusion to it is found in Crescas himself.[21] All he does, it would seem, is only

[17] Prop. VI (p. 237).
[18] See n. 11 (p. 535) on Prop. VI.
[19] Cf. J. F. W. Herschel, *Preliminary Discourse on the Study of Natural Philosophy*, Part III, Ch. III, (294); G. H. Lewes, *Aristotle*, p. 125.
[20] Prop. X, Part II (p. 263).
[21] See n. 4 (p. 569) on Prop. X.

to modify the accepted interpretation of Aristotle's theory of matter and form. Still if we look closely into Crescas' reasoning we shall find that underlying it is really an attempt to revive Atomism. For the atom is distinguished from the Aristotelian matter not only by its indivisiblity but also—and this is of greater importance—by the actuality of its existence. As a result of this latter characteristic of the atom, all the forms that the atom may assume are considered by the Atomists as being only what Aristotle would call accidents. The essential fact, therefore, about atomism, as a view opposed to Aristotle's theory of matter and form, is not that it does away with the infinite divisibility of matter but rather that it does away with the potentiality of matter and consequently also with form as a principle of actualization. That this was considered the essential fact about atomism is attested by the various restatements of the atomistic theory which have come down to us from Maimonides and others.[22] Now, this is exactly what Crescas has done to matter. He has deprived it of its potentiality. He has made it to have actual existence. He has thus also abolished form as a principle of actualization. Form, therefore, becomes only an accident. Crescas himself was aware of these far-reaching consequences of his view, but wishing to retain the Aristotelian vocabulary he argues that form, though no longer a principle of actualization and hence only an accident, may still retain its Aristotelian name, because of some other differences that may be discovered between it and all the other accidents.[23]

The unification of the forces of nature which Crescas established by bringing together celestial and terrestrial bodies under the same kind of motion was extended by him still further by his including under it the phenomenon of magnetic attraction. This phenomenon was felt to be in need of an explanation in

[22] See n. 4 (p. 569) on Prop. X.
[23] Prop. X, Part II (p. 263) and n. 31 (p. 601).

view of the fact that it seemed to contradict the Aristotelian law that every efficient cause of motion must be moved itself while producing motion in something else. Different explanations were offered, all of which, however, proceeded on the assumption that magnetic attraction was controlled by a different force from that which controlled the natural motions of the elements.[24] Logically that position is analogous to the position of modern physics which assumes that the laws which govern the electromagnetic field are different from the laws which govern the field of gravitation. Crescas, however, attempts to remove that difference. He contends that the magnet attracts the iron by a motion which is the same as the natural motion of the elements.[25] Logically, a modern analogy of Crescas' explanation would be a theory which would unite the laws of electro-magnetism and those of gravitation under one law.

In the system of Aristotle, the break which he conceived to exist within nature itself was insignificant in comparison with the break he conceived to exist between nature and that which is beyond nature, or between the universe and God. Though the cause of the universe's motion, God was in no other way related to the universe, except by the relation of absolute contrast. He was the immaterial as contrasted with the material, the immovable as opposed to the movable. Again, though the cause of the universe's motion, He was neither its immanent cause nor its external cause. He was its transcendent cause, or, to use the Greek, Arabic and Hebrew term, its 'separate'[26] cause. If we were to look in the history of philosophy for an extreme contrast to this view of Aristotle, we would probaly find it in Spinoza's conception of God as immanent in the universe, and it would be possible for us, by only exchanging Aristotle's matter and form for Spinoza's extension and thought, to express the con-

[24] Prop. IX (p. 253) and n. 10 (p. 565).
[25] Ibid.
[26] Cf Moreh Nebukim II, 1 and 12; n. 36 (p. 422) on Prop. I, Part II.

trast between them by saying that according to Aristotle God has only the attribute of thought whereas according to Spinoza God has the attributes of both thought and extension.

Now, there is a suggestion in Crescas which logically could lead one to Spinoza's position of attributing extension to God. It occurs in his discussion of space. After defining space as incorporeal extension and assuming the existence of such an infinite incorporeal extension within which the world is situated, he quotes in support of his view the old rabbinic dictum that God is the place of the world. The dictum is also known to non-Jewish authors from a non-rabbinic source[27] and its significance is usually that which it is given by those who use it. In its original sense, as used by the rabbis, it is only a pious assertion of the omnipresence of God. There is in it, however, the germ of another and radically different idea. Interpreted freely, it could be taken by one who, like Crescas, believed in the existence of an infinite space, to signify the identity of God with that infinite space or rather with the wholeness of the universe, and it would be only necessary to introduce into it the element of thought to arrive at Spinoza's novel conception of God. Crescas, however, stops short of drawing this new conclusion from the old dictum. Indeed he starts out quite promisingly by saying that God as the place of the universe implies that He is the essence and the form of the universe, which really means that God is inseparable from the universe, but without evidently realizing the significance of his own words he concludes by restoring to the dictum its original and historical sense as an assertion of the omnipresence of God within a universe from which He is separated and which He transcends.[28] God to him continues to play the traditional part of a transcendent

[27] Philo, De Somniis I, II; cf. Leibnitz, Nauveaux Essais II, xiii, §17 and Duhem, Le Système du Monde, V, pp. 231–232. Cf. Joël, Don Chasdai etc., p. 24.
[28] Prop. I, Part II (p. 201).

being unlike anything within the universe, contrasted with it as spirit with body, as the simple with the manifold, as the actual with the potential and as the necessary with the possible. Like all other philosophers who started with such premises Crescas consequently found himself compelled, in order to bridge that gulf between God and the universe, to endow this transcendent God with a will and power and all the other attributes of personality, and by doing so he got himself involved in all the traditional problems of theology which form the subjects of discussion of the remaining parts of his work.

In the history of philosophy, the opposition to Aristotle had at various times assumed different forms. Aristotle was opposed, because some of his views were found to contradict certain Biblical traditions; he was also opposed, because his reasoning on many important points was found to be logically unsustainable; and finally he was opposed, because the method of his approach to the study of nature was found to be empirically inadequate. All these modes of opposition may be discerned in Crescas. On his own asseveration, his chief motive in opposing Aristotle was his desire to vindicate the sovereignty of tradition, not so much to render it immune from the attacks of speculation as to free it of the necessity of its support.[29] Still he does not follow the tried and convenient method of hurling Biblical verses, in their crude, literal meaning, at the heads of the philosophers. As a Jew, well versed in the lore of his religion, he knew full well that Biblical verses were not to be taken in their crude, literal meaning, for having early in its history adopted a liberal method in interpreting the laws of the Bible and having explained away the verse "an eye for an eye" to mean compensation, Judaism could not with any show of consistency insist upon taking any other verse in its strictly literal sense. If some mediaeval rabbis did insist upon a literal interpretation of non-

[29] See Introduction to Book I (p. 135).

legal portions of the Bible, it was rather in utter disregard of such logical consistency. In one place, in fact, he argues quite to the contrary that the philosophers cannot derive any support for one of their views from certain literal expressions of the Bible, for those expressions, he says, are to be understood in a figurative sense.[30] Tradition, according to him, is a guide only in matters theological; he does not employ it in deciding problems concerning the nature of things. Only once, in connection with the nature of space, does he quote Biblical and rabbinic passages in support of his view,[31] and then, too, he does it rather hesitatingly and uses them only as corroborative evidence and not as a basis for his knowledge.

The method employed by Crescas in his opposition to Aristotle is of a more subtle and more effective kind. He carries the battle to the enemy's own ground. Like one Bible hero of old, he tries to slay his Egyptian with a spear plucked out of his adversary's own hand. He employs reason to show up the errors of reason. And yet for himself he is not convinced of the unlimited power of reason. Reason was well enough as a tool to be used in his attempt to upset Aristotle's scientific dogmas, but he does not consider it sufficiently reliable as a means of setting up new dogmas of his own. He is thus quite willing to employ reason in order to prove, in opposition to Aristotle, that the existence of many worlds is not impossible, but he doubts the power of reason to help us in attaining any knowledge of what is beyond this world of our experience and therefore counsels us, by suggestion, to suspend judgment and keep our mind open.[32]

With reason thus limited in its function, Crescas sometimes calls upon empirical observation for aid. He does so toward the

[30] *Or Adonai* IV, 3, in connection with the verse "The heavens declare the glory of God" (Ps. 19, 2) commonly taken by mediaeval Jewish philosophers as implying that the celestial spheres are animate and rational beings.

[31] Prop. I, Part II (p. 199).

[32] Prop. I, Part II (p. 217).

end of his discussion of infinity.[33] Again, in the discussion of magnetic attraction, in a passage the reading of which is doubtful but of which the meaning is quite clear, he says something to the effect that any rational explanation of that phenomenon is at best only hypothetical; what is certain about it is only that which is vouchsafed by observation and experience.[34] But experience as a guide to knowledge was to him still a new and untried venture. While forced to turn to its aid occasionally by his own skepticism as to the validity of speculative reasoning, he knew not what use to make of it and what its far-reaching possibilities were, and unlike the two Bacons, he did not attempt to build upon it a new method of science. Every experience to him was a single experience and was to prove only a single fact. It was never to give rise to a universal law. Again, an experience to him was something given, not something that was to be produced. It never became with him an experiment. Crescas, for instance, doubted the truth of Aristotle's theory as to the existence of naturally light objects and of a natural motion upward, and thus when he observed that air goes down into a ditch without the application of any external force, he concluded that air was not naturally light and had no natural motion upward.[35] But when Newton began to doubt these Aristotelian laws of motion, while he may not have received his original inspiration from the falling of the celebrated apple, he certainly did observe and study the falling of other bodies and after long and painstaking research established the universal law of gravitation. Again, when Crescas wanted to prove that something was wrong with a certain conclusion which was supposed to follow from Aristotle's theory that heavier bodies fall faster than lighter

[33] Prop. I, Part II (p. 213).

[34] Prop. IX, Part II (p. 257). Another reading of the same passage would imply that Crescas did not consider his explanation of magnetic attraction as conclusive until it had been verified by experience. See n. 11 (p. 568).

[35] Prop. VI (p. 239).

bodies, he resorted to a hypothesis of an original time of motion.[36] It was subtle, but it led nowhere. But when Galileo wanted to prove that Aristotle's theory was totally wrong, he climbed up to the top of the tower of Pisa, and let two unequal weights fall down at the same time and watched their landing. It was simple, but it led to an epoch-making discovery in the history of science.

In a larger sense, we may see in Crescas' critique of Aristotle the fluctuation of the human mind at the point when it began to realize that reason, which had once helped man to understand nature, to free himself from superstition and to raise his desultory observations to some kind of unity and wholeness, had itself in the system of Aristotle gone off into the wilds of speculation and built up an artificial structure entirely divorced from nature. A new way of returning to nature was sought, but none was as yet to be found. Crescas had passed the stage when man condemned reason; he had reached the stage when man began to doubt reason, but he had not yet entered upon that stage when man learned to control reason by facts.

[36] Prop. XII, Part II (p. 271). Cf. n. 13 (p. 403) on Prop. I, Part II.

EXPLANATION OF SYMBOLS

פ—Ferrara edition, 1555.

ס—MS. Sulzberger, Jewish Theological Seminary.

מ—MS. Munich.

ל—MS. Jews' College.

י—MS. Paris, Bibliothèque Nationale.

ו—MS. Vienna.

ר—MS. Rome, Vatican.

ד—MS. De-Rossi, Parma.

ק—MS. Oxford, Bodleian.

ב—MS. Bloch, Berlin.

א—MS. Adler, Jewish Theological Seminary.

נ—MS. Bamberger, Jewish Theological Seminary.

( ) = omission.

[ ] = addition.

] = different reading.

TEXT AND TRANSLATION

of the

Twenty-five Propositions

of

Book I of the Or Adonai

המאמר הראשון

בשרש הראשון שהוא התחלה לכל האמונות התוריות והוא
אמונת מציאות האל יתברך.

אמנם למה שההקדמה יתבאר עניה בשתי עניינים: הראשון,
ביאור הגבולים הנופלים בה, והשני, יחס האחד אל האחר, כאלו
תאמר, חיוב הנשוא לנושא או שלילתו ממנו; והוא מבואר מעניין
ההקדמה הזאת, רוצה לומר אמרנו שהאלוה נמצא, שהגבול הנושא
בה הוא האלוה, והנשוא הוא הנמצא; והוא מבואר שהאל יתברך
נעלם תכלית ההעלם, כמו שיבא בגזרת השם; הנה אין ענין זאת
ההקדמה אלא שהסבה וההתחלה לכלל הנמצאות נמצאת. ולזה
היה העיון בשרש הזה בדרך השני לבד, והוא אופן עמידתנו
באמתתו. ולזה ראוי שנחקור אם עמדנו על אמתת השרש הזה מפאת
הקבלה לבד והוא התורה האלהית, או אם עמדנו בה מפאת העיון
והחקירה גם כן.

ולפי שהראשון ממי שהרחיב הדבור מפאת החקירה הוא ארסטו
בספריו, בטבעיות ובמה שאחר הטבע, ומפרשי ספריו, כמו תמסטיוס
ואלכסנדר, והאחרונים, כמו אבונצר ואבן רשד, והמחברים אחריו,
כמו אבן סינא ואבוחמד ור' אברהם אבן דאוד, והנה הרב המחבר
בספרו הנקרא מורה הנבוכים נשתמש ברוב הקדמותיהם על צד

6 הנושא לנשוא ד. ‏ 7 (רוצה לומר) סלזורדבאנ‏ — שאלה פ. ‏ 8 (בה) פג — (מבואר) לוא —שהאל]
האל לוא. ‏ 9 (נעלם) ב. ‏ 10 נמצא בג. ‏ 13 האל יתברך פ — (אם) ד — בו פ. ‏ 14—13 החקירה
העיון צזורדבאנ. ‏ 15 ממה בג—שהתחיל בדבור פד. ‏ 16 בספרו ב—תמאסטיוס ב נתמסטיוס א
תמאסתיו פ. ‏ 17 או אלכסנדר זורקבנ—אבונצרי—ובן רשד סזורדבג ון' רשד פא. ‏ 18 בן
סינא פבאנ ן' סנא ז—ואבוחאמד יד ואבואמר פ—בן דאוד פולדבנ ן' דאוד פ בן עזרא* —
הנהן] היה פג. ‏ 19 (הנקרא) פ — הקדמותיהם] הקדמונים פ. ‏ 19—1 לבאר על צד הקצור לד.

INTRODUCTION TO BOOK I

OF THE first of those principles of belief designated by us as Roots, which is the source of all the other principles designated by us as Scriptural Beliefs,[1] namely, the belief in the existence of God.

The purport of any proposition can be made clear and the proof thereof established by the explanation of two things:[2] first, the meaning of the terms which constitute the proposition, and, second, the relation of the terms to each other, that is to say, whether the predicate is to be affirmed of the subject or whether it is to be denied. In the proposition under consideration, i. e., 'God is existent,' it need hardly be said that the subject is 'God' and the predicate is 'existent.' Furthermore, it is generally admitted, as will be shown later,[3] God willing, that God is absolutely inscrutable. It follows, therefore, that the proposition is nothing but an affirmation that the Cause or Principle of all beings is existent. The study of this principle of belief must thus be confined to the second kind of inquiry, namely, to show how we know that the predicate is to be affirmed of the subject.[4] The task before us then is to inquire whether our knowledge of the truth of this principle of belief rests upon tradition[5] alone, that is to say, upon the authority of the Scripture, or whether we may also attain to it by way of reason and speculation.

Of those who discoursed in detail upon the question of God's existence from the point of view of speculative reason, the first was Aristotle in his works the *Physics*[6] and the *Metaphysics*; then his commentators, such as Themistius and Alexander, and the later[7] commentators, such as Alfarabi and Averroes; then the authors after Aristotle, such as Avicenna, Algazali and Abraham ibn Daud.[8] Finally Maimonides, in his work called *The Guide of the Perplexed*, has made use of the main teachings of

131

הקצור לבאר השרש הזה בדרכים מתחלפים, וראה הרב לצרף עם

זה שני שרשים יקרים, והם היותו יתברך אחד והיותו לא גוף ולא כח

בגוף, הנה ראינו לחקור על מופתיו, אם הם נותנים האמת על כל

פנים בשלשת השרשים האלה אם לא, לפי שהם לקוחים מכלל דברי

5 הפלוסופים הראשונים, וכל מה שנאמר בהם מזולתו אין לשום לב

עליו.

ולפי שמופתיו בנויים על שש ועשרים הקדמות שהניח בראש

החלק השני מספרו, הנה יהיה סדר העיון בזה בשני דברים האלה.

האחד, אם ההקדמות ההם אשר נשתמש בהם בבאור השרשים האלה

10 מבוארות האמת ביאור מופתי, שהוא אם לא היו ההקדמות הצריכות

אל ביאור השרשים מבוארות באור מופתי, הנה השרשים לא

התבארו באור מופתי. והשני, כשנניח ההקדמות ההם אמתיות,

מבוארות באור מופתי, אם התבארו מהם השרשים באור מופתי.

והעיון הזה יהיה כפי מאמר האומר.

15 ולזה ראוי שנחלק המאמר הזה לשלשה כללים.

הכלל הראשון. בביאור ההקדמות, כפי מה שבאו מבוארות

בדברי הפילוסופים, ובאור מופתי הרב, כי אם נחקור בהם, ראוי

these men,[9] restating them briefly in the form of propositions, out of which he constructed various proofs to establish this principle of God's existence. Furthermore, the Master has deemed it fit to add thereunto two other precious principles, namely, that God is one and that He is not a body nor a force inherent in a body.[10] By reason of all this, we have selected the proofs advanced by Maimonides as the subject of our investigation, with a view to determining whether they establish the truth of these three principles in every respect[11] or not, for his proofs alone are derived from the generality of the teachings of the first philosophers, and therefore nothing that has been said by others on this subject deserves consideration.[12]

Inasmuch as Maimonides' proofs are all based upon twenty-six propositions which he has placed at the beginning of the second part of his work, our investigation of the subject will have to deal with the following two questions: First, whether the propositions which he has made use of in proving the principles are themselves established by demonstrative reasoning,[13] for if the propositions necessary for the proof of the principles have not been established by demonstrative reasoning, the principles, too, will not have been conclusively established. Second, granting those propositions to be true and to have been established by demonstrative reasoning, whether the principles can be shown conclusively to follow therefrom. In this twofold kind of investigation we shall reason from the opinion of the affirmer.[14]

In accordance with this plan it seems to us proper to divide Book I into three parts.

Part I. A commentary wherein the propositions are proved in accordance with the arguments employed by the philosophers in their own writings, and also a restatement of the Master's proofs [for the existence, unity and incorporeality of God], for intending as we do to subject both the propositions and the proofs to a

שיהיו מובנים לנו מבוארים וגלויים ונקיים מכל ספק, לפי כוונת
הרב.

הכלל השני, נחקור בו במקצת ההקדמות ובמופתי הרב, אם
נתבארו באור מופתי.

הכלל השלישי, בביאור השרשים כפי מה שתגזרהו התורה,ובאופן
עמידתנו בהם. ושם יתבאר כוונת המאמר הזה, והוא שאין דרך
לעמוד על השרשים האלו בשלמות אלא מצד הנבואה, במה
שהעידה עליו התורה ונתאמת בקבלה. ואמנם יתבאר עם זה
שיסכים בו העיון.

הכלל הראשון

בבאור ההקדמות, כפי מה שבאו מבוארות בדברי הפלוסופים,
ובמופתי הרב הלקוחים ממאמרי הפילוסופים. ולזה חלקנו הכלל
הזה לשנים ושלשים פרקים, השׁשה ועשרים לבאר השׁש ועשרים
ההקדמות, וששה עוד לבאר מופתי הרב שהם ששה.

הפרק הראשון

בביאור ההקדמה הראשונה האומרת שמציאות בעל שעור אחד
אין תכלית לו שקר.

והנה ההקדמה הזאת חקר עליה ארסטו במקומות מתחלפים
מספריו, בשמע, ובשמים והעולם, ובמה שאחר; והביא מופתים
עליה, אם בבאור המנעות מציאות גודל נבדל בלתי בעל תכלית,
ואם בבאור המנעות מציאות גודל גשמי בלתי בעל תכלית, ואם
בבאור המנעות מציאות מתנועע בלתי בעל תכלית תנועה סבובית
או ישרה, ואם בבאור כולל בהמנעות מציאות גשם בלתי בעל תכלית
בפעל. ולזה חלקנו הפרק הזה לארבעה מינים כמספר מיני
המופתים.

3 מקצת ר – הקדמות זד.   4 באור מופתי] במופת ר.   7 במה] כמו פ.   12 (ובמופתי...
הפילוסופים) פז – ממאמר ג.   14 הקדמות בג.   19 (מספריו) ד בספריו זר – והשמים בא –
ובעולם פ.   20 עליה] על זה לוורדקבנ – (אם) סלורדזקבנ – (גודל) ר – בב״ת גודל נבדל זק בנ.
21 גשמי] נבדל לדו – בב״ת גודל גשמי זק..   23 בהמנע סלוקבנ.   24 מינים סלוקבנ.   עיונים פצ׳ס.

critical examination we must first endeavor to understand them in a manner clear and thorough and free from any ambiguity, even as the Master himself would have wished them to be understood.

Part II. Wherein we shall inquire into some of the propositions and also into the Master's proofs with a view to determining whether they have been conclusively demonstrated.

Part III. An exposition of the same principles in accordance with the strict teachings of the Scripture and also a statement of the method by which we arrive at them. Therein the main contention of Book I will be made clear, namely, that it is impossible[15] to arrive at a perfect understanding of these principles except by way of prophecy, in so far as the teachings of prophecy are directly testified of in the Scripture and indirectly corroborated in tradition, though it will also be shown that reason is not necessarily at variance wth the teachings thus arrived at.

PROPOSITION I

PART I.

PROOF OF the first proposition, which reads:[1] 'The existence of any infinite[2] magnitude whatsoever is impossible.'

An inquiry into this proposition has been made by Aristotle in several places of his works, in the *Physics*, *De Caelo et Mundo*, and the *Metaphysics*,[3] and in support of it he has advanced arguments to show the impossibility of an incorporeal[4] infinite magnitude, or the impossibility of a corporeal infinite magnitude, or the impossibility of an infinite body having either circular or rectilinear motion, or again to show, by means of a general proof,[5] the impossibility of any actually infinite body. In correspondence to these four classes of arguments, we have divided this chapter into four sections.[6]

המין הראשון

בביאור המנעות מציאות גודל נבדל בלתי בעל תכלית.

וסדר המופת כן. אמר לא ימלט העניין מחלוקה, אם שיהיה
הגודל הזה הנבדל מקבל החלוקה או בלתי מקבל החלוקה. ואם
לא היה מקבל החלוקה, הנה לא יתואר בשהוא בלתי בעל תכלית
אלא כמו שיאמר בנקודה שהיא בלתי בעל תכלית ובמראה שהוא
בלתי נשמע. נשאר אם כן שיהיה מקבל החלוקה. ולא ימלט אם כן
משיהיה כמה נבדל או עצם מן העצמים הנבדלים, כנפש והשכל.
ובטל שיהיה עצם נבדל, למה שהנבדל במה שהוא נבדל אינו מקבל
החלוקה, וכבר הונח מקבל החלוקה.

ועוד שלא ימלט אם שנאמר שהוא מתחלק או שאינו מתחלק. ואם
הוא מתחלק, אחר שהיה נבדל פשוט מתדמה החלקים, חוייב שיהיה
גדר החלק והכל אחד; ולפי שהניח הכל בלתי בעל תכלית, יחוייב
שיהיה החלק בלתי בעל תכלית, והוא בתכלית הבטול שיהיו הכל
והחלק אחד. ואם אינו מתחלק, כמו שיחוייב בנבדל, הנה אמרנו
בו שהוא בלתי בעל תכלית כמו שיאמר בנקודה שהיא בלתי בעל
תכלית.

נשאר אם כן שיהיה כמה. ולא ימלט אם כן שיהיה אם כמה נמצא
בנושא ואם כמה נבדל, ובטל שיהיה כמה נבדל, אחר שהיו המספר
והשיעור, אשר עליהם יאמר הבלתי בעל תכלית, בלתי נבדלים
מן המוחש. ואם היה כמה נמצא בנושא, אחר שהיו המקרים בלתי
נבדלים מנושאם והיו התכלית והבלתי תכלית מקרים נושאם הכמה,
חוייב שיהיו בלתי נבדלים, אחר שהכמה בלתי נבדל.

<hr>

1 (המין] ב׳ג׳.   3 המופתים לא – כן] בו ס׳ק׳ב׳א׳ – ימלט פ׳צ׳פ׳א – העין לא ימלט ד׳ – מענין]
החלוקה י׳.   4 הזה] ההוא ר׳ – (הנבדל) ר׳.   5 יתואר] יתבאר ר׳.   7 לא ולרדוק׳ב׳א׳ – ימלט]
ימנע פ׳צ׳א.   8 שיהיה ז׳ר׳ – [יהיה] עצם צ׳ל׳ד׳ [שיהיה] עצם א׳.   11 ימלט] ימנע פ׳צ׳א.   12 ופשוט
צ׳פ׳לרדוק׳ב׳א׳ – ומתרמה צ׳פ׳ב׳א׳ – חייב ס׳.   13 החלק והכל ס׳לזור׳דק׳ב׳א׳.   16 [רק] בב״ת צ׳.
18 ימלט] ימנע פ׳צ׳א – שיהיה אם] אם שיהיה פ׳א.   22 נושאם] נושאים פ׳.

The First Class of Arguments

Proof for the impossibility of an incorporeal infinite magnitude. Aristotle has framed the argument in the following manner:[7] There is no escape from the disjunctive proposition[8] that this incorporeal magnitude is either divisible or indivisible. Now, if it were indivisible, it could not be described as infinite, except in the sense in which a point is said to be infinite or color inaudible. It must, therefore, be divisible. If so, however, it must inevitably be either an incorporeal quantity or one of the incorporeal substances, as, for instance, soul and intellect. But to say that it is an incorporeal substance is impossible, for the incorporeal *qua* incorporeal is not subject to division, whereas the infinite is now assumed to be capable of division.[9]

Again, that incorporeal substance would inevitably have to be either divisible or indivisible. If it be divisible, since it is also incorporeal, simple and homoeomerous, it would follow that the definition of any of its parts would be identical with that of the whole, and since the whole is now assumed to be infinite, any part thereof would likewise have to be infinite. But it is of the utmost absurdity that the whole and a part of the whole should be alike [in infinity]. And if it is indivisible, which, indeed, as an incorporeal, it must be, we can no longer call it infinite except as a point is said to be infinite.[10]

Hence, by the process of elimination, the infinite must be a quantity. But then, it must inevitably be either a quantity subsisting in a subject or an incorporeal quantity.[11] It cannot be an incorporeal quantity, for number and magnitude, of which two infinity is predicated, are never themselves separable from sensible objects. And if the infinite were a quantity subsisting in a subject, it would have to be inseparable from corporeal objects, for since quantity itself is inseparable and finitude and infinity are accidents whose subject is quantity, like all other accidents, finitude and infinity could not exist apart from their subject.[12]

ולהיות המופת הזה בנוי על ההקדמה המחייבת המנעות שיעור

נבדל למוחשות, והאומר ברחק נבדל מקיים מציאותו, כבר יהיה

נערך על הדרוש. ולזה יראה שהוא סומך על סברתו בהמנעות

הרקות. וזה שאם הודינו במציאותו לא ימנע מציאות שיעור נבדל

5 למוחשות, אבל אולי יחוייב מציאותו, למה שכבר אפשר שישוער,

ויתאמת אמרנו בו גדול או קטן ויתר משיני הכמה. אבל למה

שהרחיק מציאותו, בנה עליו המופת הזה. ולזה ראינו להביא מופתיו

על צד הקצור במין הזה, כדי שנחקור בהם, בכלל השני, אם הם

נותנים האמת בו על כל פנים, בגזרת השם.

10 והנה לפי שהאומרים ברקות דמו שתנועת ההעתק בלתי אפשרית

אם לא היה הרקות נמצא, הוא לקח תחילה בביאור שקרות הדמוי

ההוא. עוד סדר ארבעה מופתים בביטול מציאות הרקות.

והנה באור שקרות הדמוי הוא כן. אם היה הרקות סבת התנועה,

יחוייב שיהיה פועל או תכלית. אבל אינו פועל או תכלית, יוליד

15 סותר הקודם. והנה חיוב התדבקות הנמשך לקודם מבואר, למה

שהתבאר שסבות הדברים ארבעה, והם החומר והצורה והפועל

והתכלית. והוא מבואר שאין הרקות חומר התנועה ולא צורתה.

---

2 יהיה] היה יר.　　4 הרקות] הרחוק פ – הודהו לד.　　8 במין] בעין פ.　　9 (בגזרת השם) פ.

10 ההעתק] ההתקבלות פ.　　13 והנה] הוא בג – הוא,　והוא בג ותנועה] התנועה ההעתק לד.

15 ההתדבקות נמשך יר ק בג הדבקות הנמצא א.　　17 צורתה] זולתה יר.

Inasmuch as this last argument is based upon a proposition which negates the possibility of a magnitude existing apart from sensible objects, the existence of which, however, is not impossible if one admits the existence of an incorporeal distance, the argument will thus be[13] a begging of the question.[14] It seems, therefore, that Aristotle is relying here upon his own opinion as to the impossibility of a vacuum. For were we to admit the existence of a vacuum, the existence of an incorporeal magnitude would no longer be impossible; nay, its existence would of necessity be implied, since a vacuum is capable of being measured, and can thus be appropriately described by the terms great and small and the other properties of quantity. [15] It is only by rejecting first the existence of a vacuum that he was enabled to build up that argument of his. This being the case, it appears to us peculiarly fitting to give here a brief summary of all his arguments against the existence of a vacuum, so that we may inquire afterwards, in the second part, God willing, as to whether they establish the truth of his contention in every respect.

Since those who affirmed the existence of a vacuum supposed[16] that locomotion would be impossible[17] without the existence of a vacuum, Aristotle first undertook to prove the falsity of this supposition. Then, he framed four[18] other arguments to show that the existence of a vacuum is impossible.

His proof of the falsity of the assumption runs as follows:[19] If a vacuum were the cause of motion, it would have to be either its efficient or its final cause. But the vacuum can be neither an efficient nor a final cause. Hence it leads to a conclusion which denies the antecedent. The cogency of the connection between the consequent and the antecedent is evident, for it has been shown that causes are four in number, the material, the formal, the efficient, and the final; and since the vacuum can evidently be neither the material nor the formal cause of motion, it must

נשאר אם כן שיהיה פועל או תכלית. וחיוב סותר הנמשך יתבאר
כן. לפי שאנחנו נראה גשמים מתחלפים מתנועעים תנועת ההעתק,
מקצתם אל המעלה ומקצתם אל המטה, וכבר יראה שסבת
ההתחלפות אם טבע הדבר הנעתק, והוא המניע והפועל, ואם טבע
5 המקום אשר אליו התנועה, והוא התכלית. ולזה לפי שהיה הרקות
מתדמה החלקים, ואי אפשר שיתחלף בו בעניין שיהיה לקצתו טבע
מה שממנו וקצתו מה שאליו, לא ימלט העניין מחלוקה, אם שיהיה
לו טבע מה שממנו, או טבע מה שאליו, או שלא יהיה לו לא טבע
מה שממנו ולא טבע מה שאליו. ואם הנחנו לו טבע מה שממנו, כאשר
10 הונח גשם מה ברקות, חייב שיהיה נח לעולם. ואם הנחנו לו טבע
מה שאליו, חייב שיתנועע אל כל הצדדים יחד, או שיהיה נח לעולם
לפי שאין התנועה לצד אחד ראויה יותר מהתנועה לצד אחר. ואם
הנחנו שאין לו טבע מה שממנו ולא מה שאליו, כמו שהוא האמת
בעצמו, למה שהוא רוחק נבדל מהדברים הטבעיים, חייב גם כן
15 שיהיה הדבר נח לעולם. ולזה התבאר שאין הרקות פועל ולא
תכלית. וזהו מה שכוון באורו במופת הזה.

עוד עשה א ר ב ע ה מופתים לבטל מציאות הרקות.

המופת הר א ש ו ן סדורו כן. אם היה הרקות נמצא, התנועה
בלתי נמצאת. אבל התנועה נמצאת, אם כן הרקות בלתי נמצא. והנה

2 כן] בו פ – הגשמים ז ר.   3 אל] על פ – אל]על פ – המטה] המנוחה פ – כבר פלורדקבאי.
5 (ולזה) ר וזה פ – ולפי ר.   6 ואין] ואם ב י – בו] כן כ.   7 וקצתו (טבע] קאי – (העניין) פ.
9‑7 לא ימלט ... ולא טבע מה שאליו] או שלא יהיה לו טבע מה שממנו ומה שאליו פ או שלא יהיה
לו טבע מה שממנו או טבע מה שאליו או שלא יהיה לו טבע מה שממנו לא יבא מה שאליו בי.   7 ואם]
או פ – שיהיה] שהיה ר.   8 או (טבע] ז – שלא] שאין ז ר – לו (ולא] ר – (יהיה) ר – (לא) פלרדק (לו לא) א.
9 ולא (טבע) פ זא (ולא טבע) י.   10 הונח] הנחנו פלזורדבי – מה] שמה פ (מה) ר.   11 אל]
על פ.   14 הטבעיים, ולפי שהפועל והתכלית לא יחייבו חלוף התנועות אלא מצד חלוף טבעם,
והרקות אין לו טבע ולא חלופו, הנה א״כ לא חייב בתנועה ולא יהיה לא פועל ולא תכלית, חייב
ל הדברים האלוינשנים גם כן בנליון הכת״י.   17 (עוד... הרקות) פ.
15 (ולזה) זה פ.   18 (המופת) פ ז דבאי – סדורו] סדורן חברו פ – כן] כך פ.   19 (אבל) קאי – אם כן הנה לזדקבאי.

necessarily be either its efficient or its final cause. As for the validity of the proposition which denies the consequent, it can be established as follows. We observe that different elements[20] are all moved with locomotion, but some in an upward direction and others in a downward direction.[21] It is quite evident that the cause of this divergence of direction lies in the nature of the moving object, which might be called the motive and efficient cause, and in the nature of the place toward which the motion is tending, which might be said to operate as a final goal.[22] But inasmuch as the vacuum, being homoeomerous, cannot have dissimilar parts, so that some of it would have the nature of a *terminus a quo*, and others that of a *terminus ad quem*, it must inevitably either possess only one nature, *a quo* or *ad quem*, or be devoid of either. [In the first case], if we suppose all the parts of the vacuum to be *termini a quo*, then a body placed in it would have to remain always at rest; and if we suppose them to be all *termini ad quem*, then an object placed in it would either have to move in all directions at the same time or to remain always at rest, since in such a vacuum motion in one direction would not be more likely than in another. [In the second case], if we suppose the vacuum to be endowed with neither of these natures, which indeed must be the case, since the vacuum is nothing but dimension devoid of all physical contents,[23] it would again follow that an object [placed in it] would have to remain always at rest. Thus it has been demonstrated that the vacuum can be neither an efficient nor a final cause. This is what he intended to prove by this argument.[24]

He further framed *four* arguments in denial of the existence of a vacuum.

The *first* of these arguments runs as follows:[25]

If a vacuum exists, motion does not exist. But motion exists. Hence a vacuum does not exist. The proposition which denies

סותר הנמשך מבואר מן החוש. יחיוב התדבקות הנמשך אל הקודם

יתבאר כן. לפי שהתנועה אם טבעית ואם הכרחית, והתנועה

הטבעית תתחלף לפי טבע מה שממנו ומה שאליו, והיה הרקות אין

בו התחלפות, אם כן אין בו תנועה טבעית. ולפי שהההכרחית תאמר

5 בצירוף אל הטבעית, והטבעית קודמת לה בטבע, וזה שהמתנועע

בהכרח יתנועע בהכרח למה שיפרד ממקומו אשר אליו התנועה

בטבע, הנה אם כן כאשר לא תמצא הטבעית לא תמצא ההכרחית.

ועוד שאלו היתה התנועה ההכרחית ברקות, יתחייב שינוח המתנועע

בהפרד המניע ממנו. וזה שהחץ כאשר יתנועע מהמניע, והוא היתר,

10 והיתר נח, הנה הוא למה שבאויר כח על קבול התנועה לקלותו,

ידחה החץ עד שיפול למקומו הטבעי. ולמה שהוא מבואר ברקות

שאין בו כח על קבול התנועה, הנה יחוייב שינוח המתנועע בהפרד

מן המכריח, והוא הפך מה שיראה בחוש.

המופת ה ש נ י ו ה ש ל י ש י בנויים על שתי הקדמות, והוא שסבת

15 המהירות והאיחור במתנועעים, הוא חלוף המניע, או חלוף המקבל,

או שניהם. ובאור זה, שאם המניע יותר חזק יהיה יותר מהיר, וכן אם

המקבל, והוא הממוצע אשר בו התנועה, יותר חזק הקבול – כאויר

---

1 (הנמשך אל) בג. 2כן) נ"כ יר. 3והיה] והנה ד. 4אם כן] הנה א"כ סזורק באו הנה ל ד
6יתנועע] מתנועע ר – (בהכרח) ר – אשר] ולאשר בג. 7 תמצא] נמצא פ. 9יתנועע] יפרד פ א.
10 (והיתר) הנח פ – (הוא) פא – (על) קבול] לקבל י. 11וידחה פ פ בג. 12 בו] לו פ. 13 (מן)
ר – המכריח] המניע פ מהמכריע ר – מה] ממה ל דא – לחוש א. 14שבסבת פ.

the consequent can be established by sense perception; and as for the cogency of the connection between the consequent and the antecedent, it may be shown in this way. Motion is either natural or violent. Natural motion must differ in direction, and this is possible only through a difference in the nature of the places from which and toward which it tends.[26] Since the vacuum admits of no difference in the nature of its parts, there can of course be no natural motion in it. And as violent motion is so called only with reference to natural motion, which is prior to it in nature,[27] for an object set in motion by some external force is said to be moving by violence only because it moves away from the place toward which it has a natural tendency,[28] it follows that by proving natural motion to be impossible in a vacuum violent motion becomes likewise impossible. Furthermore, if there existed violent motion in a vacuum, the *motum* would have to come to rest as soon as the motor which had set it in motion was removed. In the case of a shooting arrow,[29] for instance, it is only because the air on account of its lightness is endowed with the capacity of retaining this impelling force [imparted by the motor] that the arrow, having once been set in motion by its impellent, namely, the string, [will continue in its motion], even though the string has come to rest, for the air will continue to propel it until it comes to its natural locality.[30] But as it is clear that the vacuum has no capacity of retaining the impelling force of motion, an object moving in it would necessarily have to come to rest as soon as it has parted from the motor. But this is contrary to sense perception.

The *second* and *third* arguments[31] are based upon two propositions.[32] First, the swiftness and slowness of moving objects are due to the difference in the motive force[33] or in the receptacle[34] or in both, that is to say,[35] the stronger the motive force the greater the velocity; likewise, the stronger the receptacle, i. e., the medium in which the motion takes place—as, for instance,

על דרך משל שהוא יותר חזק הקבול מהמים–יהיה גם כן יותר מהיר.
והשני, שיחס התנועה אל התנועה כיחס הכח המניע אל הכח המניע,
כשהממוצע אחד; או כיחס כח הקבול אל כח הקבול, כשהמניע
אחד; או כיחס מחובר מכח המניע אל כח המניע ומכח הקבול אל
5 כח הקבול, כשהמניעים והממוצעים מתחלפים, וכבר התבאר בספר
היסודות לאוקלידס דרך לקיחת היחס המחובר. ואחר שהונחו
אלו ההקדמות כמבוארות בעצמן, סדר המופת האחד מצד המקבל
והאחד מצד המניע.

אם אשר מצד המקבל סדורו כן. אם היה הרקות נמצא, יתחייב
10 שהמתנועע בו יתנועע בזולת זמן, והתנועה בזולת זמן הוא שקר,
יתחייב סותר הקודם. והנה התדבקות הנמשך לקודם יתבאר
בהניחנו מתנועע אחד מניע אחד, גודל ידוע, באויר וברקות. הנה
לפי שסבת המהירות והאחור בזה הוא חילוף המקבל, כמו שהתבאר
בהקדמה הראשונה, ויחס המהירות והאיחור בזה הוא כיחס האויר
15 אל הרקות, כמו שהתבאר בשנית, והוא מבואר בשני המקבלים
שיחסם כיחס הבעל תכלית אל הבלתי בעל תכלית, חוייב אם כן
שתהיה התנועה ברקות בזולת זמן. והוא שקר, למה שלא תצוייר
תנועה בגודל בזולת זמן, להיות הגודל מתחלק, ויתחייב שיתחלק
הזמן בהחלק התנועה בו.

20 ואמר אבן רשד, שהמופת הזה כחו כח המופת אשר יולד ממנו,

air which has a stronger receptive power[36] than water—the more rapid the motion. Second, the ratio of two motions is equal to the ratio of the powers of their respective motive forces, when the medium is the same, or to the ratio of the receptive powers [of their respective media], when the motive force is the same; or to the compound ratio of the powers of their respective motive forces and receptivities, when both motive force and medium are different—the rule for manipulating compound ratios having already been explained in Euclid's *Elements*.[37] With these two propositions assumed as self-evident, he has framed one argument with respect to the receptacle and another with respect to the motive force.

As to the one with respect to the receptacle, it runs as follows.[38] If a vacuum exists, an object moving in it will have to move in no-time. But motion in no-time is inconceivable. Hence it leads to a conclusion which denies the antecedent. The connection of the consequent with the antecedent may be explained by assuming an object moved by the same motor—a certain magnitude—both in air and in a vacuum. Since according to the first proposition a difference in the velocity would have to arise in consequence of the difference in its respective receptacles, and according to the second proposition the ratio between its respective velocities would be equal to the ratio between the air and the vacuum, and as it is furthermore clear that the ratio between these two receptacles would be equal to the ratio between a finite and an infinite,[39] it would thus follow that motion in a vacuum would take place in no-time.[40] But that is impossible, for no magnitude can be conceived as being moved in no-time, since every magnitude must be divisible, and the time of its motion must consequently be divisible along with its motion.[41]

Averroes has remarked here that the force of this argument is like that of the argument by which it is sought to prove

שאם היה כח מניע בלתי בעל תכלית היולני, שיחוייב שיתנועע

המתנועע ממנו בזולת זמן.

ואמנם המופת אשר מצד המניע סדורו כן. אם היה הרקות

נמצא, יתחייב שקרות ההקדמה הראשונה, עם היותה מבוארת

5 בעצמה. וזה בהניחנו שני מתנועעים, משני מניעים, מתחלפים בגודל

ידוע, ברקות; והנה יתחייב מההקדמה הראשונה שהאחד יותרמהיר

מהשני; ולפי שהוא מבואר בכל מתנועע ברקות, לפי מה שקדם,

שיתנועע בעתה, הנה יתחייב שבחלוף המניע לא תתחלף התנועה.

והוא שקר לפי ההקדמה הראשונה. והשקר הזה יתחייב מאמרנו

10 שהרקות נמצא.

המופת ה ר ב י ע י סדורו כן. אם היה הרקות נמצא, היה מתחייב

אפשרות הכנס גשם גשם בגשם. ואבל הכנס גשם בגשם הוא נמנע, שאם

לא, היה אפשר שיכנס העולם בגרגיר חרדל. יוליד שהרקות בלתי

נמצא. והנה חיוב התדבקות הנמשך לקודם יתבאר כן. לפי

15 שמציאות הרקות אינו דבר רק מציאות השלשה רחקים נבדלים,

מופשטים מן הגשם; הנה אם כן, למה שאינם גשמים ולא מקרים

נשואים בדבר, הנה אי אפשר בהם שימירו מקומם כשיכנס בהם

הגשם, כמו שיעשו המים אשר בשוקת כשיושלך בתוכה אבן. הנה אם

כן כבר נכנסו רחקי הגשם ברחקי הרקות. ואם הוא אפשרי, הנה

20 הכנס גשם בגשם אפשרי. וזה כי ההמנעות אשר יראה בהכנס גשם

בגשם איננו מצד היותו עצם, ולא מצד היותו בעל מראה, ולא בעל

that if there existed a corporeal infinite moving force, the object set in motion by it would have to move in no-time.[42]

The argument with respect to the motive force runs as follows:[43] If a vacuum existed, it would lead to the falsity of the first proposition, despite its being self-evident. For suppose two objects in a vacuum were moved by two unequal motors, differing from each other by a given magnitude. According to the first proposition the velocity of one of those moving objects would have to be greater than that of the other. But an object moving in a vacuum, as has been shown before, would have to perform its motion in an instant. It would thus follow that though the motors differed, the velocity of the motion would not differ. This, however, is impossible according to the first proposition. And this impossibility will of necessity arise once we admit the existence of a vacuum.

The *fourth* argument runs as follows:[44] If a vacuum existed, it would follow that one body could enter into another. But the interpenetration of bodies is impossible, for, were it not so, the world could enter into a grain of mustard seed.[45] Hence it follows that a vacuum does not exist. The cogency of the connection between the consequent and the antecedent may be explained as follows: The existence of a vacuum means nothing but the existence of three abstract dimensions, divested of body. Since those dimensions are not bodies, nor accidents inherent in a subject,[46] they could not leave their place if another body were entered into them, as would happen, for instance, in the case of a trough full of water, if a stone were thrown into it. Hence the dimensions of the body would have to be considered as penetrating the dimensions of the vacuum. But if that were possible, the penetration of one body into another would likewise have to be possible, for the interpenetration of bodies is considered impossible not because of their being substances or of their being endowed with color and other qualities, but rather

איכות, אלא מצד רחקיו השלשה. הנה אם כן, אם הכנס גשם

ברחקים אפשרי, הכנס גשם בגשם אפשרי. והוא שקר בטל. הנה

אם כן אין הרקות נמצא תוך העולם ולא חוצה לו.

והנה חזק זה הדעת עוד מאשר הגשם יצטרך אל מקום מצד מה

5 שהוא בעל רחקים שלשה ינוח בם, ואם כן יצטרכו גם כן הרחקים

אל רחקים, וזה לבלתי תכלית. ועוד שהרחקים תכליות הגשמים,

והתכלית, במה שהוא בלתי מתחלק, אי אפשר בו שיובדל ממה

שהוא תכלית, יתחייב אם כן המנעות מציאות רוחק נבדל.

והוא היסוד אשר סמך עליו בביאור המנעות מציאות גודל בלתי

10 בעל תכלית. והוא אשר כוון במין המופת הזה, והוא המין הראשון.

עוד סדר אלתבריזי מופת בביאור המנעות מציאות גודל בלתי

בעל תכלית, והוא מופת הדבקות. וזה שכאשר הנחנו קו בלתי בעל

תכלית מצד אחד, ודבקנו עליו קו בלתי בעל תכלית, והתחלנו

מנקודה אחת בקצה הקו אשר הוא בעל תכלית, יתחייב שיהיה קו

15 בלתי בעל תכלית גדול מקו בלתי בעל תכלית. והוא שקר, שהוא

מן הידוע שאין בלתי בעל תכלית גדול מבלתי בעל תכלית.

1—2 נשם ברחקים] ברחקי הגשם רחק פ׳ג׳ נשם בנשם ברחקים ד.   2 ובטל ג׳ – (הנה) ג׳.
4 חחק ל – (עוד) ר.   5 ג׳ רחקים פ.   6 לבלתי תכלית] ב״ת י בעל תכלית ר – ועוד] וזה ג.
7 (והתכלית במה שהוא) ג׳ והתכלית במה שהוא (תכלית) ל׳ד.   10 (הזה) ז׳פ׳ב׳ג.   11 תבריז
׳ אל תבריז ז׳אל חביריז ל׳ד.   13 בעל תכלית והתחלנו ל׳ד.   14 יתחייב] חוייב ד׳ר –
שיהוא פ.   14—15 קו ב״ת גדול ר.   16 (מן הידוע) ידוע י׳ס׳ב׳ג – (מן הידוע שאין] בב״ת) ב״ת ר.

because of the three dimensions which they possess. If it be, therefore, maintained, that these dimensions, [i. e., a vacuum], can be penetrated by a corporeal object, all other corporeal objects would likewise have to be penetrable by one another. But this is an impossible falsehood.[47]

Hence a vacuum does not exist either within the world or outside thereof.[48]

He has further strengthened his view [by two additional arguments].[49] (1) If a body requires a place for its existence, it is only because of the three dimensions in which it is posited. [Now, if incorporeal dimensions or a vacuum existed], these dimensions, too, would require dimensions, and so on to infinity.[50] (2) Then, again, dimensions are the limits of bodies, and a limit, in so far as it [is a limit], is indivisible. It is therefore inseparable from the object of which it is a limit. Hence the existence of an incorporeal extension is impossible.[51]

This is the premise upon which he depended in trying to prove the impossibility of an infinite magnitude, and this is what he intended to prove by this class of arguments, namely, the first class.

Another argument to prove the impossibility of an infinite magnitude has been advanced by Altabrizi, namely, the argument of application.[52] Suppose we have a line infinite only in one direction. To this line we apply an infinite line [which is likewise infinite only in one direction], having the finite end of the second line fall on some point near the finite end of the first line.[53] It would then follow that one infinite, [i. e., the first line], would be greater than another,[54] [i. e., the second line]. But this is impossible, for it is well known that one infinite cannot be greater than another.

המין השני

בביאור המנעות מציאות גודל גשמי בלתי בעל תכלית.

והנה התחיל תחילה בביאור כולל היות מציאות גודל בלתי בעל
תכלית בפועל, גשמי היה או למודי, נמנע. וסדר המופת כן. כל
גשם הנה יקיף בו שטח אחד או שטחים, וכל מה שיקיף בו שטח או
שטחים הנה הוא בעל תכלית; הנה אם כן כל גשם בעל תכלית
בהכרח. וכאשר התבאר לו היות כל גשם בעל תכלית, הנה אם כן
כל שטח וכל קו בעל תכלית, לפי שהם לא יובדלו מן הגשם. וכן
התבאר לו במספר בפעל, שהוא בעל תכלית בהכרח, לפי שכל
מספר בפעל הוא ספור בפעל, וכל ספור בפעל אם זוג ואם נפרד,
הנה אם כן כל מספר בעל תכלית.

עוד סדר א ר ב ע ה מופתים טבעיים בבאור המנע מציאות גודל
גשמי בלתי בעל תכלית.

המופת הר א ש ו ן סדורו כן. אם היה גשם ממושש בלתי בעל
תכלית, הנה הוא בהכרח פשוט או מורכב. ואיך שיהיה, היה בהכרח
אחד מיסודותיו בלתי בעל תכלית בגודל, אחר שהתבאר המנע
מציאות יסודות בלתי בעל תכלית בראשון מהשמע. ואם היה אחד
מהיסודות בלתי בעל תכלית בגודל, אחר שהוא ממושש ובעל
איכות, הנה ברוב הזמן היה משנה ומפסיד שאר היסודות, למה
שהיסודות הם יסודות באמצעיות איכותיהם, ולא תתמיד ההויה,
והוא הפך ממה שנראה בחוש. ועוד שאם היה האחד בלתי בעל

The Second Class of Arguments

Proof for the impossibility of the existence of an infinite corporeal magnitude.

Starting out with a general proof,[55] he first tried to show that the existence of an actually infinite magnitude, whether coporeal or mathematical,[56] is impossible. The argument runs as follows:[57] Every body is contained by a surface or surfaces, and that which is contained by a surface or surfaces is finite. Hence every body must be finite. Having convinced himself that every body must be finite, it has also become clear to him that surfaces and lines must likewise be finite, inasmuch as they cannot be separated from body. In a similar manner he has proved to himself the case of actual number, showing that number, too, must be finite, inasmuch as every actual number is that which is actually numbered, and that which is actually numbered is either even or odd. Hence every number is finite.[58]

He then proceeded to frame *four* physical[59] arguments to prove the impossibility of an infinite corporeal magnitude.

The *first* argument runs as follows:[60] If there existed an infinite tangible body, it would have to be either simple or composite. In either case, and however that simple or composite infinite body is conceived to be,[61] one of its elements would have to be infinite in magnitude, inasmuch as it has been demonstrated in the first book of the *Physics*[62] that an infinite number of elements is impossible. This element, infinite in magnitude, if it were so, and being also tangible and endowed with qualities, would in course of time bring change and corruption to other elements, [for that infinite element would have to be of a nature opposite to the others], inasmuch as elements are elements only by virtue of their own peculiar qualities,[63] and so there would be no continuance of generation. But this is contrary to sense perception. Again, if one[64]

תכלית, הנה יהיה בלתי בעל תכלית בכל רחקיו, למה שהרחקים
במה שהם רחקי גשם פשוט מתדמים, ולא ישאר מקום לשאר.

המופת הש נ י סדורו כן. כל גשם ממושש הנה לו קלות או כבדות.
והנה אם היה לו כבדות, היה במקום השפל ונבדל מן המקום העליון,
5 ואם היה לו קלות, היה בעליון ונבדל מן התחתון, זה כלו שקר
בבלתי בעל תכלית.

המופת הש ל י שי סדורו כן. אם היה כל גשם מוחש במקום, והיו
המקומות בעלי תכלית במין ובשיעור, הנה יחוייב שיהיה הגשם בעל
תכלית, אחר שהתבאר שהמקום הוא התכלית המקיף בגשם. ואולם
10 שהמקומות בעלי תכלית במין, זה מבואר, למה שהבדליהם
מוגבלים, והם המעלה והמטה והפנים והאחור והימין והשמאל, ושהם
בעלי תכלית בשיעור, הוא מחוייב, למה שאם לא היו בעלי תכלית,
לא היה בכאן מעלה מוחלט ולא מטה מוחלט, אלא בהצטרף,
ואנחנו נראה הדברים הטבעיים מוגבלים.

15 המופת הר ב י עי סדורו כן. אם היה כל גשם מוחש במקום,
והמקום הוא התכלית המקיף, יתחייב. שיהיה הגשם המתקומם בעל
תכלית. והנה חיוב התדבקות הנמשך מבואר בעצמו, למה שהמוקף
בעל תכלית בהכרח. ואולם איך יתבאר שהמקום הוא המקיף, בזה
סדר חמש הקדמות מבוארות בעצמם. האחת, שהמקום יקיף הדבר
20 אשר הוא לו מקום. והשנית, שהוא נבדל ואינו חלק ממנו. והשלישית,
שהמקום הראשון, והוא המיוחד, שוה לבעל המקום. והרביעית,
שהמקום ממנו מעלה וממנו מטה. והחמישית, שהגשמים ינוחו בזה
המקום ואליו יעתקו. אלו הן ההקדמות אשר יעמידונו על עצם

---

1 והנה] הוא זו ר.   8 הגשם] כל גשם זר כל הגשם פ נשם י.   10 שהמקומות] שיהיו המקומות
צפלזורדקבאנ – למה] במה זר.   11 ואחור פפ – ושהם] ושיהיו צפלזורדקבאנ.   12 היו (בעלי
תכלית) ירקבנ היו (לו) ב"ח א.   13 (ולא מטה מוחלט) פ.   16 המקומם פלזורדקבאנ.
17 והנה סק – (חיוב) פ   חוייב – ההתדבקות לדבנ.   20 (הוא) פ.

of the elements were infinite, it would be infinite in all its dimensions, for, being a simple substance, all its dimensions would have to be equal, and so there would be no room left for the other elements.

The *second* argument runs as follows:[65] Every tangible body must have either weight or lightness. Consequently, if the infinite had weight, it would have to be in the lower region and separated from the upper,[66] and if it had lightness it would have to be in the upper region and separated from the lower. But all this is impossible in an infinite.[67]

The *third* argument runs as follows: Since[68] every sensible body is in a place,[69] and since places are finite in both kind and magnitude,[70] it follows that every body must be finite, for place has been shown to be the limit that surrounds a body.[71] That places are finite in kind is evident, for their differences are limited in number, namely, above and below, before and behind, right and left. That they must also be finite in magnitude follows as a logical conclusion, for if they were not finite, there would be no absolute up and no absolute down, but only relative. But we observe that the natural places are limited.[72]

The *fourth* argument runs as follows:[73] Since every sensible body is in place, and place is the surrounding limit, it follows that the body which occupies place must[74] be finite. The cogency of the connection of the consequent is self-evident, for that which is surrounded must of necessity be finite. But how can it be proved that place is that which surrounds? To do this he has laid down five self-evident propositions:[75] First, that place surrounds the object of which it is the place. Second, that place is separated [from its occupant] and is not a part thereof. Third, that first place,[76] i. e., proper place, is equal to its occupant. Fourth, that place has the distinction of up and down. Fifth, that the elements are at rest in their respective places and toward those places they tend to return. These are the propositions which

המקום. עוד עשה הקש תנאי מתחלק, סדורו כן. המקום בהכרח

יראה שהוא אחד מארבעה, אם הצורה, ואם ההיולי, ואם התכלית

המקיף, ואם הרוחק אשר בין תכליות המקיף, והוא אשר יקרא

חללות. ואם לא יהיה אחד מהשלשה, רוצה לומר הצורה וההיולי

5 והחללות, יחוייב בהכרח שיהיה התכלית המקיף. ואינו אחד

מהשלשה, הנה הוא אם כן התכלית המקיף. ואולם איך יתבאר

שאינו אחד מהשלשה? אמנם שאיננו הצורה וההיולי הוא מבואר,

לפי שהם מעצמות הדבר, ואינם נבדלים ממנו, ולא תתאמת בהם

ההקדמה השנית. ואם הנחנו שהצורה הוא תכלית, הוא תכלית

10 המוקף לא תכלית המקיף. והאמת שאינו תכלית, ולא יאמר בו

תכלית אלא למה שהוא תכלית להיולי ותגבילהו.

הנה נשאר שנבאר שאיננו החללות. וארסטו יאמר בזה, שהמאמר

בשיש הנה רחקים עומדים בעצמם, יתחייבו ממנו שני שקרים.

הראשון, שיהיה לדבר האחד בעצמו מקומות רבים יחד בלתי בעלי

15 תכלית. והשני, שיהיו המקומות מתנועעים ושיהיה המקום במקום.

והנה איך יחוייב זה? כפי מה שאומר. וזה שאם היה הרוחק אשׁר בין

תכליות הגשם הוא המקום, חוייב שיהיו חלקי הגשם במקום בעצם,

וזה כי כמו שהגשם בכללו הוא במקום, להיותו ברוחק שוה לו, הנה

כל אחד מחלקיו במקום, להיותו ברוחק שוה לו. וכאשר הנחנו כלי

3 תכליות] תכלית צזרד – (והוא) ר.   4 חללות] גבניות א – (לא) פ – (אחד) פ.   5 ואיננו ב.

6 הנה הוא אם כן הוא ו הנה אם כן הוא זר ק ב נ.   8 תתאמת] יחובר א.   9 הצורה (הוא) פ

שהצורה (היא תכלית) א.   10 לא] ולא לר.   12 שהמאמר] המאמר פ שהמאמר [מבואר,

בנליון: שנבאר, בזה] בשיש פ.   13 ממנו] מזה צ מלזור ק ב א נ בזה ד.   14 שיהיה] שהיה י – אחד

זור.   17 תכליות] תכלית צזדרק ב נ.   18 להיות זר ב נ.

enable us to understand the essence of place. He has furthermore framed a hypothetical disjunctive syllogism which runs as follows:[77] Place must inevitably be thought of as one of four things: form, matter,[78] the surrounding limit, or the interval between the limits of that which surrounds,[79] i. e., that which is known as the vacuum.[80] If it cannot be any of the three, namely, form, matter and the vacuum, it necessarily follows that it is the surrounding limit. But it is none of those three. Consequently it is the surrounding limit. But how can it be shown that it is none of these three? That place cannot be identified with either form or matter is evident, for both of these belong to that which is essential to a thing and are inseparable therefrom,[81] and thus they cannot satisfy the conditions laid down in the second proposition. If we have assumed that form is a limit,[82] it is a limit only of the thing surrounded but not of the thing surrounding.[83] The truth of the matter is, form is not a limit. It is said to be a limit only in the sense that it is the final cause of matter and the limit which defines it.[84]

It therefore remains for us to prove that place is not identical with the vacuum. With regard to this Aristotle says[85] that the assertion that there are dimensions existing by themselves [without a body] would give rise to two untenable conclusions. First, that one and the same thing would have an infinite number of places at the same time. Second, that the places would be movable and that one place would exist in another place.[86] How such conclusions would ensue, will become clear from what I am to say. If the interval between the boundary lines of a body be its place, the parts of that body would have to be essentially each in its own place, for just as the body as a whole is said to be in place because of its occupancy of an interval equal to itself, so also every one of its parts would have to be assumed as existing each in its own place, since each of them occupies an interval of its own size. Supposing now that a vessel full of water is moved from

מלא מים יתנועע ממקום אל מקום, הנה כמו שהמים יעתק בכלי
עם הרוחק השוה לו, אשר יטרידהו, ויהיה ברחק אחר, כאשר המיר
הכלי בכללו מקומו, כן יעשו חלקי המים, רצוני, שהם יעתקו עם
הרחקים המיוחדים להם אל רחקים אחרים, אשר הם מקומות להם.

5   וכאשר חלקנו החלקים אל חלקים אחרים תמיד, הנה יתחייבו השני
שקרים, אם שיהיו להם מקומות בלתי בעלי תכלית, ואם שיהיו
המקומות מתנועעים ושיהיה המקום במקום.

יתחייב אם כן היות המקום השטח המקיף השוה הנבדל. וכאשר
התבאר זה, התאמת בלא ספק שהגשם המקומם בעל תכלית. וזה
10   אשר כוון במין הזה מן המופתים.

המין השלישי

בבאור המנעות מציאות מתנועע בלתי בעל תכלית תנועה ישרה
או סבובית.

אמנם המנעות תנועה ישרה במתנועע בלתי בעל תכלית, סדר
15   בזה ש ל ש ה מופתים.

הר א ש ו ן, הציע בו שתי הקדמות ידועות בעצמן. האחת, שכל
גשם מוחש יש לו אנה תיחדהו ומקום מתנועע אליו וינוח בו. השנית,
שמקום החלק והכל אחד, כאלו תאמר שמקום גוש אחד מן הארץ
הוא מקום הארץ בכללה. ואחר שהתייישבו אלו השתי הקדמות,
20   סדר המופת כן.

1 (מלא] ם ז – [אם] יתנועע ם.   3 רצוני] לנו לומר ם.   4 הרחקים ם.   5 שני ם.   8 כל שוה ב ג.

9 בלא] בלי ב ג – המקומם] המקומיי ל – מהו ד.   10 זה ם – [ההוא] הזה ל.   16 בשתי ו.

17 (בה) ם.   18 שהמקום ם ק – גוש] איש ל ר ד ג.   19 בכלל ם – שיתיישבו] שהתיישבו] לך

ק – שתי ו א.

one place to another, it would follow that just as the entire volume of water, when the vessel as a whole changes its place, is
translated by that vessel, together with its own equal interval
which it occupies,[87] and is placed in another interval, so also the
parts of the water would be affected in the same way, that is to
say, they, too, would all individually be translated together with
their particular intervals to other intervals, the latter intervals
thus becoming the places of the parts of the water as well as of
their former intervals.[88] By infinitely continuing to divide the
parts of the water, we would thus finally arrive at the two aforementioned untenable conclusions: first, that they [i. e., the parts]
would have an infinite number of places, and second, that places
would be movable and that one place would exist in another place.

Consequently, place must be the surrounding, equal and separate surface.[89] This having been demonstrated, it is now
established beyond any doubt that any space-filling body must
be finite. This is what he intended to show by this class of
arguments.

THE THIRD CLASS OF ARGUMENTS

Proof for the impossibility of an infinite object having either
rectilinear or circular motion.[90]

With respect to the impossibility of *rectilinear* motion in an
infinite movable body, he has framed *three* arguments.

The *first*[91] of these arguments is introduced by him by two
self-evident propositions. First, every sensible body has a whereness which properly belongs to it[92] and a place toward which it
moves and wherein it abides. Second, the [proper] place of the
part and the whole [of a homoeomerous body[93]] is one [in kind],[94]
as, e. g., the [proper] place of a clod of earth is the same as that
of the whole earth. Having laid down these two propositions,
he proceeds with his argument as follows:

אם היה הגשם בלתי בעל תכלית, לא ימנע משיהיה מתדמה
החלקים או בלתי מתדמה החלקים. ואם היה מתדמה החלקים,
הנה לפי שמקום הכל והחלק אחד, כמו שהתבאר בהקדמה השנית,
לא יתנועע כלל, למה שמקומו צריך שישוה לו, ואם כן כשחלק הגשם
5 הוא בחלק מקום הכל, הנה הוא אם כן במקומו, והדבר לא יתנועע
כשהוא במקומו. ואם לא היה מתדמה החלקים, הנה החלקים אם
שיהיו בעלי תכלית במספר ואם שיהיו בלתי בעלי תכלית, ואם היו
בעלי תכלית במספר, חויב שיהיה אחד מהם בלתי בעל תכלית
בגודל, וחויב שלא יתנועע תנועה ישרה כמו שקדם. ואם היו בלתי
10 בעלי תכלית במספר, חוייב שיהיו מיני האנה בלתי בעלי תכלית
במספר, כמו שהתבאר בהקדמה הראשונה. והנה מיני האנה
מוגבלים, וזה שהאנה הטבעית הוא לקוח אם מהתנועה הישרה אם
מהסבובית, והתנועה הישרה היא מן האמצע או אל האמצע,
והסבובית היא סביב האמצע, ואם היה בכאן גודל בלתי בעל
15 תכלית בין חלקי הגשם לא יהיה בכאן אמצע.

ואין לאומר שיאמר שמקום כל אחד זה למעלה מזה, וזה אל לא
תכלית; שאם היה הדבר כן, לא יהיה בכאן מעלה ומטה במוחלט.
ולפי שאנחנו נראה היסודות הארבעה מתנועעים, מהם אל המעלה
במוחלט, ומהם אל המטה במוחלט, ומהם אל המעלה ואל המטה
20 בצרוף, ואנחנו נראה שהמטה במוחלט מוגבל, הנה הפכו, שהוא
המעלה במוחלט, מוגבל, אחר שההפכים הם בתכלית המרחקן.

התבאר אם כן, איך שיהיה, שבמציאות גשם בלתי בעל תכלית

If an infinite body existed, it would inevitably have to be either of similar[95] or of dissimilar parts. [In the first case], if it were of similar parts, it could not have [rectilinear] motion; for according to the second self-evident proposition, the place of the part and the whole is [generically] one, and furthermore the proper place must be equal to its occupant; consequently in whatever part of the [infinite] place of the whole any part of the body finds itself, it will always be in its proper place, and no object can have [rectilinear] motion while in its proper place.[96] [In the second case], if it were of dissimilar parts, those parts would have to be either finite or infinite in number.[97] If they were finite in number, one of them would have to be infinite in magnitude, and, as in the preceding case, would be incapable of motion.[98] If they were infinite in number, the kinds of places would have to be infinite in number,[99] in accordance with the first self-evident proposition. But [100] the kinds of places must be limited, for the existence of natural places is derived from the existence of rectilinear and circular motion, and rectilinear motion is from or toward the centre and circular motion is around the centre[101]; but there would be no centre if the sum of the parts of the body formed an infinite magnitude.[102]

It cannot be said that the places of the elements are one above the other and so on to infinity; for if that were the case, there would be no absolute up and down.[103] [But[104] we observe that the four elements are moved, one absolutely upward, another absolutely downward, and of the remaining two, one relatively upward and the other relatively downward. We also observe that absolute lowness is limited; consequently its contrary, absolute height, must likewise be limited, inasmuch as contraries are those things which are most distant from each other.[105]]

Thus it has been shown that in either case the existence of an infinite body would exclude the possibility of rectilinear motion.

תסתלק התנועה הישרה. אבל התנועה הישרה נראית בחוש; גשם
בלתי בעל תכלית אם כן בלתי נמצא.

המופת השני סדורו כן. אם היה גשם בלתי בעל תכלית נמצא,
הנה ימצא בהכרח כובד בלתי בעל תכלית או קלות בלתי בעל
5 תכלית, אבל כובד בלתי בעל תכלית וקלות בלתי בעל תכלית
נמנע. אם כן גשם בלתי בעל תכלית נמנע. והנה התדבקות הנמשך
בקודם בהקש הזה, יתבאר על הדרך הזה. (לפי שאנחנו נראה
היסודות הארבעה מתנועעים, מהם אל המעלה במוחלט, ומהם אל
המטה במוחלט, ומהם אל המעלה ואל המטה בצרוף, ואנחנו נראה
10 שהמטה במוחלט מוגבל, הנה הפכו, שהוא המעלה במוחלט, מוגבל,
אחר שההפכים הם בתכלית המרחק). ונאמר שהוא מחוייב, אם
היה גשם בלתי בעל תכלית נמצא, שיהיה כובד בלתי בעל תכלית
נמצא. שאם לא ימצא לו כובד בלתי בעל תכלית, יהיה אם כן בעל
תכלית, ונניחהו עוד נבדל ממנו גשם בעל תכלית, והוא מבואר
15 שיהיה כובדו קטן מכובד הבלתי בעל תכלית. עוד נכפול זה הגשם
עד שיהיה כובדו גדול ככובד הבלתי בעל תכלית, אחר שכובדו
בעל תכלית. והוא מבואר שההכפל בגשם הבעל תכלית הוא
אפשר עד שיהיה יותר גדול מכובד בעל תכלית הראשון שהיה
כובד לגשם הבלתי בעל תכלית. וכל זה בתכלית הבטול, שיהיה
20 כובד חלק מהגשם, והוא בעל תכלית, גדול ככובד כל הגשם הבלתי
בעל תכלית, ויותר גדול ממנו. התבאר אם כן התדבקות הנמשך
בקודם בהקש הזה, שאם היה גשם בלתי בעל תכלית נמצא, כובד
בלתי בעל תכלית בהכרח נמצא.

ואמנם סותר הנמשך, והוא שאי אפשר שימצא כובד בלתי בעל
25 תכלית או קלות בלתי בעל תכלית, זה יתבאר אחר שנניח שלש
הקדמות. האחת, שהמתנועע שיש לו כובד יותר גדול, יתנועע תנועתו

2-1 וא"כ נשם בב"ת ז.   6 (אם כן ... נמנע) פר – ההתדבקות לד.   7 (בקודם) לרד –
התבאר ז – דרך פ – ולפי א.   9 (ואל) המטה] והמטה ל.   10 המטה זסלזור קבאג – מעלה
זסלזורדבאג – מוחלט בג.   13 ימצא] יהיה זר.   15 כובדו [של נשם ב"ת] קטן פ הכובד של
הנשם הב"ת והקטן קטן ז.   17 (הוא) פ.   19–18 שהיה כובד] שיהיה לכובד פ.   21 (התבאר)
פ.   22 מקודם פ.   23–22 כובד ... נמצא] הנה ימצא בהכרח כובד בב"ת וקלות בב"ת
פור.   23 (בהכרח) באג – נמצא [וקלות] נמצא בג ימצא וקלות פ.   26 ההקדמות פ – שהמתנועע
[שלו] פ – יש י.

But rectilinear motion is a matter of sense perception. Hence an infinite body does not exist.

The *second* argument runs as follows:[106] If an infinite body existed, infinite weight or lightness would likewise exist. But infinite weight and infinite lightness are impossible. Hence an infinite body does not exist. The connection of the consequent with the antecedent in this syllogism may be made clear as follows: (For[107] we observe that the four elements are moved, one absolutely upward, another absolutely downward, and of the remaining two, one relatively upward and the other relatively downward. We also observe that absolute lowness is limited, consequently its contrary, absolute height, must likewise be limited, inasmuch as contraries are those things which are most distant from each other.[108]) We say it must follow that if an infinite body existed, infinite weight would also exist, for if the infinite body could not have infinite weight, then its weight would have to be finite. Let us then assume a finite part taken from that infinite body.[109] The weight of this finite part would of course be less than that of the infinite. Let us then increase the magnitude of the finite part until its weight equals that of the infinite, since the weight of that infinite is now assumed to be finite. It is also evident that the finite part could be continually increased until its weight became even greater than the first finite weight of the infinite body. But all this is absolutely impossible, namely, that the weight of only a finite part of the body should be as great as that of the infinite whole of the same body, nay, even greater than it. Hence the connection of the consequent with the antecedent in this syllogism, namely, that if an infinite body existed, infinite weight and lightness would likewise have to exist.

As for the proposition which denies the consequent, namely, that infinite weight or infinite lightness cannot exist, it will become evident after we have laid down three propositions. First, an object of greater weight, in the course of its natural motion,

הטבעית, מרחק אחד, בזמן יותר מועט ממה שיתנועע המתנועע,
שיש לו כובד יותר קטן, המרחק ההוא בעינו. השנית, שיחס השני
זמנים יחס הכובד אל הכובד. והשלישית, שכל תנועה בזמן. וכאשר
נתישבו אלו ההקדמות, נניח שכובד בלתי בעל תכלית וכובד בעל
5 תכלית יתנועעו מרחק אחד בעינו. יחוייב שיהיה יחס הזמן אל הזמן
יחס הכובד אל הכובד. ולפי שאין יחס בין הבלתי בעל תכלית
והבעל תכלית אלא כנקודה אל הקו וכעתה אל הזמן, יתחייב
שיתנועע בעתה, והוא בלתי אפשר. ויתחייב עוד שיחתוך מרחק גדול
וקטן בשוה, והוא בעתה אחד. ואם הנחנו זמן מה מועט לבלתי בעל
10 תכלית, היה אפשר שימצא כובד אחד יחסו אל הכובד הקטן יחס
הזמן אל הזמן, ויהיה זה הכובד הבעל תכלית יתנועע בזמן שוה
לכובד הבלתי בעל תכלית. וכשנכפול אותו יתנועע הכובד הבעל
תכלית בזמן יותר מועט מהכובד הבלתי בעל תכלית. וכל זה
בתכלית הבטול. והבטולים נתחייבו מהנחתנו כובד בלתי בעל
15 תכלית נמצא. וכאשר התבאר המנעות מציאות כובד בלתי בעל
תכלית, התבאר אם כן המנעות מציאות גשם בלתי בעל תכלית
בגשמים הפשוטים.

ואולם במורכבים, המנעות מציאות גשם בלתי בעל תכלית
מבואר מצד החלוקה, והוא שלא ימנע אם שיהיה מדברים בלתי
20 בעלי תכלית בשיעור, או במספר, או בצורה. ואי אפשר בשעור,
שכבר התבאר המנעות שעור הגשמים הפשוטים בלתי בעל תכלית.

1 ועל] מרחק לו"ר.   4 התיישבו לד.   5 יחייב יר.   6—7 בין הבב"ת והב"ת יחס פצפא.
10 יחוסו ס.   11 כובד פ.   12 יתנועע [בזמן שוה לכובד הבב"ת] פ.   13—12 הבעל תכלית
ב"ת י.   14 יתחייבו י.   15—14 הבב"ת י.   16 אם כן נ"כ י.   18 המורכבים יר.
19 שהיה י.

will traverse a given distance in shorter time than would be required by an object of lesser weight moving over the very same distance. Second, that the ratio between the [shorter] time and the [longer] time is equal to the ratio between the [smaller] weight and the [greater] weight. Third, every motion is in time.[110] Having laid down the propositions, let us now suppose two weights, one infinite and the other finite, to be moving over the same given distance. It would follow that the ratio of the time required by the infinite to that required by the finite would be equal to the ratio of the weight of the finite to that of the infinite. But infinity has no ratio to finitude except as a point to a line and as an instant to time. It would consequently follow that the infinite weight would traverse a long and a short distance without any difference in time, that is to say, in an instant.[111] Even if we were to allow in the case of the infinite weight a certain fraction of time, some finite weight might still be assumed whose ratio to the former finite weight would be equal to the ratio between the time of the infinite weight and that of the former finite weight. The time of this new finite weight would then be equal to the time of the infinite weight. Furthermore, by increasing the new finite weight it would follow that that finite weight would perform its motion in shorter time than the infinite weight. But all this is most absurd. And these absurdities have arisen from our assumption that an infinite weight existed. Having thus shown the impossibility of an infinite weight, we have thereby also shown that there can be no infinite body among the simple bodies.

In the case of composite bodies,[112] however, the impossibility of an infinite body can be demonstrated by a disjunctive syllogism. An infinite compound body would inevitably have to be composed of elements which were infinite in one of these three respects: magnitude, number, or form. They could not be infinite in magnitude, for it has already been shown that the magnitude of simple bodies cannot be infinite. Nor could they be infinite in

וכן אי אפשר להיותם בלתי בעלי תכלית במספר, כי מצד שיתמששו
יהיו כולם שיעור בלתי בעל תכלית, שהתבאר המנעו, אחרי שהם
אחדים בצורה. וכן אי אפשר שיהיו בלתי בעלי תכלית בצורה,
שיתחייב שיהיו המקומות בלתי בעלי תכלית. ועוד שאנחנו נראה
5 התנועות בעלי תכלית. ולזה הוא מבואר שלא ימצא גשם בלתי בעל
תכלית פשוט ולא מורכב. וזה אמנם מצד התנועה.

המופת השלישי סדורו כן. אם היה גשם בלתי בעל תכלית
נמצא, הנה אי אפשר לו שיפעל ושיתפעל. אבל כל גשם מוחש אם
פועל ואם מתפעל. יוליד סותר הקודם, והוא שגשם בלתי בעל
10 תכלית בלתי נמצא. ואמנם נרצה בהפעלות ההפעלות אשר בזמן.
והנה שכל גשם מוחש פועל או מתפעל, זה מבואר בחפוש, למה
שכל גשם מוחש אם פועל לבד, כמו הגרמים השמימים, אם פועל
ומתפעל, כמו היסודות והגשמים המורכבים. ואולם שהגשם הבלתי
בעל תכלית אי אפשר לו שיפעל ושיתפעל, יתבאר בשנניח שלש
15 הקדמות מבוארות בעצמן. האחת, ששני מתפעלים שוים יתפעלו
מפועל אחד בזמן שוה, ושהמתפעל הקטן יתפעל ממנו בזמן יותר
קטן. והשנית, שכשיפעלו פועלים מתחלפים בשני מתפעלים, יחס
המתפעל אל המתפעל כיחס הפועל אל הפועל. והשלישית,
שהפועל יפעל בזמן בעל תכלית. ואחר שנתיישבו אלו ההקדמות,
20 הוא מבואר שהבלתי בעל תכלית אי אפשר לו שיפעל ושיתפעל.
וזה שהבעל תכלית אי אפשר לו שיפעל בבלתי בעל תכלית, ולא
הבלתי בעל תכלית בבעל תכלית, ולא הבלתי בעל תכלית בבלתי
בעל תכלית.

אמנם שהבעל תכלית לא יפעל בבלתי בעל תכלית הוא מבואר,
25 שאם היה פועל בו, נניח שיהיה פועל בו בזמן מה מונח, ונניח בעל

1 שישתמששו מ ז שישתמששו ק.   2 [שכבר] התבאר לזר דקבאג.   5 שלא ימצא] שאי"א להמצא
זור ק בג שאי"א למצא ד שאי"א לנמצא ל.   6 זה ס.   7 כך ד.   8 ושיתפעל] ושיפעל א —
אם] אור.   9 ואם] או מ לזר דרא — (ואם) ומתפעל בג.   11 והנה] והוא ל.   12 ואם] או לא —בלבד
צלדא – (כמו כגרמים זור ק בג – אם] או ר [און אם ק בג.   13 ומתפעל] או מתפעל צור ק באג.
15 (ששני] שמתפעלים בג.   17 השנית] הב' בג.   18 השלישית] הג' בג.   19 שהפועל [יתברך]
ס.   21 שיפעל [ושיתפעל] ל ר.   25 בו] כן ל – בו] כן ל.

number, for being contiguous[113] to each other and one in form, their aggregate would make [a continuous, simple], infinite magnitude, which has been shown to be impossible. Finally, they could not be infinite in form, for were they to be so, they would require an infinite number of places. Moreover, we observe that the motions are finite.[114]

It is thus clear that an infinite body, whether simple or compound, has no existence, and all these are indeed arguments from motion [proper].[115]

The *third* argument runs as follows:[116] If an infinite body existed, it could neither act nor suffer action. But every sensible body must either act or suffer action. Hence a conclusion which denies the antecedent, that is to say, an infinite body does not exist. By acting and suffering action we mean here an action or passion that is [completely realized] in time.[117] That every sensible body must either act or suffer action may be made clear by induction. Every sensible body either only acts, as, e. g., the celestial bodies, or both acts and suffers action, as, e. g., the elements and the composite bodies. That unlike these, an infinite body could neither act nor suffer action will be shown after we have laid down three self-evident propositions. First, two equal objects are affected by the action of one and the same agent in equal time, and a smaller object will be affected by the same agent in shorter time. Second, when two unequal agents affect two objects [in equal time], the ratio between the two objects is equal to the ratio between their respective agents.[118] Third, every agent must complete its action in finite time.[119] These propositions having been laid down, it becomes clear that an infinite could neither act nor suffer action, for it can be shown that a finite could not impart action to an infinite, nor an infinite to a finite, nor, finally, one infinite to another.

That no finite could impart action to an infinite is evident, for were that possible, let a finite act upon the infinite in some given

תכלית פועל בבעל תכלית בזמן אחר, ויהיה קטן מהראשון בהכרח.
ונכפול הבעל תכלית המתפעל עד שיפעל בזמן שוה לזמן הראשון
המונח, שהוא אפשר זה, כמו שהתבאר בהקדמה השנית. ויתחייב
אם כן שיתפעל הבלתי בעל תכלית מהבעל תכלית בזמן שוה
5 להפעלות הבעל תכלית מהבעל תכלית. והוא שקר. ואם נכפול
יותר המתפעל, יתחייב שיתפעל הבלתי בעל תכלית מהבעל
תכלית בזמן יותר קטן מהפעלות הבעל תכלית מהבעל תכלית.
וזה מגונה מאד.

וכן יתחייב שלא יפעל הבלתי בעל תכלית בבעל תכלית, שאם
10 היה פועל בו, נניח בלתי בעל תכלית פועל בבעל תכלית בזמן מה
מונח, ונניח בעל תכלית פועל בבעל תכלית בזמן אחר גדול
מהראשון. ונכפול הבעל תכלית הפועל עד שיפעל בזמן שוה לאותו
זמן, שזה אפשר כמו שהתבאר בשנית. ויתחייב אם כן שיפעל הבעל
תכלית בבעל תכלית בזמן שוה למה שיפעל הבלתי בעל תכלית
15 בבעל תכלית, הפך מה שהונח. ואם נכפול עוד הבעל תכלית,
יתחייב שיפעל בזמן יותר מועט מהבלתי בעל תכלית. והוא מגונה
מאד.

וכן יתחייב שלא יפעל הבלתי בעל תכלית בבלתי בעל תכלית.
שאם היה פועל בו, נניח בלתי בעל תכלית פועל בבלתי בעל תכלית
20 בזמן מה מונח, ונניח חלק מהמתפעל מתפעל מהבלתי בעל תכלית
הפועל בזמן, ויהיה בהכרח יותר קטן. ונכפול המתפעל עד שיהיה
בזמן שוה לזמן המונח, וזה איפשר מכח ההקדמה השנית. ויתחייב

time, and let again another finite act upon a finite object in some other given time. The time in the latter case would, of course, be shorter than that in the former. Let us now increase the finite object so that its time would be equal to the given time of the infinite object. This, according to the second proposition, could be done. It will hence follow that an infinite body would be affected by a finite agent in the same time as would be required by a finite body to be affected by a finite agent. This is contrary to truth. Furthermore,[120] if the finite object were still further increased, the result would be that an infinite would be affected by a finite in less time than a finite by a finite. But this is very absurd.

It can likewise be proved that an infinite agent could not impart action to a finite object, for if it could, let the infinite act upon a finite in a certain given time and let again a finite act upon another finite in some greater time than the former. Let us now increase the finite agent so that it would complete its action in a time equal to that of the infinite agent. This, according to the second proposition, could be done. The result would be that a finite would impart action to another finite in the same time as would be required by an infinite acting upon a finite—contrary to what has been assumed. Furthermore,[121] if the finite [agent] were still further increased, the result would be that it would perform its action in less time than the infinite agent. This is very absurd.

Finally, it can similarly be proved that an infinite could not impart action to another infinite, for if it could, let an infinite act upon another infinite in some given time, and let again a finite part of the infinite object be acted upon by the infinite agent in some other given time. The second given time would, of course, be less than the former. Let us now increase the finite object until it would receive the action in the same time as the infinite object. This, on the strength of the second proposition, could be done. The result would be that an infinite and a finite would be

שיתפעל הבלתי בעל תכלית והבעל תכלית מפועל אחד בזמן
אחד, והוא הפך מה שהונח. ואם נכפול עוד המתפעל, יתחייב
שיתפעל הבלתי בעל תכלית מהבלתי בעל תכלית בזמן מועט
מהפעליותו מהבלתי בעל תכלית. והוא מגונה מאד.

5 ואחר שהתבאר שאי אפשר לבלתי בעל תכלית שיפעל ולא
שיתפעל, הנה חוייב שאין בלתי בעל תכלית נמצא. וזה אמנם
התבאר מפאת המנעות התנועה, וזה שהשינוי הוא מין מן התנועה,
וכבר השתתף לתנועה הישרה למה ששניהם מהפך אל הפך, ולזה
סדרנו המופת הזה במופתים שהונחו מצד המנעות התנועה הישרה.

10 ואולם מפאת התנועה הסבובית, הנה הוא סדר ששה מופתים
לבאר שהיא נמנעת בגשם בלתי בעל תכלית.

המופת ה ר א ש ו ן סדורו כן. אם הגשם הבלתי בעל תכלית
הסבובי, המתנועע בסבוב, נמצא, יתחייב, שבהיות חצי קטרו
מתנועע בסבוב, שידבק על חצי קטרו הנח, כשיגיע אליו. והנה זה
15 נמנע; יוליד שהגשם הבלתי בעל תכלית הסבובי בלתי מתנועע
בסבוב. והנה התדבקות הנמשך בקודם מבואר בעצמו, להיות
הקוים היוצאים מן המרכז אל המקיף בכל הכדור שוים. ואמנם
סותר הנמשך מחוייב, למה שהוא גלוי שהמרחק שבין כל שני קוים
היוצאים מן המרכז אל המקיף מתוסף בתוספת הקוים. ולפי שהיו
20 הקוים בלתי בעלי תכלית, היה המרחק אשר ביניהם בלתי בעל
תכלית. ולפי שהוא מבואר שאי אפשר למתנועע שיחתוך מרחק
בלתי בעל תכלית, הוא מבואר שאי אפשר לו להדבק בחצי הקטר

---

affected by the same agent in equal time. This is contrary to what has been assumed. Furthermore, if the [finite] object were still further increased, the result would be that an infinite object would be affected by an infinite agent in less time than a finite object by the same infinite agent.[122] This is very absurd.

Having thus demonstrated that an infinite could neither act nor suffer action, we must consequently conclude that an infinite has no existence, and this indeed has been proved from the impossibility of [rectilinear] motion [in an infinite], for change is a species of motion, and, furthermore, it is analogous to rectilinear motion, inasmuch as they both take place between opposites.[123] It is in view of this consideration that we have included this argument among those derived from the incompatibility of rectilinear motion with the existence of an infinite.[124]

As to *circular* motion, he has framed *six* arguments to show that it would be impossible in an infinite body.[125]

The *first* argument runs as follows:[126] If an infinite, spherical body moving in a circle existed, it would follow that one of its radii[127], assumed to revolve on the centre, on reaching the position of another radius, assumed to be at rest, would have to coincide with the latter.[128] But this is impossible. Hence an infinite spherical body could not have circular motion. The connection of the consequent with the antecedent is self-evident, for the lines extending from the centre of a sphere to its circumference are all equal. As for the proposition which denies the consequent, its validity can be demonstrated as follows: It is well-known that the distance between any two lines emerging from the centre to the circumference increases in proportion to the elongation of those lines.[129] Since in the case under consideration the lines would be infinite,[130] the distance between them would likewise have to be infinite. As it is obvious, however, that no moving object can traverse an infinite distance[131], it must follow that the revolving radius could never coincide with the fixed radius. But we have

הנח. וכבר הנחנוהו דבק בו. והוא מבואר שהשקר הזה יצא מהניחנו
אותו מתנועע.

ומהאחרונים מי שחזק המופת הזה, בשאמר: ואיך ידבק בחצי
הקטר? והנה כאשר דמינו שני קוים יוצאים מהמרכז, ויחדשו זוית
בענין שמיתרו יחדש משולש שוה הצלעות, הנה אם הקוים בלתי
בעלי תכלית, המרחק אשר ביניהם בלתי בעל תכלית. אם כן הקו
האחד המתנועע אי אפשר לו להדבק עם הקו האחר, למה שיצטרך
לחתוך מרחק בלתי בעל תכלית, עם שהוא נמנע בעצמו היות בלתי
בעל תכלית מוקף משני קוים משתי קצותיו, כי המאמר בהיותו מוקף
ובלתי בעל תכלית מאמר סותר נפשו. והנה יתחייב זה בכל שני קוים
היוצאים מהמרכז אם היו בלתי בעלי תכלית, שאין ספק שכל מה
שיתוספו הקוים נוסף המרחק, אשר הוא להם מקום מיתר, ולהיות
הקוים בלתי בעלי תכלית היה המרחק אשר ביניהם בלתי בעל
תכלית בהכרח, והוא מבואר הבטול.

המופת השני סדורו כן. אם הגשם הסבובי המתנועע בסבוב
בלתי בעל תכלית נמצא, יתחייב שיתנועע בזמן בעל תכלית מרחק
בלתי בעל תכלית, והוא נמנע, יתחייב שלא ימצא מתנועע בסבוב
בלתי בעל תכלית. והנה סותר הנמשך מבואר בעצמו. והתדבקותו
לקודם יתבאר כשנציע קו בלתי בעל תכלית יוצא ממרכז. ונציע
גם כן מיתר בו. והוא מבואר שיהיה בלתי בעל תכלית, אחר שהגשם
בלתי בעל תכלית. ונציעהו נח. הנה כשיתנועע הקו היוצא מן

shown that they would coincide. It is thus clear that if we assume the infinite to have circular motion, this false conclusion
would have to follow.[132]

One of the later thinkers[133] has clinched this argument by asking: How could the two radii coincide? Let us suppose, he argues,
two lines emerging from the centre at such an angle that its
opposite chord would complete an equilateral triangle. Since the
lines are infinite, the distance between them [i. e., their intersecting chord] must be infinite. Consequently, the revolving radius
could never coincide with the other [i. e., the fixed radius], as it
would have to traverse an infinite distance, quite apart from the
consideration that it is impossible to conceive of an infinite as
bounded by two lines on its two ends, for to say that something
is both bounded and infinite is a self-contradictory proposition.[134]
The same difficulty, [according to this version of the argument],
would arise in the case of any two lines emerging from a common
point,[135] if they were conceived to be infinite. The distance between any two such lines at the point where they are intersected
by a common chord would undoubtedly increase in proportion to
the extension of the lines, and as the lines are assumed to be
infinite, the distance between them would likewise have to be
infinite. But this clearly is an impossibility.

The *second* argument runs as follows:[136] If an infinite, spherical
body moving in a circle existed, it would have to traverse an
infinite distance in finite time. But this is impossible. Hence
the existence of an infinite endowed with circular motion is impossible. The proposition which denies the consequent is self-
evident.[137] As for the connection of the consequent with the
antecedent, it may be made clear as follows: Let an infinite line
emerge from the centre; and let also a chord intersect the sphere.
Since the sphere is assumed to be infinite, it is clear that the chord
will have to be infinite.[138] Let that chord be at rest. Now,
if we suppose the radius to revolve on its centre, it will at some

המרכז בסבוב, יהיה בו זמן יפגוש המיתר ויחתכהו, וזמן לא יפגשהו.
ולהיות הגשם הסבובי המתנועע בסבוב יתנועע בזמן בעל תכלית,
יתחייב שיחתוך הקו היוצא מן המרכז מרחק בלתי בעל תכלית,
והוא המיתר המונח, בזמן בעל תכלית. והוא שקר מבואר, להיות
5 התנועה אשר בזמן בעל תכלית מחוייב שיהיה במרחק בעל תכלית.

המופת ה ש ל י ש י סדורו כן. אם הגשם המתנועע בסבוב בלתי
בעל תכלית נמצא, יתחייב אפשרות הנחת שני קוים נכחיים האחד
מתנועע נכח חברו בסבוב, והאחר [נח], שיחתכהו ויפגשהו קודם
פגישתו קצה הקו. וזה נמנע. יחוייב אם כן המנעות הקודם. והנה
10 המנעות הנמשך מחוייב, למה שהוא מבואר בעצמו, שכשהונחו שני
קוים על זה התאר, יתחייב שיפגוש הנקודה הראשונה אשר בקצה
הקו קודם שיפגוש אמצעיותו. ואמנם התדבקותו לקודם גם כן
מבואר, למה שהקו הבלתי בעל תכלית אין לו קצה והתחלה, ואין
בו נקודה שלא יהיה לפניה נקודה.

15 המופת ה ר ב י ע י סדורו כן. אם הגשם הבלתי בעל תכלית
מתנועע תנועה סבובית, הנה יש לו תמונה סבובית בלתי בעל
תכלית. והוא נמנע. יוליד שאין הגשם הבלתי בעל תכלית מתנועע
בסבוב. אמנם התדבקות הנמשך לקודם מבואר בעצמו. ואמנם
המנעות תמונה סבובית בלתי בעל תכלית, זה יראה מרושם התמונה,
20 אשר יאמר בה המהנדס ברשמה, שהיא אשר יקיף בה גבול או
גבולים. והוא מבואר, שאשר יקיף בו הגבול הוא בעל תכלית.
ובכלל התכלית הוא מצד הצורה בכל הדברים, והעדר התכלית
מצד החמר, ואחר שהיתה התמונה היא הצורה, אי אפשר שהוא
בלתי בעל תכלית.

4 בזמן (בב׳ת) ג.   5 במרחק הב׳ת ר.   7 אפשרות] באפשרות לרדקבא׳.   8 [נח] הוספתי על
פי השערה, עיין פירושי האנגלי. – שיחתכו ס.   9 והנה] ויהיה ס והוא א.   10 בעצמו] בנפשו
זורקבא׳.   12 אמצעיתו זרדקבא׳ אמצעותו ס׳.   13 בב׳ת בג – לו] בו ורא.   16 תמונה]
תנועה ף.   17 [והוא] ואם א – בב׳ת בג.   19 המנע] תנועה זרא (תמונה) בג –
(התמונה) ף תמונה י.   20 (המהנדס) ס – יקיפוף – בו] בה פזרקבא׳.

time meet the chord and intersect it while at another time it will not meet it. As a spherical body rotating upon itself must complete its rotation in finite time,[139] it follows that the radius would traverse an infinite distance, namely, the given chord, in finite time. But this is a flagrant absurdity, inasmuch as motion completed in finite time must take place over a finite distance.[140]

The *third* argument runs as follows:[141] If an infinite body moving in a circle existed, it would be possible by assuming two [infinite] parallel lines,[142] of which one turns on a pivot towards the other and the other [is at rest],[143] that the former should intersect the latter and meet it first at some point [in the middle] without having met it before at its extremity. But this is impossible. Hence the impossibility of the antecedent. The impossibility of the consequent can be established as follows: It is self-evident that when two lines are assumed to act in the manner described, the moving line must first meet the [permanent] line at its extreme point before meeting it in the middle. The connection of the consequent with the antecedent is likewise clear, for an infinite line has neither end nor beginning and there is not a point in it which has not another point before it.

The *fourth* arguments runs as follows:[144] If an infinite body could have circular motion, it would have an infinite spherical figure. But that is impossible. Hence an infinite body could not have circular motion. The connection of the consequent with the antecedent is self-evident.[145] As for the impossibility of an infinite spherical figure, it is clearly evident from the meaning[146] of the term figure, which is defined by the geometrician[147] as that which is contained by any boundary or boundaries.[148] But that which is contained by a boundary is certainly finite. Besides, it is a general truism that all finitude in things is due to form and all lack of finitude is due to matter.[149] As the mathematical figure of a thing is the form of the thing, it cannot be infinite.

המופת ה ח מ י ש י סדורו כן. אם היה הגשם הבלתי בעל תכלית
מתנועע בסבוב, היה אפשר בו, כשנוציא קו מהמרכז יתנועע בסבוב,
שיחתוך קו בלתי בעל תכלית משתי קצותיו, אם הונח עמוד על
הקטר בלתי בעל תכלית. והוא נמנע, למה שהעמוד בלתי בעל
5 תכלית, ואי אפשר שיחתוך קו בלתי בעל תכלית בזמן בעל תכלית.
יוליד שאי אפשר לגשם הבלתי בעל תכלית שיתנועע בסבוב.

המופת ה ש ש י סדורו כן. אם נניח הגשם המתנועע בסבוב בלתי
בעל תכלית, כאילו תאמר הגשם הרקיעי, יתחייב שיחתוך מרחק
בלתי בעל תכלית בזמן בעל תכלית. והוא שקר. יוליד שאין גשם
10 מתנועע בסבוב בלתי בעל תכלית. והנה סותר הנמשך, מבואר
בעצמו. וחיובו לקודם מבואר מן החוש, שאנחנו נראה באיזו נקודה
שנרשום בו שתשוב למקומה בזמן בעל תכלית.

התבאר מכל אלו המופתים שההתנועה הסבובית נמנעת בגשם
הבלתי בעל תכלית. וכבר התבאר במה שקדם שהתנועה הישרה
15 נמצעת גם כן בו. אבל התנועה הישרה והסבובית נראית בחוש.
הנה אם כן הגשם הבלתי בעל תכלית בלתי נמצא. וזהו אשר כוון
במין הזה השלישי.

המין הרביעי

בבאור כולל, בהמנע מציאות גשם בלתי בעל תכלית בפעל,
20 והוא בכח המופתים הקודמים. וסדר בזה שני מופתים.
ה ר א ש ו ן סדורו כן. אם היה הגשם בלתי בעל תכלית נמצא,
הנה אם שיתנועע תנועה סבובית או ישרה. ואם סבובית, הנה
בהכרח יש לו אמצע, כי הסבובי הוא אשר יסוב סביב האמצע.

The *fifth* argument runs as follows:[150] If an infinite body could have circular motion, it would be possible that any radius moving in a circle would traverse an infinite line from one end to the other, if, e. g., a line drawn perpendicular to the diameter were assumed to be infinite.[151] But that is impossible, for that perpendicular line is assumed to be infinite, and an infinite line cannot be traversed in finite time.[152] Hence an infinite body cannot have circular motion.[153]

The *sixth* argument runs as follows:[154] If any body endowed with circular motion, as, e. g., the celestial element, were assumed to be infinite, it would have to traverse an infinite distance in finite time. But this is impossible. Hence no substance endowed with circular motion can be infinite. The minor premise which denies the consequent is self-evident[155]. As for the connection of the consequent with the antecedent, it can be made clear from observation, for we observe that any point we may take in that sphere will reappear in the same position after the lapse of some finite time.

All these arguments have clearly shown that circular motion would be impossible in an infinite body. Nor, as has already been shown before, could it have rectilinear motion. But both rectilinear and circular motions are facts vouchsafed by sense perception. Hence an infinite body has no existence. This is what he intended to show by this third class of arguments.

THE FOURTH CLASS OF ARGUMENTS

A GENERAL proof[156] to show the impossibility of an actually infinite body, based upon the reasoning of the preceding arguments. Under this proof he has framed *two* arguments.[157]

The *first* runs as follows:[158] If an infinite body existed, it would have either circular or rectilinear motion.[159] If circular, it would necessarily have a centre, circular motion being the motion of a

ואם יש לו אמצע, יש לו קצוות, ולבלתי בעל תכלית אין לו קצוות.
הנה לא יתנועע אם כן תנועה סבובית. נשאר אם כן שיתנועע תנועה
ישרה. והנה יצטרך בהכרח שני מקומות, כל אחד מהם בלתי בעל
תכלית, האחד לתנועה הטבעית ומה שאליו, והשני להכרחית ומה
5 שממנו. ואם המקומות שנים, יהיו בעלי תכלית בהכרח, למה
שהבלתי בעלי תכלית אי אפשר שיהיו שנים במספר. וכבר הונחו
בלתי בעלי תכלית. לא יתנועע אם כן תנועה ישרה. ועוד שהמקום
אי אפשר לו שיהיה בלתי בעל תכלית, למה שהוא מוגבל, אחר
שהתבאר מעניינו שהוא התכלית המקיף.

10 ה ש נ י סדורו כן. אם ימצא גשם בלתי בעל תכלית, אם שיתנועע
מעצמו או מזולתו. ואם יתנועע מעצמו יהיה בעל חי מרגיש, וכל
מרגיש יש לו מוחשים מחוץ מקיפים בו, ואשר בזה התאר הוא בעל
תכלית. ואם יתנועע מזולתו מחוץ, יהיה בהכרח גשם בלתי בעל
תכלית, ויהיו שנים בלתי בעלי תכלית. וזה שקר, למה שיהיה מקובצם
15 יותר גדול מכל אחד מהם, ויהיה מה שאין תכלית לו גדול ממה שאין
תכלית לו, עם שיתחייב מזה מניעים ומתנועעים בלתי בעלי תכלית
במספר, כל אחד מהם בלתי בעל תכלית בגודל.

ועוד חזק זה בדברים הם בכח המופתים אשר קדם זכרם.

אלו הם המופתים שבאו בדרוש הזה בספרי ארסטו וזולתו

body around a centre, and if it had a centre it would also have extremities. But an infinite has no extremities. Hence it could not have circular motion. It must, therefore, have rectilinear motion. But if so, it would need two places, both of infinite magnitude, one to account for natural motion and to serve as a *terminus ad quem* and the other to account for violent motion and to serve as a *terminus a quo*. Now, since these places are to be two in number, they must be finite in size, for two infinites cannot exist together. But they were assumed to be infinite. Hence it must be concluded that an infinite body could not have rectilinear motion. Moreover, place cannot be infinite, since it must be bounded, for it has been shown concerning it that it is the surrounding limit.

The *second* argument is as follows:[160] If an infinite body existed, it would have either to move itself or to be moved by something not itself. If it were to move itself, it would then be an animate being endowed with sense perception. But a body endowed with sense perception must have perceptible objects outside itself to surround it,[161] and anything of such a description must be finite. If it is moved by something external to itself, the motive agent would likewise have to be an infinite body. Thus there would be two infinites. This is impossible, for since the sum of the two will be greater than either one of them, it would follow that one infinite would be greater than another. Besides, if the infinite were moved by something external to itself, there would also follow the possibility of an infinite number of movers and things moved each infinite in magnitude.[162]

He has further strengthened this class of arguments by the application of the reasoning contained in the arguments already mentioned.[163]

Such then are the arguments with regard to this problem which are to be found in the works of Aristotle and of other authors as well as in the works of Aristotle's commentators, but lacking in

מהמחברים ומפרשי ספריו, אלא שבאו מבולבלים להבהיל המעיין,
אשר הוא אחד מהמקומות המטעים, ולזה סדרנו אותם בצורתם,
בקצור מופלג, וחזקנו מקצתם בדברים לא זכרום, הכוונה ממנו
שיהיה מוכן ומזומן לברר האמת מהטעות ומקומות ההמעדה,
5 ולבלתי נשוא פנים רק לאמת.
וזה מה שכוונו בזה הפרק.

הכלל השני

נחקור בו במקצת ההקדמות, ובמופתי הרב, אם התבארו באור
מופתי אם לא. ולפי שההקדמות אשר יפול הספק באמתתם הם
10 הא' והב' והג' והז' והח' והט' והי' והי"ב והי"ו והט"ו והי"ו והכ"ב
והכ"נ והכ"ד והכ"ה, כי הכ"ו נחקור בה במאמר השלישי בגזרת השם,
ובכלל ההקדמות אשר תפול בהם החקירה בכלל הזה הם ארבע
עשרה, ומופתי הרב אשר תפול בהם החקירה ששה, חלקנו הכלל
הזה לעשרים פרקים.

הפרק הראשון

15

נחקור בו במופתים שסדר לאמת ההקדמה הראשונה, אם הם
נותנים האמת בה על כל פנים, ונחלק הפרק הזה לארבעה עיונים,
כמספר מיני המופתים הנעשים שם.

העיון הראשון

20   בחקירה במופת שסדר בבאור המנעות מציאות גודל נבדל בלתי
בעל תכלית.
ונאמר שהמופת ההוא הוא הטעאיי ונערך על הדרוש. וזה
שהמניח גודל נבדל בלתי בעל תכלית, אומר במציאות שיעור נבדל,
ולזה גם כן לא יתחייב שגדר הבלתי בעל תכלית יצדק על חלקיו,
25   כמו שלא יתחייב זה בקו הלמודי, ולא יתחייב הרכבה בו כלל אלא
מחלקיו.
אלא שזה, לפי מה שיראה, בנוי על יסוד המנעות הרקות, כמו

---

1 [המעיין] קי מהמעיין ז.   2 המטעים] המנועים ס׳ ב – בצורתם פ.   3 בקצתם ד ׀ במקצתם
ל – מדברים רפוס חוסא – כוונה פ.   5 ולבלתי פ.   6 (מה) א – שכוונו לוורק – בפרק הזה ס לוורק בג.
8 מההקדמות ז – יתבארו פ.   9 יפול [בהם] לדד – (הם) פ.   12 החקירה בהם בג.   13 תפול]
תחול לד – החקירה בהם ס לוורק בג – [הם] ששה פ.   16 אם] שה פ.   17 (בה) ס׳ד – ונחלוק
צור.   18 (מיני) פ.   20 החקירה לוורק כאנ – ובמופת פ – (נבדל) ג.   23 (נבדל) פ – אמר
פ – ונבדל פ.   25 ולא יתחייב] ולא יחייב סורק בא.   27 (לפי) פ – (מה) ס לוורק לטה פ.

orderly arrangement they tend merely to bewilder the reader in what is one of those topics[164] that easily lend themselves to misunderstanding.[165] In view of this, we have recast these arguments in their logical form,[166] restating them in exceeding brief language, strengthening some of them with points not mentioned by any of those authors, our main object being to have all their arguments well arranged and classified, in order to be able afterwards to distinguish truth from error and to detect the loci of the fallacy—and this without regard for anything but the truth.

This is what we intended to accomplish in this chapter.

Part II.

WHEREIN we shall inquire into the arguments which he has framed in support of the first proposition with a view to determining whether they establish the truth thereof in every respect. We shall divide this chapter into four Speculations, corresponding to the four classes of arguments which have been set forth in the corresponding chapter of Part I.

The First Speculation

Examination of the argument which he has framed to prove the impossibility of an incorporeal infinite magnitude.

We say that the argument is fallacious and a begging of the question. For he who assumes the existence of an incorporeal infinite magnitude likewise affirms the existence of an incorporeal quantity. By the same token, it does not follow that the definition of the infinite would have to be applicable to all its parts, just as such reasoning does not follow in the case of a mathematical line. Nor would there have to be any composition in it except of its own parts.[1]

The argument, however, as has already been pointed out in Part I, is obviously based upon the negation of a vacuum, for if

שקדם לנו בכלל הראשון. וזה שאם הודינו במציאותו לא ימנע
מציאות שעור נבדל למוחשות, אבל אולי יחוייב מציאותו, למה
שכבר אפשר שישוער, ויתאמת אמרנו בו גדול או קטן ויתר משיגי
הכמה. אבל למה שהרחיק מציאותו בנה עליו המופת הזה. ולפי
5 שאין בכל מה שחתר מופת מספיק בבטול מציאותו, ראינו להשיב
עליהם, ולבאר שקרות המופתים ההם, לפי שבזה תועלת אינו מעט
בחכמה הזאת.

והנה לפי שהאומרים ברקות דמו, לפי דעתו, שהרקות היא סבת
התנועה, אומר שהמופת הנעשה לבאר שקרות הדימוי הוא הטעאיי.
10 וזה שהאומרים ברקות לא דמו שיהיה הרקות סבת התנועה אלא
במקרה. וזה שהם חשבו שאם לא יהיה הרקות נמצא לא תהיה
תנועת ההעתק אפשרית, להמנע הכנס גשם בגשם, ונעזרו בזה גם
כן מהצמיחה וההתוך והספוגיות והמקשיות ומדמויים אחרים, כמו
שבא זה כלו בספר השמע. והיות הרקות סבת התנועה במקרה על
15 הדרך הזה לא יחייב היות הרקות פועל או תכלית.

ואמנם המופת הראשון שעשה לבטל מציאות הרקות מצד
מציאות התנועה הוא מבואר הבטול. וזה שאם היו האומרים ברקות
מחייבים היותו סבה בעצם לתנועה היה מקום למופת ההוא, אבל

4 כמה פ – שירחיק קנ.   6 לבאר פ – (לפי) פ – (אינו מעט) גדול פ אינו מעטי לדꞋ.
8 (והנה) פ – ולפי פ – שהאומר קנ – דמה קנ.   9 (הוא) פ ההוא הואꞋרꞋ.   10 (וזה) פ.   11 (שהם)
כי פ שאם רꞋ – (שאם) רꞋ – שלאꞋרꞋ – יהיה] יהיה יהיה] היה לווꞋרꞋ דꞋ בנꞋ.   12 (בזה) לדꞋ נם כן בזהꞋי.   13–12 גם
כן] אꞋכ צסꞋא.   13 (וההתוך] וההחיתוך פ והתוך יꞋרꞋ – (והספוגיות] והספונית פ –
והמקשית פ.   14 שבא] שבאר פ.   16 (מציאות) הרקות זרקפꞋבנꞋ.   17 (וזה) פ.   18 אל תנועה
בנ – ואבל פ לרקבꞋאנꞋ ואבל [במה יתבטל מופת המנעות התנועה ההכרחית ברקות] לא לꞋ.

we admit the existence of a vacuum, it would not be impossible to assume a quantity existing apart from sensible objects; nay, its existence would of necessity be implied, since a vacuum is capable of being measured and can thus be appropriately described by the terms great and small and by the other properties of quantity. It is only because of his rejection of the existence of a vacuum that he was enabled to build up his argument. As it is our belief, however, that in all his efforts there is not a single convincing[2] argument to disprove the existence of a vacuum, we have deemed it fit to set forth in great detail our refutation of his alleged arguments and to expose their absurdities, for such an inquiry will prove to be of no small benefit in the pursuit of this intellectual discipline.[3]

Since according to his opinion those who affirmed the existence of a vacuum supposed that the vacuum is the cause of motion, I shall endeavor to show that the argument advanced by him to prove the falsity of that supposition is fallacious. Those who affirmed the existence of a vacuum did not consider it to be the cause of motion except in an accidental sense,[4] that is to say, they thought that without the assumption of a vacuum, locomotion would be impossible on account of the impossibility of bodies penetrating into one another, for which contention they found support in the phenomena of increase and diminution, rareness and denseness,[5] and other examples,[6] as is all set forth in the *Physics*. Since, therefore, the vacuum was conceived by them only as an accidental cause of motion after the manner described, it does not follow that it would have to be either an efficient or a final cause.

As for the *first* argument which he has adduced to disprove the existence of a vacuum, namely, the argument from the existence of motion, its inconclusiveness is evident. There would be some room for the argument, if the vacuum were considered by those who affirmed its existence to be the essential cause of motion, but,

לא דמו לעולם אלא היותו סבה במקרה, כמו שקדם. ולזה לא

ימנע ליסודות, ואם היו מעורבים ברקות, היות להם האותות

במקומם הטבעי, והלוף טבע מה שממנו ומה שאליו, לסבת קרובו

או רחוקו מהמקיף או מהמרכז. ולזה לא ימנע מציאות התנועה

5   הטבעית וההכרחית במציאות הרקות, וכל שכן שלא יחוייב בזה

המופת המנעות מציאות הרקות חוץ לעולם, למה שאם היה הרקות

שאין לו טבע מה שממנו ומה שאליו, לא יתחייב המנעות תנועה

סבובית לגשם כדורי. וזה מבואר בנפשו.

ואמנם המופת ה ש נ י ו ה ש ל י ש י בנויים על שתי הקדמות,

10   שהאחת מהן כוזבת, והיא האומרת שיחס התנועה אל התנועה כיחס

המקבל אל המקבל כשהיו המקבלים מתחלפים. וזה כי למה

שהתנועה תחייב זמן לעצמותה, יתחייב שבהסתלק המקבל ישאר

זמן שרשי לתנועה, ידוע אצל הטבע, לפי חזק המניע. ולזה יתאמת

שיחס איחור התנועה השרשית אל איחור התנועה השרשית כיחס

15   המקבל אל המקבל, כמו שתאמר על דרך משל שיחס איחור התנועה

באיש היגע אל איחור התנועה באיש ההוא בהיותו יותר יגע כיחס

היגיעה אל היגיעה, ואם סלקנו היגיעה תשאר התנועה השרשית.

as has been stated, it was never considered by them as a cause except in an accidental sense. It would not be impossible, therefore, for the [sublunar] elements, though interspersed with a vacuum,[7] still to possess an affinity[8] to their respective natural places, nor [would it be impossible for the vacuum to possess within itself] a distinction of parts, one having the nature of a *terminus a quo* and the other of a *terminus ad quem*, this distinction to be determined by the proximity of the vacuum[9] to the circumference or the centre, or by its remoteness therefrom.[10] Hence, with the assumption of a vacuum, neither natural nor violent motion would be impossible. Much less does this argument prove 'the impossibility of a vacuum outside the world,[11] for even if there existed outside the world a vacuum in which there were no distinction of *terminus a quo* and *terminus ad quem*, it would not be impossible for a spherical body [existing in it] to have circular motion.[12] This is self-evident.

As for the *second* and *third* arguments, they are based upon two propositions, one of which is false, namely, the one which states that the ratio of one motion to another is equal to the ratio of their respective receptacles, when these latter are unlike. For since every motion by its very essence involves time in its process, it will follow that even by eliminating the receptacle there will still remain an original time of motion,[13] required by the nature of motion itself,[14] varying only according to the power of the motive force. It is only true, therefore, to say that the ratio of the retardation of one original motion to that of another is equal to the ratio between their respective receptacles, as, e. g., the ratio of the diminution of the natural speed of a person when he is fatigued to the diminution in the natural speed of the same person when he is more fatigued is equal to the ratio between the two states of fatigue, in which case, if the fatigue were to be eliminated, there would still remain an original speed. Averroes, to

והנה אבן רשד חתר להתיר הספק, בשכבר העיר עליו אבובכר

במקצת, והרבה דברים מרבים הבל.

ומהאחרונים מי שחשב לבאר המנעות הרקות בשאמר שהממוצע

תנאי במציאות התנועה, וזה להאותות טבעו למה שאליו. והוא דבר

5 לא התבאר ולא יתבאר, בשכבר אפשר שיאמר שהכובד והקלות

למתנועעים בטבע, ואין צורך בהם לממוצעים. ואולי שאפשר

שיאמר שלכלם כובד מה אלא שיתחלפו בפחות ויתר. ולפי זה

המתנועעים למעלה יהיה מהכרח היותר כבדים, כאלו תאמר

שהאויר בהיותו תוך המים יעלה מצד הכרח כובד המים הדורשים

10 המטה, להיותם יותר כבדים. וכבר יראה זה, כי אנחנו אם פנינו

מקום הארץ, ואולי עד המרכז, כבר יתמלא מים או אויר. אם זה

להכרח המנעות הרקות תוך העולם, או לסבת כובד האויר, לא

התבאר עדיין ולא יתבאר.

ועוד שאם היה שהודיענו שהממוצע תנאי במציאות התנועה, הנה

15 לא ימנע משיהיה חוץ לעולם רקות, ויתנועע בתוכו גשם כדורי

בסבוב, כי המופתים ההם לא ימנעו אלא תנועה ישרה לגשם מונח

ברקות, אבל הגשם הכדורי כבר יתנועע בתוכו מבלתי שימיר

מקומו. וזה מבואר מאד.

ואמנם המופת ה ר ב י ע י, יסודו ההקדמה האומרת שהמנע הכנס

1 הספק [בזה] ג^א – אבן] בן פרק בן ן' פ^א – אבובכר] אבונצר ל' אבו כבר פ.   3 ומאחרונים

פ – (מי) זר – [נמצא] מי ב – בשיאמר פ לזורדקב^א-ן – בשהממוצע פ.   4 ולזה האותות י.

6 צריך ל דג – (שאפשר) בנ.   13 (עדיין ולא יתבאר) י.   19-1 הכנס גשם בגשם] גשם בגשם הכנסו פ

be sure, attempted to answer this objection, which in part[15] had already been anticipated by Avempace, but his answer rather answers to the description: 'Many words that increase vanity'.[16]

Among the later thinkers there is one[17] who proposed to prove the impossibility of a vacuum by maintaining that the medium is a necessary condition in the existence of motion,[18] and this because the medium has in its nature something akin to a *terminus ad quem*.[19] But this is an assertion which has never been demonstrated and never will be, for it may be claimed, on the contrary, that the movable bodies have weight and lightness by nature, and have no need for media.[20] Or, it may also be said that all the movable bodies have a certain amount of weight, differing only *secundum minus et majus*.[21] Accordingly, those bodies which move upward are so moved only by reason of the pressure exerted upon them by bodies of heavier weight,[22] as, e. g., air, when compressed in water, will tend to rise on account of the pressure of the weight of the water, which, being heavier, will seek the below. That this is so will appear from the fact that when we make a hollow in the earth, even as far as the centre, it will immediately fill up with water or air, though, [it must be admitted], whether this is due to the impossibility of a vacuum within the world or to the weight of the air has not so far been demonstrated and never will be.[23]

Furthermore, even if we were to admit that the medium is a necessary condition in the existence of motion, it is still not impossible for a vacuum to exist outside the world[24], and in it for a spherical body to move with circular motion; for all these arguments show only the impossibility of rectilinear motion in a body assumed to be in a vacuum, whereas a spherical body may have motion in a vacuum without changing its place.[25] This is very evident.

As for the *fourth* argument, it is based upon the assumption that the impenetrability of bodies is due exclusively to their

נשם בנשם הוא מצד מרחקיו השלשה בלבד. והוא שקר מבואר
לאומרים ברקות, שאין ההמנע מפני הרחקים מופשטים, אבל מפני
הרחקים במה שהם בעלי חומר. ואם היה שאין ההמנעות מפני
החמר לבדו, למה שאם לא היה לו רחקים לא יטריד מקום, הנה
5 גם כן הרחקים, אם לא היו בעלי חומר, לא יטרידו מקום, ואז לא
היו צריכים אל מקומות בלתי בעלי תכלית. אלא שאם היה שלא
יצדקו נפרדים, רוצה לומר שאין באחד מהם די להמנעות הכנס
נשם בנשם, הנה יצדק מורכב, שהרחקים בעלי חמר יטרידו
המקום, אשר מזה הצד היא נמנע הכנס גשם בנשם. ולזה לא יתאמת
10 שהרחקים מופשטים יצטרכו אל מקום. וזה מבואר מאד.

ומה שחזק דעתו עוד בשאמר שהרחקים תכליות הגשמים, הנה
האומר ברחק נבדל אינו מודה בו, והוא מערכה על הדרוש.

הנה כבר התבאר, שאין בכל מה שאמר דבר ראוי לשום לב
עליו בבטול רוחק נבדל. והוא מה שכווננו לביאורו.

15 וכבר יראה שמציאותו מחוייב לפי סברתם, האומרים בהמנעות
מציאות גשם בלתי בעל תכלית, וזה שהוא מחוייב שלא יהיה חוץ
לעולם גשם, ואם אין שם גשם הנה אין שם מלוי, ואם אין שם מלוי,
מי יתן ואדע מה זה אשר ימנעהו לקבל רחקים גשמיים? והנה רחקים

1 מרחקיו] רחקיו ורד* רחוקו פ רחקו ז – (השלשה) קו – (בלבד) קו. 2 (מופשטים) ז
המופשטים ו – אבל] אלא זרקבנ – אבל [הם] א. 4 יטריד] יטרידו פרקבנ. 5 ואז] וזה ב.
6 (שאם) בנ – שהיה בנ. 7 (די) לזורקבנ. 10 (מאד) בנ. 11 עוד] הוא פ – (עוד) סוא.
13 התבאר] החאמת זר – שאי[ן] כי אין ר – [כבר] דבר ז – (ראוי) פצפלרקבאי. 14 לבטול א –
(מה) ד – בביאורו ר ביאורו סלזודקבאי. 15 שמציאותו] מציאותו לד – [הם] האומרים ב.
16 (מציאות) לד – הנשם ז – (חוץ) בנ.

tridimensionality. But this, according to those who believe in a vacuum, is obviously not true, for according to them, the impenetrability of bodies is due not to dimensions existing apart from matter, but rather to dimensions in so far as they are possessed of matter.[26] Matter alone, to be sure, could not account for impenetrability, for were it not for its dimensionality, matter alone would not occupy place, but neither would the dimensions alone occupy a place were it not for their materiality. This being the case, one could not argue, [as does Aristotle], that the dimensions would require an infinite number of places. The fact of the matter is, while neither of the reasons mentioned is sufficient when taken separately, that is to say, neither of them by itself is sufficient to render the penetrability of bodies impossible, they are sufficient when taken together,[27] that is to say, in view of the fact that material dimensions occupy place, it is impossible for bodies to enter into one another.[28] Hence it does not follow that the dimensions even when they are immaterial, [as in his argument], would require a place for their existence. This is very evident.

As for the statement by which he reinforced his view, namely, that dimensions are the limits of bodies, this, too, will not be admitted by him who affirms the existence of an incorporeal interval.[29] It is thus a begging of the question.

It has thus been shown that in all he has said there is nothing which merits attention as an argument to disprove the existence of an incorporeal interval. This is what we intended to do to his proof.

Furthermore, it would seem that the existence of an incorporeal interval is implied even in the view of those who deny the possibility of an infinite body. For according to their view there can be no body outside the world, and if there is no body, there is no plenum, and if there is no plenum, would that I knew[30] what should prevent that which is outside the world from being capable of receiving corporeal dimensions. But incorporeal dimensions

נבדלים עניינם המקום הפנוי לקבל רחקי גשם. ואמרנו המקום
הפנוי, למה שיראה שהמקום האמתי לגשם הוא הפנאי השוה לגשם,
אשר יטרידנו הגשם, כמו שנבאר במקומו בגזרת השם.

ולזה התבאר שגודל נבדל אינו נמנע בעצמו, אבל אולי מחוייב.
5 ואיך לא? והפנאי בעצמו כבר יאמר בו גדול או קטן, והוא משוער
בחלק ממנו, ואלו תדמה כלי קערורי הורק מן האויר, ולא נתמלא
אויר במקומו, הנה הפנאי ההוא יאמר בו גדול או קטן, והוא משוער
בחלק ממנו. ואחר שיצדק עליו גדר הכמה המתדבק, הנה הוא
בהכרח גודל, אחר שאיננו זמן.

10 ואם כן אחר שאין חוץ לעולם גשם לפי סברתם, האומרים
בהמנעות גשם בלתי בעל תכלית, יש שם בהכרח פנאי. ואחר
שהתבאר שהוא גודל, התבאר אם כן מציאות גודל נבדל. ולפי
שהוא נמנע מציאות התכלית לו, למה שהוא מחוייב שיכלה אל גשם
או אל פנוי, ואי אפשר שיכלה אל גשם, הנה אם כן יכלה אל פנוי,
15 וכן לבלתי תכלית. והתבאר אם כן לפי סברתם מציאות גודל נבדל
בלתי בעל תכלית.

ואיך שיהיה, התבאר בהכרח מציאות גודל בלתי בעל תכלית,
גשם היה או נבדל. וזהו מה שראינו לחתום בו העיון הראשון.

ואולם במופת אלתבריזי, אשר קראו מופת ההתדבקות, הוא
20 מבואר שלא יתחייב מה שחשב. וזה שהמנעות היות בלתי בעל

---

2 פנוי פ – (למה שיראה פ – הפנאין הפנוי פ.   3 יטרידו ב – בע"ה לרד.   5 (והפנאי בעצמו
פ לוורדקב.נ.   6 בחלקי ב.נ – קערורית פ קערור ב.נ.   7 (הנה הפנאי ההוא) ו – (או קטן)
ז – משוער] בשער פ.   8 לחלק פוק ב.א.נ – עליו) עלין פ – (הכמה) פ לורדק ב.א.נ.   14 פנאי
פ – נשם (א') פ – (הנה) זרק ב.נ.   18 זהו מה (שראוי וראינו לבארן לחתום ב.   19 לתבריזי פ –
התרבקות זק א.נ הדרבקות לד – (הוא) הנה לרד.   20 מה] מי פ.

mean nothing but empty place capable of receiving corporeal dimensions.[31] We have advisedly used the words 'empty place' because it is evident that the true place of a body is the void, equal to the body and filled by the body, as we shall prove in its proper place,[32] God willing.

Thus it has been shown that an incorporeal magnitude is by its own nature not impossible; nay, its existence must inevitably be implied. And why should it not? when the void itself, [without any content], may be described as great and small[33] and may be measured by a part of itself,[34] for when, for instance, you imagine a closed vessel from which the air has been cleared and into which no other air was admitted, the void within it will be described as great and small, and will be measured by a part of itself. Since the definition of a continuous quantity can thus be applied to the void, and since it is not time, it must of necessity be a magnitude.[35]

We thus conclude: Since according to the view of those who maintain the impossibility of an infinite body, there is no body outside the world, there must necessarily be there a void.[36] Since the void has been shown to be a magnitude, it has thus been shown that an incorporeal magnitude exists. But this incorporeal magnitude outside the world cannot have a limit, for if it had a limit it would have to terminate either at a body or at another void. That it should terminate at a body, however, is impossible. It must therefore terminate at another void, and so it will go on to infinity. It has thus been shown that on their own premises an infinite incorporeal magnitude must exist.

However that may be, it has been conclusively shown that an infinite magnitude, be it a body or something incorporeal, must exist. With this we deem fit to conclude the first Speculation.

As for Altabrizi's proof, which he terms the proof of application, it is obvious that his alleged conclusion does not follow. The impossibility of one infinite to be greater than another is true

תכלית גדול מבלתי בעל תכלית הוא מצד השיעור, שכשנניחהו
גדול בצירוף הכוונה בו גודל השיעור, ומה שאין תכלית לו הוא
בלתי משוער. ולזה לא היה הקו האחד גדול מהאחר, לפי שכל
אחד בלתי מקבל השיעור בכללו. ולזה איננו גדול מהאחר, ואם

5   היה נוסף מהצד שהוא בעל תכלית.   וזה מבואר בעצמו.

וכבר יתאמת זה מן החוש, למה שהוא מבואר מעניין הזמן, לאומר
בנצחותו, שזה עניינו, שהזמן הוא מתוסף מהצד שהוא בו בעל תכלית
עם היותו בלתי בעל תכלית מהצד האחר, לאומר בקדמותו. ועוד
יתבאר במה שיבא בגזרת השם שאף לאמונתנו האמתית בחדוש

10 יתחייב זה במה שאין ספק בו.

העיון השני

בחקירה במופתים שסדר בבאור המנעות מציאות גודל גשמי
בלתי בעל תכלית.

והנה הבאור הכולל שהתחיל בו תחלה הוא מבואר הנפילה,

15 שההקדמה הקטנה, האומרת שכל גשם יקיף בו שטח או שטחים,
חולק עליה בעל הריב האומר במציאות גשם בלתי בעל תכלית,
והנה סדר מערכה על הדרוש.   וכן אם הודינו לו בהמנעות גודל
גשמי בלתי בעל תכלית, לא יתחייב מה שדמה בגודל, למה שכבר
אפשר שיבדלו מן הגשם, כמו שבארנו במה שעבר. והנה במספר

20 נדבר בו במה שיבא בגזרת השם.

only with respect to measurability, that is to say, when we use the term greater in the sense of being greater by a certain measure, and that indeed is impossible because an infinite is immeasurable. In this sense, to be sure, the first one-side infinite line [in Altabrizi's proof] cannot be greater than the second one-side infinite line, inasmuch as neither of them is measurable in its totality. Thus indeed the former line is not greater than the latter, even though it extends beyond the latter on the side which is finite.[37] This is self-evident.

That this is so may be demonstrated from observation, from the case of time, which according to those who believe in its eternity, must be conceived in a similar way, that is to say, it must be conceived as capable of increase on the side on which it is limited even though it is infinite on the other side.[38] Furthermore, it will be shown subsequently, God willing, that this distinction will have to be accepted beyond any doubt even according to our own true belief in creation.[39]

THE SECOND SPECULATION

Examination of the arguments which he has framed to prove the impossibility of a corporeal infinite magnitude.

As for the general argument with which he begins his proof, its unsoundness is obvious, for the minor premise, namely, that every body is contained by a surface or surfaces is contradicted by the opponent who affirms the existence of an infinite body.[40] He is thus arguing in a circle. Furthermore, even if we agree with his conclusion as to the impossibility of a corporeal infinite magnitude, that conclusion of his must not necessarily be true with respect to magnitude in general, for dimensions, as we have already shown, are capable of existence apart from body. As to number, we shall discuss it in a subsequent chapter,[41] God willing.

ואולם המופתים הטבעיים, הנה ה ר א ש ו ן נפסד החומר והצורה.
זה שהוא מחובר מהקדמות בלתי מודות, ושהתדבקות הנמשך בלתי
מחוייב. וזה שההקדמה האומרת בהמנעות מציאות יסודות בלתי
בעלי תכלית, לא התבארה בראשון מהשמע אלא בשתי טענות.

5 האחת, כי הבלתי בעל תכלית לא תקיף בו ידיעה. והנה אין
מהכרח ההתחלות במה שהם התחלות להיותן ידועות. והוא מבואר
בעצמו. והשנית, שאם היו היסודות בלתי בעלי תכלית היה
מורכב בלתי בעל תכלית. והוא הדרוש. ולזה בהניחנו מורכב
בלתי בעל תכלית לא יתבאר המנעות מציאות יסודות בלתי בעלי

10 תכלית. התבאר אם כן היות ההקש נפסד מצד חמרו. ואולם מצד
צורתו, למה שלא יתחייב בהניחנו אחד מהיסודות בלתי בעל תכלית
שיפסיד השאר, כי כבר אפשר שלא יהיה בעל איכות, למה שכבר
אפשר שיונח גשם בלתי בעל תכלית אין איכות לו, ומזה הצד הוא
מקבל כל האיכיות, מצד היותו משולל מכלם, והוא להם יסוד.

15 וכבר נמצא גשם בלתי בעל איכות, לפי סברתם, כעניין בגרמים
השמימיים, אלא שבו כח והכנה לקבל האיכיות. וכל שכן שבזה
המופת לא התבאר המנעות מציאות גשם כדורי חוץ לעולם בלתי
בעל תכלית.

ומה שחזק עוד דעתו, שאם היה בלתי בעל תכלית יהיה בלתי
20 בעל תכלית בכל רחקיו, לא יתחייב זה. שאם היה הבלתי תכלית
עצמי לרחקים, היה מקום לחיוב ההוא, אבל אם הבלתי תכלית

---

2 בהקדמות זר – מודות] מורות פ – שהתרדבקות פ ושהההתרדבקות ר.   3 מחוייב] מתחייב פלוק
באג.   4 (לא) פ.   5 בון] בה פי.   6 להיותם באג – והוא] חה צבאג.   7 בעצמו] בנפשו ודקב
נ – (היסודות) י – היה] הוא צ (היה) ל.   8 המורכב פלודקא.   9 התבאר פ – (מציאות) זד –
היסודות פ.   14 היסוד פ זר יסודי ב.   15 הענין לר.   16 שבו] שבזה סודקבאג – הכח ראֹ –
(והכנה) ר.   17 יתבאר לוקאג.   20–21 שאם הי' סבה עצמי' לרחקיו היה פ.   21הבלתי
תכלית] הבעל תכלית פר נהיה] הב"ת גי.

As for the physical arguments, the *first* is both materially and formally defective: viz., it consists of propositions which are inadmissible[42] and the connection of the consequent with the antecedent is not necessary. The proposition denying the existence of an infinite number of elements has been demonstrated in the first book of the *Physics*[43] only by two arguments. The first of them is that the infinite cannot be comprehended by knowledge. But it is not necessary that principles *qua* principles should be known.[44] This is self-evident. The second argument is that if the elements were infinite, there would be an infinite composite body. But this is what was to be proved here. If we assume, therefore, the existence of an infinite composite body, there will be no argument for the impossibility of the existence of infinite elements. It has thus been shown that the syllogism is materially defective. As for the defectiveness of its form, it does not necessarily follow, if we assume one of the elements to be infinite, that it would cause the destruction of the other elements, for that element may be conceived as being devoid of any qualities, inasmuch as it is possible to assume an infinite element without any qualities, which, on account of its being devoid of any qualities, may be the recipient of all the qualities and act as their substratum.[45] Such a body, devoid of any qualities, is to be found, according to their own admission, in the case of the celestial bodies,[46]—a body endowed only with a capacity and predisposition for the recipiency of qualities. Still less has this argument proved the impossibility of the existence of an infinite spherical body outside the world.[47]

As for the statement by which he has reinforced his contention, namely, that if an infinite existed it would have to be infinite in all its dimensions, this, too, is inconclusive. If infinity were essential to dimensions as such, there would be some ground for his conclusion; but since infinity is to be only one of the properties of

משיג ממשיגיו ובלתי עצמי לו, לא יתחייב זה בכל הרחקים. וזה מבואר מאד.

ואמנם ה ש נ י, אשר יסודו הכובד והקלות, הוא לקוח מהגשמים המוחשים אשר תחת הגלגל. ואולם האומר בגשם הבלתי בעל תכלית יאמר שאין לו כובד ולא קלות, כמו שיאמר בגרמים השמימיים לדעת ארסטו.

ואמנם ה ש ל י ש י ו ה ר ב י ע י, אשר מצד המקום, הנה אם הודינו גדר המקום אשר אמרו, הנה לא יתנו האמת כמו שחשב. וזה שהאומר בגשם הבלתי בעל תכלית יאמר שמקומו הוא [מצד] שטח קעריריותו, והוא השטח המקיף המרכז, ומצד גבינותו הוא בלתי בעל תכלית, ואין לו מקום בפאה ההיא. ואיך לא? והגשם השמימי המקיף בכל לפי דעת ארסטו זה תוארו, רצוני שאין לו מקום מקיף אלא מוקף.

אלא שהאמת בעצמו, לפי מה שיראה, שהמקום האמתי לדבר הוא הרחק אשר בין תכליות המקיף. והשקרים אשר חייב ארסטו לזה הדעת אין עניין להם, שהם מיוסדים על שהרחקים אשר בתוך הכלי מלא מים נעתקים בהעתק הכלי, ואז היו מתחייבים השקרים ההם. והוא בדוי, ואינו אמת, שהרחקים לאומרים בפנוי ורקות בלתי מתנועעים. ולזה לא יתחייבו הבטולים ההם.

והנה לסברת ארסטו במקום יתחייבו גנויות.

מהם, שהגרמים השמימיים יתחלפו במקום. וזה שלכלם מקום

the infinite and unessential to it, it would not follow that all the dimensions would have to be infinite.[48] This is very evident.

As for the *second* argument, based upon the consideration of weight and lightness, it is derived from an analogy of sublunar sensible bodies. But he who affirms the existence of an infinite body conceives it to be without either weight or lightness, as is said to be the case of the celestial bodies according to the view of Aristotle himself.[49]

As for the *third* and *fourth* arguments, based upon place, even if we accept his definition of place, they do not sustain his alleged conclusion. For he who affirms the existence of an infinite body would maintain that the infinite has place only with reference to[50] the surface of its concavity,[51] that is, the surface which surrounds the centre,[52] whereas with reference to its convexity[53] it is infinite and therefore has no place on that side. Why should it not be so? when the all-encompassing celestial sphere answers exactly to this description, according to Aristotle's own theory, namely, that it has no place which surrounds, but one which is surrounded.[54]

The truth of the matter, as it seems, is that the true place of a thing is the interval between the limits of that which surrounds.[55] The impossibilities which, according to Aristotle, would have to ensue from this view,[56] are beside the mark, resting as they do upon the assumption that the dimensions within a vessel full of water will be moved together with the vessel, whence indeed, were this true, the alleged possibilities would have to follow. But the assumption is a figment of the imagination and is not true. The dimensions, according to those who believe in an empty space and a vacuum, are immovable, and so none of those supposed impossibilities would follow.[57]

Furthermore, Aristotle's definition of place will give rise to many absurdities:

First, the celestial bodies will differ with regard to place. All the [internal] spheres will have essential place, that is, the sur-

בעצם, רצוני השטח המקיף, והמקיף בכל לא יהיה לו מקום בעצם,
למה שאין לו שטח מקיף שוה נבדל, כי השטח אשר בגבניות אינו
נבדל ממנו, אשר בעבור זה נלחץ לומר שאין לו מקום בעצם אלא
במקרה.

5 ומהם, שהגדר אשר אמרו, בשהוא שטח מקיף שוה נבדל, איננו
מסכים גם למתנועעים תנועה ישרה. וזה שהמקום המיוחד לחלקים,
המתנועעים בעצם בתנועת הכל, איננו מקיף שוה נבדל באופן שיהיה
לו ערבות ודמיון לכל חלקי המקום כאשר חתר. וזה שמקום האויר,
על דרך משל, לפי סברתו הוא השטח המקיף בקערירות האש,
10 למה שיש לו שם ערבות ודמיון. ואמנם החלק האמצעי מן האויר
לא נמלט אם שהוא במקומו הטבעי, אם שאינו במקומו הטבעי, אשר
לו ההאותות אשר אמרו. ואם הוא במקומו הטבעי, יתחייב שמקומו
הטבעי אשר לחלק יתחלף למקום הטבעי אשר לכל, והוא בתכלית
הגנות.

15 ומהם, שאם המקום אשר לגרם השמיי, בעצם היה או במקרה,
הוא מקיף המרכז, לא יתכן בו ההאותות אשר אמרו במקוממים
בכלל. וזה שלא יצוייר בגרמים השמימים האותותם אל המטה.
וכל שכן שיסוד האש ידרוש המעלה, אשר מזה הצד יש לו ערבות
ודמיון במקיף, ושהגרם השמימי איך יהיה לו ערבות ודמיון אל
20 המטה.

1 (המקיף⁶. ⁶. 2 בגבניות ב נ. 3 אשר] אבל ⁶ לוורק ב נ. 5 ונבדל ל ד. 7 באופן (שוה] ק ב נ.
8 שהמקום פקא. 10 (מן האויר) י. 11 (לא נמלט אם שהוא במקומו הטבעי) דוק ב א נ – הטבעי
[למה] אם ×— (אם שאינו במקומו הטבעי) ר – [למה] אשר ×. 12 [יש] לו לורקב א נ – האותות ⁶–
שאמרו ב. 15 שמימי ב. 16 הוא] היה ⁶ – האותות ⁶ – במקומם ⁶. 17 (בכלל) ר – למטה
ל ד. 18 שהיסוד ⁶ – [אל] המעלה ל ד ק ב נ. 19 (איך) לורק ב א נ – יהיה] יש × – (לו) ק ב נ – אל]
עם ⁶ לורק ב א נ.

faces [of the other spheres which surround them respectively], whereas the outermost sphere, having no surrounding, equal and separate surface, for its own convex surface is inseparable from it, cannot have any essential place,[58] on which account Aristotle was compelled to say that it has no essential place but only accidental.[59]

Second, the definition he gave of place, that it is a surrounding surface, equal to the body surrounded, and separate therefrom, is not applicable in the same sense even with regard to the elements which have rectilinear motion.[60] For in the case of parts that move essentially[61] with the motion of the whole the proper place of each part cannot be described as *surrounding, equal* and *separate*, and at the same time satisfy another condition which Aristotle insists upon, namely, that each part of the object should have an agreeableness and likeness[62] to a respective part of the place.[63] The place of air, for instance, is according to his theory the surrounding surface identical with the concavity of fire, because air finds there that to which it has an agreeableness and likeness.[64] Now any part from the middle of the air must inevitably either be in its natural place, to which it is claimed to have the alleged natural affinity,[65] or not be in its natural place.[66] But if it is in its natural place, it will follow that the natural place of the part is different from that of the whole. But this is most absurd.

Third, if the place of the celestial body, be it essential or accidental,[67] were the surface surrounding the centre, the celestial sphere could not have that affinity [with its place], which they claim to be characteristic of all place-filling objects, for it is inconceivable that celestial bodies should have an affinity to the below.[68] If the element fire has an agreeableness and likeness only to that which surrounds it,[69] as is evidenced by the fact that it always tends upward, *a fortiori* how could a celestial body have an agreeableness and likeness to the below?

גם מה שדמה שהכדור המתנועע יצטרך לדבר נח, ומזה הצד

היה אפשר לומר בו שהוא במקום, הוא שקר בדוי. וזה שיתחייב

מזה שסביב קטבי הכדור דבר נח, ויתפוצצו אם כן חלקיו. אלא

שהנקודה אשר במרכז או בקטבים לא תתואר בתנועה ולא במנוחה

5 בעצם, ואם היה שתתנועע, במקרה, מצד היותה תכלית למתנועע.

ולזה לא יאמרו בעבורה שהכדור המקיף במקום.

ואמנם כשהנחנו הפנוי הוא המקום, הוא המסכים לכל המתנועעים

תנועה ישרה או סבובית, ולכל חלקיהם, מבלי שלא נבקש להם

האותות.

10 והתמה, שכאשר בקשנו ליסוד הארץ מקום, הנה אמרנו שהוא

המטה במוחלט, והנה המטה במוחלט איננו שטח כי אם נקודה, ואי

אפשר שתתואר במקום.

ולזה היה האמת עד לעצמו, ומסכים מכל צד, כשהמקום האמתי

הוא הפנוי. וכבר היה ראוי להיות כן, כי המקום היה ראוי שיהיה

15 שוה למקומם כלו וחלקיו.

ולזה המופת שסדרו איננו נותן האמת בדרוש. והוא מה שכוונּו

בזה העיון השני.

והנה להתפרסם זה הענין מהמקום, היה הרבה מהקדמונים שהיו

רואים כי מקום הדבר האמתי צורתו, בשהיא תגבילהו ותיחדהו

20 כלו וחלקיו, עד שרבותינו עליהם השלום השאילו השם הזה לצורת

Likewise, his assumption that a rotating sphere must have a stationary centre, with reference to which the sphere could be said to exist in place,[70] is a fictitious falsehood. For it would imply that around the poles of the sphere there was something stationary. But if so, the parts of the sphere will have to separate themselves from each other[71] [during its rotation]. The fact of the matter is that the point at the centre or at the poles cannot be described as being essentially either at rest or in motion,[72] and if it is moved, it is moved only accidentally by virtue of its being the extremity of something moving.[73] In view of this, the centre cannot be taken as that on account of which the surrounding [celestial] sphere is to be described as being in place.

If we assume, however, place to be identical with the void,[74] the definition will be equally applicable to all the elements, whether moving rectilinearly or circularly, and also to all their parts,[75] without our having to postulate for them any affinity.[76]

There is also this difficulty: When we were looking for[77] a place for the element earth, we decided that it is the absolute below, but the absolute below is not a surface but rather a point, and cannot be described as place.[78]

Consequently, it will be in accordance with the nature of truth, which is evident by itself and consistent with itself in all points,[79] if true place is identified with the void. That it should be so can be also shown from the consideration that place must be equal to the whole of its occupant as well as to [the sum of] its parts.[80]

Hence the argument which he has framed does not prove the thesis in question.[81] This is what we intended to show in this second Speculation.

It is because this was generally known to be the meaning of place that there were many among the ancients who identified the true place of a thing with its form, for place like form determines and individuates the thing, the whole as well as its parts,[82] so that our rabbis, peace be upon them, applied the term place figura-

הדבר ועצמותו, אמרם ממקומו והוא מוכרע, ממקום שבאת,
כלומר מאותו דבר עצמו. ממלא מקום אבותיו. והסתכל איך
העידו שהמקום הוא הפנוי, אשר ימלא בעל המקום, ולזה אמרו
ממלא, ואילו היה מכוון מדרגה לבד, היו אומרים במקום

5 אבותיו היה, כלומר, במדרגת אבותיו.

ולזה להיות השם יתברך הוא הצורה לכלל המציאות, כי הוא
מחדשו ומיחדו ומגבילו, השאילו לו השם הזה, באמרם תמיד ברוך
המקום, לא על דעתך אנו משביעים אותך אלא על
דעתנו ועל דעת המקום, הוא מקומו של עולם. והיה

10 הדמיון הזה נפלא, כי כאשר רחקי הפנוי נכנסים ברחקי הגשם
ומלואו, כן כבודו יתברך בכל חלקי העולם ומלואו, כאמרו
נקדוש קדוש קדוש השם צבאות מלא כל הארץ
כבודו, ירצה כי עם היותו קדוש ונבדל בשלש קדושות, שירמוז
בהם אל היותו נבדל משלש עולמות, הנה מלא כל הארץ

15 כבודו, שהוא יסוד העיבור שביסודות כבודו.

ומזה העניין אמרו ברוך כבוד השם ממקומו, כלומר
שתואר הברכה והשפע ממקומו, רוצה לומר מעצמותו ולא מזולתו.
ויהיה הכנוי ממקומו שב אל הכבוד. ואם תרצה שיהיה הכבוד
נאצל, יהיה העניין כפשוטו, ויהיה הכנוי שב אל השם, כלומר

1 ממקומו] שמקומו נ – מוכרע] מוכרח לירד א.   2 בעצמו לירד – מקום] כבוד ד – אבותיו פ –
והשתכלי א והשכל ר – אין] אש' פ.   3 שהעידו קבנ – הוא] היה פ.   4 (לבד) בלבד
ובאו.   5 אבותינו פ – (היה) – במדרגות פ – (אבותיו) פ.   6 להיותו פ – (הוא) פלוירד
באו – צורה לרדקבנ.   7 מיחדו ומחדשו ומגבילו ז מחדשו ומגבילו ומיחדו לזורדקבנ – זה
השם ל.   8 ולא ר – דעתך] דעתנו וקבאנ – (אותך) פבאנ.   9 דעתנו] דעתם וא – המקום נב'ה
פ – עולם [ואין העולם מקומו] נ.   11 כאמרו] כאמרם זרד כאומרם ל.   13 ירצה [בן] לרדק
בנ – שירמז] שיראה פ שיחריז ד.   14 אל היותו] אלהותו ר להיותו פ.   16–17 כלומר
שתואר] כאלו שתאמר אל תואר פ.   17 [הוא] ממקומו צא – מעצמותו] לעצמו פ.   18 כנוי
פ סורד באו – שב ממקומו פ.

tively to the form and essence of a thing, as, when they say: 'It is proved from its own place;'[83] 'From the place from which you come,'[84] that is to say, from the very thing itself; 'He fills his ancestors' place.'[85] You may note how in the last-quoted expression they have indirectly testified that place is identical with the void which an object occupies, thus accounting for their use of the word 'fills,' for if by 'place' in this quotation were meant 'grade,'[86] they would have said, 'He was in his ancestors' place,' which would mean, 'in the exalted position of his ancestors.'

Accordingly, since the Blessed One is the form of the entire universe, having created, individuated and determined it, He is figuratively called Place, as in their oft-repeated expressions, 'Blessed be the Place;'[87] 'We cause thee to swear not in thy sense, but in our sense and in the sense of the Place;'[88] 'He is the Place of the world.'[89] This last metaphor is remarkably apt, for as the dimensions of the void permeate through those of the body and its fullness, so His glory, blessed be He, is present in all the parts of the world and the fullness thereof, as it is said, '[Holy, holy, holy is the Lord of Hosts], the whole earth is full of his glory',[90] the meaning of which may be stated as follows: Though God is holy and separated by a threefold holiness,[91] alluding thereby to His separation from three worlds, still the whole earth is full of His glory, which is an allusion to the element of impregnation, which is one of the elements of Glory.[92]

Of the same tenor is the conclusion of the verse, 'Blessed be the glory of the Lord from His place,' that is to say, the 'Blessedness' and 'Affluence,' ascribed to God is from His place, that is, to say, from God's own essence and not from something outside Himself, and so the pronominal suffix 'His' in 'from His place' will refer to 'glory.'[93] If, however, you prefer to consider 'Glory' as an emanation, the verse will be taken according to its more literal meaning, the pronominal suffix referring to God, the meaning of the verse thus being, the 'Glory of God' is 'blessed' and is

שכבוד ה' ברוך ומושפע ממקום השם, רוצה לומר עצמותו,
להיותו נאצל ממנו, ולא יצטרך לפירוש הרב אשר פירש מקומו
מדרגתו, כי אין ראוי ליחס מדרגה אצל השם.

וזה מה שראינו לחתום בו זה העיון השני.

העיון השלישי

בחקירה במופתים שסדר בהמנעות מתנועע בלתי בעל תכלית
תנועה ישרה או סבובית.

אולם המופתים שסדר בהמנעות תנועה ישרה לגשם הבלתי בעל
תכלית, ויחייב מזה המנעות מציאות גשם בלתי בעל תכלית, הנה
הם בנויים כלם על הגשם המוחש, ולזה יהיה החיוב חלקי, ולא
יתבאר עדיין המנעות מציאות גשם בלתי בעל תכלית בלתי מוחש.
אלא שכשנחקור בהם נמצאם בלתי נותנים האמת על כל פנים, אף
בגשם מוחש.

וזה שהמופת הראשון המיוסד על האנה, יש לאומר שיאמר
שמקומות האנה, עם היותם מוגבלים במין, רוצה לומר המעלה
והמטה, הם בלתי מוגבלים באיש, וזה שהמקומות הם זה למעלה
מזה לבלתי תכלית. ואם אין שם מעלה במוחלט, לא יקרה מזה
ביטול, ואם היה שהתנועה הישרה נראית בחוש.

ואמנם המופת השני, המיוסד על הכובד והקלות, הנה כשנניח
הגשם בלתי בעל תכלית בעל כובד וקלות, לא יתחייבו החיובים

---

poured forth in abundance 'from the place of God,' i.e., from His essence,[94] inasmuch as it is an emanation. There is no need, therefore, for the Master's interpretation of 'His Place' to mean 'His grade,'[95] for it is an impropriety to ascribe to God any distinction of grade.

This is wherewith we deem it fit to conclude this second Speculation.

THE THIRD SPECULATION

Examination of the arguments which he has framed to prove the impossibility of an infinite body having either rectilinear or circular motion.

As for the arguments which he has framed to prove the impossibility of *rectilinear* motion in an infinite body, whence he infers the impossibility of an infinite body, they are all based upon the analogy of a sensible body. His reasoning, therefore, proves only one particular case,[96] but there still remains to be proved the impossibility of an infinite body which is imperceptible by the senses. Moreover, upon further inquiry we shall find that his arguments are not conclusive in any respect, even with regard to a sensible body.

In the case of the *first* argument, based upon whereness, his opponent may contend that the places toward which the elements tend, though limited in kind, that is, the above and the below, are still unlimited individually, that is to say, those places exist one above the other *ad infinitum*.[97] The fact that there would be no absolute above will give rise to no impossibility, even though rectilinear motion is perceptible by the senses.[98]

As for the *second* argument, based upon weight and lightness, even if we admit the infinite body to be endowed with weight and lightness, the consequences he saw in his imagination will not

שדמה. וזה שלכל כובד וקלות זמן שרשי, אם מפאת האמצעי אשר
בו יתנועע, ואם להכרח היות התנועה בזמן. ולא יתחייב אם כן
כובד בעל תכלית מתנועע בזמן קטן מכובד בלתי בעל תכלית,
אבל יתחייב היות כובד נשם בעל תכלית מתנועע בזמן שוה לכובד
5 נשם בלתי בעל תכלית. ולא יקרה מזה בטול, למה שזה קרה מפאת
הכרח שמירת הזמן השרשי, אשר מפאת האמצעי ומפאת התנועה.
ולזה לא יתחייב שיתנועע הכובד הבלתי בעל תכלית בעתה כאשר
חשב.

והנה המופת השלישי, המיוסד על הפעל והפעלות, החיוב
10 אשר חשב, שאי אפשר לנשם הבלתי בעל תכלית שיניע מה שיש לו
תכלית, למה שאין יחס ביניהם, והיה ראוי שתהיה פעולתו בבלתי
זמן, אינו. וזה שלמה שאי אפשר לתנועה אלא בזמן, הוא מן ההכרח
שיהיה לתנועה זמן שרשי, אם נניח התנועה באנה. ואם נניח התנועה
באיך, הנה מהיות הבלתי בעל תכלית פועל ומשנה בזולת זמן, לא
15 יקרה ממנו בטול, ולא יהיה כנגד המוחש.

ולזה הוא מבואר שאין בכל מה שחתר לבאר המנעות נשם בלתי
בעל תכלית מפאת התנועה הישרה מחוייב.

ואמנם מפאת התנועה הסבובית, הנה הוא גם כן בלתי מחוייב,
להיותם בנויים גם כן על הנשם המוחש, ולאומר שיאמר, שיש שם
20 נשם בלתי בעל תכלית, והוא בלתי מתנועע בסבוב לסבות שזכר.
והנה כשנחקור בהם, נמצאם בלתי נותנים האמת בהם אף בגשם
מוחש.

follow. For every object that is described as heavy or light has some original time [in which to perform its motion], due either to the medium in which its motion takes place[99] or to the necessity of motion taking place in time.[100] It will not, therefore, follow that a finite weight will perform its motion in less time than an infinite weight. It will only follow that a body of finite weight and one of infinite weight will perform the same motion in equal time. But no impossibility will happen as a result of this, for this may be explained to come about as a result of the inevitable persistence of the original time, which, [as said above], is due either to the medium or to the nature of motion itself. Hence, neither will it follow, as he imagined, that an infinite weight will move in an instant.

As for the *third* argument, based upon acting and suffering action, the consequence he thought would follow, namely, that because there is no ratio between infinity and finitude, an infinite body could not produce motion in a finite body unless that motion was in no-time, does not follow. If the motion in question is that of place, it will always have that original time without which, as has been said, no motion is possible. And if the motion in question is that of quality, the inference that an infinite would act and produce change in no-time will lead to no impossibility,[101] nor is it contrary to sense perception.

It is thus clear that in all his attempts to prove the impossibility of an infinite body from rectilinear motion there is not a single argument that is conclusive.

As for the arguments from *circular* motion, they are likewise inconclusive, being again based upon the analogy of a [finite] sensible body. His opponent may, therefore, argue that while indeed there is an infinite body, it is incapable of circular motion for those very reasons given by Aristotle.[102] Upon further reflection, however, we shall find that the arguments do not prove his contention even with regard to sensible bodies.

זה שהמופת הראשון, מה שחייב בו סותר הנמשך, והוא
שהמרחק אשר בין שני הקוים בצד המקיף בלתי בעל תכלית, להיות
המרחק נוסף בתוספת הקו, ואחר שהקו מתוסף לבלתי תכלית
המרחק אם כן נוסף לבלתי תכלית, יש למערער שיאמר, המרחק

5 מתוסף כתוספת המספר, ושהתכלית בו לעולם שמור. וכבר יראה
זה, מפני שהידיעה בהפכים אחת, והנה כבר התבאר בספר
החרוטים אפשרות התקצר המרחק לבלתי תכלית, ויהיה המרחק
בו שמור בו לעולם. וזה שאפשר שיונחו שני קוים, שכל מה שיתרחקו
יתקרבו, ולא יתכן הפנשם לעולם, ואפילו יוצאו לבלתי תכלית.

10 הנה יש שם מרחק שמור לא יפסד, וכל שכן בתוספת, שאפשר
שיתוסף לעולם, ושיהיה התכלית שמור בו.

והוא האמת הגמור, שהמרחק בלתי בעל תכלית שבין שני קוים,
ואם הם בלתי בעלי תכלית, אין מציאות לו, להיות המרחק לעולם
מוקף, וכמו שיתבאר עוד מדברינו בגזרת השם. אלא שתחלה נבאר,

15 שאם היה היה החיוב שיסד בו סותר הנמשך אמת, היה מתחייב שיהיה
המרחק בלתי בעל תכלית ובעל תכלית יחד, ואף לא נניחהו
מתנועע. וזה שהמופתים שסדר הם בנויים על בטול התנועה
הסבובית לגשם הבלתי בעל תכלית, אבל אם נניחהו בלתי בעל

In the *first* argument, he proves the proposition which denies the consequent [by contending] that the distance at the circumference between any two radii [of an infinite sphere] must be infinite on the ground that the distance between radii increases in proportion to the elongation of those radii, concluding from this that wherever there is an infinite elongation of the radii there must be an infinite distance between them. To this the opponent may answer that distance increases [infinitely] in the same way as number[103] is said to increase [infinitely], namely, without ever ceasing to be limited. That the possibility of infinite increase is not incompatible with being actually limited may appear from the case of infinite decrease, for the examination into contraries is by one and the same science.[104] It has been demonstrated in the book on *Conic Sections*[105] that it is possible for a distance infinitely to decrease and still never completely to disappear. It is possible to assume, for instance, two lines, which, by how much farther they are extended, are brought by so much nearer to each other and still will never meet, even if they are produced[106] to infinity. If, in the case of decrease, there is [107] always a certain residual distance which does not disappear, *a fortiori* in the case of increase it should be possible for a distance, though infinitely increased, always to remain limited.

What we have just said is wholly in accordance with the truth, for an infinite distance between lines has no existence even when the lines themselves are infinite, inasmuch as a distance must always be bounded, as will appear in the sequel, God willing. But first we shall endeavor to show that if the reasoning by which he established the minor premise which denies the consequent were true, it would follow that the distance in question would be both infinite and finite at the same time—and this even if we do not assume that the infinite is capable of motion. For, according to him, the arguments are only meant to show that an infinite body could not have circular motion, whereas were we to assume an

תכלית בלתי מתנועע לא יקרה ממנו בטול, וכל שכן אחר שנתבאר
שחוץ לעולם בהכרח מילוי או רקות, ואיך שיהיה רחק בלתי בעל
תכלית נמצא, ואף אם לא ימצא, עלינו שנניחהו על צד שישתמש בו
המהנדס בגדר הקוים הנוכחיים, ובזולתו מהשרשים. ואולם איך
יתאמת שאם היה החיוב שעשאו אמת שיהיה המרחק בלתי בעל
תכלית ובעל תכלית יחד, הנה כפי מה שאומר. הנה אם יתחייב
בקוים בלתי בעלי תכלית היוצאים מהמרכז שיהיה המרחק ביניהם
בצד המקיף בלתי בעל תכלית, להיות המרחק נוסף בתוספת הקו,
הנה יתחייב זה בכל שני קוים היוצאים מהמרכז, ובאיזו זוית הזדמן.

וכאשר נצייר בצד המקיף אשר המרחק אשר ביניהם בלתי בעל תכלית,
ונרשום אצל הקו האחד בשיעור ידוע נקודה, אין ספק שאפשר לנו
להוציא קו מהנקודה הרשומה אל נקודת המרכז, למה שהוא מן
הידיעות הראשונות שאפשר להוציא קו ישר מכל נקודה אל כל
נקודה, ויחדש אם כן זוית ידוע, ואם היו בצד המקיף במרחק בעל
תכלית, וכבר הונח שכל הקוים היוצאים מאיזו זוית הזדמן יחדשו
בצד המקיף מרחק בלתי בעל תכלית, אם כן היה בעל תכלית
ובלתי בעל תכלית יחד. והשקר הזה יתחייב מהנחתינו החיוב אמתי.

אלא שהאמת הגמור שעם היות הקו בלתי בעל תכלית, לא
יתחייב מציאות מרחק בלתי בעל תכלית בין שני קוים. וזה שהוא

2 ואיך שיהיה] ואחר שהיה פ – רחק] שרוחק ליוקבאי.   3 ואף] או אף פ – (אם] פ – שהשתמש פ.
4 (המהנדס) פ – בנדרי א – הנכוחיים ב נ.   5 התאמת לד – שיהיה] שהיה לד.   6 (הנה) כפי בי.
7 הבב״ת ל.   8 מהמקיף ק – ולהיות פ.   9 הנה (אם] פ – (מה בכל שני] פ – קוים בקוים פ.
11 ונרשום] ונחשוב ל – האחד [ירצה בזה הטיתר] לד.   12 להוציא הקו] לד.   13–12 (מהנקודה...
קו] ב נ.   15 מאיזה צב – יתחדשו ב.   16 (כן] פ. – (היה] הוא לי.   17 בב״ת פ והבב״ת ר –
האמתי פ האמת זק בנ.   18 היותו פ.   19 שני] הב׳ לורדא – הקוים פ.

infinite body incapable of motion, he would find nothing impossible in the assumption of an infinite body. Moreover, according to what has been shown already, there must be outside the world either a plenum or a vacuum, in either of which cases there must exist an infinite distance. Or, if it does not actually exist, we may still assume its existence after the manner of the geometer who makes use of infinity in the definition of parallel lines,[108] and in the other hypotheses.[109] But how it could be shown, as we have suggested, that if his reasoning were correct it would result that the distance would have to be both infinite and finite at the same time, I will now explain by the following: If it were true that the distance between two infinite radii at their intersection with the circumference were infinite, on the ground that the distance between two emerging lines must increase in proportion to the elongation of those lines, that, of course, would have to be true in the case of any two radii emerging from the centre at any central angle whatsoever. Let us now imagine that, on the circumference between the radii which are infinitely distant from each other, we take a point at a certain distance from one of the radii. A line can undoubtedly be drawn from that point to the centre, for it is one of the postulates[110] that a straight line can be drawn between any two points. This line will make a certain central angle with the aforesaid radius, and at the same time the two lines will be at a finite distance from each other at the circumference. But the assumption is that any two radii, making any central angle whatsoever, would be infinitely distant from each other at the circumference. Hence the distance would be both finite and infinite at the same time. This absurdity will follow if we assume his reasoning to be true.

The real truth of the matter is that even if the radius in an infinite sphere is assumed to be infinite, it need not necessarily follow that there would have to be an infinite distance between two such radii. For it is evident that whatever point we may take

ידוע שהקו הבלתי בעל תכלית היוצא מן המרכז, אי אפשר שנרשום
בו נקודה, שלא יהיה הקו שבין הנקודה והמרכז בעל תכלית. ואחר
שהמרחק שבין הקוים אי אפשר להיות בלתי בעל תכלית אלא
אצל נקודה שיהיה בה הקו בלתי בעל תכלית, והנקודה ההיא אין
5 מציאות לה, אין מציאות אם כן למרחק הבלתי בעל תכלית שבין
שני הקוים. ובכלל שכשנאמר בקו שהוא בלתי בעל תכלית, כבר
אמרנו בו שאין לו קצה ותכלית, ואלו היה נמצא מרחק בלתי בעל
תכלית, היה ראוי שיהיה בקצה, והוא משולל הקצה. הנה מרחק
בלתי בעל תכלית בין הקוים אין מציאות לו. ואם היה שהגשם
10 בכללו יתנועע, והוא בלתי בעל תכלית, הנה לא יתנועע חלק ממנו
אלא על קו בעל תכלית. ואם היה זה רחוק מן הציור, הנה השכל
מחייבו.

וראוי שתדע שהחיוב הזה שחייבנו, היות המרחק שבין שני הקוים
הבלתי בעלי תכלית היוצאים מן המרכז בעל תכלית, יחייב היות
15 כל הסבוב שימצא במתנועע הזה בעל תכלית. וזה יתבאר בקלות.
למה שהזוית הבעל תכלית אשר אצל המרכז, כאשר חדשנו זויות
שוות לו אצלו, הנה היו בעלי תכלית במספר בהכרח, להיות
המרחק אשר אצל המרכז בעל תכלית, והיה המספר בעל תכלית,
חויב שיהיה המרחק בעל תכלית בהכרח.

20 וכאשר היה זה כן, התבאר שהחיוב שחשב לחייב בו סותר הנמשך
במופת הזה אינו אמת.

ובזה נתבטל המופת ‫ה ח מ י ש י‬.

והנה המופת ‫ה ש נ י‬ ו‫ה ש ל י ש י‬ ‫ו ה ש ש י‬, מיוסדים על חתוך

---

2 אחר פ.   3 הב״ת פ.   4 נקודה] הנקודה לד – שיהיה] שיש ר – (הקו ר – (הקו] ר – הבב״ת ל.   5 בין
צפזרדק.   6 קוים פ   – ובכלל] ויבטל ד – כשנאמר לורק ג: כשאם׳ ד.   7 שאין לח שהוא פ
(שאין לח) ז – המרחק פ באני.   8 הבב״ת לזרדק באני.   9 [שני] הקוים פר –
הגשם פ.   13 אשר חייבנו ב ג.   14 מהמרכז באני.   17 (בהכרח) בג – ולהיות לו.   20 לחייב]
לחיוב פ להייב ז – (בו) פ ג.   21 אינן אינגו באני.   22 החמישין הו׳ פצפזרדק באני הג׳ ל
שניתי על פי השערה, עיין פירושי האנגלי.

in the infinite radius, the line between that point and the centre will always be finite. Consequently, since the distance between two radii cannot be infinite unless it be between two points in those radii at which the radii themselves are infinite, and since there are no such points, it must, therefore, follow that there can be no infinite distance between those radii. Generally speaking, when we say of a line that it is infinite, we mean that the line has no extremity or limit, whereas an infinite distance [between infinite radii], if it existed, would have to mean the distance between the extremities of the infinite radii. But an infinite radius has no extremity. Hence there can be no infinite distance between the radii. And even though the sphere as a whole is capable of rotation, notwithstanding its being infinite, any given part of it performs its rotation on a finite axis.¹¹¹ This, to be sure, is remote from the imagination, but reason compels us to assume it.¹¹²

You may further know that the conclusion we arrived at, namely, that the distance between two infinite radii must always be finite, leads also to the conclusion that any distance which these radii may traverse in their revolution must likewise be finite. This can be easily demonstrated. If [in the argument in question] we draw around the centre a certain number of angles, each of them being equal to the finite central angle [formed by the infinite radii], the number of these new angles will have to be finite, inasmuch as the distance around the centre is finite. Now, since the number of the angles is finite, the distance [traversed by the radii] must likewise be finite.

This being the case, it is evident that the reasoning by which he tried to establish the minor premise in order to deny the consequent in this argument [i. e., the *first*] is unsound.

This also disposes of the *fifth*¹¹³ argument.

As for the *second*, *third* and *sixth*¹¹⁴ arguments, they are based upon the intersection of the infinite line by a revolving line,

קו מתנועע בסבוב, נכחי היה או לא לקו הבלתי בעל תכלית. והנה
למה שהתבאר המנעות חלק ראשון בתנועה, למה שחוייב כל
מתנועע כבר התנועע, הנה לא יתחייב מציאות נקודה ראשונה
מהפנישה. ולזה איננו רחוק שיפגוש הקו בשיעור בעל תכלית
5 בתנועה בעל תכלית, וזה להכרח קצה התחלת התנועה בזולת זמן.

והנה המופת ה ר ב י ע י מיוסד על ההקדמה האומרת שהגשם
הבלתי בעל תכלית המתנועע בסבוב יש לו תמונה סבובית, והוא
שקר, שאחר שהגשם בלתי בעל תכלית, הנה הוא נעדר הקצוות,
ולזה אין לו תמונה. וזה כי אם היה מהכרח התנועה בסבוב תמונה
10 סבובית, היה לזה מקום ספק, אבל כבר אפשר בכל תמונה
להתנועע בסבוב. וכאשר סלקנו מהגשם גבוליו, הנה סלקנו ממנו
התמונה, ולא יתחייב אם כן היותו בעל תכלית.

כבר התבאר מזה שאין בכל המופתים שסדר דבר יחייב סלוק
התנועה הסבובית בגשם הבלתי בעל תכלית, אבל התבאר מדברנו
15 אפשרות התנועה בגשם הבלתי בעל תכלית. וכבר יתבאר עוד חיוב
אפשרותה מהחוש. וזה שאנחנו נראה הגשם הניצוצי יתנועע בסבוב
בזמן בעל תכלית. והנה כאשר נדמה הקו הניצוצי בלתי בעל
תכלית, ונשתמש בו כאשר ישתמש המהנדס בו, הנה לא ימנע
משיתנועע תנועתו הבלתי בעל תכלית בזמן בעל תכלית, ואם כבר
20 ימשך הניצוץ לבלתי בעל תכלית. ואם היה שאין מציאות לבלתי בעל
תכלית, לפי סברת בעל הריב, הנה השכל יגזור שלא ימנע הניצוץ

1 בקו פ.   2 המנעת פ – החלק הראשון ל' חלק הראשון רד – שכל לד.   3 תחייב פ התחייב
א – (מציאות) ל.   4 קו פ.   5 ולזה לזד זה ר – קצת רג.   6 הרביעי] הה' פ צזוורדק כאן
הוא ל' שניתי על פי השערה, עיין פירושי האנגלי.   7 תמונה] תנועה פ.   8 הבב"ת זרד.
9 תמונה [וצורה] ל – בסבוב בתמונה רבג.   10 מקום ספק לזה ג – (כבר) ל.   11 סלקנו] חלקנו
סזרקו – סלקנו] חלקנו זרג.   13 וכבר לד – (מזה) פ.   14 לגשם צסלוורדק רבג – בבב"ת רד –
יתבאר א.   15 בב"ת לרדבג – התבאר פזרא – (עוד) פ.   18 ישתמש] נשתמש א – בו המהנדס
צלורדק כאן.   19 הבלתי בעל תכלית] הסבובית פ.   20 לבלתי] לבעל פ.

whether that line be assumed to be parallel[115] to the infinite line at the start or not.[116] Since, however, it has been shown that there can be no first part of motion, because every object that is moved must have already been moved, it does not follow, as he claimed, that there would have to be a first point of meeting.[117] It is not inconceivable, therefore, that the infinite line [in question] should meet the other line in a finite distance[118] with a finite motion,[119]— and this may be accounted for by the fact that the extreme beginning of motion must take place in no-time.[120]

As for the *fourth*[121] argument, it is based upon the proposition which states that an infinite body moving in a circle must necessarily have a spherical figure. This, however, is untrue, for if a body is conceived to be infinite it has no extremities, and thus it has no figure.[122] There would be some ground for his objection if circular motion required a spherical figure, but an object of any figure may have circular motion.[123] By conceiving, therefore, a body devoid of any boundaries, we conceive it also to be devoid of any figure, and so it does not follow that it would have to be finite.

All this has shown that among all the arguments he has adduced there is nothing which proves conclusively the impossibility of circular motion in an infinite body. Quite the contrary, our discussion has made it clear that motion is possible in an infinite body. This possibility may be further demonstrated by an argument from observation. We observe that a luminous body may complete a revolution in finite time. If we assume a ray of that luminous body to be infinite, allowing ourselves to make use of such an assumption after the manner of the geometer, we may conclude that it would not be impossible for that ray, though infinitely extended, to complete its infinite motion in finite time. Though according to the view of our opponent an infinite has no

מלהתנועע, אם היה אפשרות להיותו בלתי בעל תכלית. וזה מבואר
בנפשו.

ועוד כי אם היה שלא נדמה הניצוץ בלתי בעל תכלית, הנה לא
ימלט שלא ירשום נקודה בתנועתו בגודל הבלתי בעל תכלית,
5 שהתבאר מדברינו היותו מחוייב במלוי או ברקות, ולזה כאשר נדמה
בגודל ההוא קו בלתי בעל תכלית, נכחי לניצוץ מונח, הנה קצה
הניצוץ, כשיתנועע, ירשום נקודה בקו נכחית לקו הניצוצי.

ויתבאר מזה בקלות הפך מה שחייבהו במופתים אשר סדר.

ודי בזה העיון השלישי.

10 # העיון הרביעי

בחקירה במופתים שסדר לבאר באור כולל המנע מציאות גשם
בלתי בעל תכלית בפעל.

ואם הם בכח המופתים הקודמים, הנה המופת ה ר א ש ו ן, לא
יתחייב מהתנועה בסבוב שיש לו אמצע. וזה כי למה שהוא משולל
15 הקצוות, אין לו אמצע. והמופת ה ש נ י, כבר אפשר שיתנועע
בעצמו, ולא יתחייב שיהיו לו מוחשים מחוץ מקיפים. ושאר מה
שנאמר בו, התרם מבואר במה שנאמר.

התבאר מכל זה, שאין בכל מה שחשב לאמת ההקדמה הזאת
דבר מספיק. ולפי שהטעות שבהתחלות מביא אל הטעות שאחר

actual existence, still reason decrees that had it been possible for the ray to be infinitely extended, it would not thereby become incapacitated from having motion.[124] This is self-evident.

Furthermore, supposing that the ray were not infinite, still in the course of its revolution it would have to come in contact at a certain point with that infinite magnitude which, as has been shown in our discussion, must exist [outside the world] either as a plenum or as a vacuum. If we now imagine a certain infinite line in that magnitude parallel to the ray when at rest, the extremity of the ray, in its rotation, will have to meet that parallel line at a certain point. By this observation, then, we may easily establish the contrary of what he has been trying to show by the arguments which he has adduced.

This will suffice for the third Speculation.

THE FOURTH SPECULATION

Examination of the arguments which he has framed to demonstrate by a general proof the impossibility of an actually infinite body.

Though these arguments derive their force from the reasoning of the preceding arguments, it may be further urged in refutation of the *first* argument that circular motion does not imply the existence of a centre, for an infinite, having no extremities, likewise has no centre.[125] Again, in refutation of the *second* argument, it may be urged that the infinite may be moved by itself and still it will not follow that it would have to be surrounded by sensible objects from without. As for the remaining assertions made by him in this class of arguments, their refutation is evident from what has already been said before.

All this, then, shows clearly that in all his devices to prove this proposition [i. e., that an infinite magnitude is impossible] there is not a single argument which is convincing. And as an error in first principles leads to error in what follows on the first

ההתחלות, הביא זה לחייב שאין שם עולמות אחרים. וזה שהוא
חייב תחילה שאין חוץ לעולם מילוי ולא ריקות, וחייב שאלו היו
שם עולמות אחרים, היו היסודות מתנועעים מעולם אל עולם.
והוסיף הזיות ודברים מרבים הבל. ולמה שהטעות בהתחלה

5 מבואר, וזה שכבר התבאר במה שקדם חיוב מציאות גודל בלתי
בעל תכלית, וחיוב רקות או מלוי בלתי בעל תכלית חוץ לעולם,
הוא מבואר שמציאות עולמים רבים אפשרי. ולא יתחייב תנועת
היסודות מעולם אל עולם, וזה שכל אחד מהיסודות מתנועע תוך
מקיפו אל המקום הנאות לו. וכל מה שנאמר בזה לחייב ההמנעות

10 הבל ורעות רוח.

ולהיות האפשרות הזה אמת, אין ספק בו, ואין דרך אצלנו ומבוא
דרך החקירה לדעת אמתת מה שחוץ לעולם, מנעו חכמינו עליהם
השלום לדרוש ולחקור מה למעלה, מה למטה, מה לפנים,
מה לאחור.

15 וזה מה שראינו לחתום בו העיון הזה הרביעי בפרק הראשון.

הכלל הראשון, הפרק השני

בבאור ההקדמה השנית האומרת שמציאות גודלים אין תכלית
למספרם שקר, והוא שיהיו נמצאים יחד.

הנה אחר שבאר בהקדמה הראשונה המנעות מציאות גודלים
20 בלתי בעלי תכלית בשעור, באר בהקדמה הזאת השנית המנעות
מציאות גודלים בלתי בעלי תכלית במספר.

principles,[126] the implication of this proposition has led him to conclude that there are not any other worlds.[127] For having first proved to his own satisfaction that outside the world there is neither a plenum nor a vacuum, [he argued therefrom that there cannot be many worlds], and he [further] argued that if there were many worlds the elements would move from one world to another,[128] to which arguments he added many other fanciful speculations and 'words that increase vanity.'[129] But since the error of his initial premise is manifest, for it has already been shown before that an infinite magnitude must exist and that outside the world there must exist an infinite plenum or vacuum, it clearly follows that the existence of many worlds is possible. Nor can it be contended that the elements would move from one world to another, for it is quite possible that each element would move within the periphery of its own sphere towards its own suitable place.[130] Thus everything said in negation of the possibility of many worlds is 'vanity and a striving after wind.'[131]

Inasmuch as the existence of many worlds is a possibility true and unimpeachable, yet as we are unable by means of mere speculation to ascertain the true nature of what is outside this world, our sages, peace be upon them, have seen fit to warn against searching and inquiring into 'what is above and what is below, what is before and what is behind.'[132]

With this we deem fit to close the fourth Speculation of the first chapter.

PROPOSITION II

Part I.

PROOF OF the second proposition, which reads: 'The existence of an infinite number of magnitudes is impossible, that is, if they exist together'.[1]

Having shown in the first proposition that magnitudes cannot be infinite in measure, he now shows in this second proposition that they cannot be infinite in number.

ואמנם אמתות זאת ההקדמה יגיע במופתי ההקדמה הראשונה,

וזה שכל גודל יש לו שיעור מה, וכאשר הוספנו עליו גודל אחר, היה

מקובץ שעורם יותר גדול, וכאשר יוסיף גודלים בלתי בעלי תכלית

במספר, יהיה השיעור בלתי בעל תכלית, אשר התבאר המנעו.

5

הכלל השני, הפרק השני

בחקירה בהקדמה השנית האומרת שמציאות גודלים אין תכלית

למספרם שקר.

והוא מבואר שיסוד ההקדמה הזאת היא אמות ההקדמה הראשונה,

וכאשר התבאר בטול הראשונה, יתבאר בקלות בטול ההקדמה

10 הזאת השנית. אלא שיש לאומר שיאמר שאף בשלא תתאמת

הראשונה, תתאמת השנית מצד המנעות מספר בלתי בעל תכלית,

וזה בשנאמר כל מספר אם זוג ואם נפרד, והזוג והנפרד כל אחד

מוגבל ובעל תכלית, אם כן כל מספר בעל תכלית. והנה כבר קדם

לנו בפרק השלישי מהכלל הראשון שאין זה דעת הרב, גם אבוחמד

15 ואבן סינא מסכימים עמו.

והנה אבן רשד נתעורר בזה בביאורו לספר השמע. ומה שראוי

שיאמר בזה הוא שהמספר בפעל, רוצה לומר הספורים בשם מספר,

הנה הם מוגבלים, וכל מוגבל בעל תכלית בהכרח, אבל בעלי

As for the truth of this proposition, it can be established by the arguments employed in the proof of the first proposition. The reasoning may be stated as follows: Every magnitude is of a certain size. Now, if to any given magnitude we add another magnitude, their combined size will be greater. Consequently, if an infinite number of magnitudes were added together, their total size would be infinite. But a magnitude of infinite size has already been shown to be impossible.[2]

PART II.

EXAMINATION OF the second proposition, which reads: 'The co-existence of an infinite number of magnitudes is impossible'.

It is obvious that this proposition rests upon the proof of the first proposition. But inasmuch as the falsity of the first proposition has been demonstrated, this proposition, too, can be easily shown to be false.

One may, however, argue that even if the first proposition cannot be conclusively established, the second may still be demonstrated independently on the ground of the impossibility of an infinite number. That number cannot be infinite may be shown by the following reasoning: Every number is either even or odd; even and odd are each limited and finite; hence every number must be finite.[3] In answer to this we may refer to what has been shown above, in the third chapter of the first part, [Proposition III, Part I], namely, that this absolute negation of infinite number does not represent the view of the Master and that both Algazali and Avicenna are in agreement with him.[4]

The argument from odd and even has indeed been advanced by Averroes in his commentary on the Physics.[5] But in refutation of it, the following may be urged with telling effect: Actual number, i. e., things counted and numbered, is indeed limited, and every thing limited must needs be finite. But things which only

המספר, רוצה לומר אשר מדרכם שיספרו אבל אינם ספורים
בפעל, אין הבלתי בעל תכלית נמנע בהם, ולו הונח שיהיה זוג או
נפרד, וזה שכבר אפשר שיאמר זונים בלתי בעלי תכלית או נפרדים
בלתי בעלי תכלית.

5   אלא שהאמת הגמור הוא שהחלוקה למספר אל זוג ואל נפרד
הוא במספר הבעל תכלית המוגבל, אבל במספר הבלתי בעל
תכלית, למה שאינו מוגבל, הוא בלתי מתואר בזוג ונפרד. וכבר
העירונו בזה בפרק הנזכר.

הכלל הראשון, הפרק השלישי

10   בבאור ההקדמה השלישית האומרת שמציאות עלות ועלולים
אין תכלית למספרם שקר, ואם לא יהיו בעלי גודל, משל זה, שיהיה
זה השכל דרך משל סבתו שכל שני, וסבת השני שלישי, וכן אל בלתי
תכלית, זה גם כן מבואר הבטול.

        הנה אחר שבאר בהקדמה השנית המנעות מציאות בלתי בעל
15   תכלית בדברים אשר להם סדר במצב, בגודלים, באר המנעות
מציאותו בדברים אשר להם סדר בטבע, בעלות ועלולים, כי העלה
היא אשר בהמצאה ימצא העלול, ואם יצוייר העדרה לא יצוייר
מציאות העלול.

        ולזה השתלשלות עלה ועלול לבלתי תכלית נמנע. וזה שהעלול
20   הוא אפשרי המציאות בבחינת עצמו, והוא צריך אל מכריע

3 שיאמר] שיהיו פ – זונית פלורד קג א, זונות זו – נבלתי בעלי תכלית] פ – נאו פי.   4–3 נפרדים
בלתי בעלי תכלית] ז.   7 או בנפרד לד, נפרד ס, או נפרד זר.   8 העירונו] הנחנו ל.   11 יהיו]
היו פ.   12 השלישי] סזוג נ.   14 שבאר] שנמר י.   19 ועלול] לעלול לר.   20 מכריח זרא.

possess number, that is to say, which have the capacity of being numbered but are not actually numbered,[6] even though assumed to have the distinction of even and odd, are not excluded from the possibility of being infinite, for infinity may be predicated of even numbers or of odd numbers.[7]

The real truth of the matter, however, is that the division of number into even and odd applies only to a finite and hence limited number; but infinite number, inasmuch as it is unlimited, does not admit of the description of even and odd.[8] We have already discussed this distinction in the aforementioned chapter.

PROPOSITION III

Part I.

Proof of the third proposition, which reads: 'The existence of an infinite number of causes and effects is impossible, even if they are not magnitudes. To assume, for instance, that the cause of a given Intelligence be a second Intelligence, and the cause of the second a third, and so on to infinity, can be likewise demonstrated to be impossible'.[1]

Having shown in the second proposition the impossibility of an infinite [number] with reference to objects which have order in position, namely, magnitudes, he now shows that it is likewise impossible with reference to objects which have order in nature, namely, causes and effects,[2] for by a cause is meant that the existence of which implies the existence of an effect and should the cause be conceived not to exist the effect could not be conceived to exist.[3]

It is because of this relation between cause and effect that an infinite series of causes and effects is impossible. The argument may be stated as follows: An effect by its own nature has only possible existence, requiring therefore a determinant to bring about

יכריע מציאותו על העדרו, אשר המכריע ההוא הוא עלתו. ולזה
השתלשלות עלות ועלולים לבלתי תכלית לא ימלט כללם מהיותם
כלם עלולים אם לא. ואם היו כלם עלולים, הנה הם אפשרי
המציאות, ולפי שהיו צריכים אל מכריע יכריע מציאותם על
5 העדרם, הנה להם עלה בלתי עלולה בהכרח. ואם לא היו
עלולים כלם, הנה אחד מהם עלה בלתי עלולה, אשר הוא תכלית
ההשתלשלות, וכבר הונח שלא היה לו תכלית. זה שקר בטל.
והשקר הזה התחייב בהניחנו עלות ועלולים אין תכלית למספרם.

וצריך שנתעורר, שלא חייב המנעות בלתי בעל תכלית אלא
10 לדברים שיש להם סדר במצב, בגודלים, או בטבע, בעלות
ועלולים, אבל בדברים אשר אין להם סדר במצב ולא בטבע.
בשכלים או בנפשות, הנה לא ימנע מציאותם בלתי בעל תכלית.
וזה הוא דעת אבן סינא ואבוחמד. ואולם אבן רשד יראה ההמנעות
גם בדברים שאין להם סדר, כי הוא אמר שהמספר בפעל הוא בעל
15 תכלית בהכרח. וזה שכל מספר בפעל הוא ספור בפעל, וכל ספור
בפעל הוא אם זוג ואם נפרד, ומה שהוא זוג או נפרד הוא בעל תכלית
בהכרח.

ומה שיראה לנו בזה הוא, שהחלוקה הזאת למספר היא אמתית,
אין המלט ממנה, אבל המספר הבלתי בעל תכלית, אחר היותו
20 בלתי מוגבל, לא יתואר בזוגיות והפרדה, ולזה אין הבלתי בע׳

the preponderance of existence over non-existence, which determinant constitutes its cause. Now, it must inevitably follow that in the aggregate of an infinite series of causes and effects either all the members of the series would be effects or some of them would not be effects. If they were all effects, they would all have possible existence. They would require some determinant to bring about the preponderance of existence over non-existence, and so they would necessarily presuppose the existence of a causeless cause [outside the series]. And if they were not all effects, one of them at least would then be a causeless cause, which one would thus mark the end of the series. But the series is assumed to be endless. Hence an impossible contradiction. And this contradiction ensues because we have assumed the existence of an infinite number of causes and effects.[4]

We must observe, however, that the possibility of infinite number is denied by the author only with reference to objects which have order either in position, as magnitudes, or in nature, as causes and effects; he does not deny its possibility with reference to objects which have no order either in position or in nature, as, for instance, intellects or souls.[5] This is in accordance with the view of Avicenna and Algazali.[6] Averroes, however, finds it to be impossible even with reference to objects which have no order whatsoever,[7] for he maintains that actual number must necessarily be finite. He reasons as follows: Every actual number is something actually numbered, and that which is actually numbered must be either even or odd, and that which is even or odd must necessarily be finite.[8]

For our own part, we will say this with regard to Averroes' argument: While indeed the division of number into odd and even is true and unavoidable, still infinite number, not being limited, is not to be described by either evenness or oddness.[9] And so an infinite number is not impossible in the case of intellects and souls. It is for this reason that in his propositions about the im-

תכלית נמנע בו. ולזה מה שדקדק הרב בהמנעות המספר הבלתי
בעל תכלית בדברים שיש להם סדר במצב, בגודלים, או בטבע,
בעלות ועלולים, בשיהיה האחד עלה לשני והשני לשלישי, וכן
לבלתי תכלית.

הכלל השני, הפרק השלישי

בחקירה בהקדמה השלישית האומרת שמציאות עלות ועלולים
אין תכלית למספרם שקר.

ואומר שהמופת אשר סדר אלתבריזי בזה, אשר העירונו עליו
בפרק השלישי מהכלל הראשון, והרמוז במאמר השמיני מספר
השמע ובמה שאחר, בלתי מספיק לפי דעת הרב. וזה שהוא לא יחייב
המנעות מספר בלתי בעל תכלית אלא לדברים שיש להם סדר
והדרגה במצב או בטבע, ולזה אפשר בשכל אחד שיהיה עלת
שכלים בלתי בעלי תכלית במספר. ובכלל אין המנעות מציאות
עלולים בלתי בעלי תכלית מעלה אחת, אם היה אפשר לעלה אחת
אצילות יותר מעלול מעלול אחד. ואחר שאין המנעות לעלולים להיות
בלתי בעלי תכלית, ואם להם עלה לכללם, הנה אם כן לא יחייב
מציאות העלה לכללם המנעות הבלתי בעל תכלית לעלולים. ולזה
כאשר נניח עלות ועלולים, בשיהיה האחד עלה לשני, והשני לשלישי,
וכן לעולם, מי יתן ואדע, בשנניח לכל אלו עלה אחת, איך יחייב
מציאותה המנעות הבלתי בעל תכלית לעלות ועלולים. וזה שלא

possibility of infinite number the Master has specifically confined himself to objects that have order either in position, as magnitudes, or in nature, as causes and effects, when these are so arranged that the first is the cause of the second, the second of the third, and so on to infinity.

PART II.

EXAMINATION OF the third proposition, which reads: 'The existence of an infinite number of causes and effects is impossible.'

I say that the argument framed here by Altabrizi, which has been discussed by us in the third chapter of the first part, and of which there is a suggestion in the eighth book of the *Physics*[10] and in the *Metaphysics*,[11] is not altogether sufficient, considering the particular view espoused by the Master. For the Master, as has been shown, does not preclude the possibility of an infinite number except in the case of things which have order and gradation either in position or in nature. According to this, it will be possible for one Intelligence to be the cause of an infinite number of other Intelligences. On general principles, it must be admitted that the emanation of an infinite number of effects from one single cause would not be impossible, if it were only possible for a single cause to be the source of emanation of more than one effect.[12] And so, inasmuch as it is evident that there can be an infinite number of effects, despite their all being dependent upon a common cause, it must follow that the assumption of a common cause for more than one effect would not make it impossible for those effects to be infinite in number. This being the case, assuming now a series of causes and effects wherein the first is the cause of the second and the second of the third and so on for ever, would that I knew why, by the mere assumption of a common cause for the series as a whole, the number of causes and effects within that series could not be infinite? That their infinity is impossible on

יחוייב זה מצד היות עלה ראשונה לכלם, שהוא בהנחתנו עלולים

בלתי בעלי תכלית כבר נודה בעלה ראשונה לכלם, והוא מבואר

שלא ימנע היותם בלתי בעלי תכלית, אחר שאין המנע בלתי בעל

תכלית במספר בדברים שאין להם סדר במצב או בטבע. והנה

5 כשנניח גם כן העלולים ההם הבלתי בעלי תכלית כל אחד עלה

לחברו לא יקרה מזה שום בטול, אלא שאנו צריכים לדבר יכריע

מציאותם על העדרם, אחר שכלם אפשרי המציאות, ואנחנו כבר

נודה בעלה הראשונה אשר לא יתחייב התכלית לזולתה מהעלולים,

והיא המכרעת מציאותם.

10 וכבר חתר קצת המפרשים לאמת ההקדמה הזאת בשאמר, זה

לשונו: כי מה שלא יגיע בעצם אם לא בקדימת מה שאין לו סוף הנה

לא יגיע, ואי אפשר שימצא, עד כאן. והנה אם היתה הקדימה זמנית,

היה מקום לטענה הזאת, ואם כבר תקבל המחלוקת, למה שאנחנו

נראה שמה שלא יגיע אם לא בהקדמת מה שאין סוף לו הנה יגיע,

15 כאלו תאמר, על דרך משל, שהיום הזה שאנחנו בו הגיע, ואם לא

הגיע אלא בקדימת מה שאין סוף לו, לאומרים בקדמות העולם,

אלא שזה במקרה, ושנודה באפשרות שבמקרה ובהמנעות אשר

1 יחייב פזוכנ, יתחייב א — (זה) צזורק בזאנ — העלה צספלזוורד — הראשונה לדנ — לכלם] לכללם פ — הראשונה לודג.   5 (ההם) פ — כל] בכל ג.   6 (שום) זר — יכריח פ.   9 המברחת זר.

10 חתרו פ בנ — מהמפרשים לודגא — בשאמרו פ.   11 סוף לו באור נרבוני.   12 הנה לד — היתה] תהיה ד.   14 (לה) לוק בנ.   16 בקדימת] בקדמות פ בהקדמת א.

the ground of the dependence of the entire series upon a first
cause is without any justification, for assuming, as we did before,
the existence of an infinite number of effects, [which are not interrelated among themselves as cause and effect], we likewise posit a
first common cause for all the effects, and yet, we have shown,
that those effects can be infinite, inasmuch as an infinite number
is not impossible in the case of things which have no order in
position or nature. By the same token, no impossibility will
happen if we assume those infinite effects to be each successively
the cause of the other. To be sure, it will be necessary for us
[to posit at the beginning of the series] something [uncaused] to
bring about the preponderance of the existence over the non-
existence [of the causes and effects within the series], since [by
themselves] they all have only contingent existence. But still, we
have already admitted the possibility of a first common cause
which would not necessitate that the effects proceeding from it
should be finite, even though it would bring about the existence
of those effects.[13]

A certain one[14] of the commentators has attempted to prove
this proposition by an argument which we quote verbatim: 'That
which cannot be realized[15] by itself, unless it be preceded by
something infinite, will never be realized and cannot come into
existence.'[16]

Now,[17] if the 'precedence' [implied in Maimonides' proposition]
were of a temporal nature, there might be some room for this reasoning,[18] though, I must say, even in temporal precedence the
argument is not wholly immune from criticism. For we see that
that which cannot arrive except by the precedence of what is
infinite does actually arrive: thus, for instance, the present day
in which we are is here, even though its arrival, according to the
view of those who believe in the eternity of the universe, had to
be preceded by something infinite. Indeed, it may be rejoined
that in that case the precedence was only accidental.[19] But still,

בעצם צריך האמתה. אבל כשנודה בחלוק הזה בקדימה אשר
בזמן, אין מקום לו בקדימה אשר בסבה, אחר שהם יחד בזמן, כי
אחר שהדברים יחד בזמן אחד, מי חייב המנעות בשיהיה כל אחד
עלה לאחר ואפשרות בהיותם כלם עלולים, אחר שנודה באפשרות
5 היותם בלתי בעלי תכלית יחד.

אלא שהמכוון מזאת ההקדמה, ומה שאנו צריכין ממנה, הוא
מציאות עלה ראשונה בלתי עלולה, היו העלולים בלתי בעלי
תכלית וכל אחד עלה לחברו או בעלי תכלית.

הכלל הראשון, הפרק הרביעי

10 בבאור ההקדמה הרביעית האומרת שהשנוי ימצא בארבעה
מאמרות, במאמר העצם, והוא ההויה וההפסד, ובמאמר הכמה,
והוא הצמיחה וההחסרון, ובמאמר האיך, והוא ההשתנות, וימצא
במאמר האנה, והוא תנועת ההעתק, ועל זה השנוי באנה תאמר
התנועה בפרט.

15 הנה למה שהשנוי ממנו בזמן וממנו בזולת זמן, כשילקח השנוי
סתמי בשלוח, תתאמת זאת ההקדמה. והיא כמבוארת בעצמה,
כי השנוי אשר בכמה ובאיך ובאנה הוא בזמן, והשנוי אשר בעצם
הוא בזולת זמן, כמו שהתבאר בספר ההויה וההפסד.

ומה שצריך שנתעורר עליו, למה ייחד אלו הארבעה מאמרות,

to admit that something is possible when accidental and to deny its possibility when essential, needs to be demonstrated.[20] Granted, however, that the distinction between accidental and essential holds true in the case of things which precede one another in time, it has no place in the case of things which precede one another only as causes, but co-exist in time. Admitting, therefore, as we must, that things which co-exist in time can be infinite in number, by what show of reason can we confine that possibility only to things that are all equally the effects of one cause and deny that possibility of the same effects when they are arranged among themselves as the effects of each other?

But what this proposition really means to bring out, and what conclusion thereof is actually needful for our purpose, is the fact that there must exist a first cause, which is uncaused by anything else, regardless of the view whether its effects, when they are one the cause of the other, are infinite or finite.[21]

PROPOSITION IV

PROOF of the fourth proposition which reads: 'Change exists in four categories: in the category of substance, which is generation and corruption; in the category of quantity, which is growth and diminution; in the category of quality, which is alteration; and in the category of place, which is the movement of translation. It is this change in place that is called motion proper'.[1]

Inasmuch as some kinds of change are in time while others are in no-time, by taking the term change in an unrestricted, absolute[2] sense, the proposition will have been proved to be true. [That the term change is to be here so understood] is quite self-evident, for change in the categories of quantity, quality, and place is in time, whereas that in the category of substance is in no-time,[3] as has been shown in the book *De Generatione et Corruptione*.[4]

The following argument, however, may be urged against the author. Why did he enumerate only these four categories, when as

והוא מבואר שהשינוי כבר ימצא בשאר המאמרות, כאלו תאמר,

במאמר המצב ושיפעל ושיתפעל. אלא שלמה שלכל שינוי שתי

בחינות, אם מצד הנושא, והוא העתק המשתנה מתאר אל תאר,

ובבחינה הזאת היא בשאר המאמרות, והוא שינוי בזולת זמן, ואם

5 מצד חמר השינוי, כאלו תאמר, בכמות ובאיכות ובאנה, ובבחינה

הזאת היא במאמר אשר בו חומר השינוי, והוא פונה בעיין הזה אל

הבחינה הזאת. והיה השינוי אשר בעצם נמשך לתנועה אשר באלו

המאמרות, ייחד הרב אלו הארבעה מאמרות. ודרך בזה דרך

ארסטו בספרו במה שאחר. והוא הנכון מה שיאמר בזה לפי

10 מה שיראה.

אלא שנשאר עלינו לבאר, למה ייחד השינוי באנה, שהוא ההעתק,

לתנועה בפרט, אחר שהתנועה בכמה היא השינוי באנה גם כן, אחר

שבו העתק מה. וכבר נתעורר אלתבריזי מזה ואמר, כי להיות

ההעתק באנה מוחש, יחד לו התנועה, ולא יחד אותה לצמיחה, כי

15 ההעתק בה איננו מוחש. ולפי מה שיראה בצמיחה, אין בו העתק

באנה, למה שידוע שהצמיחה בצומח הוא בכל קטריו, ולזה לא

נשאר חלק רמז אליו שיתאמת בו העתק מאנה אל אנה. ולזה יחד

הרב התנועה אל ההעתק באנה.

a matter of common knowledge change exists as well in the other categories[5], as e. g., position[6], action and passion?[7] [The solution of this difficulty may be given as follows]: Every change has two aspects[8]. First, it may be regarded with respect to the sub-stratum, in which case change means the transition of that which underlies the change from one accident to another[9]. In this respect, change exists in the other categories[10], and is in no-time. Second, change may also be regarded with respect to the matter of the change, that matter being, e. g., quantity, quality, and place[11]. In this respect it exists in that category in which the matter of the change is to be found[12]. It is change in this latter respect that the author has in mind in this proposition[13]. But inasmuch as change in the category of substance is consequent upon the motion existing in those [three] categories[14], the author has enumerated those four categories. In this he has followed the path trod by Aristotle in the *Metaphysics*.[15] This would seem to be the right[16] solution of the difficulty.

There still remains for us to explain why he has restricted the use of the term motion proper to change in the category of place, that is, to translation, when, as a matter of fact, motion in the category of quantity is likewise a change in place, inasmuch as it always entails some act of translation.[17] This question has already been raised by Altabrizi,[18] in answer to which he says that the term motion proper is applied by the author to loco-motion because the act of translation therein is perceptible; but he does not apply it to growth because the act of translation therein is not perceptible. It would seem, however, that in growth there is no translation in place at all, for plants, as is well known, grow in all directions, and consequently there is no definite part therein of which translation from one place to another can be truly affirmed.[19] It is for this reason that the Master has re-stricted the use of the term motion proper to translation in place.

הכלל הראשון, הפרק החמישי

בביאור ההקדמה החמישית האומרת שכל תנועה שינוי ויציאה
מן הכח אל הפעל.

הנה אמרו שכל תנועה שינוי הוא מבואר לפי מה שקדם, אבל לא
5   יתהפך זה. וזה שאין כל שינוי תנועה, למה שהשינוי ממנו שיהיה
בזולת זמן, כמו ההויה וההפסד והעתק הנושא מתואר אל תואר, אשר
מזה הצד יכנס במאמר שיפעל וישתפעל. אבל השנוי ממנו אשר
הוא בבחינת חמר השינוי אשר בו יצדק שם התנועה לבד. ודעהו,
כי לא הרגישו בזה החלוק מהמון המתפלספים.

10   ואמנם אמרו שהוא יציאה מן הכח אל הפעל, הוא נמשך למה
שגדרו התנועה שהיא שלמות מה שבכח מצד מה שהוא בכח. והנה
יצדק עליו שהוא שלמות, למה שהתנועה בין מה שממנו ומה שאליו,
וכשהיה במה שממנו, היה בכח גמור והוא נח, וכשיהיה במה שאליו,
היה לו שלמות גמור והוא נח, וכשהוא במה שבין, הנה הוא שלמות
15   מה, אבל מצד שהוא עדיין בכח, ולזה אין לו שלמות גמור. ולזה
התאמת שהתנועה יציאה מן הכח אל הפעל.

ואולם כבר יראה שהגדר הזה איננו אמתי לתנועה, למה שמסגולת
הגדר ההתהפך על הנגדר, כמו שהתבאר בספר המופת, ולפי
שהגדר הזה כבר יצדק גם כן בהנעה, יתחייב אם כן שיהיה תנועה,

7 אבל השנוי (ממנו) אשר לורקבאני אבל (השנוי ממנו) אשר סד (אבל) השנוי ממנו אשר פ
(אבל) השנוי וממנו אשר צז.   8 (אשר) פלורדקבאני – (בו) פ– יצדק) שיצדק פ יצטרך ד.
9 ההחלוק] החלוף פ.   10 (הוא] והוא לקבג.   12 שאליו (התנועה)] ג.   13 וכשהיה] וכשהיא לד
וכשיהיה צקבג – היה] היא לד – (והוא נח) הוא נח) פר – שאליו (התנועה)] ג.   14 היה] הנה לד – (לה] פ כ.
19 גם (כן) בג – אם כן] אם כן גם כן ב.

PROPOSITION V

PROOF of the fifth proposition which reads: 'Every motion is a change and transition from potentiality to actuality'.[1]

His statement that every motion is a change is evident from what has been said before. The proposition, however, is not convertible[2], for not every change is motion, inasmuch as there is a kind of change that takes place in no-time, as, e. g., generation and corruption and the transition of the substratum from one accident to another, in which latter respect, change is to be included under the categories of action and passion.[3] But still change may also be regarded with respect to the matter of the change, to which alone applies the term motion proper. Bear this in mind, for none of the host of philosophizers has noted this distinction.[4]

As for his statement[5] that motion is a transition from potentiality to actuality, he follows the definition generally given of motion, namely, that it is the actuality[6] of that which is in potentiality in so far as it is in potentiality.[7] There is a justification for describing motion as an actuality. For motion takes place between a *terminus a quo* and a *terminus ad quem*. Accordingly, when it is yet in the *a quo*, it is in a state of complete potentiality, and is thus at rest; when it is already in the *ad quem*, it has a complete actuality, and is again at rest. It is only when it is in the interval that it is an actuality in some respect, but that only in so far as it is still potential. Thus it has no complete actuality.[8] Hence it has been demonstrated that motion is a transition from potentiality to actuality.

It would seem, however, that this is not a true definition of motion. For one of the characteristics of a definition is that it is convertible into the *definiendum*, as has been shown in the *Posterior Analytics*.[9] Since the foregoing definition will also apply to motivity, it will follow that motivity is motion, and will thus

ויצטרך גם כן אל הנעת מניע, וזאת ההנעה השנית גם כן תנועה, וזה לבלתי תכלית.

ולזה היה הגדר האמתי לפי מה שיראה לנו הגדר האחר אשר זכרו, והוא שלמות המתנוע במה שהוא מתנועע. והנה אמרו שלמות

5 יורה שאיננו בכח גמור, אבל שיש לו פעל ושלמות מה. ואמרו במה שהוא מתנועע יורה שאין לו פעל ושלמות גמור.

ואיך שיהיה הגדר, ההקדמה אמתית, שכל תנועה שינוי ויציאה מן הכח אל הפעל.

הכלל הראשון, הפרק הששי

10 בבאור ההקדמה הששית האומרת שהתנועות מהם בעצמות, ומהם במקרה, ומהם בהכרח, ומהם בחלק. אולם אשר בעצמות, כהעתק הגשם ממקום למקום, ואולם אשר במקרה, כמו שיאמר בשחרות שהוא בגשם שהועתק ממקום אל מקום, ואולם אשר בהכרח, כתנועת האבן אל המעלה במכריח יכריחה על זה, ואולם

15 אשר בחלק, כתנועת המסמר בספינה, כי כאשר התנועעה הספינה, נאמר שכבר התנועע המסמר גם כן, וכל מחובר יתנועע בכללו יאמר שחלקו כבר התנועע.

הנה המכוון בהקדמה הזאת שהתנועה לה בחינות. וזה אם עצמותית, טבעית היתה או הכרחית, ויכנס בזה הרצונית, כ ה ע ת ק

20 ה ג ש ם מ מ ק ו ם א ל מ ק ו ם; או מקרית, שניחס התנועה לדבר שאין

---

3 שיראה] שיורה ר – אחר פ.   4זכרנו פזכרה קבנ – שהוא] הוא ובנ – (והנה) פ – (ואמרו פ –
שלמות [המתנועע] פ.   5 שאיננו בכח גמור אבל] פ – (שיש) פ – לוורדקבא – (אבל... מה) נ.
6שהוא] הוא בנ.   11 מהם באנ – מהם באנ – מהם באנ.   12גשם פבאנ.   14 מעלה ו –
יכריחנה פ.   15 (כתנועת) פ – כמסמר ד – [אשר] בספינה פ – (כאשר) פ – כשנתנועעה פ.
16 וכן כל פ – המחובר זרא   17 בחלקו פ – שכבר חלקו ל – כבר] נ"כ פ.   20 שניחס] שהניח צ.

require a motive agent for its motion. But that second motivity will likewise be motion, and this will have to go on to infinity.[10]

It seems to us, therefore, that the true definition of motion is the other definition mentioned by Aristotle, namely, that it is the actuality of that which is movable in so far as it is movable.[11] His use of the term 'actuality' is meant to indicate that motion is not complete potentiality, but that it has some degree of energeia and entelecheia.[12] His use of the qualification 'in so far as it is movable' is likewise meant to indicate that it has not a complete energeia and entelecheia.

But, however the definition may be phrased, the proposition remains true, namely, that 'every motion is a change and transition from potentiality to actuality.'

PROPOSITION VI

PROOF of the sixth proposition which reads: 'Of motions some are according to essence, some are according to accident, some are according to violence, and some are according to part[1]. Motion is according to essence, as when a body is translated from one place to another. It is according to accident, when, e. g., blackness which exists in a body is said to be translated from one place to another. It is according to violence, as, e. g., the motion of a stone upward brought about by a certain force applied to it in that direction. It is according to part, as, e. g., the motion of a nail in a boat, for when the boat is moved we say that the nail is likewise moved; and similarly, when something composed of several parts is moved as a whole, every part of it is likewise said to be moved.'[2]

The purpose of this proposition is to show that motion is classifiable.[3] First, essential, 'as when a body is translated from one place to another'[4], which may be either natural or violent, and voluntary motion, too, is to be included in this class. Second,

מדרכו שיתנועע מעצמו אלא שיתנועע במקרה, כמו שיתנועע

השחרות אשר בגשם בתנועת הגשם; ואם הכרחית, עצמותית היתה

או מקרית, כתנועת האבן למעלה; ואם אשר בחלק, הכרחית

היתה או טבעית. וההפרש אשר בין המקרית ואשר בחלק, שהמקרית

5 היא כשניחס התנועה אשר במקרה לדבר שאין מדרכו שיתנועע, ואשר

בחלק הוא שניחס התנועה אשר בחלק לדבר שמדרכו שיתנועע.

אבל מה שצריך להתעורר עליו, אמרו במשל התנועה אשר

בעצמות, כהעתק הגשם ממקום למקום. ולפי שבתנועת

הגלגל לא יעתק גשם הגלגל ממקום למקום, למה שלא ימיר המקום

10 בכללו, ואמנם יעתקו חלקיו, הנה לא תהיה התנועה עצמית לכללו

כי אם לחלקיו. וזה חילוף מה שיראה. כי התנועה לגלגל היא אם

רצונית תשוקיית לפי דעת ארסטו ואם טבעית למה שיראה לנו. וזה

שלמה שאנחנו נראה שהתנועה טבעית בגשמים בכלל, והיו הגשמים

הפשוטים אשר תחת הגלגל היסודיים בעלי כובד וקלות מתנועעים

15 תנועה ישרה, הנה גשם הגלגל בכללו, שאינו מתואר בכובד וקלות,

התנועה הטבעית לו הסבובית. ולזה היתה התנועה הסבובית לגלגל

עצמותיית, ואם היה שלא יעתק הגלגל ממקום אל מקום בכללו,

בחלוף מה שיראה מדברי הרב.

1 בעצמו ס׳.   5 דרכו ס׳.   7 שנתעורר לד שיעורר להתעורר א.   11 וזה] ולזה ל – (אם) ס׳.
12 למה] לפי מה סלודבא.   14 יסודיים א.   16 הטבעית לו היא הסבובית לד הטבעית
הסבובית ג׳ הטבעית לסבובית י.

accidental, as when we attribute motion to something which cannot be moved essentially, but is moved accidentally, as, e. g., the blackness in a body which is moved by the motion of the body.[5] Third, violent, which may be either essential or accidental, 'as, e. g., the motion of a stone upward'.[6] Finally, according to part, which may be either violent or natural.[7] The difference between 'accidental' and 'according to part' may be stated as follows: It is 'accidental,' when we attribute motion as something accidental to an object which ordinarily is incapable of independent motion. It is 'according to part,' when we attribute motion as something participated by an object which ordinarily is capable of independent motion.[8]

What we ought to animadvert upon him for is his statement in the illustration of essential motion, namely, 'as when a body is translated from one place to another.' According to this illustration, in the case of the motion of the [celestial] sphere, where the body of the sphere is not translated from one place to another, inasmuch as it is only[9] its parts that are so translated whereas the sphere as a whole does not change its place, it will follow that only the parts will thus have essential motion but not the whole.[10] This is contrary to what seems to be the truth. For the motion of the sphere is voluntary [or] appetent, as is Aristotle's view, or natural, as seems to us. For we are of the opinion that motion of whatever description is natural to all the elements [whether sublunar or translunar]. That the simple translunar elements are moved with rectilinear motion is due only to the fact of their having weight and lightness. The common substance of the celestial spheres, therefore, not being endowed with either weight or lightness, has motion in a circular direction as its natural motion. Thus [according to either view] the circular motion of the sphere must be essential, even though the sphere as a whole is not translated from one place to another, contrary to what would seem to be implied in the Master's statement.[11]

וכן אמרו במשל אשר במקרה, בשחרות אשר בגשם, אשר
הוא בגודל מה, והוא נעתק מגודל אל גודל. כבר יאמר התנועה
אשר במקרה גם בנקודה אשר בתכלית הגשם, ואם איננה בגודל כי
אם בתכלית.

5 ואולם אמרו במשל אשר בהכרח, כתנועת האבן אל
המעלה, נמשך לדעת היוני המפורסם, אשר ליסודות תנועות
טבעיות הפכיות, כתנועת האבן אל מטה ותנועת האש למעלה,
ושפטו מפני זה, שהיסודות הארבעה, לאחד מהם, והוא הארץ, כובד
מוחלט, ולאש קלות מוחלט, ולאויר ולמים כובד וקלות צרופי.
10 והדעת הזה לפי מה שיראה לא התבאר ולא יתבאר. וזה שיש לאומר
שיאמר, שלכל אחד מהיסודות כובד מה, אלא שהם מתחלפים
בפחות ויתר. ואמנם היתה תנועת האש למעלה, לכובד האויר
אשר ידחה אותו למעלה, כאשר יקרה לאבן תוך הכור אשר
בו זהב או עופרת מותך או כסף חי שיתנועע אל המעלה, למה
15 שכובד המתכות ידחהו. ובדמות זה יקרה אל האויר ואל המים.
וכבר יראה זה, למה שכאשר חפרנו בארץ, ירד בחפירה האויר,
ונתמלאת ממנו. ואם היה שאפשר לטוען שיטעון שזה ממנו להמנע
הרקות תוך הגלגל, אבל איננו נמנע שיהיה זה לכובד היסוד ואולם
איך שתהיה תנועת האבן למעלה, הנה על כל פנים מפאת מכריח,
20 כמו שבא במשל.

5 [הוא] בהכרח ס ל ו ו ר ד ק ב א נ. 6 [הוא] נמשך ק ב א נ. 7 אל מטה] למטה ל ו ד אל המטה
ר ק ב א נ – (ותנועת) ל ד – והאש ל ד – למעלה] אל המעלה ז ר ק ב א נ. 9 ולאש] והאש ל ו ו ר ד ק ב א נ –
והאויר והמים ז ר ק ב א נ והמים והאויר ס ל ו ד – צרופי] בצרוף ס. 12 ויותר ס – היתה] היות
ס ז ר א. 14 (חי) פ – מעלה פ. 16 (מה) ז ר ק. 17 ונתמלאה י ותמלא ב א נ ונמצא ז –
אפשר ר א – (לטוען) ז ר ק ב נ – שיטען] למעון ז ר ד ק ב א נ. 18 מתוך ז ר ק א נ – היסודי א. 19 הנה]
הוא י – על] עם ג – מפאת] מופת ס ז ק ב א נ – מכריע ק י.

Again, in his illustration of accidental motion, he uses the phrase 'blackness which exists in a body.' This would seem to imply that there can be no accidental motion except of something residing in some magnitude and capable of being translated from one magnitude to another.[12] But as a matter of fact accidental motion may apply to the point at the extremity of a body, even though it does not exist in a body but at the extremity thereof.[13]

As for his illustration of violent motion, which he finds in 'the motion of a stone upward,' he follows the well-known theory of the Greek,[14] namely, that the elements are endowed with natural motion in opposite directions, as, e. g., the motion of a stone downward and the motion of fire upward, whence it is inferred that of the four elements, one, i. e., earth, has absolute weight, fire has absolute lightness, while air and water have only relative weight and lightness.[15] But this theory seems never to have been demonstrated and never will be. On the contrary, one may argue, that all the elements possess a certain amount of weight, but some possess more of it and some less.[16] That fire tends upwards may be due to the pressure of the air which pushes it upwards,[17] as happens in the case of a stone which, upon being dropped into a crucible in which there is molten gold or lead or mercury, comes up to the top, because of the pressure of the metals which push it upward. The same may also be said to happen in the case of the elements air and water. That [air possesses some weight] is more-over supported by observation. For when we make a digging in the ground, the air immediately descends into the hollow and fills it up.[18] Though the opponent might claim that this last phenomenon is due to the fact that a vacuum is impossible within the world, still it is not impossible that the descent of the air into the hollow is due to the weight which that element possesses.[19] But, whatever may be the explanation [of natural motion], it is clear that the upward motion of a stone is due, as has been shown in the illustration, to some external force.

ודאי בזה ההערה בזה הפרק.

הכלל הראשון, הפרק השביעי

בבאור ההקדמה השביעית, האומרת שכל משתנה מתחלק, ולזה
כל מתנועע מתחלק, והוא גשם בהכרח, וכל מה שלא יתחלק לא
יתנועע, ולא יהיה גשם כלל.

הנה ההקדמה הזאת כוללת חמש הקדמות. האחת שכל משתנה
מתחלק. השנית שכל מתנועע מתחלק. השלישית שכל מתנועע
הוא גשם בהכרח. הרביעית שכל מה שלא יתחלק לא יתנועע·
החמישית שכל מה שלא יתחלק אינו גשם.

ואמנם הרביעית והחמישית הן מבוארות מעצמן. אם הרביעית,
מבוארת בהפך הסותר מהשנית, וזה שכאשר התבאר שכל מתנועע
מתחלק, והיא ההקדמה השנית, יתחייב, מהפוך הסותר, שמה שלא
יתחלק לא יתנועע, והוא הרביעית. ואם החמישית, מבוארת מגדר
הגשם והיותו מכמה המתדבק.

ואמנם הראשונות צריכות באור.

אולם הראשונה, נתחבטו בה המפרשים, לפי שארסטו יחד בה
המופת בששי מהשמע, למה שהמשתנה מחוייב שיהיה מקצתו במה
שממנו ומקצתו במה שאליו, וזה כי בהיותו במה שממנו הוא נח בלתי
משתנה עדיין, וכשהוא במה שאליו הוא נח כבר השתנה, ואי אפשר

The critical comments contained in this chapter will suffice [for this proposition].[20]

PROPOSITION VII

PART I

PROOF of the seventh proposition, which reads: 'Everything changeable is divisible. Hence everything movable is divisible, and is necessarily a body. But that which is indivisible cannot have motion, and cannot therefore be a body at all'.[1]

This proposition contains five theses[2]: First, everything changeable is divisible. Second, everything movable is divisible. Third, everything movable is necessarily a body. Fourth, that which is indivisible cannot have motion. Fifth, that which is indivisible cannot be a body.

The fourth and fifth theses are self-evident. The fourth may be proved by the conversion of the obverse[3] of the second, for having stated that everything movable is divisible, which is the second thesis, it naturally follows, by the conversion of the obverse, that that which is indivisible cannot have motion, which is the fourth thesis. [By the same method of the conversion of the obverse] the fifth may be inferred from the definition of body, and from the fact that body is described as a continuous quantity.[4]

The first [three] theses, however, must needs have some explanation.

With regard to the first thesis the commentators [of Aristotle] have been debating with themselves as to its meaning,[5] for the demonstration thereof is given by Aristotle in the sixth book of the *Physics*[6] as follows: An object in change, he says, must be partly in the *terminus a quo* and partly in the *terminus ad quem*, for when it is wholly in the *terminus a quo* it is at rest, not having as yet begun to change; and when it is in its *terminus ad quem*, it is likewise in a state of rest, having already been com-

לו להיות כלו במה שממנו וכלו במה שאליו יחד, יחוייב אם כן שיהיה

מקצתו במה שממנו ומקצתו במה שאליו, ומה שזה דרכו הוא מתחלק

בהכרח.

ולפי שהבאור הזה לא יכלול אלא המשתנה בזמן, אבל המשתנה

5 בזולת זמן, כתכליות השנויים והתנועות, לא יצדק עליו זה, והיה

הבאור אם כן חלקיי, היה אלכסנדר יראה שכל משתנה בזמן,

והמשתנה בזולת זמן הוא בחוש לבד, אבל הוא בזמן, ולא יורגש

למעוטו. והיא סברא נפסדת מבוארת הבטול.

ואולם תמסטיוס קבל מציאות משתנה בזולת זמן, אלא למה

10 שהמשתנה בזולת זמן הוא נמשך למשתנה בזמן, היה הבאור אצלו

כולל. ואולם אבובכר אבן אלצאיג, עם שקבל גם כן מציאות

משתנה בזולת זמן, והוא המשתנה מהעדר אל מציאות, כחול הצורה

בחמר, פירש המשתנה באיך, כחם שיתקרר, וכקר שיתחמם, שזה

יהיה בזמן בהכרח.

15   ואולם אבן רשד דקדק עוד, כי למה שתכליות השנויים אינם

שנויים באמת, כי אז הם נחים, באור ארסטו כולל המשתנה באמת,

והיה המשתנה כולל כל סוגי השנוי.

ולא אדע מה הרויח אבובכר במה שפירש המשתנה באיך, כי

---

pletely changed, and as the whole thing cannot be at once both in the *terminus a quo* and in the *terminus ad quem*, it follows that it must be partly in the one and partly in the other. Whatsoever is thus conceived must necessarily be divisible.

Inasmuch as this demonstration assumes only things that change in time but cannot be applied to things that change without time, as, e. g., the terminations of the processes of change and motion, the demonstration will thus be only of particular application.[7] Compelled by this difficulty, Alexander was led to believe that everything that is changed is changed in time; and that if anything appears to be changed in no-time it is only an illusion; in reality it is in time, but the time is imperceptible on account of its brevity.[8] This view of Alexander, however, is erroneous and self-evidently false.[9]

Themistius, on the other hand, admits the existence of timeless change, but, inasmuch as change in no-time is always consequent upon change in time, he finds the demonstration to be of general application.[10]

A different interpretation is given by Avempace. While admitting the existence of timeless change, as, e. g., the change from non-being to being, which occurs instantaneously when form settles on matter,[11] he takes the term 'changeable' [in the proposition] to refer only to change in the category of quality, as, e. g., the refrigeration of a hot object or the calefaction of a cold object, which changes must always take place in time.[12]

Averroes makes a still nicer distinction. The final points of the various changes, he says, are not changes in the true sense of the term, for by that time they have already come to rest. Aristotle's demonstration, however, deals only with cases of true change, and in that sense it is of general application. Thus, according to this interpretation, the term 'changeable' [in the proposition] will include all the categories of change.[13]

I am, however, at a loss to know what Avempace has gained by

הוא מבואר שהשינויים אשר באיך להם תכליות שינויים בזולת זמן,

וזה שהשחור המתלבן, בתכלית תנועתו, היה לבן בזולת זמן.

ואיך שהיה, הנה יראה שהרב לקחו כפי דעת אבן רשד. ולזה

חייב מאמרו כל משתנה מתחלק שכל מתנועע מתחלק, וזה למה

5 שהמשתנה כבר יכלול כל מיני השינויים, וכמו שביאר בהקדמה

הרביעית.

ולזה נתאמתו שתי ההקדמות הראשונות.

ואולם השלישית, אמרו שכל מתנועע גשם, היא מבוארת מאד.

וזה שאם נקח התנועה בפרט, כמו שפירש הרב שהיא התנועה באנה,

10 הנה למה שהאנה ייחד מקום, והמקום הוא מיוחד לגשם, הוא מבואר

שהמתנועע גשם. ואם נקח התנועה כוללת כל מיני השינוי, להיות

כלם צריכין אל נושא גשמי, הוא מבואר בהם שהמשתנה הוא גשם.

נתאמתו אם כן אלו השלש הקדמות הראשונות.

אלא שצריך שיותנה באמרו כל מתנועע, המתנועע בעצם. וזה

15 שאנחנו נמצא אשר יתנועע במקרה לא יתחלק, כי הנקודה שהיא

תכלית הקו. כבר תתנועע בהתנועע הקו שהיא תכלית לו, והקו

בהתנועע השטח והגשם, והנקודה לא תתחלק ואינה גשם. אבל

הכוונה במתנועע בעצם.

ונתבארה אם כן ההקדמה השביעית הכוללת ההקדמות החמש.

1כי השינויים לרדק – בזולת זרא.   2המתלבן] שיתלבן ד – בתכלית] בתכליות גנ – היה] יהיה
זרא.   3 בן רשד באנ ב"ר פס.   4כל] שכל ר – כל [מה] פ – (שכל מתנועע מתחלק] פ.
5 (מיני) פ – כל מיני השנויים כבר יכלול ג – שנויים ד.   8 [הוא] גשם י.   9 (נקח) י.   12 (בהם)
ז – שהמשתנה] שהמתנועע זד – (הוא) לזרדקבנ.   13 התאמתו רא והתאמתו פ – אם כן נ"כ בנ.
14 (שצריך) פצ צריך ז – מתנועע [מתחלק] ד – המתנועע] מתנועע פ.   16 כברן לא פ –
תתנועען התנועעה א – לו תכלית פ.   17 (והגשם) פ.

restricting the application of the term 'changeable' to the category of quality, for in quality, too, the final points of its various changes are timeless. When a black object, for instance, turns white, it becomes completely white only at the end of its motion, and that is in no-time.[14]

However Aristotle's proposition may be interpreted, it is quite evident that the Master has taken it in Averroes' sense. Consequently, from the premise that 'everything changeable is divisible' he logically infers that 'everything movable is divisible', inasmuch as he takes the term 'changeable' to include all the kinds of change that he has enumerated in the fourth proposition.

Thus have been proved the first two theses.

As for the third[15], namely, everything movable is a body, it is very clear. For if we take motion in its proper sense, which the Master has explained to be locomotion, then, since locomotion implies a certain place, and place is peculiar to bodies[16], it must necessarily follow that whatever is movable is a body. And if we take the term motion to include all the kinds of change, again, since they all require some corporeal subject[17], it also follows that in their case, too, whatever is changeable is a body.

Thus have been proved those first three theses.

The following qualification must, however, be stipulated: When the author uses the phrase 'everything movable' he means only that which is moved essentially, for that which has only accidental motion we sometimes find to be indivisible. Take, for instance, the point at the extremity of a line. It is moved with the motion of the line of which it is the extremity, the line in its turn being moved with the motion of the surface or the solid, and still the point is indivisible and is not a body. But as has been said, the term movable must be taken to refer here only to that which is moved essentially.[18]

Thus has been proved the seventh proposition containing those five theses.

הכלל השני, הפרק הרביעי

בחקירה בהקדמה השביעית האומרת שכל משתנה מתחלק.

וזה שאנחנו נמצא בנפש המדברת, שהיא משתנה בקנין המושכלות

מהמוחשות והמדומות אשר יהיו בזולת זמן, והתנועות הנפשיות,

5 כשמחה והדאגה, אשר יהיו בזמן.

והנה אלתבריזי נתעורר מהספק הזה, ואמר בהתרו שהכוונה

בזה באיכיות גשמיים. ויראה שנמשך לדעת אבובכר בבאור דברי

ארסטו, כאשר העירונו בפרק שביעי מהכלל הראשון, ואולם לדעת

אבן רשד נאמר, לפי פירושו, שהכוונה בזה באיכיות ותנועות גשמיות,

10 ויהיה אם כן כל ההקדמה הזאת כפל ומותר, וביחוד אמרו

שהמתנועע בתנועות גשמיות הוא גשם. ועוד שאם ההקדמה הזאת

חלקית, ומיוחדת באיכיות הגשמיות, הנה לא יוכל להשתמש ממנה

במה שיבא במשתנה בכלל.

אלא שהתר הספק לפי מה שיראה הוא כפי התנאי שהעירונו

15 במתנועע, וזה שאנו צריכין להתנות בו המתנועע בעצם. וכן נאמר

PART II

EXAMINATION of the seventh proposition which reads: 'Everything changeable is divisible.'

[Against this proposition the following criticism may be urged]:

We find in the case of the rational soul that it suffers a change in the process of its acquisition of intellectual conceptions out of sensible perceptions and forms of the imagination[19]—a change which is in no-time.[20] Likewise, the motions of the soul,[21] as pleasure and care, imply a change which is in time.[22] [And yet the soul is indivisible.]

Altabrizi has already called attention to this difficulty, to solve which he has suggested that the term 'changeable' in this proposition should be taken to refer only to corporeal qualities[23]. It would seem that Altabrizi has followed Avempace's interpretation of Aristotle's words, the nature of which we have discussed in the seventh chapter of the first part. But even if we accept Averroes' interpretation, we may still say with Altabrizi that the term 'changeable' should be taken to refer to corporeal qualities and motions. As a result of Altabrizi's explanation, however, the entire proposition will be tautological and redundant,[24] and especially redundant will be that part of the proposition which, according to his explanation, will be tantamount to saying that that which is moved by corporeal motions is a body. Furthermore, if this proposition were to be of particular application, referring only to [change] of corporeal qualities, Maimonides could not have used it in a subsequent chapter with reference to changeableness in general.[25]

It seems, therefore, that the solution of the difficulty must needs have recourse to the condition we have stipulated with reference to the term 'movable,' according to which we have qualified its meaning as referring only to that which is moved essentially. Likewise here, with reference to the term 'change-

אנחנו במשתנה, רוצה לומר, המשתנה בעצם. ולהיות הנפש
המדברת בלתי משתנה בעצם, אלא למה שיקרה היותה היולנית,
לא יבטל אמות ההקדמה הזאת. אלא שהתבאר אם השינוי הקורה
לה, אם אפשר להיותו עצמי אם לא, יתבאר במה שיבא בגזרת השם.

הכלל הראשון, הפרק השמיני

בבאור ההקדמה השמינית האומרת שכל מה שיתנועע במקרה
ינוח בהכרח, אחר שאין תנועתו בעצמותו, ולזה לא יתנועע התנועה
המקרית תמיד.

יסוד ההקדמה הזאת, לפי מה שיראה, מה שהניח ארסטו בשמיני
מהשמע, שמה שיהיה במקרה אפשר בו שימצא ושלא ימצא,
והאפשרי אין ראוי בו שלא יצא אל הפועל בזמן בלתי בעל תכלית.
ולזה כבר יחוייב במתנועע במקרה שינוח.

הכלל השני, הפרק החמישי

בחקירה בהקדמה השמינית האומרת שכל מה שיתנועע במקרה
ינוח בהכרח.

וזה שמה שימצא במקרה יעבור שלא ימצא, כשלא יהיה מתחייב
לנמצא בעצם. ולזה כבר אפשר בנשם שיתנועע במקרה תמיד,

able,' we may say that it refers only to that which is changed essentially. Consequently, since the rational soul is never changed essentially, but only through the contingency of its being material, it in no way contradicts the truth of this proposition. The question, however, whether the change that is contingent to the soul can be essential or not, will be discussed in some subsequent chapter,[26] God willing.

PROPOSITION VIII

PART I

PROOF of the eighth proposition, which reads: 'Everything that is moved accidentally must of necessity come to rest, inasmuch as its motion is not in its own essence. Hence that accidental motion cannot continue forever'.[1]

The basis of this proposition would seem to be the principle laid down by Aristotle in the eighth book of the *Physics*, namely, everything that is accidental has in itself the possibility both of being and of not being.[2] But that which is possible cannot be conceived as not becoming actually realized in infinite time.[3] Hence it follows that whatever is moved accidentally must of necessity come to rest.[4]

PART II

EXAMINATION of the eighth proposition, which reads: 'Everything that is moved accidentally must of necessity come to rest.'

[The criticism of this proposition is as follows]:

[The statement that] everything that exists by accident may possibly cease to exist is true only in the case of a thing which is not the necessary result of something whose existence is essential. It may, therefore, be possible for a body to be moved accidentally

למה שיתחייב כן ממתנועע אחר בעצם, כמו שיקרה לכדור האש

שהוא מתנועע בהכרח מצד תנועת הגלגל התמידית, וכן שטחי הגלגל

וחלקיהם מתנועעים במקרה בתנועת הגלגל העצמית, והוא מין

מהמתנועע במקרה שלקח הרב במשלו בהקדמה הששית.

5   וכבר נתעורר מזה אלתבריזי וזולתו, עד כי הנרבוני חשב ליישב

ההקדמה הזאת, באמרו שירצה בו שכל מה שיתנועע במקרה, במה

שהוא מתנועע במקרה, ינוח בהכרח, כאלו תאמר על דרך משל

שנפש האדם המניעה האדם, והיא מתנועעת במקרה בהנעתה ואינה

מתנועעת בעצם, הנה למה שבהנעתה מתנועעת במקרה יחוייב בה

10   שתנוח. וכן תאמר בנפש הגלגל המניעה לו, והיא מתנועעת במקרה

בהנעתה, יחוייב לה שתנוח, אם לא שהצטרף שם מניע אחר נבדל

בלתי מתנועע אפילו במקרה.

והנה כשנשתדל בזה נמצאהו בלתי מחוייב. וזה כי כשניחס

ההתנועעות במקרה לנפש הגלגל אינו אלא על צד הקשרה בגלגל,

15   הקשר מציאות או הקשר עירוב, אשר הוא מתנועע בעצם. ואחר

1 שיתחייב] שיתנועע ר – כן] בו כל – ממתנועע] מתנועע סלי – שקרה פ.   4 מן המתנועע פ –

במשלו בכללו.   5 כין שמשה סבר' משה לד משה צזרקנ שרבי משה א – וישב ל.   6 (בו)

פ.   8 שנפש אדם פ – ואינה] ואם אינה לורדקבאנ.   9 הנה [יחוייב בה] צסא – למה] לפי פ –

מחוייב פ.   10 היא] והנה פ – המתנועעת פ – המתנועעת סלזורדבאנ.   11 (בהנעתה) ר להנעתה סלזורדקבאנ –

שיצטרף קנ – (שם) ר – האחר קנ.   13 כשנשתדל] כשנסתכל לזרד כשנשתכל קבאנ – נמצא

פ – (כי) לזורקבאנ – כשנתיחס לור שכניחס ו שכשניחס א.   14 ההתנועעות] התנועעות ד

המתנועעות ג התנועות ר – (הגלגל) סלזורקבאנ.

forever, inasmuch as its accidental motion may have to be continued forever as the necessary result of something that is moved
essentially. An example of this is to be found in the case of the
globe[5] of fire whose motion is violent, being brought about by the
perpetual motion of the [celestial] sphere[6]; or in the case of the
superficies of the [celestial] sphere, and the parts thereof,[7] which
are moved accidentally by the essential motion of the sphere [as a
whole].[8] Motion of this [latter] kind is a species of accidental
motion according to the illustration used by the Master in the
sixth proposition.[9]

This difficulty has already been raised by Altabrizi and others[10],
with the result that he of Narbonne thought of setting the
proposition aright by putting upon it the following construction: Everything that is moved accidentally in so far as it
is moved accidentally, must of necessity come to rest, as, e. g.,
the human soul, which is the principle of motion in man and
which, though unmoved essentially, is moved accidentally in the
process of its causing motion. This motion it is which according to
the proposition must come to rest, inasmuch as it is only the
accidental result of its own action in producing motion. By the
same token, the soul that moves the celestial sphere would likewise have to come to rest, for it, too, is moved accidentally as a
result of its own action in producing motion in the sphere, were it
not for the fact that there is an additional cause for the motion of
the soul of the sphere, namely, an absolutely separate mover
which is not moved even accidentally.[11]

If we examine[12], however, Narboni's reasoning with regard to
the soul of the sphere, we shall find it inconclusive. For if we
ascribe to the soul of the sphere any accidental motion at all, it is
only in consequence of its union—a union either of inexistence or
of admixture[13]—with the sphere, which is itself moved essentially.
Since the motion of the soul of the sphere is thus brought about
only through its union with the sphere, it is obvious that this

שאין התנועה לה אלא על זה הצד, הוא מבואר שלא תגיע ממנה

לה לאות מזה הצד. וזה שכאשר נניחה מניעה לגלגל תנועה נצחית

בעצם, הנה שם התנועה המקרית אשר ניחס לה כבר תמשך אל

העצמית, וכבר הנחנו שאפשר שתתנועע תמיד. ולא יקרה מזה

5 בטול, אבל נמצא דברים מקריים, מתחייבים לעצמיים, תמידיים

בהתמדת העצמיים.

הכלל הראשון, הפרק התשיעי

בבאור ההקדמה התשיעית האומרת כי כל גשם שיניע גשם אמנם

יניעהו בשיתנועע גם הוא בעת הנעתו.

10 ההקדמה הזאת מבוארת בעצמה. אמנם צריך שיותנה בה שיהיה

המניע הפועל, אבל המניע על דרך התכלית, כאלו תאמר, שהאש

מניע האויר שיעלה אל שטחו, להאותות המקום ההוא אל האויר,

כבר יניעהו והוא לא יתנועע. ולזה היה אמרו גשם שיניע גשם, ירצה

שיניעהו אם בדחייה או במשיכה.

15 וכבר הקשו על זה ממה שנראה בחוש, שהאבן המגניטס שיניע

הברזל כשימשכהו אצלו ולא יתנועע. והנה השיבו בזה בשני פנים:

union could not create in it an incapacity to continue that motion. Consequently, admitting, as we do, that it is the soul which causes the sphere to move with an essential and eternal motion, that accidental motion which we ascribe to the soul as a result of its own action must of necessity be co-extensive with the essential motion which it causes, and thus we must also admit that it would be possible for the soul to continue its accidental motion forever.[14] Still to admit this possibility will in no way invalidate the principle of this proposition, for it may very well be granted, that things accidental which proceed as necessary results from things essential will continue eternally when the essential things continue eternally.[15]

PROPOSITION IX

PART I

PROOF of the ninth proposition, which reads: 'Every body that moves another body moves that other body only by being itself moved at the time it moves the other.'[1]

This proposition is self-evident. The following qualification, however, must be stipulated, namely, that the proposition refers only to a mover which acts as an efficient cause, but in the case of a mover which acts as a final cause, it may cause motion without being itself moved. An instance of such a mover is to be found in fire which moves air and causes it to rise to the [concave] surface of the former, by reason of the affinity between that place and air. Consequently, in saying 'every body that moves another body,' he means that the former body moves the latter either by pushing or by drawing.[2]

Against this proposition an objection has been raised from the fact commonly observed that the Magnesian stone[3] causes iron to move, by drawing it in its direction, without being itself moved.[4] In reply to this, two explanations have been offered.

האחד, כי לאומר שיאמר שהברזל הוא שיתנועע בעצמו, וזה אמנם

מהמזג אשר יקנה מהאבן. והשני, שאם הודינו שהאבן ימשכהו, הנה

יהיה זה כשיותכו מהאבן גשמים ימששו הנמשך וימשכוהו, אם בדרך

משיכה או בדרך דחייה.

הכלל השני, הפרק הששי

בחקירה בהקדמה התשיעית האומרת כי כל גשם שיניע גשם אמנם

יניעהו בשיתנועע גם הוא בעת הנעתו.

הנה השני פנים אשר זכרו ממה שיראה ממשיכת אבן המגניטס

הברזל, מבוארי הנפילה בעצמם. כי שיקנה הברזל מזג משכונת

המגניטס, אשר לכל אחד כח טבעי שעור גדול, למה שהוא גלוי

מעניינם היותם קשי ההפעלות מאד, הוא רחוק קרוב לנמנע. ומזה

הצד הוא רחוק מאד שיותכו גשמים מהמגניטס ימשכו הברזל ויניעוהו.

ועוד שלא ימלט העניין מהיות הגשמים ההם המניעים, היוצאים

מהמגניטס, שיפעלו במשיכה או בדחייה. והנה בדחייה צריך

שיתנועעו הגשמים תנועות הפכיות בעת שידחו הברזל ויביאהו אל

המגניטס. ואם במשיכה גם כן צריך שיתנועעו הגשמים תנועות

First, one may say that the iron is set in motion by itself, and this indeed is due to a certain disposition it acquires from the stone. Second, even if we admit that it is the stone that sets the iron in motion, it may still be explained as being due to the effluxion of certain corporeal particles from the stone which come in actual contact with the iron and set it in motion either by drawing or by pushing.[5]

Part II

Examination of the ninth proposition, which reads: 'Every body that moves another body moves that other body only by being itself moved at the time it moves the other'.

The two explanations mentioned by the commentators with regard to the phenomenon of the power of the Magnesian stone to attract iron are self-evidently groundless. That the iron should acquire from the magnet, through its proximity to the latter,[6] a new disposition [and thereby move itself toward the magnet], either one of which acts would imply a natural force of considerable strength,[7] it being clear from the nature of the case that both these acts are very difficult of performance,[8] is a far-fetched assumption and well-nigh impossible. For the same reason, it is likewise past comprehension that corporeal effluvia should flow out of the magnet and pull the iron and thus set it in motion. Furthermore, we cannot escape the conclusion that the particles issuing forth from the magnet and causing motion must inevitably act either by drawing or by pushing. If by pushing, then those particles, when they begin to push the iron in order to bring it to the magnet, will have to move in a direction opposite to [that which they took when moving from the magnet to the iron]. If by drawing, then the particles will likewise have to move alternately in opposite directions, namely, [first], toward the iron,

הפכיות, אל הברזל, ואחר כך ימשכוהו ויתנועעו עמו לצד המגניטס.

ואיך יהיה זה, מי יתן ואשער. וכל זה בתכלית הגנות.

ולזה יראה שהתשובה הנכונה במה שיראה מאבן המגניטס,

שלברזל תנועה טבעית אל המגניטס, ביחס ידוע אצל הטבע, כמו

5 שיש לה תנועה טבעית אל המטה, אם להאותות אשר לו אל המקום

ואם בסגולה בו אשר לא נשער אלא שאמתהו החוש.

הכלל הראשון, הפרק העשירי

בבאור ההקדמה העשירית האומרת כי כל מה שיאמר שהוא

בגשם יחלק אל שני חלקים, אם שתהיה עמידתו בגשם, כמקרים, אם

10 שתהיה עמידת הגשם בו, כצורה הטבעית, ושניהם כח בגשם.

כבר היה מן הקדמונים מי שיראה שהגשם אין בו הרכבה כלל,

אבל הוא אחד בעצמו וגדרו, ואם היה שנרגיש בהם הרכבה, הנה

במקרים ומשיגים בלתי עצמיים. והנה ארסטו ומפרשי ספריו הכו

על קדקד הסברא הזאת, בשאמרו שאין המלט בכל גשם משני

15 דברים עצמיים לו, והם החמר והצורה. וזה שאנחנו נראה הגשמים

שבכאן הוים ונפסדים. ולפי שהדבר הנפסד לא יקבל הדבר ההוה,

---

and then drawing the iron and moving along with it toward the magnet. How that would be possible, would that I knew.[9] All this is of the utmost absurdity.

It seems, therefore, that the true explanation of the phenomenon of the Magnesian stone is that iron possesses, according to a certain relation to nature, a natural tendency toward the magnet, just as it possesses a natural tendency toward the below, which tendency is due either to its affinity with its appropriate locality or to some natural property inherent within it[10] of which we do not know anything except that it is warranted by sense perception.[11]

PROPOSITION X

Part I

Proof of the tenth proposition, which reads: 'Everything that is said to be in a body falls under either of two classes.[1] It is either something that exists through the body, as accidents, or something through which the body exists, as the natural form. Both accidents and the natural form are to be conceived as a force in a body'.[2]

Among the ancients[3] there were some who held that body has no composition in any sense whatsoever, but that it is one in essence and in definition. If we observe in bodies, they say, some kind of composition, it is only with reference to accidents and [other] unessential properties[4]. Aristotle and the commentators upon his works,[5] however, knocked this view on the head,[6] by demonstrating conclusively that every body must inevitably consist of two essential parts, matter and form. For we observe that all the mundane bodies are subject to generation and corruption; and as that which no longer is cannot be the recipient of that which is coming to be, it is necessary to postulate the

יצטרך להניח נושא יקבל את שניהם, והוא החמר הנקרא היולי.

והוא מבואר שהוא עצמי להוה, כי הוא נושאו. ולפי שהמקבל הוא

דבר זולת המקובל, הנה יחוייב שיהיו בו שני דברים.

ולפי שהמקובל בו יאמר שהדבר הווה ומוגבל ובו נתעצם, הוא

5 מבואר שהוא עצמי להוה. ולפי שהנושא אי אפשר שיהיה בעצמו

בפעל, שאם היה בפעל לא תהיה הויה אלא שינוי, הוא מבואר

שקיום הדבר ועמידתו הוא בדבר המקובל, והוא הצורה הטבעית.

ואולם המקרים, אשר אין המלט מהם בכל גשם, הוא מבואר

שעמידתם בגשם הכולל החמר והצורה הגשמית, שאם היה להם

10 קיום ועמידה בעצמם היו עצמים.

ולפי שכל אחד משני אלו, רוצה לומר הצורה והמקרה, אין לו

מציאות בעצמו, ושניהם צריכין אל נושא, כמו שהתבאר, תפס

בתיבת כח ואמר ששניהם כח בגשם.

וצריך שתתבונן אמרו שעמידת הגשם בצורה הטבעית, שהוא

15 לקח הגשם, שכולל החומר והצורה הגשמית, ביחס אל הצורה

הטבעית המיוחדת, כיחס החמר אל הצורה בכללו, שקיומו ועמידתו

בה.

1(את) לד.   8(המקרים) קאנ – בהם לד.   9(שאם) בא.   12שתופס פבאנ שתפס ל זור.

14 בצורת י בו כצורה ד.   15שהוא כולל פלזורדקבאנ   16המיחדת פ.

existence of a substratum which is to be the common underlying recipient of both of them. This substratum is matter, the so-called hyle.[7] That matter must be essential to that which comes to be,[8] is self-evident, inasmuch as it is its substratum. But still the recipient must be something distinct from that which is received, it follows therefore that in every body there must be two principles.

Again, as it is that which is received through which a thing is said to come into being, by which it is limited and in which it has its essence, it is evident that this, too, must be essential to that which comes to be[9]. But the substratum, it is quite clear, cannot have actual existence by itself[10], for if it had actual existence, the process of coming-to-be would be an alteration rather than a generation.[11] Hence it must follow that the being and existence of a thing must depend upon that which is received, that is to say, upon the natural form.[12]

As for accidents, which no body is destitute of, it goes without saying that they can exist only in bodies composed of matter and corporeal form,[13] for if accidents could have being and existence by themselves, they would be substances.[14]

Since neither of these two, namely, form and accidents, have independent existence, both, as has been shown, requiring some substratum, the author, making use of the term 'force' in a special sense, says that 'both accidents and the natural form are to be conceived as a force in a body'.[15]

You must note that the assertion that body exists through the natural form indicates that Maimonides has taken the term body, which includes both matter and corporeal form, in its relation to the natural proper form as analogous to the relation of matter to form in general, the former of which has its being and existence in the latter.[16]

הכלל השני, הפרק השביעי

בחקירה בהקדמה העשירית האומרת כי כל מה שיאמר שהוא
בגשם יחלק אל שני חלקים, אם שתהיה עמידתו בגשם, כמקרים,
ואם שתהיה עמידת הגשם בו, כצורה הטבעית.

5 ראוי שתדע שאבן סינא ואבוחמד והנמשכים אחריהם היו רואים
שמציאות החמר והצורה בכל גשם, ואף בגרמים השמיים, למה
שהצורה הגשמית אצלם אינה זולת דבקות השלשה רחקים מתחתכים
על זויות נצבות, ולפי שהדבקות זולת המתדבק, למה שהמתדבק
מקבל החלוק והדבקות אינו מקבל החילוק, צריך אם כן אל נושא
10 יקבל החלוק והדבקות. השכל אם כן יגזור בכל גשם שני דברים
עצמיים לו, והם החמר והצורה. ואולם אבן רשד, למה שהגרם
השמיימי לא יקבל החילוק בפעל, יראה שאין בו רבוי והרכבה
כלל. וזה כי הגשם אחד במציאות, אלא שהשכל יחייב בו הרכבה
מנושא ונשוא מצד ההויה וההפסד, לפי שהנפסד לא יקבל ההויה,
15 וכמו שקדם לנו ביאורו בפרק העשירי מהכלל הראשון, הגשם הנצחי
אם כן, שלא יפול תחת ההויה וההפסד, לא יגזור השכל בו הרכבה
כלל מחומר וצורה.

והנה לפי דעת אבן רשד, מה ההכרח, מי יתן ואדע, שלא נאמר
כן בגשמים ההוים והנפסדים, רוצה לומר שהחמר בהם הגשמות,
20 והצורה היא הצורה המיוחדת לכל אחד ההולכת מהלך השלמות

---

5 שבן סינא א שב״ס פ.   6גשם] הנשמים לד – ולמה ר.   8 (למה שהמתדבק) פ אמרו
שהמתדבק י.   9 (יקבל) ו – והתדבקות פא והדבוק לזזרדקבנ.   11 הם פ – ואולם] ואמנם
ג – בן רשד באגי ב״ד פ.   14 שהדבר הנפסד צפלוורדקבאג – ההוה לזור.   15 (ביאורו) קי –
(וא׳כ) הגשם לד – הנשם ג – הנצחי [השמיימי] לד.   16 (אם כן) לד – תחת] אחר לזורד –
(תחת) א.   17 (וצורה) פ באג.   18 הנה א – (לפי) ק – לדעת פ – בן רשד בג ב״ד פ נ׳ רשד א –
הכרח י ההכרח א.   19 הנפסדים לזזרד א.

Part II

EXAMINATION of the tenth proposition, which reads: 'Everything that is said to be in a body falls under either of two classes. It is either something that exists through the body, as accidents, or something through which body exists, as the natural form.'

It behooves you to know that Avicenna, Algazali, and those who follow them are of the opinion that the distinction of matter and form obtains in every body, including also the celestial spheres.[17] For believing that the corporeal form is nothing but the continuity of the three dimensions,[18] intersecting each other at right angles,[19] they reason as follows: Since continuity must be something different from the thing continuous, seeing that the latter may become divided whereas the former may not[20], there must exist a substratum capable of receiving both the continuity and the division. Reason therefore decrees[21] that in every body there must be two essential principles, namely, matter and form.[22]

Averroes, however, contends that inasmuch as the celestial sphere is not subject to actual division, it is not necessary to postulate in it any plurality and composition. For body, he argues, is one in reality. It is only on account of the phenomenon of generation and corruption, [23] seeing that that which no longer is cannot be the recipient of that which is coming to be, that reason postulates therein the distinction of subject and something borne by the subject, as we have explained it above in the tenth chapter of the first part. But as the eternal [celestial] sphere does not come under the law of generation and corruption, there is no reason why we should conceive it to be composed of matter and form.[24]

In view of Averroes' theory, however, would that I knew[25] what prevents us from maintaining the same with regard to the elements that are subject to generation and corruption, namely, that their matter be corporeality, and their form be the proper form of every one of the elements, which is related to corporeality

לנשמות, והגשמות, הנקרא אצלו צורה גשמית, שתהיה הולכת מהלך
החמר אל הצורה המיוחדת. ויהיה אם כן החמר בזולת הצורה
המיוחדת יצטרך אל מקום ונמצא בפעל. והנה שהדי במרומים,
שהגרם השמיי, שהוא גשם בלא חמר, נמצא בפעל. והנה בזה
5 יותרו קושיות חזקות ומבוכות רבות אשר בטבע ההיולי למה שהונח.

ואם כן הוא, הנה לטוען שיטעון שאין בכאן צורה מיוחדת יהיה
קיום הגשם בו, אבל הצורה הגשמית הוא הנושא בפעל והמעמדת
הצורה המיוחדת. ואם היה שאין ראוי לומר בצורות המיוחדות
היותם מקרים, למה שבהם יחודים יובדלו בהם מהמקרים, כאלו
10 תאמר שהצורות המיוחדות להם מקומות מיוחדים ושאינם מקבלים
התוספות והחסרון וכיוצא באלו, הנה אמנם יאמר בהם שהם דברים
עצמיים. אבל שיהיה עמידת הגשם וקיומו בו, לא, למה שצורת
הגשמות, שהיא הנושא, היא לעולם נמצאת בפעל, ועמידת הצורה
המשלמת אותו היא בו.

הכלל הראשון, הפרק האחד עשר

15

בבאור ההקדמה האחת עשרה האומרת כי קצת הדברים
שעמידתם בגשם יחלקו בהחלק הגשם, ויהיו נחלקים במקרה,
כמראים ושאר הכחות המתפשטות בכל הגשם, וכן קצת המעמידות
לגשם לא יחלקו בשום פנים, כנפש וכשכל.

---

3 ונמצאת באו – סהדי לודקנ סתרי ר.   5 הותרו לר יוערו ר – ונבוכות] ומרורות ר
ונבוכת סא.   9 יחודים] מיוחדים ד – יבדלו לר נבדלו ס – בהם] מהם לד.   11 יאמר
אמר ו נאמר ג – הדברים יר.   12 בקיומו סבאו – (לא) ס.   16 (כי) בקצת ג.   17 יתחלקו
לודרא.   18נשם זרא.   19 והשכל לרא ושכל פ.

as an entelechy, and that corporeality, designated by him as corporeal form, be regarded as matter in relation to the proper form.[26] As a result of this view, it would follow that even without its specific form, matter would be in place and would have actual existence.[27] Behold, my witness is in heaven,[28] for the heavenly sphere, which, [according to Averroes], is body without any matter, has actual existence. This theory would remove many a difficulty, strong and perplexing, which exists with regard to the nature of matter as it is generally understood.

This being so, an opponent may now further contend that the proper form is not that through which the body exists,[29] but, quite the contrary, it is the corporeal form which, being an actually existing substratum, sustains the existence of the proper form.[30] To be sure, the proper forms could not on that account be rightfully called accidents,[31] seeing that they possess peculiarities which distinguish them from accidents, as, e. g., they have appropriate localities of their own,[32] and are not subject to increase and decrease, and other things of a similar nature. They must, indeed, be considered as substances. Still to say that body exists and has its being in the proper form must be emphatically denied. Quite the contrary, the corporeal form, which we now propose as the substratum, always has actual existence, whereas the existence of the [proper] form, which to be sure is the entelechy of the corporeal, is dependent upon the latter.

PROPOSITION XI

PROOF of the eleventh proposition, which reads: 'Among the things which exist in a body, there are some which participate in the division of that body, and are therefore accidentally divisible, as, e. g., colors and all other forces[1] that are distributed throughout the body. In like manner, among the things which constitute the existence of a body, there are some which cannot be divided in any way, as, e. g., the soul and the intellect.'[2]

הנה חלוקת הדברים אשר עמידתם בגשם והמעמידות לגשם

מבוארת בעצמה, למה שהמקרים אשר עמידתם בגשם, מהם יחלקו

במקרה בחלוקת הגשם, כמראה וכשיעור, ומהם שלא יחלקו,

כנקודה, והקו מצד הרחב, והשטח מצד העמק. וכן המעמידות

5 לגשם, מהם שיחלקו בחלוקת הגשם, כהיולי, אשר הוא הדבר

המקבל החלוקה, למה שצורת הגשמיות, שהיא דבקות הרחקים,

לא יקבל החלוקה, שאין מדרך ההפך שיקבל ההפך.

ומה שצריך לבאר אמרו כנפש וכשכל, כי הוא יראה שהם

כח בגוף, ולמה שאין מתפשטות בכל הגוף לא יחלקו בחלוקת הגשם.

10 ועוד יתבאר זה לפנינו בגזרת השם.

כי ארסטו יראה בחלוף זה, שהשכל הנקנה נקשר בגוף הקשר

מציאות לא הקשר עירוב, ולזה לא יתנועע במקרה כשיתנועע הגוף.

ולזה יראה שהשכל הנבדל הוא המניע לגלגל ולא יתנועע במקרה.

ולהיותו מניעו הוא נפשו. ולזה יקרא הגלגל מתנועע מפאת נפשו.

The division of things which exist in a body as well as of those which constitute the existence of a body [into some which are divisible and some which are not divisible] is self-evident. For of accidents that exist in a body, some are accidentally divided with the division of the body, as, e. g., color and quantity, while others are indivisible, as, e. g., a point, or a line with respect to width, or a surface with respect to thickness. In like manner, of things which constitute the existence of a body, some participate in the division of the body, as, e. g., prime matter, which is that element in a body that is subject to division, for corporeal form, being the continuity of the dimensions, is not subject to division, inasmuch as opposites cannot be the recipients of each other.[3]

What needs explaining, however, is his statement 'as, e. g., the soul and the intellect.' For the author is of the opinion that soul and intellect are forces existing in a body, and it is only because they are not distributed throughout the whole body that they do not participate in the division of the body. We shall give full consideration to this problem in a later part of this work,[4] God willing.

For Aristotle is diametrically opposed to this view.[5] He is of the opinion, [and in this Maimonides agrees with him], that the acquired intellect is conjoined with the body by a nexus of inexistence rather than by a nexus of admixture. In consequence of this, the acquired intellect, [according to both of them], is not moved accidentally with the motion of the body. By the same token, Aristotle maintains that the Intelligence [of the sphere], which is separated [from the sphere in the same manner as the acquired intellect is separated from the body], is the [first] mover of the sphere, causing motion in the latter without itself being moved accidentally. Still that Intelligence, though separate, being the principle of the sphere's motion, is in a sense the latter's soul, and it is in that sense that the sphere is said to be moved by

והרב יראה, שׁשׂכל הגלגל הוא כח בגוף, ויתנועע במקרה בתנועת

הגלגל. ולזה יחד מופת על שאין השכל ההוא מניעו, כי למה שיתנועע

במקרה יצטרך לנוח בהכרח, כמו שביאר בהקדמה השמינית, וייחד

מופת על שהכח המתפשט אינו המניע, כי יהיה בעל תכלית,ו ויהיה

5 פעלו בעל תכלית, אחר שיתחלק בהחלקו. ולזה אמר כי מניעו הוא

השכל הנבדל, כמו שיראה במה שכתב בפרק הראשון מהחלק השני

בספרו המורה.

הכלל הראשון, הפרק השנים עשר

בבאור ההקדמה השתים עשרה האומרת שׁכל כח נמצא מתפשט

10 בגשם הנה הוא בעל תכלית, להיות הגשם בעל תכלית.

הנה ארסטו ביאר ההקדמה הזאת בשמיני מהשמע, וסדר המופת

כן. כל גשם אם שיהיה בעל תכלית או בלתי בעל תכלית, אבל

מציאות גשם בלתי בעל תכלית נמנע, כמו שהתבאר במה שקדם.

נשאר אם כן שיהיה הכח בגשם בעל תכלית. והנה מציאות כח בלתי

15 בעל תכלית בו יראה שׁהוא נמנע, אחר הניחנו הקדמה אחת

מבוארת בעצמה, והיא שׁהכחות המתפשטים אשׁר בגשמים מתחלקים

---

1 והגוף פ. 2 [מניעו] יצטרך כ. 3 [לנוח] לזה פ נוח ב – שביאר] שהתבאר ז רק באג. 4–3 [וייחד...
תכלית] הוספתי על פי השערה, עיין פירושי האנגלי. 5 פעלו בעל תכלית] פועלו ב"ת פ
פועל ב"ת ר פעולו בלתי תכלית פ – שתחלק לר שיחלק פ שהתחלק א – בהתחלקו פ – ולזה
ולכן זוו רק באג – המניעו ו זו רא. 6 ממה פ ל זוו רק באג. 7 מספרו פ לו ר דק ב – במורה צ ל זוו ר דק ב
11 הקדמה ב ג – הוו א – ז ו ר ב ג. 13 [גשם] לר.

its own soul. As against this, the Master maintains that the Intelligence of the sphere is, [like the hylic intellect in its relation to the human body], a force inherent in the body of the sphere, in consequence whereof it is moved accidentally with the motion of the sphere. It is for this reason that he advances a special argument to show that the Intelligence of the sphere cannot be the [first] mover of the sphere, for inasmuch as it has, [according to his own view], accidental motion, it would have to come to rest, as he has stated in Proposition VIII. [Previous to this he had already shown by another argument that the first mover could not be a force distributed throughout the body of the sphere, for a force like that would have to be finite], inasmuch as it must be divisible with the division of the sphere, and thus its action would have to be finite.[6] He thus concludes that the [first] cause of the motion of the sphere must be an Intelligence which is absolutely separate from the sphere, all as may be gathered from his discussion in the first chapter of the second part of his work *The Guide*.

PROPOSITION XII

PART I.

PROOF of the twelfth proposition, which reads: 'Every force that is distributed through a body is finite, that body itself being finite.'[1]

Aristotle has demonstrated this proposition in the eighth book of the *Physics*.[2] His argument runs as follows: Every body must be either finite or infinite; but, as has already been shown before, the existence of an infinite body is impossible; it follows therefore that the body in which a force exists must be finite. That in such a finite body no infinite force can exist will become manifest after we have laid down the following self-evident proposition, namely, that forces distributed through bodies must participate

בהחלק הגשמים, ושכל מה שיהיה הגשם יותר גדול יהיה כח הנעתו

יותר גדול, כאשר נראה בחלק הגדול מהארץ יותר גדול הנעה

מהחלק הקטן ממנה. וכאשר התישב זה, סדר ההקש כן. אם ימצא

כח בלתי בעל תכלית בגשם בעל תכלית, יתחייב אחד משני דברים,

5 אם שיניע מתנועע מה בעתה, או שיהיו כח בלתי בעל תכלית וכח

בעל תכלית שום בהנעה, ושניהם מבוארי הבטול.

ואיך יתחייב זה, כפי מה שאומר.

נניח הגשם אשר בו כח בלתי בעל תכלית יניע מתנועע מה בזמן

מה. הנה כבר אפשר במניע בעל תכלית שיניע המתנועע ההוא,

10 למה שעלינו להניחו בשיעור שיניעהו המניע בעל תכלית. ואין ספק

שיצטרך בהנעתו אל זמן יותר גדול מהמניע הבלתי בעל תכלית.

והנה לא ימלט המניע הבלתי בעל תכלית אם שיניעהו בעתה או

בזמן. ואם יניעהו בזמן, יהיה בהכרח חלק ידוע מהזמן היותר גדול,

והוא ידוע שאפשר לנו שנקח מהגשם הבלתי בעל תכלית חלק יהיה

15 יחסו אל הבעל תכלית האחר יחס הזמן הקטן אל הזמן הגדול,

ויהיה אם כן חלק הבלתי בעל תכלית, שהוא בעל תכלית בהכרח,

שוה בהנעה אל הכח הבלתי בעל תכלית.

התבאר אם כן חיוב התדבקות הנמשך לקודם, והוא שאם ימצא

in the division of those bodies and that the greater the size of the body the stronger its motive force,[3] as we observe, for instance, a large clod of earth to possess a stronger motive force than a smaller clod. This proposition having been established, the syllogism of the argument may be framed as follows: If in a finite body an infinite force were possible, either of the following two conclusions would ensue, namely, either the infinite force would move a certain object in an instant or an infinite force and a finite one would be equal in their power of producing motion. Both of these conclusions, however, are notoriously absurd.

How such conclusions would have to ensue, will now be explained.

Let the body in which that infinite force is assumed to abide set a certain object in motion in a certain time. Undoubtedly there could be found some finite motive force which would also be capable of setting that object in motion—for we will assume that object to be of a size that could be moved by that finite motive force. The finite force will undoubtedly require a greater time than the infinite force to effect its motion. Now, the infinite force must inevitably be able to effect its motion either in an instant or in some extended time. If it does it in time, that time will of necessity be a certain portion of the greater time [required by the finite force]. Now, it is well-known that we can take from the body [with] the infinite [force] a certain portion the ratio of whose magnitude to the magnitude of the other body [with] the finite [force] would be equal to the ratio of the lesser time to the greater time. Thus it would result that a part of the infinite, which is of necessity finite, would be equal in its motive power to the infinite force.

We have thus demonstrated the inference of the consequent from the antecedent, namely, that if in a finite body an infinite

כח בלתי בעל תכלית לגשם בעל תכלית, יתחייב אחד משני דברים,

אם שיניע המניע הבלתי בעל תכלית מתנועע מה בעתה, ואם שיהיו

כח בלתי בעל תכלית וכח בעל תכלית שוים בהנעה.

הכלל השני, הפרק השמיני

5   בחקירה בהקדמה השתים עשרה האומרת שכל כח נמצא

מתפשט בגשם הנה הוא בעל תכלית, להיות הגשם בעל תכלית.

ואומר שהסבה אשר זכרה כבר התבאר בטולה במה שקדם.

וזה שהמנעות גשם בלתי בעל תכלית לא התבאר עדיין.

אבל נניחהו, ואומר שהוא בטל. וזה שלא נודה בחיוב התדבקות

10 הנמשך לקודם בהקש. וזה שלא תתחייב התנועה בזולת זמן, למה

שלכל תנועה זמן שרשי אין המלט ממנו. ולא יתחייב גם כן שווי הזמן

לכח הבלתי בעל תכלית והבעל תכלית, למה שיחס הכח אל הכח

יהיה בזמן העודף על זמן השרשי הידוע אצל הטבע, וזה שהבלתי

בעל תכלית יניע בזולת זמן, חוץ מהזמן השרשי, והבעל תכלית

15 יצטרך בו לזמן מה. ולו הונח מניע בעל תכלית יניעהו בזמן השרשי,

לא יקרה ממנו בטול, למה שכבר ימצא החילוף ביניהם במתנועע

גדול, שהמניע בעל תכלית יצטרך זמן בהנעתו חוץ מהזמן השרשי,

force were possible, the following alternative conclusions would have to ensue, namely, either the infinite motive force would have to effect its motion in an instant or an infinite force and a finite one would be equal in their motive power.

PART II.

EXAMINATION of the twelfth proposition, which reads: 'Every force that is distributed through a body is finite, that body itself being finite.'

I say that the basis of his argument may be refuted on the ground of what has already been said,[4] namely, that the impossibility of an infinite body has not been conclusively established.

Granted, however, that an infinite body is impossible, I still maintain that his reasoning is inconclusive, for we do not admit the cogency of the connection of the consequent with the antecedent in the syllogism of the agrument. In the first place, the conclusion that there would be motion without time does not follow, inasmuch as every motion has that original time from which it is never free.[5] Nor, in the second place, does it follow that the finite and the infinite forces would produce motion in equal time, for the ratio of one force to the other would be equal to the ratio of their respective lengths of time in addition to that original time which may be assumed to exist by the nature of motion itself.[6] Thus, for instance, the infinite would effect motion within the original time only, without any other time, whereas the finite would require some additional time besides the original. Even in assuming a finite mover which would likewise cause motion in the original time only, the alleged absurdity would not ensue, since a difference might still be found between such a finite mover and the infinite mover if the size of the object moved by them were increased, in which case the finite mover would require for the effectuation of its motion some

והבלתי בעל תכלית יניעהו בזמן השרשי לבד. זהו הדרך שנתבטל
בו המופת.

ואולם צריך שתתעורר, שכשנודה במופת, צריך שיובן בלתי
בעל תכלית בחוזק. וזה שהוא מבואר שהבלתי בעל תכלית כבר
5 יאמר בשתי בחינות, אם בחוזק ואם בזמן. ולזה כשנודה חיוב המופת
בבלתי בעל תכלית בחוזק, הנה לא יתחייב בבלתי בעל תכלית
בזמן. וזה שכבר אפשר בכח אשר בנשם הבעל תכלית, שיניע תנועה
בעל תכלית בחוזק זמן בלתי בעל תכלית, כשלא יהיה לו סבת
היניעה והלאות, כאלו תאמר בתנועה הסבובית, שאינה במשיכה
10 ולא בדחייה, וכל שכן בגרם השמיימי, שכבר הוסכם מהם שאיננו
בעל איכיות, ולא יקרה לו החולשה והזקנה, כמו שבא בספר השמים
והעולם. ועוד שכבר אפשר שיאמר בתנועה הסבובית שהיא טבעית
לגרם השמיימי כאשר התנועה הישרה טבעית ליסודות. והוא מבואר.

הכלל הראשון, הפרק השלשה עשר

15 בבאור ההקדמה השלש עשרה האומרת שאי אפשר שיהיה דבר
ממיני השינוי מתדבק אלא תנועת ההעתק לבד, והסבובית ממנה.

ואמנם הכוונה בהקדמה הזאת הנה שאי אפשר במיני השינוי, רוצה
לומר בשני מינים המקבילים, שיהיה תנועה מדובקת. וזה שכבר
קדם שהשינוי בארבעה מאמרות, והם סוגים מתחלפים. והנה בשני

time in addition to the original time, whereas the infinite would cause the object to move in the original time only. Thus the proof has been shown to be refutable.

You must, however, note that even if we accept this proof, the term infinite in the proposition is to be understood to refer only to infinite in intensity. For it is evident that the term infinite may be used in a twofold respect, with regard to intensity and with regard to time.[7] Hence even if we accept the conclusiveness of the proof with regard to an infinite in intensity, the same will not follow with regard to an infinite in time.[8] In the latter case, it is quite possible that a force residing in a finite body should produce motion of finite intensity but of infinite time, providing only that the motion is of a kind in which there is no cause of lassitude and exhaustion, as, for instance, circular motion, which is caused neither by drawing nor by pushing,[9] and all the more so [the circular motion of] the celestial sphere,[10] about whose substance the philosophers are agreed that it is devoid of any qualities, and is not subject to caducity and senility, as is to be found in *De Coelo et Mundo*.[11] Furthermore, circular motion may be said to be natural to the celestial substance in the same manner as rectilinear motion is natural is to the [sublunar] elements.[12] This is evident.

PROPOSITION XIII

PART I.

PROOF of the thirteenth proposition, which reads: 'None of the several species of change can be continuous, except locomotion, and of this, too, only that which is circular.'[1]

The purpose of this proposition is to show that there can be no continuous motion between two species of change, that is to say, between two opposite species. For as has already been stated, change exists in four categories, and these constitute different genera.[2] Now, that between two of such genera, as, e. g., be-

סוגים הדבר בהם מבואר שאין שם תנועה אחת מדובקת, כאלו

תאמר המשתנה מהלובן אל השחרות, והמתנועע מאנה אל אנה,

אבל בסוג אחד בעצמו, כאלו תאמר באיך מהלובן אל השחרות

ומהשחרות אל הלובן, גם כן איננו שינוי מדובק. וזהו מה שרצה

5 באמרו, ד ב ר מ מ י נ י ה ש י נ ו י. כי אין לאומר שיאמר במין אחד

מן השינוי שאי אפשר שיהיה מתדבק, וזה שהשיני ממנו בזמן וממנו

בזולת זמן, והשינוי אשר בזמן הוא בהכרח מתדבק, להיות הזמן

מתדבק; ואם לא, היה הזמן מחובר מעתות. אלא שהכוונה בזה הוא

בשני מיני השינוי המקבילים. או שרצה באמרו מתדבק תמיד נצחי.

10 והנה ההקדמה הזאת בארה ארסטו בשאמר, כי למה שהתנועה

תקרא בשם מה שאליו התנועה, כי אנחנו נאמר במתנועע מהשחרות

אל הלובן מתלבן, ובתנועה חלק מה ממה שאליו גמור, חוייב שיהיה

נח במה שאליו; ואם לא, היה השלמות האחרון בכח, ולא היה מה

שאליו גמור, והיו התנועות המקבילות תנועה אחת, והיה הדבר

15 ישתחר ויתלבן יחד. אלא שהעניין בו כעניין בהויה. וזה שהתנועה

4 מדובק] מתדבק נ – שנרצה ⁵ שירצה ר.   5 באומרו פⁱ במאמרו ד – (כי) ⁱ – האחד לורדק

נ ג.   7 בזולת] בלתי י.   8 (בזה) פ – (הוא) ⁱ – (הוא) ר   9 המקבלים רדאג – רצה א – מתדבק [מתדבק]

לזנא.   12 ומתלבן ⁱ – (מה) ⁱ.   14 והיה] והנה כ והיא א.   15 ישתחרר סלורדקבאו

ישתחרת פ.

tween one object changing from whiteness to blackness and another object moving from one place to another, there can be no continuous motion is quite evident. But even [between two changes] within one genus, as, e. g., the changes within the genus quality, from whiteness to blackness and from blackness to whiteness [of the same object], it must likewise be evident that there can be no continuous change.³ That is what the author means by his statement 'none of the several species of change.' For to say that he means thereby to deny the possibility of continuous change even within one species is impossible, and for the following reason: Change is either in time or timeless, and change in time must of necessity be continuous,⁴ inasmuch as time is continuous, for if change in time were not continuous, time would be composed of instants.⁵ Hence the proposition must be assumed to refer only to change between two opposite species. Or, [if the proposition is to refer also to change within one species], the term "continuous" must be understood to have been used here by the author in the sense of *perpetual, eternal.*⁶

Aristotle⁷ has demonstrated this proposition by the following argument:⁸ Motion is named after the terminus toward which it tends; thus we say, for instance, with regard to an object that is moved from blackness toward whiteness, that it is whitening.⁹ Furthermore, in motion there must be a certain part which is an absolute *terminus ad quem*. It therefore follows that motion must come to rest on its arrival at the *terminus ad quem*, for if that were not so, the ultimate completion of motion would be potential, and there would never be a perfect *terminus ad quem*, whence it would follow that opposite motions would be one motion, and a thing would be whitening and blackening at one and the same time. The case of qualitative motion must therefore be analagous to that of generation. For in the motion of

אשר בעצם, כאשר נתהוה, נח, ויתנועע אחרי כן אל ההפסד. ואמנם

בין ההויה וההפסד אמצעי, שלא יצוייר בו שיתהוה ויפסד יחד.

ואולם בתנועת ההעתק גם כן הדבר בו מבואר, למה שתנועת

ההעתק, אם שתהיה ישרה או סבובית או מורכבת משתיהן. והנה

5 בתנועה הישרה הדבר מבואר שיתחייב בין כל שתי תנועות הפכיות

מנוחה; ואם לא, היה מתנועע אל המעלה ואל המטה יחד. ועוד

שהאמצע בכל גודל כבר ימצא בשני צדדים, אם בכח ואם בפעל:

כי הוא כאשר התנועע בו מתנועע מה בהתדבקות לא ירשום בו

נקודה או קו בפעל, למה שהקו אינו מחובר מנקודות ולא השטח

10 מקוים; וכאשר עמד, רשם בו נקודה או קו בפעל; ואם היה רושם

נקודה או קו בפעל, כשהוא מתנועע בהתדבקות, היה מחוייב שיהיה

בהם זמן יעמוד באמצע; וזה שהוא מבואר שהיותו מתנועע אל

האמצע והיותו מתנועע מן האמצע הם שתי תכונות מתחלפות; ואם

היה הנקודה או הקו בפעל, היה מחוייב שיהיו שתי תכליות התכונות

15 בפעל, ושיהיה הזמן מחובר בעתות. וכאשר התבאר זה בקו הישר,

הוא מתחייב בקו המורכב מהישר והסבובי, אשר הוא החלזוני. וזה

שכאשר ניחהו מתדבק, כבר יתנועע בפעל אל המעלה והמטה

בהתדבקות, ויתחייבו ממנו הבטולים הקודמים.

---

the category of substance, the object comes to rest when its generation is complete, and then begins to move backward towards corruption. But between these motions of generation and corruption there is an intervening instant in which the object cannot be conceived to be both generated and corrupted.[10]

That the like takes place also in locomotion is equally manifest.[11] Locomotion is rectilinear, circular, or composed of both of these.[12] With respect to rectilinear motion it is obvious[13] that between the motion in two opposite directions there must be an interval of rest, for if not, the same object would be moved upward and downward at the same time. Furthermore,[13] the middle of any magnitude is to be understood in two senses, as actual and as potential, of which the following is an illustration. When a certain object is moved with a continuous motion over any magnitude, it does not mark on it any actual point or line, inasmuch as a line is not composed of points nor a surface of lines; it is only when the moving object stops that it marks an actual point or line. Hence, [conversely], if an object which is moved with a continuous motion has marked an actual point or line, it must be inferred that at a certain time it had stopped at some point in the middle. Now, it is manifest that the motion of that object towards that middle and its motion away from it are in opposite directions, and since the point or line marked by that object is, [as we have said], actual, it must follow that the extremities of these opposite motions are likewise actual, and thus, [if we do not postulate an interval of rest between them], time would be composed of instants.[14] This having been shown to be the case of [motion in] a straight line, the same must also hold true with regard to [motion in] a line composed of straight and circular parts,[15] that is, a spiral,[16] for if we suppose it to be continuous, it would be actually moved upward and downward with one continuous motion, whence the aforesaid absurdities would ensue.

ולזה היה מבואר שההתדבקות איננו אפשרי אלא בתנועת

ההעתק, והסבובית ממנה, שמה שממנו ומה שאליו אחד, ומזה הצד

אפשר בה ההתדבקות והנצחיות.

הכלל השני, הפרק התשיעי

בחקירה בהקדמה השלש עשרה האומרת שאי אפשר שיהיה דבר

ממיני השינוי מתדבק אלא תנועת ההעתק לבד, והסבובית ממנה.

והנה כאשר ידוקדקו טענות ארסטו בזה יראה שהם דמויים לבד

והזיות. וזה שהשחור כאשר יתנועע אל הלובן, ואם היה שלא ינוח

בלובן אבל ישתחר, הנה לא יחוייב שיתלבן וישתחר יחד אלא בשתי

בחינות, שהוא במה שיתלבן ראשונה יצדק עליו שיתלבן, ובמה

שיתנועע אחר כן אל השחרות יצדק עליו שישתחר, ולא יקרה מזה

בטול.

וכל שכן בתנועה הישרה, שלא יתחייב מנוחה בין שתי התנועות,

אבל אפשר שתהיה מדובקת, ואי אפשר לעמוד עליה מהחוש, כמו

שאמר ארסטו. אבל יחוייב, שאם נדמה מתנועע קל בתכלית הקלות

מתנועע אל המעלה והר נופל עליו בתכלית הגודל, שאין ספק עליו

From all that has been said, it is evident that continuity is impossible except in locomotion, and of this, too, only that which is circular,[17] in which case both the *terminus a quo* and the *terminus ad quem* are identical,[18] for which reason continuity and eternity are possible in it.[19]

PART II.

EXAMINATION of the thirteenth proposition, which reads: 'None of the several kinds of change can be continuous except locomotion, and of this, too, only that which is circular.'

When Aristotle's arguments in proof of this proposition are closely examined, it becomes evident that they are all mere fancies and conceits. For even if the black object which is moved toward whiteness returned in the direction of blackness without first stopping at whiteness, it would not necessarily follow that at the juncture of the two motions the object would be both whitening and blackening at the same time. No, its whitening and blackening would be only two aspects of the same motion, that is to say, in so far as its motion is first toward whiteness, it is appropriately described as whitening, and in so far as its motion afterwards turns towards blackness, it is appropriately described as blackening. And so, no absurdity would ensue therefrom.[20]

In the case of rectilinear motion, it is still less conclusive that there must be a pause between the two [opposite] motions, for they may as well be one continuous motion, though they are not perceived as such by the senses, as has been said by Aristotle.[21] Nay, opposite motions must necessarily be continuous. Suppose, for instance, that an extremely light object is moved upward, and an extremely large object of the size of a mountain comes down upon it. There is no doubt that the latter will cause

שיניעהו אל המטה; ואם היה בין שתי התנועות ההפכיות מנוחה,

יתחייב שיעמוד ההר נח עם תכלית גדלו.

והחיוב שדמה הטעאיי, שלא יתחייב מהיות התנועות מתחלפות

שימצא שם עתה בפעל. זה יתבאר בעתה אשר הוא תכלית ההפסד

5 והתחלת ההויה, או תכלית ההויה קודמת והתחלת ההויה מתאחרת,

שהוא מחוייב שלא ימצא עתה בפעל. ואיך לא? והנה תנועת ההויה

נמשכת לתנועת האיך, והעתה שבין האיכיות איננו נמצא בפעל, ואם

האיך הראשון תכלית הויה קודמת והשני התחלה למתאחרת. וזה

מבואר מאד.

10 הכלל הראשון, הפרק הארבעה עשר

בבאור ההקדמה הארבע עשרה האומרת שתנועת ההעתקה יותר

קודמת שבתנועות, והראשונה מהם בטבע, כי ההויה וההפסד יקדם

לה ההשתנות, וההשתנות יקדם לה קריבת המשנה מן המשתנה, ואין

צמיחה ולא חסרון אם לא שיקדם להם הויה והפסד.

15 הנה ההקדמה הזאת בארה ארסטו בחפוש, וכוון בה הקדימה

בטבע ובזמן. והוסיף בה ביאור, שהתנועה הסבובית קודמת לשאר

---

2 ההר נח] ההכרח פ.   3 התנועות] התכונות צ'לנדרא תכונות קב'.   5 ההויה] המציאות
לזר מציאותי – הקודמת קב' הקודם לזור – והתחלת] ותכלית' והתחלה פ – ההויה] ההויה
ובא הוית קי.   8 התחלה] הויה א'.   9 (מאד) ב'.   12 יקדם] יקרים פ.   13 (בתחלת]
ההשתנות כ.   14 ואין חסרון ב.   16 שהתנועה] שתנועה ב.

the former to change its motion to the downward direction. Now, if there were a pause between these two [opposite] motions [of the lighter object], it would follow that the mountainous object, too, with all its size, would have to stop in the middle of its downward motion.[22]

Again, the conclusion which he has fancifully deduced is fallacious; for from the assumption that the motions are opposite, it must not necessarily follow that there is an actual instant [of rest] between them. It can be shown from an analogy of the instant which marks the end of corruption and the beginning of generation, or rather the end of an anterior generation and the beginning of a posterior generation, that there must not necessarily be an actual instant. Why should it not be so? Motion of generation is always consequent on motion of quality, and still the instant between the opposite qualities does not exist actually,[23] even though the first quality is the end of the anterior generation and the second the beginning of the posterior. This is very evident.

PROPOSITION XIV

PART I.

PROOF of the fourteenth proposition, which reads: 'Locomotion is prior to all the other kinds of motion and is the first of them in nature, for generation and corruption are preceded by alteration, which in its turn is preceded by the approach of that which alters to that which is to be altered, and, similarly, growth and diminution are impossible without previous generation and corruption.'[1]

Aristotle has demonstrated this proposition by the method of induction,[2] and has made it clear that he meant to establish the priority of locomotion both in nature and in time.[3] He has furthermore proved that circular motion is prior to all other

התנועות למה שאינה מהפך אל הפך, ולא ישיגנה שינוי, והמתנועע
בה אין לו כח על השינוי, אבל ענינו דומה אל הפעל הגמור.

הכלל השני, הפרק העשירי

בחקירה בהקדמה הארבע עשרה האומרת שתנועת ההעתקה
יותר קודמת שבתנועות, והראשונה מהם בטבע, כי ההויה וההפסד
יקדם לה ההשתנות, וההשתנות יקדם לה קריבת המשנה מן המשתנה,
ואין צמיחה ולא חסרון אם לא שיקדם להם הויה והפסד.

הנה על דרך ההויה הנמשכת תתאמת ההקדמה הזאת, אבל על
דרך התחלת ההויה, אם היתה מלא דבר, כאשר יתבאר, הנה
יתאמת שההויה קודמת לשאר התנועות, ושתנועות הכמה והאיך
קודמות להעתק, למה שהיו בעלי איכות וכמות קודם שהתנועעו,
והכמה בשלוח קודם לאיך.

הכלל הראשון, הפרק החמשה עשר

בבאור ההקדמה החמש עשרה האומרת כי הזמן מקרה נמשך
לתנועה ודבק עמה, לא ימצא אחד משניהם מבלתי האחר, לא
תמצא תנועה כי אם בזמן, ולא יושכל זמן אלא עם תנועה, וכל מה
שלא תמצא לו תנועה אינו נופל תחת הזמן.

1 וישינה ליזורדבא‎ג.   7 להם] לה פ – ההויה וההפסד פ.   9 היתה] תהיה ד.   10 התאמתי
יתבאר ו – שיתנועעו ור.

motions,[4] by reason of the fact that it does not take place between opposite boundaries,[5] that its velocity is not subject to variation,[6] that the substance to which it is peculiar is incapable of change,[7] nay, that in everything it maintains the character of perfect actuality.[8]

PART II.

EXAMINATION of the fourteenth proposition, which reads: 'Locomotion is prior to all the other kinds of motion and is the first of them in nature, for generation and corruption are preceded by alteration, which in its turn, is preceded by the approach of that which alters to that which is to be altered, and, similarly, growth and diminution are impossible without previous generation and corruption.'

With reference to relative generation,[9] the proposition may be accepted as true. With reference, however, to the first generation, if it is *ex nihilo*, in the manner that will be explained,[10] it can be shown that it is generation which precedes all the other motions,[11] and that qualitative and quantitative motions precede locomotion, for things must have possessed qualitative and quantitative properties before they began to be moved [in place],[12] and, finally, that absolute quantity precedes quality.[13]

PROPOSITION XV

PART I.

PROOF of the fifteenth proposition, which reads: 'Time is an accident that is consequent on motion and is conjoined with it. Neither one of them exists without the other. Motion does not exist except in time, and time cannot be conceived except with motion, and whatsoever is not in motion does not fall under the category of time.'[1]

ההקדמה הזאת כוללת ארבע הקדמות. האחת, היות הזמן מקרה.
והשנית, היותו דבק לתנועה באופן שלא ימצא אחד מהם בלתי
האחר. והשלישית, שלא יושכל זמן אלא עם תנועה. והרביעית,
שמה שלא תמצא בו תנועה אינו נופל תחת הזמן. והנה יתבארו
5 בבאור גדר הזמן.

ואמנם ארסטו, ואם היה שהתחלפו בו הקדמונים בסברתם חלוף
רב, אין צורך לזכרם להיותם מבוארי ההפסד, הנה גדרו בשהוא
מספר הקודם והמתאחר בתנועה.

וזה שאין ספק הצטרכו אל נושא, להיותו בלתי עומד כלל, וכל
10 שכן שיהיה עומד בעצמו כמו הדברים שלא יצטרכו אל נושא. וזה
שהזמן יחלק אל עבר ואל עתיד, כי ההוה הוא עתה, והוא בלתי
נמצא, ואיננו זמן, והעבר כבר נפסד, והעתיד איננו עדיין. ולזה
הנה הצטרכו אל נושא מבואר בעצמו. והיא ההקדמה הראשונה
מאלו הארבע.

15 ולפי שאנחנו נראה שאנחנו נשער התנועה המהירה והמאוחרת
בזמן, וזה שהתנועה המהירה היא אשר יתנועע המתנועע בה שיעור
ידוע בזמן יותר קצר מהמאוחרת, הנה התבאר שהזמן איננו תנועה,
כי לא ילקח הזמן בגדר עצמו. ולהיות המהירות והאיחור בתנועה
מקרה דבק בה ובלתי נפרד ממנה, והיה שנשער אותם בזמן, נתאמת
20 שהוא מקרה דבק לתנועה, והוא ההקדמה השנית.

וכאשר היה זה כן, והיה הזמן משער לעולם התנועה איך
שלוקחה, אם בבחינת מהירות ואיחור אם בבחינת הקודם והמתאחר
ממנה, כבר יצדק אמרנו בגדרו שהוא מספר הקודם והמתאחר

2 מבלתי לו דקבאין. 4 ימצא יא — בן לו לירדא — התבארו פ. 6 (בו) סי. 8 המספר ר.
11 כי הזמן ירא — ואל עתיד [ואל ההוה] פ [ואל ההוה] ואל עתיד א. 13 (הנה) יצטרכו פ.
17 (בזמן) יד. 19 (בה) ירקאיי — בלתי לד. 21 (משער) לד.

This proposition contains four premises.[2] First, time is an accident.[3] Second, time is conjoined with motion in such a manner that neither one of them exists without the other.[4] Third, time cannot be conceived except with motion.[5] Fourth, whatsoever is not in motion does not fall under the category of time.[6] All these premises may be proved by the following discussion of the definition of time.

In contradistinction to all the ancients, who held widely different views with regard to time[7]—views which may be disregarded on account of their notorious untenability[8]—Aristotle defines time as the number of priority and posteriority of motion.[9]

Time no doubt needs a subject, for time itself has no existence whatsoever, still less can it exist in itself after the manner of things which are in no need of a subject.[10] For time is divided into past and future, inasmuch as the present is only an instant, which has no existence, and is not time. Now the past is always gone, and the future is never yet arrived; whence it is self-evident that time needs a subject.[11] Hence the *first* of the four premises.

Since we are accustomed to measure swift and slow motion by time, for swift motion is [defined as] that by which an object traverses a certain distance in less time than by motion called slow, time cannot be identical with motion, for time cannot be included in the definition of [that which is identical with] itself.[12] Yet,[13] on the other hand, since swiftness and slowness, which are measured by time, are accidents adjoined to motion and inseparable from it,[14] it follows that time must also be an accident adjoined to motion. Hence the *second* premise.

This being the case, namely, that time is always the measure[15] of motion, whether taken with respect to swiftness and slowness or with respect to priority and posteriority,[16] we are therefore justified in framing the definition of time by saying that it is number of priority and posteriority of motion. The term motion

בתנועה. ולפי שלוקחה התנועה בגדרו, נתבארה ההקדמה
השלישית והוא שלא יושכל הזמן אלא עם תנועה.

ואמנם ההקדמה הרביעית, שהיא אמרנו שמה שלא תמצא בו
תנועה אינו נופל תחת הזמן, היא מבוארת בעצמה, כשיתבאר ענין
5 הנפילה תחת הזמן, והוא הדבר שינבילהו הזמן ויעדיף עליו משתי
קצותיו. ולזה היו הדברים הנצחיים אינם נופלים תחת הזמן בעצם,
כי לא ינבילם הזמן ולא יעדיף עליהם. ואם היה שיהיו נופלים תחת
הזמן, הוא במקרה, והם אשר היו מהם מתנועעים, כי למה שהתנועה
כבר ינבילה הזמן, כשנקח חלק ממנה, כבר יהיו המתנועעים נופלים
10 תחת הזמן במקרה, מצד תנועתם. ואמנם הנבדלים. להיותם בלתי
מתנועעים, אינם נופלים תחת הזמן לא בעצם ולא במקרה.

הכלל השני, הפרק האחד עשר

בחקירה בההקדמה החמש עשרה האומרת כי הזמן מקרה נמשך
לתנועה ודבק עמה, לא ימצא אחד משניהם מבלתי האחר, לא
15 תמצא תנועה כי אם בזמן, ולא יושכל זמן אלא עם התנועה, וכל
מה שלא תמצא לו תנועה אינו נופל תחת הזמן.

ואומר שכאשר נדקדק בגדר הזמן נמצא ההקדמות הארבע
הנכללות בהקדמה הזאת, כמו שקדם לנו בכלל הראשון, כוזבות.
כי למה שהוא מבואר בעצמו, שכבר יאמר במנוחה גדולה כאשר
20 נח דבר מה זמן גדול, וקטנה כאשר נח זמן מועט, הנה מבואר שהזמן
ישוער במנוחה מזולת מציאות התנועה בפעל. ואם היה שנשער

is thus included in the definition; hence it proves the *third* premise, namely, that time cannot be conceived except with motion.

As for the *fourth* premise, namely, whatsoever is not in motion does not fall under the category of time, it will become self-evident when it is made clear that the expression "falling under the category of time" applies only to an object which is comprehended by time and transcended by it on both ends.[17] Consequently, the eternal beings are not essentially in time,[18] inasmuch as they are not comprehended and transcended by time. If they are sometimes said to be in time, it is only accidentally, and that, too, is true only of some of them, namely, of those that are endowed with motion,[19] Thus the movable [eternal] beings, on account of their motion, may be duly said to be in time, inasmuch as motion can always be made to be comprehended by time, as when, for instance, we take any finite part thereof.[20] The separate [Intelligences], however, having no motion whatsoever, are neither essentially nor accidentally in time.[21]

Part II.

Examination of the fifteenth proposition, which reads: 'Time is an accident that is consequent on motion and is conjoined with it. Neither one of them exists without the other. Motion does not exist except in time, and time cannot be conceived except with motion, and whatsoever is not in motion does not fall under the category of time.'

I say that when we closely examine the definition of time, we shall find that the four premises which this proposition contains, as has been shown in the first part, are all false. For it is self-evident that rest is described as long when an object remains at rest for a long time, and as short when it remains so only for a short time, whence it must follow that time is measured by rest without the presence of actual motion. Even if it were admitted

המנוחה בציורנו שיעור המתנועע בה, הנה יתאמת שאין צורך
מציאות התנועה בפעל בזמן. וכל שכן שהמנוחה, בזולת ציורנו
בתנועה, כבר תתחלף בפעל ברב ובמעט. וכאשר היה זה כן, הנה
מי יתן ואדע למה לא ישוער הזמן בה בזולת ציורנו התנועה. ולזה
5 הגדר הנכון בזמן יראה, שהוא שיעור התדבקות התנועה או המנוחה
שבין שתי עתות. וכבר יראה שהסוג היותר עצמי לזמן הוא שיעור,
כי להיותו מהחכמה המתדבק והמספר מהמתחלק, היה הניחנו אותו
מספר סוג בלתי עצמי וראשון. ואמנם שוער בתנועה ובמנוחה, למה
שציורנו בשיעור התדבקותם הוא הזמן. ולזה יראה היות מציאות
10 הזמן בנפש. וכאשר היה זה כן, הנה ההקדמה ה ר א ש ו נ ה, והיא
האומרת היות הזמן מקרה, כשרצינו בו שאיננו עצם, היא אמתית, ואם
רצינו בו היותו מקרה נמצא חוץ לנפש, היא כוזבת, למה שהוא נתלה
במנוחה כמו בתנועה, והמנוחה היא העדר התנועה, ואין מציאות
להעדר. ולזה יתחייב שיהיה הזמן נתלה בציורנו שיעור התדבקות
15 אם בתנועה ואם במנוחה, אחר שיאמר בכל אחת גדולה או קטנה.

ואולם ה ש נ י ת, והיא האומרת היות הזמן דבק לתנועה באופן
שלא ימצא האחד מהם בלתי האחר, כוזבת גם כן, שכבר ימצא זמן
בזולת תנועה, והוא המשוער במנוחה, או בציור התנועה ואם היה
שלא תמצא בפעל.

that we measure rest only by supposing a corresponding measure of the motion of an object moved during the same interval,[22] it would still follow that actual motion is not necessary in the conception of time. The argument is all the stronger in view of the fact that rest, without any supposition on our part of a corresponding [actual] motion, can actually be distinguished as long and short. Such being the case, would that I knew, why time should not be measured by rest alone, without our supposing a corresponding motion? Hence it is evident that the correct definition of time is that it is the measure of the duration of motion or of rest between two instants.[23] It is, moreover, evident that the genus most essentially appropriate of time is magnitude,[24] for as time belongs to continuous[25] quantity and number to discrete,[26] if we describe time as number, we describe it by a genus which is not essential nor primary.[27] It is indeed measured by both motion and rest, because it is our supposition of the measure of their duration that is time. It seems therefore that the existence of time is only in the soul.[28] Such being the case, the *first* of these premises, stating that 'time is an accident,' is true only if we thereby mean that it is not a substance;[29] but if we mean thereby that time is an accident existing outside the soul, it is false,[30] for time depends as much upon rest as upon motion, and rest is the privation of motion and privation has no existence. It thus follows that time depends upon our supposition of the measure of the duration of either motion or rest, inasmuch as either of them may be described as great and small.

As for the *second*, stating that time is joined to motion in such a manner that neither one of them exists without the other, it is likewise false, for time may exist without motion, namely, that time which is measured by rest or by the supposition of motion without its actual existence.

ואולם השלישית, והיא האומרת שלא יושכל זמן אלא עם

תנועה, גם כן כוזבת מזה הצד, אלא שנאמר, כי למה שהמנוחה היא

העדר התנועה, כשנשער הזמן במנוחה, נשכיל התנועה. אבל שלא

יושכל זמן אלא בשיהיה עם תנועה, הנה לא.

5   ואולם הרביעית, והיא האומרת שמה שלא תמצא בו תנועה

אינו נופל תחת הזמן, הנה הנבדלים. ואם היו בלתי מתנועעים, כבר

נפלו תחת הזמן, כאשר יתאמת שנתהוו כשהזמן היה קודם להם, למה

שאין מהכרח הזמן מציאות התנועה בפעל, אלא ציור שיעור התנועה

או המנוחה. ולזה יתאמת מאמר רבי יהודה בר רבי סימון כפשוטו,

10   והוא אמרו מלמד שהיה סדר זמנים קודם לכן. וגם לא

יצטרך לרחוק בפירוש הרב בכתוב הראשון שבתורה, והוא אמרו

בראשית ברא שיהיה ענינו בהתחלה, שכבר יהיה כפל ומותר,

שאם בראו הנה היה התחלה וסבה לו; ולומר שתאר הבריאה היה

בענין שהיה התחלה וסבה בלבד, חלילה לו לרב מהדעת הזה,

15   בשכבר האריך והרחיב הדבור בבטול ראיות ארסטו על הקדמות,

וחדש טענות מספיקות לאמת אמונת החדוש, כמו שיבא בגזרת הצור.

As for the *third*, stating that 'time cannot be conceived except with motion,' it is equally false and for the same reason. What we may reasonably maintain is that, since rest is the privation of motion, when we measure time by rest, we inevitably conceive of motion; but to say that the idea of time cannot be conceived except it be connected with motion must be denied.

As for the *fourth*, stating that 'whatsoever is not in motion does not fall under the category of time,' the Intelligences, though immovable, may still have existence in time,[31] inasmuch as it can be demonstrated that time existed prior to their creation on the ground that time does not require the actual existence of motion, but only the supposition of the measure of motion or rest.[32] In view of this, the passage of Rabbi Jehudah, son of Rabbi Simon,[33] which reads: 'It teaches us that the order of time had existed previous to that,' may be taken in its literal sense. Nor will there be any more need, [if we admit the existence of time prior to creation], to go as far afield as the Master in the interpretation of the first verse of Genesis and take the words *Bereshit bara* [*Elohim*] to mean that 'In being Himself the principle, [i. e., the cause], God created heaven and earth,[34]—an interpretation which renders the verse tautological and redundant, for, if He created the world, He surely was its cause and principle. To say that [what the Master means is that] the manner of creation was suchwise that God was nothing but a principle and cause[35]—far be it from him to entertain such a view, for previously[36] he has already discoursed at great length and in full detail upon the refutability of Aristotle's proofs for eternity and has also adduced convincing arguments in support of the belief in creation, as will be shown later,[37] God willing.

הכלל הראשון, הפרק הששה עשר

בבאור ההקדמה השש עשרה האומרת כי כל מה שאינו גוף לא
יושכל בו מניין, אלא אם יהיה כח בגוף, וימנו אישי הכחות ההם
בהמנות החמרים שלהם או נושאיהם, ובעבור זה העניינים הנבדלים,
אשר אינם גוף ולא כח בגוף, לא יושכל בהם מניין כלל אלא בהיותם
עלות ועלולים.

הנה להיות מהות המין הכולל אישים מתחלפים במספר הוא
אחד במין רבים במספר, הוא מבואר שלא יושכל בו מספר אלא
להלוף המקום או הזמן או מקרה מהמקרים הנמצאים בו.

ולהיות מה שאינו גוף ולא כח בגוף נבדל, והוא בלתי נופל תחת
הזמן, במה שקדם, ובלתי מוגבל במקום, ולא ייוחס לו מקרה
מהמקרים, הנה הוא מבואר שלא יושכל בנבדלים מניין אלא מחלוף
הנמצא בם, והוא בהיותם עלות ועלולים.

הכלל השני, הפרק השנים עשר

בחקירה בהקדמה השש עשרה האומרת כי כל מה שאינו גוף לא
יושכל בו מניין, אלא אם יהיה כח בגוף, וימנו אישי הכחות ההם
בהמנות החמרים שלהם או נושאיהם, ובעבור זה העניינים הנבדלים,

3 היה פ. 5 (בהם) ר מהם י. 7 (המין) פ – (במספר) פ׳. 8 במין כמו פצ – (במין רבים)
וק בג. 9 מקרה [מה) פ. 11 במה] כמו לור – ייוחס] יוחד א – לח בו זרא. 12 בחלוף
לרדבאנ. 13 בהם לור דבי.

PROPOSITION XVI

Part I.

Proof of the sixteenth proposition, which reads: 'Whatsoever is not a body does not admit of the idea of number except it be a force in a body, for then the individual forces may be numbered together with the matters or subjects in which they exist. It follows, therefore, that separate beings, which are neither bodies nor forces in bodies, do not admit of any idea of number except when they are related to each other as cause and effect.'[1]

Inasmuch as the quiddity of a species which embraces numerically different individuals is one in species but many in number, it is self-evident that no number can be conceived in that quiddity except with reference to some distinction arising from time, place, or some other accident which may happen to exist in the particular.[2]

Now, that which is neither a body nor a force in a body is called a separate being,[3] and this, according to the preceding proposition, does not fall under the category of time,[4] nor is it bounded by place,[5] nor can any of the accidents be attributed to it.[6] Hence it follows that no numerical plurality can be conceived in separate beings except with reference to some distinction which is appropriate to them, and such a distinction may be found among them when they are related to each other as cause and effect.[7]

Part II.

Examination of the sixteenth proposition, which reads:' Whatsoever is not a body does not admit of the idea of number except it be a force in a body, for then the individual forces may be numbered together with the matters or subjects in which they exist. It follows, therefore, that separate beings, which are

שאינם גוף ולא כח בגוף, לא יושכל בהם מניין כלל אלא בהיותם

עלות ועלולים.

הנה כבר יראה שההקדמה הזאת גם כן כחבת, למה שהנפשות

הנשארות אחר המות כבר יושכל בהם מניין בהכרח. וזה שלא ימלט

5 מחלוקה, והוא אם שיהיה הנשאר אחר המות עצם הנפש השכלי,

או שיהיה השכל הנקנה לאדם באמצעות חושיו וכחותיו. ואם הוא

עצם הנפש, כבר תיוחד כל אחת מהנפשות במה שהשיגה

מהמושכלות או מהדבקות בשם יתברך, ומה שהשיגה האחת כבר

יתחלף במה שהשיגה האחרת. ולזה כבר ימנו כאשר ימנו אישי

10 העצם, למה שלכל אחד מקרים ייחדוהו עם היות המהות אחד.

ואם הנשאר הוא השכל הנקנה, הוא מבואר שהמושכלות הנקנות

לנפש האחת כבר תתחלפנה לנפש האחרת. ולזה כבר ימנו מבלתי

שיהיו עלות ועלולים.   ולומר שהנשאר הוא ההכנה שתדבק עם

השכל הפועל ותתאחד עמו, ולזה יהיה המניין בהם נמנע, הדעת

15 הזה כבר יתבאר במה שיבא שהוא דעת נפסד, וחלילה לו לרב

מהיותו בעל זה הדעת. אלא שיראה שכוון הרב באמרו ה ענ יינים

ה נ ב ד ל י ם שהיו לעולם נבדלים, ולא היו כחות בגוף במה שעבר.

1 שאינם] אשר אינם ב אי.   4 ימלט [הענין] י.   5 [אם] י.   6 אם (שיהיה) לד.   8 (מהמושכלות...
שהשיגה] ז – מהדבקות] הדבקות פ – (יתברך) לד.   9–8 (כבר ... האחרת) י.   9 יתחלף]
יתחלקו לד תתחלף ר – שהשיג האחר ב אי.   10 ולמה פ – ייחדוהו פ אי – המהות] המיחדות י –
אחד] אחת פ.   12 אחת פ לד רק ב אי.   13 [ולומר] ואין לומר לד – הכנה פ – שתתדבק ל
תתדבק ר.   14 [כי] הדעת לד.   15 שהוא] הוא צ.   16 באמרו א במאמרו פ.   17 [שהיו]
שהיו א צ פ לו רד.

neither bodies nor forces in bodies, do not admit of any idea of
number except when they are related to each other as cause and
effect.'

This proposition, too, can be shown to be false, in view of the
fact that the souls which remain immortal after death must
necessarily admit of the idea of number. For the following disjunctive reasoning is unavoidable, namely, that the part immortal is either the substance of the rational soul itself[8] or the
intellect acquired[9] by man by means of his senses and faculties.[10]
Now, if it is the substance of the rational soul itself, then each
soul is possessed of an individulaity according to its attainments
in intellectual conceptions or in its union with God,[11] blessed be
He, for the attainments of one soul must differ from those of
another. This being the case, souls should be numerable in the
same manner as individual corporeal substances,[12] which, though
being all one in essence, are numerable on account of their each
having accidents by which they are individualized. And if the
immortal part is the acquired intellect, the case is still clearer,
for the intellectual conceptions acquired by one soul are different
from those acquired by another. Thus the souls of the departed
may be numbered even though they are not related to each other
as cause and effect. To say that the part immortal is only the
predisposition which unites with the Active Intellect and becomes
one with it,[13] whence indeed the souls of the departed could not
be subject to number—to say this would be to maintain a view
which will be shown later[14] to be erroneous, and far be it from the
Master to espouse it. It must, therefore, be concluded that in
using the expression "separate beings," the Master means only
to refer to such beings as have always existed apart from matter
and had not been previously forces in a body.[15]

הכלל הראשון, הפרק השבעה עשר

בבאור ההקדמה השבע עשרה האומרת שכל מתנועע לו מניע
בהכרח, אם חוץ ממנו, כאבן תניעה היד, או יהיה מניעו בו, כגשם
החי, כי הוא מחובר ממניע ומתנועע, ולזה כאשר מת ונעדר ממנו
המניע, והוא הנפש, ישאר המתנועע, והוא הגשם, במקום כמו שהיה,
5 אלא שהוא לא יתנועע אותה התנועה. ולמה שהיה המניע הנמצא
במתנועע נעלם בלתי נראה לחוש, נחשב בחי שהוא מתנועע בלתי
מניע. וכל מתנועע יהיה מניעו בו, הנה הוא אשר יקרא מתנועע
מצדו, עניינו שהכח המניע למה שיתנועע ממנו בעצמות, נמצא
10 בכללו.

יסוד ההקדמה הזאת לבאר שכל מתנועע יש לו מניע. והנה לפי
שהמתנועע, אם שיתנועע בטבע, כתנועת האבן אל המטה, ואם
בהכרח, כתנועת האבן אל המעלה, ואם בבחירה, כתנועת הבעל
חי, הנה המתנועעים בהכרח ובבחירה הדבר בהם מבואר שהמניע
15 בהם זולת המתנועע. ואולם המתנועע בטבע יתבאר מזה, למה
שנמצאו המתנועעים בטבע מתחלפים בצד, וזה שתנועת האבן אל
המטה ותנועת האש אל המעלה, הוא מחוייב שאין התנועה לו במה
שהוא גשם בשלוח, שאם היה כן לא היו מתנועעים בצדדים מקבילים,

---

2 ויש] לו לרד. 3 מחוץ קבג – [אשר] תניעה זרא תניענו ד – שיהיה פ. 4 מהמניע פ לרבג –
והמתנועע ד – נעדר פ ויעדר י. 7 יחשבי נחשוב קפא. 9 מצדו] מעצמו ס זרא – ועניינו קי –
שהכח] שהונח פ. 11 [הזאת] פ זאת בג – (יש) פא. 12 כתנוע [האש למעלה] והאבן זרא.
13 טעלה פ. 15 מזה] ממה שאומר בג למה שאומר א (מזה) זר. 16 שתנועת האבן] שהאבן
התנועע זר ק בג שהאבן מתנועע א. 17 מטה ו – מעלה י.

PROPOSITION XVII

PROOF of the seventeenth proposition, which reads: 'Everything moved must needs have a mover, which mover may be either without the object moved, as, e. g., in the case of a stone set in motion by the hand, or within the object moved, as, e. g., the body of a living being, for a living being is composed of a part which moves and a part which is moved. It is for this reason that when an animal dies and the mover, namely, the soul, is departed from it, the part that is moved, namely, the body, remains for some time in the same condition as before and yet cannot be moved in the manner it has been moved previously. But inasmuch as the mover, when existing within the object moved, is hidden from the senses and cannot be perceived by them, an animal is thought to be something that is moved without a mover. Everything moved which has its mover within itself is said to be moved by itself, which means that the force by which the object moved is moved essentially exists in the whole of that object.'[1]

The main purpose of this proposition is to show that everything moved has a mover.[2] For every object in motion, is moved either by nature, as, e. g., the motion of a stone downward, or by violence, as, e. g., the motion of a stone upwards, or by volition, as, e. g., the motion of a living being.[3] Now, in the case of objects moved either by violence or by volition, it is evident that the motive agent is something different from the object moved.[4] But that the same holds true in the case of an object that is moved by nature will become clear from the following consideration:[5] Objects which are moved by nature are found to vary with respect to the direction of their motion; thus, e. g., the tendency of a stone is downward, whereas that of fire is upward. This seems to indicate that the motion of each element is not simply due to the fact that it is a body in the absolute, for, were it so, the elements would not each move in an opposite direction.

אלא שהתנועה המיוחדת לכל אחד, במה שהוא זה הגשם. ולהיותם

שוים ומשותפים בנשמות, הנה אם כן צורת כל אחד המיוחדת היא

המניעה התנועה ההיא, באמצעות הכח אשר שם בה, והוא הנקרא

טבע. ולזה היה טבע כל אחד הוא המניע.

הכלל הראשון, הפרק השמונה עשר

בבאור ההקדמה השמונה עשרה האומרת שכל מה שיצא מן הכח

אל הפעל מוציאו זולתו, והוא חוץ ממנו בהכרח, כי לו היה המוציא

בו ולא יהיה שם מונע, לא היה נמצא בכח עת אחד, אבל היה בפעל

תמיד. ואם היה מוציאו בו, והיה לו מונע והוסר, אין ספק שמסיר

המונע הוא אשר הוציא אותו מן הכח אל הפעל. וחתם ההקדמה

הזאת באמרו, והבן זה.

ההקדמה הזאת כבר תתאמת בחפוש. וזה כי מה שיאמר עליו

שהוא בכח דבר, הנה יהיה אם בפועל ואם במתפעל. והנה

במתפעל, אם שיהיה בעצם אם במקרים. ואמנם בעצם, בהויה

והפסד, אין ספק שמוציא הכח בהם זולתם, למה שהוא מבואר

It must rather be the fact that each element is a particular kind of body that accounts for its particular motion. Now, with reference to corporeality all elements are alike and they all share it in common. Consequently, it is their respective proper forms that must be assumed to bring about their diverse natural motions,[6] and that, indeed, by means of a force implanted in form, which force is called nature.[7] The nature of an element may thus be considered as its motive cause.

PROPOSITION XVIII

PROOF of the eighteenth proposition, which reads: 'Everything that passes from potentiality to actuality has something different' from itself as the cause of its transition and that cause is necessarily outside itself, for if the cause of the transition existed in the thing itself and there was no obstacle to prevent the transition, the thing would never have been in a state of potentiality but would have always been in a state of actuality; and if the cause of the transition, while existing in the thing itself, encountered some obstacle which was afterwards removed, then the same cause which has removed the obstacle is undoubtedly to be considered as the cause which has brought about its transition from potentiality to actuality.' The author concludes this proposition by saying 'Note this.'[1]

This proposition may be proved inductively as follows:[2] Whenever it is said of anything that it is potentially a certain thing, it means that it is either potentially an agent or potentially a patient. In the latter case, again, the potentiality to suffer action may refer either to a substance or to accidents.[3] Now, in the case of substance, as, e. g., the process of generation and corruption,[4] there can be no doubt that the cause that brings about the realization of this potentiality of generation or corruption is not identical with the substances themselves, for it is well

שהדבר לא יהוה עצמו ולא יפסיד עצמו. ואמנם במקרים, בשינוי

בכמה ובאיך ושאר המאמרות, הנה להצטרכם אל נושא, אין ספק

שהכח אשר בנושא יפעלם ויוציאם מן הכח אל הפעל. ואולם

בבחינת הפועל, וזה כשנאמר בדבר שהוא פועל לדבר בכח, אין

5 ספק שהכח אם שיהיה בו או חוץ ממנו. ואם הוא חוץ ממנו, הנה

מוציאו זולתו. ואם הוא בו, הנה למה שהכח בו לפעול, אם לא יהיה

לו מונע ולא יחסר בו תנאי, הנה יהיה בפעל תמיד. ולזה אם לא

יהיה בפעל תמיד, הוא מפני שהיה לו מונע, ולזה מסיר המונע הוא

המוציא.

10 והנה צריך שנתבונן בזה הרבה, כי אמרנו בדבר שהוא בכח כך,

הנה יחייב שינוי במתפעל בהכרח. ואמנם בפועל, אם יהיה הכח

בו לפעול ויש לו מונע מצד המקבל, הנה אם היה שהמסיר המונע

---

known that nothing can generate or corrupt itself.[5] Likewise in the case of accidents, as, e. g., the change of quantity, quality, and the other categories,[6] it is clear beyond any doubt that since all these accidents must needs have a subject for their existence, it will be the force contained in that subject that will energize them and cause them to pass from potentiality into actuality.[7] In like manner, in the case of a potential agent, as, e. g., when we assert of something that it is the potential agent of something else,[8] there is no doubt that the potentiality must reside either within the agent itself or without it. If it is without the agent, then it need hardly be said that the cause which brings about the transition from potentiality to actuality is likewise without. And if the potentiality resides within the agent itself, then, if the agent is assumed to encounter no obstacle nor to be hindered in its action by the lack of some required condition, it would have to be permanently in a state of actuality, since the capacity to act resides within itself. As the agent is not, however, permanently in a state of actuality, we must assume, of course, that the cause of its inactivity is due to some kind of obstacle, and so whatsoever causes the removal of that obstacle must be considered as the cause of the transition.[9]

We must, however, bear in mind the following distinction: When we assert of anything that it possesses a certain potentiality, if that potentiality is one to receive action, then the thing in question, [upon the realization of its potentiality], must indeed undergo some change. In the case of a potentiality to act, however, it is altogether different. For when an agent has the potentiality to act, but is prevented from acting on account of some obstacle on the part of that which is to be the recipient of the action, then, though the remover of that obstacle may still

הוא המוציא מן הכח אל הפעל, אבל לא יחוייב שינוי בפועל, ולזה

מה שהעיר במקום הזה וחתם ההקדמה הזאת באמר, והבן זה.

הכלל הראשון, הפרק התשעה עשר

בבאור ההקדמה התשע עשרה האומרת שכל אשר למציאותו

5 סבה הוא אפשר המציאות בבחינת עצמותו, כי אם נמצאו סבותיו,

נמצא, ואם לא נמצאו, או נעדרו, או השתנה יחסם המחייב

למציאותו, לא ימצא.

והיא מבוארת בעצמה, כי מה שלמציאותו סבה, אם שיהיה

מחוייב בבחינת עצמו או נמנע או אפשר, כי טבע החלוקה כן חייב.

10 ואיננו מחוייב לעצמותו, כי מה שהוא מחוייב לעצמותו, לא יצוייר

העדרו בהעדר זולתו, ומה שלמציאותו סבה, הנה העדרו מחוייב

בהעדר סבתו. ואיננו גם כן נמנע לעצמותו, כי מה שהוא נמנע

מציאותו, אי אפשר שיהיה למציאותו סבה. מחוייב אם כן שיהיה

אפשר בבחינת עצמו, רוצה לומר, שמציאותו, נצחי היה או בלתי

15 נצחי, אפשר שיהיה מצוייר ההעדר בהעדר סבתו.

be called the cause of the transition from potentiality to actuality, yet this fact does not imply that the agent in question must itself undergo a change.[10] It is with reference to this distinction that the author has made his cryptic remark and concluded the proposition by saying "Note this."

PROPOSITION XIX

PROOF of the nineteenth proposition, which reads: 'Everything that has a cause for its existence is in respect to its own essence only possible of existence, for if its causes exist, the thing likewise will exist, but if its causes have never existed, or if they have ceased to exist, or if their causal relation to the thing has changed, then the thing itself will not exist.'[1]

This proposition is self-evident.[2] For a thing which has a cause for its existence must in respect to its own essence be necessary, impossible, or possible, these being the only alternatives conceivable. Now, in respect to its own essence it cannot be necessary, for whatsoever is necessary in respect to its own essence cannot be conceived as non-existent, even were there no cause in existence;[3] whereas that which has a cause for its existence would have to be non-existent were its cause not to exist. Nor can it in respect to its own essence be impossible, for whatsoever is in respect to its own essence impossible precludes the possibility of there being a cause to bring about its existence. Hence in respect to its own essence it must be only possible, that is to say, its existence, be it eternal or transient, might be conceived as non-existent were its cause not to exist.[4]

הכלל הראשון, הפרק העשרים

בבאור ההקדמה העשרים האומרת שכל מחוייב המציאות
בבחינת עצמותו, הנה אין סבה למציאותו באופן מהאופנים ולא
בעניין מהעניינים.

ההקדמה הזאת גלויית האמת משלפניה, מהפך הסותר. וזה כי
אשר למציאותו סבה איננו מחוייב המציאות, יחוייב בהכרח
שהמחוייב המציאות אין למציאותו סבה. והפלא איך לא חברה עם
התשע עשרה.

הכלל הראשון, הפרק האחד ועשרים

בבאור ההקדמה האחת ועשרים האומרת שכל מורכב משני
עניינים הנה אותה ההרכבה היא סבת מציאותו על מה שהוא עליו
בהכרח, ואינו מחוייב המציאות לעצמותו, כי מציאותו במציאות
חלקיו ובהרכבתם.

הנה למה שחלקי הדבר זולת כללות הדבר, והדבר בכללו הוא
מורכב, הנה אם כן המורכב למציאותו סבה, וכבר קדם לנו שאשר
למציאותו סבה איננו מחוייב המציאות. המורכב אם כן איננו מחוייב
המציאות.

5 (כי) סל/ ז ור ד ק ב א ו.   6 שאשר לו ד.   7—6 (איננו ... סבה) פר א.   7 חברה] זכרה ל
10 האחת ועשרים] הכוללת ב ג.   11 שהוא] שהיו י.   13 והרכבתם ז [הוא] והרכבתם ר א
בהרכבתם פ.   14 שחלק פ – והדבר] והנה ג.   15 (הנה) צ – (אם כן) ז – [הוא] סבה פ.
16—15 (וכבר ... סבה) י.   16 איננו] ואיננו י.   17—16 (המורכב ... המציאות) ל ז ור ד ב א ו.

PROPOSITION XX

PROOF of the twentieth proposition, which reads: 'Everything that is necessary of existence in respect to its own essence has no cause for its existence in any manner whatsoever or under any condition whatsoever.'[1]

This proposition may be proved from the preceding one by the conversion of the obverse,[2] for since that which has a cause for its existence is not necessary of existence, it must inevitably follow that that which is necessary of existence has no cause for its existence. I wonder why he did not combine this proposition with the nineteenth.[3]

PROPOSITION XXI

PROOF of the twenty-first proposition, which reads: 'Everything that is composed of two elements has necessarily their composition as the cause of its existence as a composite being, and consequently in respect to its own essence it is not necessary of existence, for its existence depends upon the existence of its component parts and their combination.'[1]

Inasmuch as the parts of a thing are different from the whole of the thing and the thing as a whole exists only as something composed of those parts, it follows that that which is composed of parts has a cause for its existence.[2] But it has already been shown that a thing which has a cause for its existence cannot be necessary of existence.[3] Nothing composite, therefore, can be necessary of existence.

הכלל הראשון, הפרק השנים ועשרים

בבאור ההקדמה השתים ועשרים האומרת שכל גשם הוא מורכב
משני עניינים בהכרח, וישיגוהו מקרים בהכרח. אולם השני עניינים
המעמידים אותו–חמרו וצורתו. ואולם המקרים המשינים אותו–
5 הכמה והתמונה והמצב.

הנה להכרח מציאות נושא להויה והפסד, חוייב מציאות החמר.
ולהיות החמר בעצמו משולל מכל צורה, למה שאם היה לו צורה
היה ההויה השתנות ולא הוייה, ולכן אשר ייחדהו ויגבילהו וישימהו
נמצא בפעל נרמז אליו הוא הצורה. התבאר אם כן שהדברים
10 המעמידים אותו הוא החמר והצורה.

ולהיות המקרים יצטרכו אל נושא, ומהם מתפרדים אל הנושא,
ומהם בלתי מתפרדים, הנה אשר הם בלתי מתפרדים הם הכמה,
שלא יצוייר הגשם זולתו, והתמונה, אשר במאמר האיך, שלא יפרד
מן הגשם, למה שהיה רושם התמונה שהיא אשר יגבילה קו או קוים,
15 והמצב, שהוא יחס חלקיו קצתם אל קצת ואל הגשמים אשר מחוץ.
והנה נתיחדו אלו, למה שהם בלתי מתפרדים מהגשם. והוא אשר
רצהו באמרו, וישיגוהו מקרים בהכרח, ופירש הכמה
והתמונה והמצב.

3 שני העניינים ר. 6המציאות ד. 6המציאות א – הנושא פ ב ג – ההויה וההפסד ז להויה וההפסד פ להוה
הפסד זא להויה ההפסד וד להויה ולהפסד ק. 8(ולכן פלורדקבאג – ייחדו ז ייחדהו
וג – ויגבילהו ו – וישימהו ו ישימהו פ. 9ורמח לודקבאג רסמ ז – (הוא) י והוא ל –
נתבאר ז. 11הצטרכו זרא – (אל) מהנושא פ ז. 12 (אשר) לד. 14יגבילהו פ ב. 17רצהו
פ רצה ג – המקרים פ – (בהכרח ופירש) ר.

PROPOSITION XXII

PART I.

PROOF of the twenty-second proposition, which reads: 'Every body is necessarily composed of two elements, and is necessarily subject to accidents. The two constituent elements of a body are matter and form. The accidents to which a body is subject are quantity, figure, and position.'[1]

The existence of matter is deducible from the necessity of postulating the existence of a subject underlying the process of generation and corruption. Matter, however, is itself absolutely formless, for if it had any kind of form, substantial change would not be generation but rather alteration; it follows therefore that it is form which confers upon matter individuality and definiteness and renders it a 'this' in actuality.[2] It has thus been shown that matter and form are the constituent elements of every body.[3]

Accidents are likewise in need of a subject, and there are some accidents which are separable from their subject while there are others which are inseparable.[4] Now, those which are inseparable are *quantity*, without which no body can be conceived, *figure*, which belongs to the category of quality,[5] and, being defined as something bounded by any line or lines,[6] is inseparable from body, and *position*,[7] by which is meant the relation of the respective parts of a body to each other and the relation of the body as a whole to other bodies.[8] Thus these three accidents are distinguishable from the others by reason of their being inseparable from the body, and it is these accidents that were meant by the author when he said that a body 'is necessarily subject to accidents,' as he himself immediately makes it clear by mentioning 'quality, figure, and position.'

הכלל השני, הפרק השלשה עשר

בחקירה בהקדמה השתים ועשרים שכל גשם הוא מורכב משני

עניינים בהכרח, והם שני עניינים המעמידים אותו, אשר הם חמרו

וצורתו.

5   הנה זאת חקרנוה בפרק השביעי מהכלל הזה. ולפי דעת אבן

רשד איננו מוכרח, ואבל כבר ימצא גשם בלתי מורכב מחומר וצורה,

והוא הגרם השמיימי. וכבר דברנו שם מה שבו די בהקדמה הזאת.

הכלל הראשון, הפרק השלשה ועשרים

בבאור ההקדמה השלש ועשרים האומרת שכל מה שהוא בכח,

10   ולו בעצמותו אפשרות מה, כבר אפשר בעת מה שלא ימצא בפעל.

ההקדמה הזאת נבוכו בה רבים מהמפרשים, כמו אלתבריזי

והנרבוני, ולא עלה בידם. וזה שמפשט הלשון יראה שאין המלט

מהכפל. וזה שמה שהוא בכח דבר, לו בעצמותו אפשרות מה

לדבר ההוא; ואם כן אמרו ולו בעצמותי אפשרות מה, כפל

15   ומותר. גם אמרו כבר אפשר בעת מה שלא ימצא בפעל,

אין עניין לו, וזה שאשר לו א פ ש רו ת מ ה אין עניין לו יותר מאמרנו

---

3 והם] ואם בג – אשר] אם א.   5 חקרנוהו סב – בן רשד באו ב׳ר פ.   6 המוכרח ס – אבל לד.

7(די) בג.   10 (כבר) י – שכבר י – ימצא] יצא P.   11 (רבים) לזרדקבאו – המפרשים זרקאו –

התבריזי זרא.   13 (מה) פ.   14כפול לד.   15נם] כי פ – ימצא פקבאו יצא – לפועל קבאו.

16ממאמרנו מאמרו קי.

Part II.

Examination of the twenty-second proposition which reads to
the effect that every body is necessarily composed of two elements, which two elements constitute its existence, and these are
matter and form.

This proposition has been examined by us in the seventh chapter of this part, [Prop. X, Part II]. Averroes, it may be gathered,
does not believe that every body must necessarily be composed
of matter and form, for there exists, according to him, a body
which is not composed of matter and form, namely, the celestial
sphere. But we have already discussed this question in the aforementioned chapter and what we have said there will suffice also
as a criticism of this proposition.

PROPOSITION XXIII

Part I.

Proof of the twenty-third proposition, which reads: 'Whatsoever is in potentiality, and in whose essence there is a certain
possibility, may at some time not exist in actuality.'[1]

This proposition has been the cause of perplexity to many of
the commentators, as, for instance, Altabrizi and Narboni, none
of whom, however, has succeeded in elucidating it. The wording
of the proposition seems to be inexplicably tautological. For
when a thing is potentially something else, there assuredly is in its
essence a certain possibility for that something else, and so the
additional statement 'and in whose essence there is a certain
possibility' is quite tautological and redundant.[2] Again, the
concluding statement 'may at some time not exist in actuality,'
adds nothing to the statement preceding it, for when a thing is
said to contain a certain possibility it means nothing more than
to say that at some time it may pass into actual existence and

אפשר שיצא לפעל ואפשר שלא יצא, ולזה היה המשפט הזה כמשפט
אמרנו האדם אדם.

ואם היתה הכוונה באמרו ולו בעצמותו אפשרות מה,
שנושא הכח היה לו אפשרות שימצא ושלא ימצא, ואם לא יראה כן
5 מאמרו אפשרות מה, שאם היתה הכוונה על מציאותו לא יתכן
אמרו מה; אבל נניח כן, הנה אם כן הנושא כבר יצא לפעל, ולזה
יהיה אמרו כבר אפשר בעת מה שלא ימצא בפעל בלתי
מתיחס כלל.

ומה שיראה לנו בבאור זאת ההקדמה הוא כפי מה שאומר. כל
10 מה שהוא בכח דבר, והאפשרות ההוא הוא בעצמו—וזה
שהאפשרות בכח דבר, ממנו שהאפשרות בעצמו, כאלו תאמר
שהשחור אפשר בעצמו שישתנה וישוב לבן, ואפשר שיהיה האפשרות
נתלה בדבר חוץ ממנו, כאלו תאמר שאפשר בשמש שישחיר בתנאי
שיהיה המקבל גשם לח. ולזה גזר, שכאשר יהיה האפשרות בעצמו,
15 כבר אפשר בעת מה שלא ימצא בפעל, רוצה לומר שיהיה
נעדר. וזה שהיות האפשרות בעצמו, בלתי צריך לדבר מחוץ, יחייב
היותו בחמר מקבל השינוי, ולכן אפשר שיהיה נעדר בעת מה, כי
החמר המשתנה הוא סבת ההעדר בעצם. והנה יסכים הפירוש הזה
במה שהשתמש בו הרב בזאת ההקדמה בפרק הראשון מהחלק השני
20 מהמורה.

1 שאפשר ד ב – (כמשפט) כאמרנו ז ר ק ב א נ.   4 [זה] הכח י – (שימצא) ס – ושלא] שלא ס ולא ר.
5 היתה] היה פ ב א נ – (הכוונה) ל ד.   6 מה] בה ב נ – ולזה] וזה י.   7 כבר [יהיה] פ – ימצא בפעל]
יצא לפעל פ ב א נ יצא בפעל ל י ק.   9 ומה] וכל מה ס – הוא [כי ההקדמה היא] ו ר ק א נ
הוא [כי היא] ב – (כפי) כמה ד.   10 ואפשרות ב נ.   11 שהאפשרות בכח] שהאפשר בכח ל ו ד ב
שהאפשרות [הוא] ב כ ח י.   12 שיהיה] שהיה י.   13 באפשר צ – שישחיר [שאפשר] ל ד.
16 שהיות] שהיותו י שיהיו' פ – יחייב] יחויב פ יתחייב ר.   17 יקבל ס – [יהיה] אפשר
ס ל ו ר ד ק ב א נ.   18 בעצם] בגשם י בעצם [בגשם] ל.   19 שנשתמש ו ב א נ שישתמש [בו] ר – (בו)
ד – בזו ו ר א – מחלק פ נ – שני ק.

at some time it may not. The proposition, therefore, has no more meaning than the statement that man is man.³

It may be rejoined that the statement 'and in whose essence there is a certain possibility' means to affirm that the subject of the potentiality [after its realization] has a possibility [of continuing] to exist or not. To be sure, the expression 'a certain possibility' would not seem to warrant such an interpretation, for were the statement to refer to [the continuance of] the existence of the subject of the potentiality, the use of the expression 'a certain' would be quite inappropriate. Still supposing this to be the meaning of the statement, then the conclusion 'may at some time not exist in actuality' is entirely inappropriate, inasmuch as that subject has already come into existence.⁴

What seems to us to be the correct interpretation of the proposition may be stated as follows: 'Everything that is potentially something else, and the possibility [of becoming that something else] is inherent in the thing itself...'⁵ The implication of the last statement is that the possibility involved in a thing which is potentially something else may either inhere in the thing itself, thus, e. g., black has in itself the possibility of becoming white, or be dependent upon something external to itself, thus, e. g., the sun has the possibility of turning an object black provided the recipient of the action is moist.⁶ Referring, therefore, to the case where the possibility is inherent in the thing itself, Maimonides states that at some time it may not exist in actuality, that is to say, it may be non-existent.⁷ The reason for this is as follows: When the possibility is said to be in the thing itself, and not dependent upon anything external to the thing, then it must be in matter which is susceptible of change. Consequently, it may at some time be non-existent, for changeful matter is the cause of privation in any corporeal substance.⁸ This interpretation of the proposition will agree with the use the Master makes of it in the first chapter of the second part of *The Guide*.⁹

הכלל השני, הפרק הארבעה עשר

בחקירה בהקדמה השלש ועשרים האומרת שכל מה שהוא בכח
ולו בעצמותו אפשרות מה, כבר אפשר בעת מה שלא ימצא בפעל.
הנה לפי הנאמר שם בפרק השביעי גם כן, הנה כבר אפשר שימצא
גשם בפעל בזולת צורה מיוחדת, אשר לו בעצמו אפשרות לקבל
5 צורה, ולא יתכן בענינה שלא ימצא בפעל, כי הגשמות נשאר בו
תמיד. וכבר תפול ההערה הזאת בהקדמות הארבע ועשרים והחמש
ועשרים. ואולם השש ועשרים נחקור בה במאמר השלישי, בגזרת
הצור, ונבאר שם שאין ספק בשקרותה.

הכלל הראשון, הפרק הארבעה ועשרים
10

בבאור ההקדמה הארבע ועשרים האומרת שכל מה שהוא בכח
דבר אחד הוא בעל חמר בהכרח, כי האפשרות הוא בחמר לעולם.
ההקדמה הזאת מבוארת בעצמה עם מה שקדם. וזה שמה שהוא
בכח דבר אחד יתחייב שיהיה נושא הכח וישאר עם האחד, ואם לא,
15 לא היה הוא דבר אחד, ומה שזה דרכו הוא החמר, שהצורה איננה
בכח להיות דבר אחד. ולזה יתאמת שהאפשרות הוא בחמר לעולם.
ואולם צריך שנתעורר, כי למה שהאפשרות, אם שיאמר בנושא
הנמצא, כאלו תאמר שחומר הנחשת אפשר שיהיה זנגאר, ואם שיאמר
בנושא הנעדר, כאלו תאמר הזנגאר אפשר שיחול בחמר הנחשת,
20 הנה הכוונה בזה האפשרות אשר בנמצא.

3 ימצא] יהיה לזורד נמצא א יהיה נמצא פ.   4 (הנה כבר) פ.   7 הזאת] בזאת פ –
בהקדמות] ההקדמה פ   בהקדמה לא בהקדמת זבג.   8–7 כ״ד וכ״ה פו   הכ״ד וכ״ה ג
הכ״ד והכ״ה א.   9–8 (בגזרת הצור) בג.   9 שאין שם פ.   11 מה] מי פ.   14 נשאר ד.
14–15 ואם לא היה הוא דבר אחר לוק בג ואם לא היה הדבר ההוא אחר ר ואם לא היה הדבר
הוא האחר א.   15 היה] יהיה ד.   15 והצורה א – אינה זבאג.   17 שתתעורר בג שנתאמת א.
18 זנגאר ג.   19 בנושא] בנשוא סוק באג בה שהוא זר – הזנגאר יא הזאננגאר ב.   20 אפשרות
לד.

PART II.

EXAMINATION of the twenty-third proposition, which reads: 'Whatsoever is in potentiality, and in whose essence there is a certain possibility, may at some time not exist in actuality.'

Again, in view of what has been said above in the seventh chapter, [Prop. X, Part II], a body may exist in actuality without any proper form and, though having within itself the possibility of receiving form, will never be without actual existence, inasmuch as the corporeality always stays with it.[10] The same criticism may be urged also against Propositions XXIV and XXV. As for Proposition XXVI, we shall examine it in Book III, God willing, wherein we shall show that there can be no doubt as to its falsity.

PROPOSITION XXIV

PROOF of the twenty-fourth proposition, which reads: 'Whatsoever is potentially a certain thing is necessarily material, for possibility is always in matter.'[1]

This proposition is self-evident, being the sequel of the proposition preceding. For whatsoever is potentially a certain thing, must be the subject of that potentiality,[2] and it must remain with that 'certain thing' [even after the latter has become realized], for, were it not so, it would not be the same thing.[3] Anything answering to this description is matter, inasmuch as form has not the potentiality of becoming a certain thing. It is thus true to say that possibility is always in matter.

We must, however, observe that inasmuch as the term possibility may apply either to an existent subject, thus, e. g., bronze as matter may become verdigris,[4] or to a non-existent subject, thus, e. g., verdigris may settle on the matter bronze,[5] in this proposition the term possibility is to be taken with reference to an existent subject.[6]

הכלל הראשון, הפרק החמשה ועשרים

בבאור ההקדמה החמש ועשרים האומרת שהתחלות העצם
המורכב האישי, החמר והצורה, ואי אפשר מבלתי פועל, רוצה
לומר מניע הניע הנושא עד אשר הכינו לקבל הצורה, והוא המניע
הקרוב, המכין לחומר איש מה, ויחוייב מזה העיון בתנועה והמניע
5　והמתנועע.　וכבר התבאר בכל זה מה שיחוייב לבארו. ונסח דברי
ארסטו, כי החמר לא יניע עצמותו.　וזאת היא ההקדמה הגדולה
המביאה לחקור מהמניע הראשון.

ההקדמה הזאת מבוארת בעצמה, כי להיות החמר והצורה בלתי
10　נמצאים כל אחד בפני עצמו לבדו, ואנחנו נראה שהדבר יתהוה
מדבר, ולא מאיזה דבר הזדמן, הוא מבואר שאי אפשר בזולת נושא,
נשאר לעולם, יפשוט צורה וילבש צורה.　ולכן היו התחלות איש
העצם העצמיות החמר והצורה; ואם היה ההעדר הקודם מן
ההתחלות, הוא במקרה.　אלא שלמה שהוא צריך בהכרח אל מניע
15　יכין החמר לקבל הצורה המיוחדת, הוא מבואר שאי אפשר בזולת
פועל.　אלא שלמה שאינו מעצם הדבר, אינו נמנה בהתחלות. ואולם
למה שאין המלט ממנו, למה שההחומר לא יניע עצמותו, והיה המניע
מניע בעצמותו למתנועע בתנועה, הוא מבואר שהעיון במניע מביא
אל העיון בתנועה ובמתנועע.

4 (אשר) סלודקבג.　5 חוייב פפבג יחוייב ר.　6 בכל) כל פ – שחוייב פאי שיחייב ל –
ונסח] וגניח לד.　7 חור יא.　8 מהמניע] על המניע לד.　10 אחד [ואחד] ל – (בפני) בעצמו
לורדקבאג.　11 (מדבר) פ.　13 (התחלה] מן גרא [הוא] מן קג.　14 שהוא] הוא ז יא.
17 למה] לרוב ו – (בעצמותו) לורדקבג.　18 ובתנועה וקפבג.

PROPOSITION XXV

PROOF of the twenty-fifth proposition, which reads: 'The principles of any individual compound substance are matter and form, and there must needs be an agent, that is to say, a mover which sets the substratum in motion, and thereby renders it predisposed to receive a certain form. The agent which thus predisposes the matter of a certain individual being is called the immediate mover. Here the necessity arises of inquiring into the nature of motion, the moving agent and the thing moved. But this has already been explained sufficiently; and the opinion of Aristotle may be formulated in the words that matter is not the cause of its own motion. This is the important proposition which leads to the investigation of the existence of the prime mover.'[1]

This proposition is self-evident. For inasmuch as matter and form do not each exist separately without the other, and we perceive that while one thing is generated from another thing[2] it is not generated from anything casual,[3] it is manifest that the process of generation and corruption would be impossible without the assumption of a permanently residual substratum capable of taking off one form and putting on another.[4] Consequently the essential principles of any individual corporeal substance[5] are matter and form. Though the privation which precedes[6] [form] is included among the principles, it is a principle only in an accidental sense.[7] Then, again, inasmuch as the process of generation necessarily implies the existence of a mover whose function is to render matter predisposed to receive its proper form, it is likewise manifest that the process would be impossible without the assumption of an agent.[8] As that agent, however, does not constitute an essential part of the substance, it is not numbered with the principles. Still, the assumption of such an agent is inevitable, for matter cannot be the cause of its own motion,[9] and, furthermore, it is by means of motion that the mover acts essentially upon the thing moved. Consequently, the speculation concerning the mover leads to speculation concerning motion and the thing moved.

NOTES

to the

Twenty-five Propositions

of

Book I of the Or Adonai

NOTES

INTRODUCTION TO BOOK I.

1. Hebrew בשרש הראשון שהוא התחלה לכל האמונות התוריות. "Of the first root which is the beginning of all the scriptural beliefs."

The term שרש, like its synonym עיקר and its Arabic equivalent اصل, is used in mediaeval Jewish philosophy in the general sense of fundamental principles of religious belief (cf. Neumark, *Toledot ha-'Iḳḳarim be-Yisrael* I, pp. 1–5). Crescas, however, uses it as a specific designation for the beliefs in the existence, unity and incorporeality of God, and it is contrasted by him with all the other fundamental religious beliefs which he designates by the expression "Scriptural Beliefs" אמונות תוריות. The latter is subdivided by him into (1) פנות ויסודות, *fundamentals*, (2) דעות אמתיות, *true opinions*, (3) סברות, *probabilities*. (See *Or Adonai, Haẓa'ah,* p. 3.) Hence my expanded translation of this passage.

2. Hebrew שההקדמה יתבאר ענינה בשתי עניינים. Similarly Hillel of Verona begins his commentary on the Twenty-five Propositions with the statement: ודע אחי כי צריך לך ולכל מבין בביאור אלו ההקדמות שני עניינים. "Know, my brother, that thou or any one else who wishes to understand the meaning of these propositions must needs have recourse to the explanation of two things." The two things enumerated by Hillel, however, are not the same as those mentioned here by Crescas.

3. *Or Adonai* I, iii, 1.

4. Hebrew אופן עמידתנו באמתתו. But later: לעמוד על השרשים האלו. The Talmudic expression עמד על, *to understand*, is used in mediaeval Hebrew as a translation of the similar Arabic expression وقف على, *to pause at, to pay attention to, to understand, to form an opinion of.* (Cf. Ginzberg, *Geonica*, Vol. I, p. 25). The expression עמד ב.. is used by Crescas in the same sense.

Literally: "how we know the truth of this principle."

5. The term קבלה is used by Crescas in the following three senses:

(1) *Tradition* as distinguished from *speculation*, in which sense it is used here and later in III, i, 5, p. 70a: כפי מה שבא בקבלה. והוא שהש"י חידשו והמציאו בעת ידוע כאמרו בראשית ברא. In this sense

it is the equivalent of הגדה, خبر, as used in *Emunot we-Deot*,
Introduction: ונחבר עליהם משך רביעי הוצאנו אותו בשלש ראיות ושב לנו
שם בשכלים מקום לקבול ההגדה הנאמנת, and III, 6: שרש והוא ההגדה הנאמנת.

(2) *Rabbinic tradition* as distinguished from תורה in its wider
sense of Bible, as below at the end of this preface: אלא מצד הנבואה
הוא מבואר ששלמות and in I, iii, 6: במה שהעידה עליו התורה ונתאמת בקבלה
השורש הזה מצד התורה והקבלה... ואמנם כבר בא בדברי רז״ל. In this
sense it is also used in the following passage of *Ḥobot ha-Lebabot*,
Introduction: וכאשר נתברר לי חיוב החכמה הצפונה מן השכל והכתוב והקבלה
(والمنقول).

(3) *Prophetic and Hagiographic books of the Bible* as distinguished from תורה in its narrower sense of Pentateuch, as later
in II, i, 1: ואם מפאת הקבלה, כמו שבאו הרבה כתובים על זה, אמר כי כל
לבבות דורש ה'. In this sense it is used in *Emunot we-Deot* II, 10:
וכיון שבארתי שהמושכל והכתוב והמקובל (والمنقول) הסכימו כלם על הרחקת
הדמיון. Cf. Mishnah Ta'anit II, 1: ובקבלה הוא אומר קרעו לבבכם
ואל בגדיכם.

6. Hebrew טבעיות. The term טבעיות is used by Crescas both with
general reference to Aristotle's writings on the natural sciences
and with particular reference to his *Physics*, as in the following
passages of the *Or Adonai*: (a) III, i, 1: לפי שהתבאר בטבעיות שתנועת
(b) *Ibid.* לפי שהתבאר בטבעיות שהגרם השמימי. ההעתק היא הקודמת שבתנועות.
(c) III, i, 3: שכבר יתבאר במעט עיון למי שעיין בטבעיות. אין הפך לו
(d) IV, 4: ואמנם לפי שהתבאר בטבעיות שהגרמים מניעים היסודות.
Of these four passages only the first and third may refer to the
Physics proper.   Aristotle's own terms φυσικά and τὰ περὶ
φύσεως are also sometimes to be taken as references to his general
writings on the physical sciences (cf. Zeller, *Aristotle*, Vol. I,
p. 81, n. 2). In this place it would seem that Crescas has specific
reference to Aristotle's discussion of the Prime Mover in *Physics*,
Book VIII.

7. Here Crescas seems to be using the term אחרונים, "later" (or
"modern," "recent"), to distinguish the Moslem and Jewish philosophers from their Greek predecessors.   Further down in this
passage, however, he refers to all these names as the "first" (or
"early", "ancient") philosophers: לפי שהם לקוחים מכלל דברי
הפילוסופים הראשונים, evidently in contrast to Maimonides. But the

term "ancients," הקדמונים, is elsewhere applied by him to the pre-
Aristotelian philosophers (cf. Props. X, XV) and הקודמים to
Aristotle and his followers (cf. Book IV, 2). In another place he
uses the term "later", אחרונים, with reference either to Averroes
or to Gersonides (cf. Prop. I, Part II, n. 17, p. 409). Evidently
Crescas uses all these terms in relative and variable senses.

Shahrastani applies the term *ancient*, القدماء, to the pre-Aristotelian philosophers and their followers, and the term *later*, المتأخرين,
to Aristotle and his followers among the Greek-writing philosophers. (Cf. *Kitab al-Milal wal-Niḥal*, ed. Cureton, pp. 253, 311).
The Moslem philosophers, beginning with Al-Kindi, are considered
by him as a distinct subdivision of the *later*. (Cf. *Ibid*. pp. 253,
349). Among these latter he considers Avicenna as the "first and
foremost." *Ibid*. p. 312: مقدم المتأخرين ورئيسهم.

Maimonides himself, in *Moreh* I, 71, like Shahrastani, designates
the pre-Aristotelian philosophers, especially the Atomists and the
Sophists, as *ancient* (הראשונים, הקדמונים: אלמתקדמין) and refers
to Aristotle and his followers as the *later* (האחרונים, אלמתאכרין).
Still within the Christian and Moslem theologians he distinguishes an earlier group and applies to them the same term
ancient or *first:* הראשונים (אלאקדמון) מן המדברים; המדברים הראשונים
(אלאול) מן היונים המתנצרים ומן הישמעאלים. In his letter to Samuel
ibn Tibbon Maimonides, again, uses the term *ancient* with
reference to the works attributed to Empedocles, Pythagoras and
Hermes as well as to the writings of Porphyry, all of which he
characterizes as פילוסופיא קדומה, *ancient philosophy*. See *Kobez
Teshubot ha-Rambam we-Iggerotaw* II, p. 28b: ואמנם זולתי חבורי אלה
הנזכרים, כמו ספרי בנדקלוס וספרי פיתאגוראס וספרי הרמס וספרי פורפיריוס,
כל אלה הם פילוסופיא קדומה. In Shahrastani, however, Porphyry
is included among the *later* (*op. cit.* p. 345). It is not impossible
that by קדומה, in his letter, Maimonides does not mean *ancient*
but rather *antiquated* and *obsolete*. Cf. Steinschneider, *Uebersetzungen*, p. 42, n. 297.

8. The names enumerated here by Crescas are arranged in
chronological order with the exception of Themistius which
should come after Alexander, but in this he errs in the good company of Shahrastani, Cf. *Kitab al-Milal wal-Niḥal*, pp. 343–344.
There is no ground for Joël's suggestion that the text here is

either corrupt or Crescas was not well orientated in the chronological order of the men mentioned by him (cf. *Don Chasdai
Creskas' religionsphilosophische Lehren,* p. 3, n. 1). Joël seems
to have overlooked the characteristic distinction between the
words מפרש, *commentator,* and מחבר, *author,* both of which are
advisedly used here by Crescas. They refer to two well recog-
nised methods of literary composition employed by mediaeval
authors, namely, commentaries on standard texts and independent
treatises. Maimonides, in a letter to Phinehas ben Meshullam,
speaks of these two methods as being practised from antiquity
by both Jews and non-Jews in all the branches of secular and
religious sciences. See *Ḳobeẓ Teshubot ha-Rambam we-Iggerotaw*
I, p. 25b: דע אלופי ומיודעי שכל מי שכתב ספר, בין בדברי תורה, בין בשאר
החכמות, בין מן הגוים הקדמונים בעלי החכמות, בין מן הרופאים, אחד משני דרכים
הוא אוחז, או דרך חבור או דרך פירוש.
Thus, distinguishing between commentators and authors, Crescas
names immediately after the Greek commentators, Alexander and
Themistius, the אחרונים, i. e., the *later* or *recent* or *modern,* meaning thereby the Arab commentators of whom he mentions Alfarabi
and Averroes, for Alfarabi, too, was known as a commentator as
well as an author. Thus also Maimonides refers to Alfarabi's
comments or glosses on Aristotle's *Physics. Moreh* II, 19: וכבר
זכר אבונצר בתוספותיו על ספר השמע. Then, under independent
authors he mentions in chronological order Avicenna, Algazali,
and Abraham Ibn Daud. A similar distinction between *author*
and *commentator* is again made by Crescas toward the end of his
criticism of Proposition I: בספרי אריסטו, חולתו מהמחברים, ומפרשי
ספריו.
The names given here by Crescas, with the exception of
Algazali and Abraham Ibn Daud, occur in Maimonides' letter
to Samuel Ibn Tibbon. See *Ḳobeẓ Teshubot ha-Rambam we-
Iggerotaw* II, pp. 28b–29a: וספרי אריסטו הם הם השרשים והעקרים לכל
אלו החבורים של חכמות, ולא יובנו, כמו שזכרנו, אלא בפירושיהם–פירוש אלכסנדר
או תאמסטיוס או ביאור אבן רשד.....וספרי עלי אבן סינא......אינם כספרי
אבונצר אלפראבי. It will be noted that in this letter Alexander
is correctly mentioned before Themistius, and that the works
of Alexander, Themistius and Averroes are described as *commentaries* (ביאור, פירוש), whereas those of Alfarabi and Avicenna
are called *books* (ספרי).

As for Crescas' intimation that Maimonides in writing the *Moreh* had drawn upon the works of these men, it is only partially true. The names of Alexander, Themistius and Alfarabi are all mentioned in the *Moreh*. Though Avicenna, Algazali and Abraham Ibn Daud are not mentioned in the *Moreh*, traces of their influence can be easily discovered in that work. There is no evidence, however, that Maimonides was acquainted with the works of his older contemporary Averroes at the time of his writing of the *Moreh*, though Maimonides mentions him subsequently in his letter to Samuel Ibn Tibbon. A sort of argument from silence would seem to point to the conclusion that the *Moreh* was written in complete ignorance of the works of Averroes. Throughout the *Moreh*, on all the points at issue between Avicenna and Averroes, Maimonides follows the views of the former and restates them without the slightest suggestion of his knowledge of the views of the latter. In one place Crescas infers that Maimonides must have understood a certain passage of Aristotle in accordance with Averroes' interpretation as against that of Avempace. See his criticism of Proposition VII: והנה יראה שהרב לקחו כפי דעת בן רשד. It is not clear, however, whether Crescas meant to say that Maimonides followed Averroes' interpretation or whether he meant to say that Maimonides simply happened to arrive at a similar interpretation. Similarly Shemtob, in his discussion of Prop. XVII, suggests that Maimonides was aware of a controversy between Avicenna and Averroes (cf. Prop. XVII, n. 7, p. 675). Later Jewish philosophers, Joseph Kaspi and Isaac Abravanel, definitely state that Maimonides had no knowledge of the works of Averroes when he wrote the *Moreh*. Cf. *'Amude Kesef*, p. 61: והמורה לא ראה ספרי בן רשד, and *Shamayim Hadashim* I, p. 7b: והנה הרב עם היות שלא ראה דברי אבן רשד כי בזמן אחד היו מרוחקים מאורצותם, הרב במצרים ואבן רשד בקורטובה.

9. The implication of Crescas' statement here as well as of his subsequent statement לפי שהם לקוחים מכלל דברי הפלוסופים הראשונים that Maimonides himself has constructed the proofs for the existence, unity and incorporeality of God out of the propositions is not altogether true. The proofs themselves are taken from the works of other philosophers.

10. Taken literally, the text would seem to imply that Maimonides was the first among philosophers to prove the unity and the incorporeality of God in addition to His existence. This, however, would not be true. Proofs for the unity and the incorporeality of God are already found in Aristotle's works (cf. *Metaphysics* XII, 7, and *Physics* VIII, 10), not to mention the works of early Moslem and Jewish philosophers. What Crescas probably wanted to say here is that besides the four common proofs advanced by Maimonides for existence, unity and incorporeality of God, he has also advanced several particular proofs for unity and incorporeality only (see *Moreh* II, 1). In his summary as well as in his criticism Crescas includes in his discussion also these additional proofs (cf. *Or Adonai* I, i, 31–32, and I, ii, 19–20).

11. Hebrew אם הם נותנים האמת על כל פנים. The same expression occurs again later, p. 178. I have translated it literally. The phrase, according to this literal rendering, would seem to contain an allusion to Aristotle's definition of truth as something which is "consistent with itself in all points," מסכים מכל צד (see Prop. I, Part II, n. 79, p. 456).

It is not impossible, however, that the expression על כל פנים is used by Crescas in the sense of *necessary, demonstrative, apodeictic,* as the equivalent of בהכרח or of his own ביאור מופתי. In this sense it is used by both Judah ibn Tibbon in his translation of the *Ḥobot ha-Lebabot* and by Ḥarizi in his translation of the *Moreh Nebukim*. See *Ḥobot ha-Lebabot* I, 7: על כל פנים נצטרך (اضطرارا (Arabic text, p. 51, l. 2; p. 55, l, 7; p. 58, l. 3). יש לו לומר, על כל פנים فيلزم ضرورة (Arabic text, p. 55, l. 3). על כל פנים, يلزمه (Arabic text; p. 56, l. 7). *Moreh Nebukim* III, 25; החלוק על כל פנים (Samuel ibn Tibbon: החלוקה בהכרח), Arabic חה החלוק הוא בהכרח ועל כל פנים. Cf. *ibid.* II, 1: אלתק סים צרורה, (Samuel ibn Tibbon: וזאת חלוקה הכרחית), Arabic: צרוריה. והדא קסמה צרוריה.

Similarly the term אמת here may mean not simply "truth" but "verification", "confirmation", and hence "proof". And, again, the term נותן here may have the meaning of מחייב, as in the Talmudic expressions היא הנותנת, הדין נותן. In *Ḥobot ha-Lebabot* I, 5, the Arabic يجب · · · ضرورة (p. 45, l. 7) is translated by הדין נותן. Also in *Hegyon ha-Nefesh*, p. 5a, the expression והמדע נותן undoubtedly stands for והמדע מחייב.

Thus here the expression אם הם נותנים האמת על כל פנים may be the equivalent of אם הם מחייבים באור מופתי or of Crescas' own אם הם מבארים באור מופתי, "whether they establish a demonstrative proof."

12. Hebrew וכל מה שנאמר בהם מזולתו אין לשום לב עליו. The term מזולתו may refer either to Maimonides implied in the pronominal suffix in מופתי or to כלל in מכלל דברי הפילוסופים.

The purpose of this remark by Crescas is to account for his failure to discuss the proofs of the existence of God advanced by Jewish philosophers prior to Maimonides. His explanation is that they are of no importance, inasmuch as they are not of Aristotelian origin. Similar sentiments, couched almost in the same language, as to the dispensability of views un-Aristotelian, are expressed by many Jewish and Moslem philosophers.

Maimonides, *Moreh* II, 14: ולא אשגיח למי שדבר זולת אריסטו מפני שדעותיו הם הראוים להתבונן.

Algazali, *Maḳaṣid al-Falasifah* III, p. 246: فان قيل ما حقيقة المكان قيل ما استقر عليه رأى ا'رسطاطاليس هو الذى ا'جمع عليه الكل

MS. Adler 1500: ואם יאמר מה הוא אמתת המקום נאמר מה שנתישב עליו דעת אריסטו והוא אשר ישוב אליו הכל.

Averroes, *Intermediate Physics* VI, 7: כי מה שימצא לזולתו באלו הדברים במי שהיו לפניו ממה שראוי לשומו מסופק באלו הדברים, כל שכן שנשימם התחלה.

Shahrastani, *Kitab al-Milal*, p. 312: وايس الامر على ما مالت اليه ظنونهم.

Shem-ṭob, Commentary on the *Moreh* II, 1: ואולם דעת החכם אשר כי מי שירצה לבקש האמת אשר אין גמגום עליו, and II, 4: אליו ישוב הכל ראוי שילך דרך אריסטו והוא אשר אליו ישוב הכל.

13. Hebrew באור מופתי. Crescas uses the term באור in the sense of "proof" in general, as in this expression and in the expression בבאור ההקדמה. This logical sense of באור, of which the Arabic is بيان, is to be distinquished from באור in the sense of "commentary", of which the Arabic equivalent is شرح. The term in its latter sense is used by Crescas in Prop. II, Part II: בביאורו לספר השמע. The term מופת is used by Crescas in two senses: (1) Apodeictic or demonstrative proof, as in this expres-

sion, which is the accepted meaning of that term in Hebrew.
Cf. *Millot ha-Higgayon*, ch. 8. (2) The formal process of reasoning or the argument by which the proof is established. He thus
speaks of a באור as containing several מופתים or of the מופת of
a באור, as in the expression חדו מה שכוון באורו במופת הזה, p. 140.

Etymologically, באור and بَيَان reflect the Greek ἀπόδειξις
a showing, and מופת and بُرهَان reflect the Greek τεκμήριον,
a sure sign. In Aristotle both these terms are used in the sense
of a *demonstrative proof*. Evidently the terms באור and بَيَان have
lost that forceful sense of demonstrative proof.

The term באור is also used in Hebrew as a translation of the
Arabic وضوح to designate a kind of reasoning which lies midway between pure tradition קבלה, تَقلِيد and demonstrative proof
بُرهَان, מופת. Cf. Algazali, *Mozene Zedek*, pp. 6–7: וכל זה בדרך
נעלה בו מגבול הקבלה אל גבול הבאור, אשר אלו נחקרה אמתתו והארכנו לדבר
בו היה עולה אל גבול המופת. *Mizan al-Amal*, p. 3: وكل ذلك بطريقة
يترقى عن حد طريق التقليد الى حد الوضوح لو استقصى بحقيقته
وطوّل الكلام فيه ارتقى الى حد البرهان.

14. Hebrew והעיון בזה יהיה כפי מאמר האומר. The Parma and Jews'
College MSS. have here the following marginal note: ירצה כפי
סונת החכם שהניחה. The Vatican MS. has the same note but without שהניחה.

What Crescas means to say here is that in his criticism of the
philosophers he, as interrogator or opponent, will press his re-
spondents with consequences drawn from their own premises,
even though he himself does not admit them, for his purpose is
to show the contradictions to which their own premises might
lead. This sort of *argumentum ad hominem*, as it later came to
be known (see Locke, *Essay Concerning Human Understanding*
IV, xviii, § 21), is one of the several forms of Aristotle's *dialectic*
arguments as opposed to the *didactic* (see Grote, *Aristotle* II,
p. 71). Didactic arguments are described by Aristotle as "those
which syllogize from the proper principles of each discipline, and
not from the opinions of him who answers" (*De Sophisticis
Elenchis*, ch. 2). A dialectic argument, contrariwise, must

therefore be one which reasons "from the opinions of him who answers".

The expression כפי מאמר האומר thus reflects the Greek ἐκ τῶν τοῦ ἀποκρινομένου δοξῶν (*ibid.* 165b, 2). מאמר האומר = قول القائل.

The same expression is used by Averroes in stigmatizing the *dialectic* character of Algazali's arguments against philosophy, as in the following passages in his *Happalat ha-Happalah:*

Disputation I: חה סתירה כפי מאמר האומר לא כפי העין בעצמו.

Ibid. והמחלקת השלמה אמנם היא אשר תמור בטול דעתם כפי העין בעצמו לא כפי מאמר האומר.

Disputation III: כי ממאמר האומר.

Disputation XI: חה סתירה כפי מאמריהם לא כפי העין בנפשו.

Cf. also *Intermediate Physics* IV, i, 1, 9: חה הפירוש מסכים למה שנראה מהאומר ולאמת בעצמו.

15. Hebrew שאין דרך. Similarly later, p. 216: ואין דרך אצלנו ומבוא. The equivalent Arabic expression لا سبيل, used in *Ḥobot ha-Lebabot* I, 6, p. 47, l. 2; p. 49, l. 13, *et passim*, is translated by Judah ibn Tibbon simply by אין יכולת or אי אפשר.

PROPOSITION I

PART I

1. The Hebrew version of this Proposition is taken from Samuel ibn Tibbon's translation of the *Moreh Nebukim*.

2. Hebrew אין תכלית לו. Equivalent terms for תכלית are תכלה, סוף, מכולה.

Cf. Narboni, *Ma'amar be-'Eẓem ha-Galgal le-Ibn Roshd* III: שאמרנו בלתי מכולה יאמר בשני ענינים.

Neveh Shalom VII, i, 3, p. 100b: חה כלו מחוייב מהיות צורתם פועלת פעל בלתי מכולה.

Narboni's Commentary on the *Moreh*, II, Introduction, Proposition I: והאין סוף הוא מצד הכח, כי הכלל ואין סוף שני מקבילים.

Likkutim min Sefer Meḳor Ḥayyim III, 10: כי היו עצמי כל אחד מהעצמים מוגדרים עצורים אינם נמשכים ללא תכלה.

3. *Physics* III, 4–8; *De Caelo* I, 5–7; *Metaphysics* XI, 10. The corresponding references in Averroes' *Intermediate Commentaries* which are the direct source of Crescas' summaries of Aristotle, are as follows: *Intermediate Physics* III, iii, 1–8; *Intermediate De Caelo* I, 7; *Intermediate Metaphysics* X.

4. Hebrew נבדל, i. e., נבדל למוחשות, χωριστὸν αἰσθητῶν, *separated from sensible objects*.

5. Hebrew באור כולל. The same designation of this argument is used by Crescas later, p. 174.

Aristotle himself designates this argument by the term "logical" (λογικώτερον, *De Caelo* I, 7, 275b, 12). Similarly the first of the second class of arguments in this chapter is characterized by Crescas as באור כולל (below p. 150), whereas Aristotle calls it "logical", λογικῶς, in *Physics* III, 5, 204b, 4, and "general" (or "universal"), καθόλου in *Physics* III, 5, 204a, 34, and in *Metaphysics* XI, 10, 1066b, 22). Averroes calls it "general", כולל, in *Intermediate Physics*, but "logical", הגיוני, in *Intermediate Metaphysics*. The interchanging of these two terms may be explained on the ground that among the several meanings which the expression "logical" proof has in Aristotle there is one which describes it as consisting of abstract reasoning from "universal" or "general" concepts which have no direct and appropriate bearing upon the subject in question (cf. Schwegler, *Die Metaphysik des Aristoteles*, Vol. IV, p. 48, n. 5; Ross, *Aristotle's Metaphysics*, Vol. II, p. 168; both on *Metaphysics* VII, 4, 1029b, 13). Averroes himself similarly describes "logical" proofs as those "composed of propositions which are general and true but not appropriate to the subject under consideration. And therein is the difference between such propositions and essential propositions, for essential propositions are appropriate and pertain to the subject under consideration. And the difference between logical propositions and contentious propositions consists, on the other hand, in this: Logical propositions are true in their entirety essentially, whereas the contentious are false in part, and are not true in their entirety except accidentally." *Intermediate De Caelo* I, 7, Third Proof: והם מחוברות מן ההקדמות הכוללות הצודקות, אשר אינן מיוחדות בסוג המעיין בו, וזהו ההפרש ביניהן ובין ההקדמות העצמיות, שההקדמות העצמיות מיוחדות בסוג המעיין בו ונערכות אליו. וההפרש גם כן בין אלה ההקדמות

ההגיוניות ובין ההקדמות הויכוחיות שאלו הן צודקות בכל בעצם והויכוחיות כחבות
בחלק ואינן צודקות בכל כי אם במקרה.
Cf. *Sefer ha-Gedarim*, p. 19a: הקש הגיוני, הוא אשר הקדמותיו כוללות
וצודקות, אלא שהם בלתי מיוחסות.

6. Hebrew מין, *kind, class, section*. The Sulzberger and Munich
manuscripts read here עיון, *Speculation*. The term עיון, Arabic
נטר, as a designation of a class of arguments is found in the
Hebrew translations of *Moreh* II, 1. Crescas himself uses it later
in his criticism of this proposition. Most of the MSS., however,
read here מין.

7. Hebrew כן. אמר. Literally: "in the following manner. He
said:" The word אמר, "he said", is generally used in Averroes'
Intermediate Commentaries to jntroduce the beginning of a
translation or paraphrase of a text by Aristotle.

Originally in Aristotle and Averroes the arrangement of the
argument is as follows:

(a) The infinite cannot be something immaterial, and of independent existence.

Physics III, 5, 204a, 8–14, which is restated in *Intermediate
Physics* III, iii, 4, 1 as follows: "We say that it is impossible that
there should be an infinite existing by itself apart from sensible
objects. For it would inevitably have to be either divisible or
indivisible. If it were indivisible, it could not be described as
infinite except in the sense in which a point is said to be infinite
and color is said to be inaudible. But this is not the sense which
those who affirm the existence of an infinite are agreed upon
(ישיבוהו, cf. ישוב אל = عَلى جمع! above p. 325, n. 12), nor is it that
which is the subject of our investigation." (Latin, p. 452 v b, 35).
ונאמר שאי אפשר שימצא דבר אין תכלית לו עומד בעצמו נבדל למוחשות.
וזה שלא ימנע מהיותו מקבל החלוקה או לא יקבלה. ואם היה בלתי מקבל
החלוקה, הנה לא יתואר בשהוא בב"ת אם לא כמו שיאמר בנקודה שהיא בב"ת
ובמראה שהוא בלתי נשמע. וזה דבר לא ישיבוהו האומרים כן, ואינו ממה שנחקור
עליו.

Cf. *Metaphysics* XI, 10, 1066b, 1–7, which is restated in *Intermediate Metaphysics* X.

(b) The infinite cannot be an immaterial quantity, either
magnitude or number, existing by itself. This refers to the views

of the Pythagoreans and of Plato, both of whom considered the
infinite as a certain essence subsisting by itself, the former identifying it with number, the even, and the latter identifying it with
magnitude. Their views are given by Aristotle in *Physics* III, 4.

Physics III, 5, 204a, 17–19, restated in *Intermediate Physics,
loc. cit.*, as follows: "If it is divisible, it must inevitably be either
an immaterial quantity or a quantity existing in a subject or one
of the immaterial substances. It cannot be an immaterial quantity, for inasmuch as number and magnitude are inseparable from
sensible objects it must follow that that which is an accident to
number and magnitude must likewise be inseparable; and infinity
is such an accident, for finitude and infinity are two accidents
existing in number and magnitude, inasmuch as the essence of
number and magnitude is not identical with the essence of the
infinite." (Latin, p. 452 v b, 36).

ולא ימנע, אם יקבל, שהוא כמה נבדל, או כמה נמצא בנושא, או יהיה עצם
מהעצמים הנבדלים. ובטל שיהיה כמה נבדל, אחר שהיה המספר והשעור בלתי
נבדלים למוחש, הנה מחוייב שיהיה מה שיקרה למספר והשעור בלתי נבדל ג״כ,
והוא בהעדר התכלית, כי התכלית ואין התכלית שני מקרים נמצאים במספר והשעור,
כי מהות המספר והשעור בלתי מהות מה שאין תכלית לו.

Cf. *Metaphysics* XI, 10, 1066b, 7–9, restated in *Intermediate
Metaphysics, loc. cit.*

(c) The infinite cannot be an accidental quantity existing in
something else. This refers to the views of the early Greek
Physicists and of the Atomists, all of whom considered the infinite
as an accidental quantity, either the magnitude of one of the
elements or the number of the atoms. Their views are given by
Aristotle in *Physics* III, 4.

Physics III, 5, 204a, 14–17, restated in *Intermediate Physics,
loc. cit.*, as follows: "Since it is not a separate quantity, nothing
is left for it but to be an inseparable quantity. It will then be
something existing in a subject. But if so, that subject, and not
the infinite, will be the principle, but this is something to which
they will not agree." (Latin, *ibid.*).

ואחרי שלא יהיה כמה נבדל, הנה כבר נשאר שיהיה כמה בלתי נבדל. הנה
יהיה מה שימצא בנושא. ואחר שיהיה זה כן, הנה יהיה זה הנושא הוא ההתחלה,
לא מה שאין תכלית לו, והם לא יודו בזה.

Cf. *Metaphysics* XI, 10, 1066b, 9–11, restated in *Intermediate Metaphysics, loc. cit.*

(d) The infinite cannot be an immaterial substance, having actual existence, like soul and intellect.

Physics III, 5, 204a, 20–32, restated in *Intermediate Physics, loc. cit.*, as follows: "After we have shown that the infinite cannot be an immaterial nor a material quantity, there is nothing left but that it should be an immaterial substance, of the kind we affirm of soul and intellect, so that the thing assumed to be infinite, that is, described as infinite, and infinite being itself be one in definition and essence and not different in thought. However, if we assume the infinite to be of this kind, its essence thus being at one with its definition, then, as a result of its being infinite, we shall be confronted with the question whether it is divisible or indivisible. [In the first case], if it be divisible, then the definition of a part and the whole of it will be the same in this respect, as must necessarily be the case in simple, homoeomerous things. But if this be so, then the part of the infinite will be infinite. For the parts must inevitably either be different from the infinite whole or not be different thereof. If they be different, then the infinite will be composite and not simple; if they be not different, then the definition of the part will be the same as that of the whole, for this reasoning must necessarily follow in the case of all things that are homoeomerous. Just as part of air is air and part of flesh is flesh, so part of infinite is infinite, forasmuch as the part and the whole in each of these are one in definition and essence. If a difference is found in the parts of homoeomerous bodies, it is due only to the subject, which is the recipient of the parts, and not to the form, for if we imagine the form of a homoeomerous body without a subject, the parts and the whole thereof will be the same in all respects and without any difference. [In the second case], if we say that the infinite immaterial substance is indivisible, which must be the case of an immaterial *qua* immaterial, then it cannot be called infinite except in the sense in which a point is said to be infinite. In general, the treatment of the existence of an immaterial infinite is irrelevant to the present subject of discussion". (Latin, p. 453 r a, 37).

ואחרי שבטלנו שיהיה כמה נבדל ובלתי נבדל, הנה לא נשאר אלא שיהיה עצם
נבדל, כמו שנאמר אנחנו בנפש והשכל, עד שהיה הדבר המונח אשר אין תכלית
לו, ר"ל המתואר באין תכלית, וישות בלא תכלית, דבר אחד בגדר ומהות ובלתי
מתחלק במאמר. אלא שכאשר הנחנו בענין כן, והיה עצמותו כפי גדרו, אמנם הוא
במה שאין תכלית, חייב בהכרח גם שנאמר שהוא מתחלק או בלתי מתחלק. ואם
אמרנו שהוא מתחלק, הנה יהיה גדר החלק והכל ממנו בזה הענין אחד, כענין
בדברים הפשוטים המתדמים החלקים. וכאשר היה הענין כן, כבר יהיה חלק מה
שאין תכלית לו ובב"ת]. מה שהחלקים לא ימנעו אם שיהיו מתחלפות בגדר לכל
אשר הוא אין תכלית או בלתי מתחלפים. ואם היו מתחלפים, היה מה אין תכלית לו
מורכב ולא יהיה פשוט. ואם היו בלתי מתחלפים חייב שיהיה גדר החלק והכל
אחד בגדר, לפי שזה ענין מתחייב בכל הדברים המתדמים החלקים, כמו שחלק
באויר אויר, וחלק בבשר בשר, כן חלק מה שאין תכלית לו הוא מה שאין תכלית
לו, כאשר היה החלק והכל בם אחד בגדר ובמקום. ואמנם ההלפות החלקים
בגשמים המתדמים הוא מפני הנושא המקבל החלקים לא מפני הצורה. שאלו ציירנו
צורת הגשם במתדמה החלקים בזולת נושא, היה החלק והכל בם אחד מכל הצדדים
בלתי מתחלף. ואם אמרנו שהוא לא יקבל החלוקה, והוא המתחייב לנבדל באשר
הוא נבדל, לא יאמר עליו שהוא בלתי בעל תכלית רק על צד מה שיאמר בנקודה
שהיא אין תכלית לה. ובכלל המאמר במציאות דבר נבדל אין תכלית לו, בלתי
מיוחס לזאת החכמה.

Cf. *Metaphysics* XI, 10, 1066b, 11–21, restated in *Intermediate
Metaphysics, loc. cit.*

In the *Physics*, it will have been noticed, parts (b) and (c)
come in reversed order. Averroes, however, presents them in
the *Intermediate Physics* in the order in which they appear in the
Metaphysics.

In his reproduction of these arguments (from the *Intermediate
Physics*), it should be observed, Crescas has rearranged them in
the following order: (a), (d), (c), (b), parts (a) and (d) being somewhat merged together. His reason for departing from the original
order must have been in order to conclude the arguments with the
rejection of the infinite as quantity on the ground of the insepara-
bility of quantity from material objects, which would enable him
to introduce the discussion about a vacuum. See below n. 12.

8. Hebrew חלוקה, قَسِيم, قِسْمَة, διαίρεσις. (*Analyt. Prior.* I, 31).
More fully חלוקה בשכל (*Epitome of the Physics* III, p. 11b). By
the analogy of מתחלק in the expression הקש תנאי מתחלק, it is to
be translated by *disjunction, disjunctive proposition* (*judgment
or syllogism*).

9. This is taken from part (a) of the argument as given by Averroes.

10. This is taken from part (d) of the argument as given by Averroes.

The composite nature of this passage, consisting, as we have shown, of parts (a) and (d), explains the redundancy of raising again the question whether the immaterial infinite might be divisible immediately after it has already been concluded that it must be indivisible.

The same difficulty has been pointed out by the supercommentators in the text of Averroes. But there at least the superfluity is not so obvious, since several passages intervene between (a) and (d). Cf. Narboni's supercommentary on Averroes' *Intermediate Physics*, *ad loc.* (f. 34a): "The question whether it is divisible or indivisible has already been discussed above [see above note 7 (a) and (d)], and he should have, therefore, taken up here only the possibility of its being indivisible, etc. Our answer is that the two alternatives are enumerated here again because above their enumeration was only casual, for an immaterial quantity is indeed indivisible. But here, [speaking of an immaterial substance], it is the proper place for the discussion of the question as to whether anything immaterial is divisible or not, and therefore he enumerates the two alternatives etc. Or we may say that [even here] he mentions the possibility of its being divisible [only to dispose of it], for an immaterial substance is certainly indivisible and its very essence compels us to think of it as indivisible."

שהוא מתחלק או בלתי מתחלק, שכבר עשתה עשתה למעלה, ולא היה לעשות אלא
אם הוא בלתי מתחלק וכו'. נשוב להם, שהסבה שהביא זה החלוקה שנית הוא
בעבור שלמעלה עשה בה חלוקה במקרה, לפי שהחכמה נבדל לא יקבל החלוקה,
ובעבור שבכאן הביא אמיתת הדברים אמר להביא כאן וכו'. או נאמר, שבעבור
שהעצם נבדל אינו מקבל החלוקה ומהותו יגזור שאינו מקבל החלוקה, הביא גם
כן אם הוא מקבל החלוקה.

11. A marginal note by a pupil of Crescas on the Parma and Jews' College MSS. reads as follows: "I am greatly surprised at the Master, of blessed memory, for all this redundancy. Having started above by saying that the infinite must inevitably be either an immaterial quantity or an immaterial simple substance and

having shown that it cannot be an immaterial substance and
must therefore be an immaterial quantity, he had only to show
now that it cannot be an immaterial quantity. What need was
there for raising the question whether that quantity, which he
has said must be immaterial, can be conceived to subsist in a
subject? It is possible that what the Master, of blessed memory,
meant to say here is as follows: Hence, by the process of elimination, the infinite magnitude must be a quantity. But, then, it
must be inquired concerning quantity itself whether it subsists in
a subject or is immaterial. But it cannot be immaterial. It must
therefore subsist in a subject. Hence an immaterial infinite is
impossible. According to this interpretation of the text, his statement ואם היה כמה נמצא בנושא, i. e., and if it [=the infinite] were
a quantity subsisting in a subject, should be understood as if it
read 'and since quantity must subsist in a subject' etc."

נפלאתי מהרב ז"ל בכל זה האריכות. כי אחר שהוא אמר למעלה, לא ימלט
אם שיהיה כמה נבדל או עצם נבדל פשוט ובטל שהוא עצם נבדל הנה מחוייב
שיהיה כמה נבדל, ולא נשאר לו רק לבטל היותו כמה נבדל. ואיך יהיה הגודל
הנבדל כמה נמצא בנושא. ואפשר שרצון הרב ז"ל הוא זה. הנה נשאר כי הגודל
הבב"ת הוא כמה.   נחקור מן הכמה בעצמו אם הוא נמצא בנושא או נבדל. והנה
הוא בטל שיהיה נבדל. הנה ישאר שיהיה נמצא בנושא. שקר הוא שיהיה בב"ת
נבדל. ואמרו ואם היה כמה וכו', הנה הוא כאלו אמר ואחר שהכמה הוא נמצא
בנושא וכו'.

What this pupil of Crescas is trying to do is to twist the text
and read into it a new meaning in order to remove the redundancy.
The redundancy, however, is due to the fact that Crescas has
somehow rearranged the original order of the argument as given
by Averroes and outlined above in n. 7.

12. The reason given here by Crescas for the impossibility of an
infinite quantitative accident does not agree with the one offered
here by Aristotle. Aristotle says: "Further, if the infinite is an
accident of something else, it cannot be *qua* infinite an element
in things, as the invisible is not an element in speech, though the
voice is invisible" (*Metaphysics* XI, 10, 1066b, 9–11 and cf.
Physics III, 5, 204a, 14–17).

Cf. *Intermediate Metaphysics* X: "Furthermore, if that which
they assume to be infinite is only of the accidental kind of beings,
it cannot be an element of things *qua* infinite, as is assumed by

those who affirm its existence, just as the voice is not an element
of the letters *qua* its invisibility."

ועוד אם היה זה אשר יניחהו לא ב"ת הוא במין המקרה, הנה לא יהיה יסוד
הגמצאות מצד מה שהוא בב"ת כפי מה שיניחוהו האומרים בו, כמו מה שלא יהיה
הקול יסוד האותיות מצד מה שהוא בלתי נראה.

Cf. also above n. 7 (c).

Crescas has purposely departed from the original text in order
to form a natural and easy transition from the problem of infinity
to that of vacuum.

13. Hebrew כבר יהיה. The use of כבר with the imperfect, which
does not occur in Biblical or Mishnaic Hebrew, is common in
Crescas and in other philosophic Hebrew authors. It is undoubtedly due to the influence of its Arabic equivalent قد which is
used, with a variety of subtle distinctions, both with the perfect and the imperfect. With the perfect the Arabic قد means
not only, as the Hebrew כבר, *already*, but also *now, really*, expressing the fulfillment of an expectation. With the imperfect it means
sometimes, perhaps. Some of these usages of the Arabic قد
may be discerned in the use of כבר in mediaeval Hebrew, but in
the case of Crescas its meaning has to be determined independently from the context. According to Ibn Janaḥ the basic meaning of both قد and כבר is the emphasis of certainty and the
affirmation of truth. *Sefer ha-Shorashim*, p. 211: ופרוש כבר בערבי
קד ושתי המלות, ר"ל כבר בעברי וקד בערבי, הם לקיים הדבר ולהמציאו.
This is in agreement with what is cited in the name of Arab
grammarians. See Lane's *Arabic-English Lexicon*, p. 2491.

14. Hebrew נערך על הדרוש. The expression מערכה על הדרוש (see
below p. 186) is the equivalent of المصادرة على المطلوب, τὸ ἐξ
ἀρχῆς αἰτεῖσθαι, *petitio principii*, begging the question. (Cf.
Joel, *Don Chasdai Creskas' religionsphilosophische Lehren*, p. 22,
n. 1).

The Greek expression means to assume the very thing pro-
pounded for debate at the outset. In the Latin form of the expression the term *principii* is a literal translation of ἐξ ἀρχῆς.
More accurately it should have been *quaesiti* or *probandi*, as in
the English rendering (see H. W. B. Joseph, *An Introduc-*

tion to Logic, p. 591, n. 3; Grote, *Aristotle* I, p. 225). In the Arabic and the Hebrew renderings, ἐξ ἀρχῆς is accurately rendered by مطلوب, דרוש, which are the technical terms for *quaesitum*.

As for the Arabic مصادرة, its root means, in addition to *return, proceed, issue, result*, also *demand with importunity*, and hence it is a justifiable translation of the Greek αἰτεῖσθαι, which, meaning literally *ask, beg*, is used in logic in the sense of *assume, postulate*. Thus also the Arabic مصادرة translates the Greek αἴτημα, *postulate*, (literally, *request, demand*) in Euclid's *Elements* (See below p. 466, n. 109).

But how the Hebrew מערכה came to be used as a translation of the Arabic مصادرة, both in the expression מערכה על הדרוש and in the sense of *postulate* in Euclid (see below p. 466, n. 109), is not so obvious. An attempt has been made to explain it on the ground that the Hebrew מערכה has also the connotation of *asking, demanding, begging* (see Moritz Löwy, *Drei Abhandlungen von Josef B. Jehuda*, German text, p. 16). It seems to me, however, that the use of מערכה as a translation of مصادرة is due to its synonymity with סדר. It has been shown that the Arabic صادر is often translated by its homophonous Hebrew word סדר, though the two have entirely different meanings. (Examples are given by Moritz Löwy, *op. cit.*, pp. 10 and 6. n. 1). As a result of this the Hebrew סדר has acquired all the meanings of the Arabic صادر. Such Hebrew words with Arabic meanings are numerous in philosophic Hebrew. The translation of مصادرة by סדר would thus be quite usual. But as סדר in its original Hebrew sense is synonymous with מערכה, the Arabic مصادرة thus came to be translated by מערכה. It is not impossible also that the Arabic صادر has acquired for the Hebrew readers the original meaning of the Hebrew סדר and ערך and, without knowing the underlying Greek term for مصادرة, they took the expression المصادرة على المطلوب to mean "arrangement of an argument on the question" and thus translated it by מערכה על הדרוש. That מערכה was taken in the sense of סדר may perhaps be gathered from the expression והנה סדר מערכה על הדרוש used by Crescas in I, ii, 1, p. 190.

A similar modern case of the failure to identify the Greek term underlying the Arabic مصادرة in this expression and of taking it in one of its ordinary senses is to be found in the rendering of this word by the German *Zurückgehen* (cf. Haarbrücker, *Abu-'l-Fath Muhammad asch-Schahrastânî's Religionspartheien und Philosophen-Schulen*, Vol. II, p. 225, ed. Cureton, p. 357).

15. Quantities are divided into "magnitude" and "number." "Magnitudes" are said to be "measurable" but not "numerable." Again, "magnitudes" are said to be "small" and "great" but not "much" and "few." If a vacuum is "measurable" and is said to be "small" and "great," it must be a magnitude." Cf. below p. 418, n. 33.

16. Hebrew דמו, reflecting the Greek οἴονται used in the corresponding passage in *Physics* IV, 7, 214a, 24.

17. Cf. *Physics* IV, 6.

18. Averroes divides Aristotle's arguments against the existence of a vacuum into five. Crescas, in his turn, groups these five arguments into two main classes, one which may be termed elenchic and the other deictic.

19. Cf. *Physics* IV, 8, 214b, 12–27, and Averroes: שמע אמצעי, מ"ד, כ"ב, פ"ה, המופת הראשון.

20. Hebrew גשמים, literally, *bodies*, i. e., גשמים פשוטים, *simple bodies*, by which Aristotle generally calls the elements. Cf. ἁπλᾶ σώματα in *De Caelo* III, 1, 298a, 29.

21. I. e., fire and air are moved upward whereas earth and water are moved downward.

22. That is to say, the cause of natural motion is due to the fact that the elements have proper places to which they are respectively adapted by their nature, and toward which they tend when they are separated from them. This impulsive motion of the elements is their momentum (ῥοπή), and it is called lightness (κουφότης) when it is upward but weight (βάρος) when it is downward. This momentum might be further called, as here suggested, the efficient cause of motion. But then, also, the

proper place of each element is conceived to act as an attraction. The respective proper places of the elements might, therefore, be called the final causes of motion. Cf. below n. 33.

The expression ואם.....אם is not to be translated here by "either . . . . or," for the two reasons offered are not alternatives but are to be taken together.

The passage in Averroes reads: "We say that inasmuch as there are bodies which have locomotion upward, as fire, and bodies which have locomotion downward, as earth, it seems clear that the cause of the difference in the direction of their respective locomotion must be two things: first, the difference in the nature of the objects moved, and, second, the difference in the natures of the localities toward which they are moved. This is self-evident, for fire indeed is moved in a direction opposite to that of the motion of earth, because its nature is opposite to that of earth and the nature of its place [is opposite] to the nature of the place of earth, for the respective places toward which their motions tend are assumed to be related to the motion as an entelechy and perfection and the respective objects of motion are assumed to be related to it as a motive agent."

ונאמר שלמה שהיו בכאן גשמים שתמצא להם תנועת ההעתק למעלה, כמו האש;
וגשמים תמצא להם תנועת ההעתק למטה, כמו הארץ; והיה מהגלוי שסבת התחלפותו
בזה ההעתק אמנם הוא שני דברים, אחד מהם חלוף טבע הנעתקים, והשני חלוף
טבעי המקומות אשר יעתקו עליהם. וזה ענין ידוע בעצמו, כי האש אמנם היה
נעתק אל הפך צד ההעתק הארץ, לפי שטבעו מתנגד אל טבע הארץ, וטבע מקומה
אל טבע מקומה. כי חלוף טבע המקום יונח מתנועותיהם מדרגת התמימות והשלמות
לתנועה, וחלוף המתנועעים במדרגת הפועל לתנועה.

23. The Jews' College MS. adds here within the text, after the word הטבעיים and before חייב, the following passage: "For the efficient and the final cause bring about motion in different directions only because of a difference in their own nature. But a vacuum has nothing that can be described as its own nature nor anything that is opposite to that nature. Hence it cannot cause motion nor can it be an efficient or final cause."

לפי שהפועל והתכלית לא יחייבו חלוף התנועות אלא מצד חלוף טבעם, והרקות
אין לו טבע ולא חלופו, הנה א״כ לא יחייב בתנועה ולא יהיה לא פועל ולא תכלית.

The same passage occurs also on the margin of the MS. It must have originally been a marginal note written by a pupil of

Crescas from whom we have other notes on the margin of the Parma and Jews' College MSS.

24. Hebrew וזהו מה שכוון באורו במופת הזה, which is an adoption of Averroes' וזהו מה שכוונו לבארו. This phrase is commonly used by Arab philosophers at the conclusion of their arguments. See, for instance, وذلك ما اردنا بيانه, at the end of chapters 1, 2, 3, and 9 of Avicenna's treatise on psychology published by Landauer in the *Zeitschrift der Deutschen Morgenländischen Gesellschaft*, Vol. 29, (1875), pp. 335–418. It is probably borrowed from Euclid, whose *quod erat demonstrandum* is translated into Arabic by وذلك ما اردناه. (Cf. Arabic translation of the *Elements*, Calcutta, 1824).

25. Cf. *Physics* IV, 8, 214b, 28–215a, 24, and Averroes: השמע הטבעי האמצעי, מ״ד, כ״ב, פ״ה, המופת השני.

26. Hebrew והתנועה הטבעית תתחלף לפי טבע מה שממנו ומה שאליו. Averroes has here ומה שממנו ומה שאליו יתחלפו במטבע בתנועה הטבעית. Aristotle says: "Natural lation, however, is different; so that things which are naturally moved will be different" (*Physics* IV, 8, 215a, 11–12). מה שממנו $= \xi\xi$ οὗ, מה שאליו $= \epsilon\dot{i}s$ ὅ.

27. So also Averroes כי ההכרחית אמנם תאמר בהצטרף אל הטבעית, והטבעית קודמת עליה בטבע. Aristotle says: "For compulsory motion is contrary to nature, and that which is contrary to nature is posterior to that which is according to nature" (*Physics* IV, 8, 215a, 3–4).

28. Not found in Averroes' *Intermediate Physics* nor in Aristotle.

29. The word חץ is also used by Averroes. Aristotle has τὰ πιπτούμενα.

30. Aristotle suggests two reasons for the continuation of the motion of a projectile after the removal of the exterior force. "Either through an antiperistasis, as some say, or because the air being impelled, impels with a swifter motion than that of the lation of the impelled body through which it tends to the proper place." (*Physics* IV, 8, 215a, 14–17). The explanation given by Averroes and reproduced here by Crescas corresponds to the second of Aristotle's reasons.

The term לקלותו does not occur in the *Intermediate Physics*.

31. Cf. *Physics* IV, 8, 215a, 24–216a, 26, and Averroes השמע הטבעי האמצעי, מ"ד, כ"ב, פ"ה, המופת השלישי והרביעי.

32. This formal division into two propositions is Crescas' own. Averroes has here: "It is self-evident that when of two objects in motion one is moved faster than the other the ratio of one motion to the other is equal either to the ratio of one motive force to the other, if the motive forces differ, or to the ratio of one receptacle to the other, if there is a difference only in the receptacle, or to the compound ratio of both of them, if there is a difference in both, i. e., the motive agent and the receptacle. Since the difference in the motion must inevitably be due either to the motive agent or to the receptacle or to both, he has framed one argument with respect to the swiftness and slowness due to the receptacle alone and another argument with respect to the swiftness and slowness due to the motive force alone."

וזה שלמה שהיה מן הידוע בעצמו שכל שני מתנועעים אחד מהם יותר מהיר מהשני שיחס אחת מהשתי תנועות אל השנית יהיה יהיה אם ביחס המניע אל המניע, כאשר התחלפו המניעים, או כיחס המקבל אל המקבל, כשהתחלפו במקבל לבד, או במחובר מיחסיהם, כאשר התחלפו בהם יחד, ר"ל בפועל ובמקבל. וזה שלמה שחלוף התנועה לא תמנע שתהיה אם מפני הפועל ואם מפני המקבל או משניהם, עשה המופת האחד מפני המהירות והאיחור הנמצאים מפני חלוף המקבל לבד, והשני מפני המהירות והאיחור הנמצאים מפני חלוף המניע לבד.

Cf. *Physics* IV, 8, 215a, 25–29: "We see the same weight and body more swiftly borne along, through two causes; either because there is a difference in that through which it is borne along, as when it moves through water, or earth, or air; or because that which is borne along differs, if other things remain the same, through excess of weight or levity."

33. Hebrew מניע, literally, "*movens*," or "motive force." See above n. 22.

Aristotle has here: "for we see that things which have a greater momentum (ῥοπὴν), of either weight (βάρους) or levity (κουφότητος), if in other respects they possess similar figures, are more swiftly carried through an equal space (χωρίον = מקבל), and that according to the ratio the magnitudes have to each other" (*Physics* IV, 8, 216a, 13–16).

34. Hebrew מקבל, literally, δεξαμενή, δεκτικόν. But here it probably represents the term χώρα (see above n. 33) which also in Latin is sometimes translated by *receptaclum* instead of *spatium.* Cf. *Physics* IV, 2, 209b, 11–12: διὸ καὶ Πλάτων τὴν ὕλην καὶ τὴν χώραν ταὐτὸ φησιν εἶναι ἐν τῷ τιμαίῳ. "Idcirco etiam Plato in Timaeo materiam et receptaclum ait idem esse."

35. Hebrew ובאור זה ... יותר מהיר. Not found in the *Intermediate Physics.*

36. Hebrew יותר חזק הקבול. Aristotle would have said that air being more attenuated than water will impede the motion less than water (see *Physics* IV, 8, 215a, 29).

37. Cf. *Elements*, Book V, Definition 14. This reference to Euclid is not found in the *Intermediate Physics.*

38. Cf. *Physics* IV, 8, 215a, 31–215b, 21.

39. Hebrew והוא מבואר בשני המקבלים שיחסם כיחס הב"ת אל הבב"ת, literally, "the ratio of a finite to an infinite." This statement is not found in Averroes. He only says: "But inasmuch as in a vacuum there is no recipient, motion will have to be in no-time, that is, in an instant." Aristotle has here: "But a vacuum has no ratio by which it may be surpassed by a body; just as nothing (μηδέν) has no ratio to number" (*Physics* IV, 8, 215b, 12–13). אבל למה שהיה אין ברקות מקבל, חוייב שתהיה התנועה בזולת זמן, ר"ל בעתה.

40. Hebrew זולת זמן, ἄχρονον.

41. This last statement is not found in Averroes. It is based upon the Aristotelian principle that time, motion and magnitude are continuous quantities (*Physics* IV, 11) and hence divisible (*Physics* VI, 2). Cf. also below Propositions VII and XV.

42. That is to say, both these arguments are based upon the proposition that there cannot be motion in empty time. The argument referred to is found in *De Caelo* I, 6, 273a, 21–274a, 18, and is reproduced later by Crescas in his third class of arguments. The original passage of Averroes reads as follows:

חה המופת בעצמו כחו כח המופת אשר יולד ממנו שאם ימצא כח מניע בב"ת היולאני שיחוייב שיתנועע המתנועע ממנו בזולת זמן. חה, כאשר הנחנו המקבל

אחד והמגיע מתחלף, הנה יחס התנועה אל התנועה יחס המגיע אל המגיע. הנה
כאשר הנחנו אחד משני המגיעים בב״ת בכח, לא נשאר בכאן יחס בין שני המגיעים,
הנה יחוייב ממנו שתהיה התנועה בזולת זמן. וכמו כן כאשר סלקנו המקבל באחת
שתי התנועות והנהנהו באחרת, והמגיע אחד, יחוייב שלא יהיה בין שתי תנועות יחס.

In Gersonides' supercommentary on the *Intermediate Physics*,
(*ad loc.*), Averroes' passage is paraphrased as follows: ויאמר אבן
רשד שכח זה המופת הוא כח המופת אשר נולד ממנו שאם ימצא כח מגיע בב״ת
היולאני שיחוייב שיתנועע המתנועע ממנו בבלתי זמן. חה דבר בארו אריסטו
בספר השמים והעולם.

Evidently the text here is based directly upon Gersonides.

The expression כח המופת, *vis demonstrationis*, *nervus probandi*,
refers to the formal arrangement and the cogency of the reasoning
which shows the inference of the consequent from the antecedent.
Thus the Figure of a syllogism is its כח. Cf. Averroes, *Kol
Meleket Higgayon, Niẓẓuaḥ*, p. 58a. והוא מאמר כחו כח הקשר בתמונה
הראשונה. Shem-ṭob's Commentary on the *Moreh* II, 14: וכח זה
המופת הוא על זה התואר: אם הש״י פועל העולם אחר ההעדר, לא ימלט
אם נשלמו כל התנאים להיות פועל או לא נשלמו וכו'.
See below n. 77.

43. Cf. *Physics* IV, 8, 216a, 12–21.

44. Cf. *Physics* IV, 8, 216a, 26–216b, 12, and Averroes: שמע טבעי
אמצעי, מ״ד, כ״ב, פ״ה, המופת החמישי.

45. Hebrew גרגר חרדל. Cf. Matthew 17, 20. Averroes has here
גרגר דוחן, *a grain of millet*, and refers to Aristotle: והיה נכנס העולם
בגרגר דוחן כמו שיאמר אריסטו. The expression is to be found in
the *Physics* IV, 12, 221a, 22–23: καὶ ὁ οὐρανὸς ἐν τῇ κέγχρῳ
ὅτε γὰρ ἡ κέγχρος ἐστίν, ἔστι καὶ ὁ οὐρανός.

The Greek κέγχρος, *a grain of millet*, is usually translated by
the Hebrew דוחן. It is thus rendered in the following Hebrew
translations of Averroes' *Intermediate Physics*: (1) Serahiah ben
Isaac, MS. Bodleian 1386. (2) Kalonymus ben Kalonymus, MSS.
Bibliothèque Nationale, Cod. Heb. 937 and 938. The same term
is also used in the following supercommentaries on the *Intermediate Physics*: (1) Gersonides, MS. Bibliothèque Nationale, Cod.
Heb. 964. (2) Narboni, MS. Bibliothèque Nationale Cod. Heb.
967. Cf. also Narboni on the *Moreh* II, Introduction, Proposition 2: ויהיה כל אחד כגרגר דוחן.

The expression גרר חרדל, however, is found in Ibn Tibbon's translation of the *Moreh* I, 56: כי גרניר החרדל וגלגל הככבים הקיימים מתדמים. Cf. *Emunah Ramah* II, iv, 3, p. 63: עד שיעבור שיצא, על דרך משל, מגרניר החרדל גלגל המזלות. It is also found in the following works: (1) Isaac ben Shem-ṭob's second supercommentary on the *Intermediate 'Physics* (*loc. cit.*), MSS. Munich Cod. Heb. 45 and Cambridge University Library, Mm. 6. 25; and (2) his third supercommentary on it, MS. Trinity College, Cambridge, R. 8. 19(2). (3) Abraham Shalom's translation of Albertus Magnus' *Philosophia Pauperum*, MS. Cambridge University Library, Mm. 6.32(6), p. 31a, 1.9: כגרניר החרדל טול הגלגל. (4) Joseph ben Shem-Tob's translation of Crescas' *Biṭṭul Iḳḳere ha-Noẓerim*, 5. (5) Both these expressions occur in Profiat Duran's *Iggeret Al Tehi Ka-Aboteka:* ואם אפשר הוא האמן שיכנס העולם כלו בגרניר חרדל ודוחן.

The two terms occur also in the *Intermediate Physics*, in the passage corresponding to the above-mentioned *Physics* IV, 12, 221a, 22–23: חה שאלו יהיה הדבר יאמר בו שהוא בדבר כאשר הוא נמצא עמו וכבר יהיו השמים בגרניר חרדל לפי שכבר ימצאו עם גרגר דוחן.

46. Hebrew הנה אם כן למה שאינם גשמים ולא מקרים נשואים בדבר, הנה. Averroes has here: היה מבואר שהרחקים אי אפשר בהם שימירו מקומם אשר הם הריקות אי״א בם שיפנו ולא יריקו למקום הגשם הנה בם אחר שהרחקים לא יתועעו במה שהם מקרים בעצמם. Aristotle says: "In a vacuum, however, this is impossible; for neither is a body" (*Physics* IV, 8, 216a, 33–34).

47. Hebrew שקר בטל. Again later והוא בדוי ואינו אמת (p. 194, l. 18), הוא שקר בדוי (p. 198, l. 2). Similarly in *Moreh Nebukim* I, 73, Prop. X, Note: חהו הנקרא הבדוי עו) השקר: (Ḥarizi's translation: המחשב הכחב), Arabic: אלמכתרע אלכאדב. In all these expressions there is an allusion to the difference between an "impossible falsehood" and a "possible falsehood." See Shem-ṭob on *Moreh Nebukim, loc. cit.*, and cf. the following passage in *Metaphysics* IX, 4, 1047b, 12–14: "For the false and the impossible are not the same; that you are standing now is false; but that you should be standing is not impossible."

48. This statement refers to the two views concerning the existence of a vacuum maintained respectively by the Pythagoreans

and the Atomists. According to the former, the vacuum exists outside the world. According to the latter, the vacuum exists within the world, comprehending the atoms and separating them from each other. Cf. *Physics* IV, 6.

This concluding remark does not occur in the corresponding passage in Averroes (*Intermediate Physics* IV, ii, 5), but it occurs later in IV, ii, 6, and it reads as follows: "Thus it has been established that a vacuum does not exist either within the bodies or outside of them."

הנה כבר התבאר שאין הרקות נמצא לא תוך הגשמים ולא חוץ להם.

Crescas has purposely taken it out of its original place and put it as a conclusion of the arguments against the existence of a vacuum, because he is later to contend that the arguments fail to prove the impossibility of a vacuum outside the world, whatever their validity with reference to the possibility of a vacuum within the world. See below pp. 183, 185.

49. These two additional arguments occur in Aristotle and in Averroes in reversed order.

Cf. *Intermediate Physics* IV, ii, 5, Fifth Argument: "It may also be shown that there is no vacuum from the consideration that a vacuum is an immaterial dimension. The argument is as follows: Dimensions are nothing but the extremities of bodies, an extremity *qua* extremity is indivisible, and an extremity cannot be separated from the object of which it is an extremity. This is self-evident, unless you say that accidents can be separated from the subjects in which they exist. The geometrician, indeed, does abstract a line and a plane and a body. He does this, however, only in discourse and in thought but not in reality. Furthermore, a body requires a place only because it possesses three dimensions by virtue of which it is a body. Now, since it is only because of its possession of dimensions that a body requires [other] dimensions in which to rest, then [immaterial] dimensions, [were they to exist], would require [other] dimensions, and so it would go on to infinity, thus giving rise to Zeno's difficulty about place."

וכבר יורה ג"כ שלא ימצא רקות מצד מה שהריקות רוחק נבדל. וזה שהרחקים אינם דבר יותר מתכליות הגשמים, והתכלית במה שהוא תכלית בלתי מתחלק. ותכלית אי"א שיובדל לדבר אשר הוא לו תכלית, מה ענין ידוע בעצמו, אלא אם היו אפשר שיובדלו המקרים. ואמנם יפשוט הגימטרי הקו והשטח והגשם במאמר

ובמחשבה ולא במציאות. ועוד כי הגשם יצטרך אל מקום במה שהוא בעל רחקים
שלשה אחר אשר היה אמנם הוא גשם בם. ואם הצטרך הגשם מצד שהוא בעל
רחקים אל רחקים ינוח בם, יצטרכו הרחקים אל רחקים, וילך העינן אל בלתי
תכלית, ויחוייב ספק זנין במקום.

For references to Aristotle see below notes 50, 51.

Crescas has purposely reversed the original arrangement of the
two arguments in order to be able to conclude with the statement
"Hence the existence of an immaterial extension is impossible,"
which, according to him, is the chief basis of Aristotle's rejection
of infinity.

50. This argument is based on *Physics* IV, 8, 216b, 12–21.

51. This argument is based upon the following passage: "For
these fancy there is a vacuum separate and per se . . . . But this
is just the same as to say that there is a certain separate place;
and that this is impossible, has been already shown" (*Physics* IV,
8, 216a, 23–26).

52. Crescas characterizes the argument here as מופת הדבקות.
Later in his criticism of this proposition he calls it again הדבקות,
according to the Munich and Paris MSS. and the printed editions.
The Vienna and Oxford MSS. read there התדבקות without the
definite articles. Both הדבקות and התדבקות occur in Isaac ben
Nathan's translation of Altabrizi. In the anonymous translation
the term used is מופת ההדבק. The Arabic original for these terms is
مطابقة (cf. *Makaṣid al-Falasifah* II, p. 127: اطبقنا דבקנו) which in its
turn is a translation of the Greek ἐφαρμόζω used in Euclid's
Elements. Now, the Greek term has two meanings. (1) The passive ἐφαρμόζεσθαι means "to be applied to" without any implication of fitness and equality. (2) The active ἐφαρμόζειν means
"to fit exactly," "to coincide with." (Cf. Heath, T. L. *The Thirteen Books of Euclid's Elements*, Vol. I, pp. 224–225). In the
Arabic translation of the *Elements* (Calcutta, 1824), the term
ἐφαρμόζοντα in Axiom 4 of Book I is translated by المتطابقة من
غير تفاضل, *agreeing without a remainder*.

The Hebrew דבקות and the Latin *applicatio* appear as translations of the same Arabic word, probably مطابقة, in *Fons Vitae*
II, 14: "Locus autem non est nisi applicatio superficiei corporis

ad superficiem corporis alterius." Cf. *Liķķuṭim min sefer Meķor Ḥayyim* II, 21: המקום יחייב דבקות שטח גוף בשטח גוף אחר.

53. Hebrew: והתחלנו מנקודה אחת בקצה הקו אשר הוא ב״ת. Literally, "and we begin from a point at the end of the line which is finite."

Crescas' argument as it stands would seem to imply that only one line is infinite in one direction whereas the other line is infinite in both directions. In Altabrizi, however, both lines are assumed to be infinite only in one direction (see next note).

54. The proof as fully given by Altabrizi is as follows: If an infinite were possible, let AB be infinite at B and finite at A. Take any point C in AB and draw line Cb, again infinite at b and finite at C. AB is, therefore, longer than Cb by AC.

A   C   B

b

Let us now apply Cb to AB so that C falls upon A.

The question is would b coincide with B or not. If they do coincide, it would contradict the assumption that AB is longer than Cb.

If they do not coincide, then Cb would have to be finite at b, which, again, contradicts the assumption.

Furthermore, if they do not coincide, Bb would have to be equal to AC, and so AB would have to be finite, which contradicts the assumption.

Hence, no infinite can exist.

The text of Altabrizi reads as follows:

אולם מופת הדבקות הוא זה. אלו היה מרחק מתפשט אל בלתי בעל תכלית
במלוי או רקות, אם היה לנו שניח קו יוצא מהתחלה היא נקודת א׳ בזה המרחק
הבלתי בעל תכלית וילך אל בלתי תכלית, ונקראו קו א״ב כמו זה א   ג   ב,
ועיח נקודה אחרת בזה הקו אחר נקודת א׳ בשעור אמה, והיא נקודת ג׳, הנה הגיעו
פה שני קוים, האחד מהם קו א״ב והוא מצד א׳ בעל תכלית ומצד ב׳ בלתי בעל
תכלית, והשני קו ג״ב והוא גם כן מצד ג׳ בעל תכלית ומצד ב׳ בלתי בעל תכלית.
וכאשר הנחנו במחשבתנו דבקות אחד מהם על האחר מהשני צדדים הבעלי תכלית,
ועניו זה הדבקות שנקביל במחשבה החלק הראשון מקו א״ב מצד א׳ בחלק הראשון
מקו ג״ב מצד ג׳ והחלק השני בשני והחלק השלישי בשלישי וכן אל בלתי תכלית,
הילכו מקבילים אל מה שאין תכלית לו מבלתי חתוך או יחתך אחד משניהם?
והראשון בטל, ואם לא, היה החסר כמו הנוסף, היה קו א״ב נוסף על קו ג״ב בקו
א״נ. הנה ישאר השני. וידוע שהנחתך יהיה הוא הקו החסר ויהיה בעל תכלית,

והנוסף אמנם נוסף עליו בשעור בעל תכלית, והוא שעור אמה, הנה הוא יהיה גם
כן בעל תכלית, ויהיה הקו המונח בלתי בעל תכלית בעל תכלית מצד ב', וכבר
הנחנוהו בלתי בעל תכלית, זה בטל. וכבר חויב מהנחת מרחק מתפשט אל זולת
תכלית בטל, ויהיה נמנע בטל. הנה כל גודל הוא בעל תכלית מוגבל. וזהו הדרוש.

The same proof, somewhat differently stated, is given by Algazali in his *Kawwanot*, *Metaphysics* (*Maḳaṣid al-Falasifah* II, p.
126f).

הראיה השנית: שאם אפשר קו בלי תכלית הנה יהיה זה הקו קו א"ב. ואין תכלית
לו בצד ב', ונרמז אל נקדת ג' וד', ואם היה מד' אל ב' בעל תכלית, הנה כאשר
נוסף עליו ג"ד היה ג"ב בעל תכלית, ואם היה מד' אל ב' בלתי בעל
תכלית הנה כאשר נוסף עליו ג"ד היה ג"ב בעל תכלית, ואם היה
מד' אל ב' בלתי בעל תכלית, הנה אם דבקנו במחשבה ד"ב על
ג"ב, הנה אם שילכו יחד בצד ב' בלי שנוי, חה שקר, אחר שיהיה
המעט שוה לרב, כי ד"ב יותר מעט מג"ב, ואם קצר ד"ב מג"ב
ונכרת תחתיו ונשאר ג"ב קיים עומד וכבר הגיע לתכלית ד"ב
בהכרתו מצד ב', וג"ב לא יוסיף עליו אלא בשעור ג"ד הבעל
תכלית בבעל תכלית, ומה שנוסף על הבעל תכלית בבעל
תכלית הנה הוא בעל תכלית בהכרח.

The proof is also found in Shahrastani, p. 403 (ed. Cureton),
Emunah Romah I. 4. They both seem to have taken it from
Avicenna's *Al-Najah*, p. 33, reproduced in Carra de Vaux's
Avicenne, p. 201. A similar argument is given also in *Ḥobot ha-
Lebabot* I, 5.

A similar argument by Roger Bacon is referred to by Julius
Guttmann in his "Chasdai Creskas als Kritiker der aristotelischen
Physik," *Festschrift zum siebzigsten Geburtstage Jakob Guttmanns*,
p. 51, n. 2.

55. Cf. above n. 5.

56. Hebrew גשמי היה או למודי. The *Intermediate Physics* uses here
the terms "physical" טבעי and "mathematical" למודי. Aristotle
uses the terms "intelligible" and "sensible" οὔτε νοητὸν οὔτε
αἰσθητόν. (*Physics* III, 5, 204b, 6–7; see also *Metaphysics* XI,
10, 1066b, 24). The Hebrew translation of the *Physics* with Averroes' Long Commentary (MS. Bodleian, 1388), reads in one place
בלמודים או במושכלות, i. e. "mathematical or intelligible" and in
another לא מושכל ולא מוחש, i. e., "intelligible," "sensible."

57. Cf. *Physics* III, 5, 204a, 34–204b, 10; *Metaphysics* XI, 10, 1066b, 21–26; and Averroes שמע טבעי אמצעי, מ"ג, כ"ג, פ"ד, החלק השני, מה שאחר הטבע האמצעי מ"י. Cf. also *Milḥamot Elohim* VI, i, 11.

58. Averroes has here ב"ח, הנה כל ספור כל, i. e., "everything numbered," which is quite different. See below Prop. II, Part II, p. 219. See also *Emunah Ramah* I, 4.

59. The designation of the succeeding arguments as "physical" (φυσικῶς—טבעיים) is also found in Aristotle and Averroes (cf. *Physics, loc. cit.* and *Metaphysics, loc. cit.*). Averroes designates them also as "appropriate" מיוחדים in contradistinction to the preceding argument which he calls "general" and "logical." See above notes 5, 55.

60. Cf. *Physics* III, 5, 204b, 10–205a, 7; *Metaphysics* XI, 10, 1066b, 22–1067a, 7; and Averroes: שמע טבעי אמצעי, מ"ג, כ"ג, פ"ד, החלק השני, המופת הראשון; מה שאחר הטבע האמצעי, מ"י.

61. In the original of Averroes the argument is as follows:
The infinite must be either *simple* or *composite*.

A. If *composite*, it could not be composed of an *infinite number* of elements, but would have to be composed of a finite number of elements, of which either (a) one or (b) more than one would be infinite in *magnitude*.

B. If *simple*, it would have to be either (a) one of the four elements, or (b) some neutral element outside the four.

Crescas, as will be noted, reproduces only the main alternatives, A and B, leaving out the subdivisions (a) and (b) under each of these, but he seems to allude to these subdivisions in the expression ואיך שיהיה, which accordingly is to be taken to mean not only "and in either case," i. e., whether simple or composite, but also "and however that simple or composite infinite body is supposed to be," referring to (a) and (b).

Following is the text of the *Intermediate Physics:* "First argument. Every infinite tangible object must be either simple or composite. If it were composite, inasmuch as the elements of which it is composed must be finite in number, for it has already been proved in Book I of this work that nothing composite can be made up of an infinite number of elements, it would follow that either one or more than one of its elements would be infinite in magnitude, for if not, the composite object could not be called infinite. But if one of the elements were infinite, it is clear that the other simple elements of which the composite whole is made up would become resolved into that element, inasmuch as elements are contraries, and they persist together only by that uniformity of relation [שוי, aequitas], and equilibrium [יושר, mediocritas] which exists among their forces. And even if the force inherent in one particle of that infinite element were weaker than the force inherent in a corresponding particle of the same size of the finite element, just as we may say that the force which is in a portion [משך, tractus] of air is weaker than the force which is in a similar portion of water and earth, still this would not refute [יסתור, prohibet] [our argument] that the infinite would bring corruption to the finite, for if we multiply that weaker particle to infinity the result would necessarily be something more powerful than the finite total of the stronger particles. And if more than one of the simple elements were infinite, it would follow that one of them would fill the whole place and there would remain no room for the others, for inasmuch as a body is extended in all dimensions, i. e., the six directions, it follows that an infinite body, by virtue of its being a body, is infinite in all directions. The same conclusion must necessarily also follow, if we assume that only one of the elements is infinite, namely, that no room would remain for the rest, be that finite or infinite. Since none of these alternatives is possible, there can be no infinite composite body.

He further says that there cannot exist a simple, tangible, infinite body whether it be one of the four elements or something intermediate between them,—as has been assumed by some physicists in order to avoid the difficulty confronting them that an infinite element would bring corruption to the other elements, —or be it an element additional to the four elements, even though it would seem that there is no other element outside fire, air, water and earth. The argument is as follows. If there existed in this sublunar world a fifth element, it is clear that all the composite objects would be resolved into it, for if we assume an element, qua element, to be infinite, all the other elements must suffer corruption, and thus the entire world would be changed into the nature of that element, inasmuch as an element is an element by virtue of the contrary qualities which exist in it. By the same token it would follow that that intermediate element, which is assumed by some people, would, by virtue of its being an element, have to contain something contrary, and thus, if it were infinite, the other elements would have to suffer corruption." (Latin, p. 453 r b—v b).

המופת הראשון, שכל גשם מה ממושש בלתי בעל תכלית הנה הוא אם פשוט
ואם מורכב. ואם היה מורכב, והיו היסודות אשר מהם הורכב בעלי תכלית
במספר, כפי מה שהתבאר מהמנע מציאות יסודות אין תכלית במספר במורכב
מהם במאמר הראשון מזה הספר, הנה יחוייב שיהיה אחד מהם בב"ת בגודל או
יותר מאחד; ואם לא, לא נאמר במורכב שהוא בלתי ב"ת. אבל אם היה בם אחד
בלתי ב"ת, הוא גלוי שיפסדו שאר הפשוטים אשר חובר מהם המורכב אליו מצד
מה שהיסודות הפכים. ואמנם ישארו בשווי והיושר אשר בין כחותיהם. ואם היה
הכח הנמצא בחלק אחד מהיסוד שאין תכלית לו יותר חלוש מהכח הנמצא בחלק
מן היסוד הב"ת השוה לזה החלק אשר מיסוד בלתי ב"ת, כמו שנאמר מהכח אשר
במשך אחד מהאויר יותר חלוש מהכח אשר ממשך אחד ממים וארץ, לא יסתור
זה דבר בשיהיה הבלתי ב"ת יפסיד הב"ת, לפי שאנו כאשר כפלנו זה החלק החלוש
הכח של בלתי תכלית יתקבץ ממנו מה שהוא יותר חזק מהחלק הב"ת בהכרח.
ואם היו הבלתי ב"ת מהפשוטים יותר מאחד, חוייב שיהיה אחד מהם הוא אשר ימלא
המקום, ולא ישאר לנשארים מקום, לפי שהגשם למה שהיה הוא הנמשך אל כל
הרחקים, ר"ל הפיאות השש, חוייב שיהיה הגשם הבב"ת אמנם הוא גשם בב"ת בכל
הפיאות. מה יחוייב בהנחת אחד לבד מהם בב"ת, עד שלא יהיה לנשאר מקום
בין שיהיה ב"ת בין שיהיה בב"ת. וכאשר היו כל אלו החלוקות נמנעות, הנה אי"א
שימצא גשם מורכב בלתי ב"ת.

ואומר עוד שהוא אי אפשר שימצא גשם פשוט ממושש בב"ת בין שיהיה אחד
מהיסודות הארבעה או אמצעי ביניהם, כפי מה שיניחוהו קצת הטבעיים לברוח
מאשר יתחייב להם שיהיה מפסיד הנשאר, או יהיה יסוד נוסף על היסודות הארבעה,
ואם היה נראה שאין יסוד בלתי האש והאויר והמים והארץ. מה שאלו היה בכאן
יסוד חמישי, היה נראה מענין המורכבות שהן יותכו אליו, לפי שכאשר הנחנו יסוד
בלתי ב"ת במה שהוא יסוד, חוייב שיפסדו שאר היסודות, וישתנה זה העולם אל
טבע אותו היסוד, כי היסוד אמנם הוא יסוד באיכיות ההפכיות הנמצאות בו. ולזה
יחוייב ביסוד הממוצע אשר יניחוהו אנשים שיהיו בו הפכיות מצד שהוא יסוד. ואם
היה בלתי ב"ת נפסדו הנשארים.

62. Averroes has here הספר מזה הראשון במאמר שהתבאר מה כפי. The reference is to *Physics* I, 4.

63. This is an allusion to alternative B(b) given above in note 61; that is to say, no element can be conceived as being neutral and without qualities.

64. Averroes employs this argument in refutation only of A(a) and (b) given above in n. 61. From Crescas' use of the definite האחד, which undoubtedly refers to היה בהכרח אחד מיסודותיו בב״ב בודל, it appears that he applies it to all the alternatives included under both A and B.

65. Cf. *Physics* III, 5, 205b, 24–31; *Metaphysics* XI, 10, 1067a, 23–29; and Averroes: שמע טבעי אמצעי, מ״ג, כ״ג, פ״ד, החלק השני, המופת השלישי; מה שאחר הטבע אמצעי, מ״י.

This argument, which Crescas advances as the second of the physical arguments, is the third in the original texts of Aristotle and Averroes. Crescas has omitted here the original second argument; but he has inserted it later in his third class of arguments. See below n. 91.

66. Hebrew ונבדל מן המקום העליון. In one text of Kalonymus' translation of the *Intermediate Physics* (Paris, Cod. Heb. 938) the corresponding passage reads ונבדל ממנו המקום העליון, i. e., "the upper place would be separate from it." In another text of apparently the same translation (Paris, Cod. Heb. 943), it reads ויעדיף ממנו המקום העליון, i. e., "the upper place would be greater than it." Without the original Arabic text before me, I venture to suggest that this difference must have arisen in the uncertainty of the reading فضل or فصل in the original Arabic text, the former meaning "to be greater" and the latter "to be separated." The copy used by Crescas evidently read ונבדל ממנו המקום העליון, which he has changed to ונבדל מן המקום העליון.

A similar uncertainty on the part of the same translator as to the reading of فضل or فصل may be also noted in two corresponding passages in his translations of the *Intermediate Physics* and *Intermediate Metaphysics* (quoted below in n. 71 (a). In the former it reads כי הגשם לא יבדל מהמקום, i. e., "the body cannot be separated from place." The context, however, would warrant here the reading "the body cannot be greater than place." Cf. *Physics* III, 5, 205a, 35: οὔτε τὸ σῶμα μεῖζον ἢ ὁ τόπος

In the corresponding passage in the *Intermediate Metaphysics* it correctly reads: כי אלו היה אפשר שיעדיף הגשם על המקום.

These two readings are also reflected in the Latin translation of Averroes in a passage quoted below in n. 71 (a).

67. Averroes concludes here: ואם היה בשניהם היה לו כובד וקלות, חה בטל, i. e., "and if it were in both places it would have both weight and lightness, which is impossible."

68. Cf. *Physics* III, 5, 205b, 31–206a, 8; *Metaphysics* XI, 10, 1067a, 28–33; and Averroes: שמע טבעי אמצעי, מ"ג, כ"ג, פ"ג, ח"ב, המופת הרביעי; מה שאחר הטבע האמצעי, מ"י הרביעי. Cf. also *Milḥamot Elohim* VI, i, 11, p. 339, ואולם באנה.

69. Hebrew מקום. The term מקום throughout this discussion represents the Greek τόπος in Aristotle, which is to be translated according to context by either *place* or *space*. Aristotle has one definition for both space and place, space being only place that is remote and general, as, for instance, heaven, according to Aristotle, is the remote and general place of all things that exist (cf. J. Barthélemy Saint-Hilaire, *Physique D'Aristote*, Vol. I, Préface, p. LI). Aristotle himself designates this distinction by contrasting "common (or general) place" (τόπος κοινός) with "proper place" (ἴδιος τόπος) or "first place" (πρῶτος τόπος). Cf. below n. 76. There is a reference to this distinction in *Moreh Nebukim* I, 8, where Maimonides says that the Hebrew term מקום in its original meaning applies both to a particular and to a general place. מקום. זה השם עיקר הנחתו למקום המיוחד ולכולל. (Cf. Munk, *Guide* I, 8, p. 52, n. 1). The Greek χώρα may be discerned under the Hebrew מקבל See above n. 34.

70. Hebrew במין ובשיעור.. Averroes adds here "that is, in quality and in quantity" ר"ל, והיו המקומות בעלי תכלית במין ובעלי תכלית בשעור באיכות ובכמות.

71. In the original texts this argument is divided into two parts:

(a) Everything is in place. Place has six directions. Each of these is finite. Consequently, everything is finite, for nothing can be greater than its place.

(1) *Intermediate Physics, loc. cit.:* "It may also be said that if every sensible object is in a place, and places are finite in species

and finite in magnitude, i. e., in quality and in quantity, it follows
that every body must be finite. For there is no doubt that it must
be in a certain place, and moreover in one of the several natural
places, and if the place is finite it must necessarily belong to a
body that is finite, inasmuch as the body cannot be separated
from the place (on the margin of the Latin version there is another reading: "*excedit locum.*" See above n. 66). That the places
are finite in species is clear, for their differentiae are finite, and
these are, down and up, before and behind, right and left. It can
likewise be shown that each one of these is finite in quantity, for
these differentiae cannot be of infinite dimensions, for [if they
were], those places could not be distinguished by nature, inasmuch as they would have no natural boundaries, but they would
be so 'only by relation. But it is clear from the motions of those
which move toward them and rest in them that they are limited
by nature." (Latin, p. 454 v a, 54). (Cf. *Physics* III, 5, 205b,
31–206a, 2).

ויאמר גם כן שאם היה כל גשם מוחש במקום, והיו המקומות בעלי תכלית במין
ובעלי תכלית בשעור, ר"ל באיכות ובכמות, יחוייב שיהיה כל גשם ב"ת. זה שאין
ספק שיהיה במקום, ובמקומות הטבעיים, וזה שהמקום הב"ת הנה הוא בהכרח
לגשם ב"ת, כי הגשם לא יבדל מהמקום. ואמנם שהמקומות בעלי תכלית במין
זה מבואר, לפי שהבדליהם ב"ת, והם מטה ומעלה, ופנים ואחור, וימין ושמאל.
וכמו כן יתבאר שכל אחד מהם ב"ת בכמה. זה שאלו ההבדלים אי"א שיהיו
ברחקים בב"ת, לפי שלא היו נכרים אלו המקומות בטבעם, אחר שלא היו גבולים
טבעיים, ואמנם יהיה בהצטרף. והתבאר שהם בטבע מוגבלים מתנועות המתנועעים
אליהם ונוחם בם.

(2) *Intermediate Metaphysics, loc. cit.*: "Further, every sensible
body is in a place, be that body simple or composite, and the
places are six, up and down, right and left, before and behind,
and none of these can be infinite nor can anything existing in
them be infinite. For how could anything existing in them be
infinite, unless the body could be greater than the place in which
it is." (Cf. *Metaphysics* XI, 10, 1067a, 28–30).

ועוד שכל גשם מוחש הוא במקום, בין שיהיה פשוט או מורכב, והמקומות ששה,
אם מעלה ואם מטה, אם ימין ואם שמאל, ואם פנים ואם אחור. ואי"א שיהיה אחד
מאלו בב"ת, ולא בם בב"ת. ואיך יהיה בם מה שהוא בב"ת, אלא אילו היה אפשר
שיעדיף הגשם על המקום אשר הוא בו.

(b) Since place is the limit of that which surrounds a body, the body thus surrounded and limited cannot be infinite.

(1) In the *Intermediate Physics* Averroes does not reproduce this argument in full. He only refers to it by saying that the impossibility of an infinite "will become clearer when it will have been shown that place is the boundary of that which surrounds." ויתבאר בם יותר כאשר יתבאר שהמקום הוא תכלית המקיף. (Cf. *Physics* III, 5, 206a, 2–8).

(2) *Intermediate Metaphysics*, *loc. cit.*: "In general, if there cannot be an infinite place, inasmuch as place is the surrounding limit, and this means either up or down or one of the other differentiae of place, there cannot be an infinite body, unless the occupant of the place is greater than the place in which it is." (Cf. *Metaphysics* XI, 10, 1067a, 30–33).

ובכלל אם היה נמצא שימצא מקום בלתי ב"ת, אחר שהיה המקום הוא התכלית המקיף, חה אם מעלה ואם מטה, ואם זולת זה מהבדלי המקום, הנה הוא נמצא שימצא גשם אין תכלית לו, אלא א"כ יהיה בעל מקום יעדיף על המקום אשר הוא בו.

Crescas, it should be noted, has merged these two arguments together, by quoting the definition of space within the first argument.

72. Hebrew הדברים הטבעיים, literally, "natural things." I have taken it to refer to the natural or proper places of the elements. Cf. quotations above n. 71 (a).

The reasoning of this argument is to be carried out as follows: The six species of place must be each limited in extension, for the following reason: The existence of these distinctions in place is known from an observation of the different kinds of natural motion. Natural motion is either upward, downward, or in a circle. Motion downward is limited, and so also is lower place limited. Consequently, motion upward and the upper place must be limited and absolute. See below n. 104.

73. This is not given /by Aristotle and Averroes as a separate argument. It is rather Crescas' own elaboration of the second part of the preceding argument. See above n. 71(b). It is, however, given as a separate and independent argument in *Emunah Rāmah* I, 4: "Furthermore, if an infinite body existed, it could not be in place at all, for anything that is in place is enclosed

by the surfaces of its place, and an infinite cannot be enclosed
by anything, inasmuch as that which encloses a thing must be
greater than the thing, seeing that it surrounds the thing. Consequently, if anything enclosed an infinite, it would have to be
greater than the infinite. But that is absurd."

ועוד שאם היה גשם בלתי בעל תכלית, לא יהיה במקום כלל, לפי שכל מה שהוא
במקום, שטחי מקומו כופים עליו, ולא יתכן בבלתי בעל תכלית שיהיה דבר אחר
כלל כופה עליו, כי מה שהוא כופה לדבר הוא יותר גדול ממנו, מצד שהוא מקיף
בו. ואם כפה דבר על הבלתי בעל תכלית, היה יותר גדול מבלתי בעל תכלית.
חה בטל.

74. Hebrew המתקומם. The MSS. read המקומם and so it reads also in
Part II of this proposition (p. 198, l. 15). But the form המתקומם
occurs also in 'Olam Ḳaṭan I, 3, ed. Horovitz, p. 15: לפי שאין מקום
בלי מתקומם ואין מתקומם בלי מקום, and in Albalag quoted below Prop.
I, Part II, n. 23 (p. 414). The term reflects the Arabic مُتَمَكِّن
(cf. Horovitz, ibid., p. XIV) = τὸ τόπον κατέχον, corpus locatum
(cf. Husik, *Judah Messer Leon's Commentary on the 'Vetus
Logica'*, p. 115).

75. Cf. *Physics* IV, 4, 210b, 34–211a, 5: "First, then, we should
think that place comprehends that of which it is the place, and
that it is not anything of that which it contains. And, again, that
the first place is neither less nor greater than the thing contained
in it; and also that it does not desert each particular thing, and
is not inseparable from it. Besides this we should think that
every place has upward and downward, and that every body
naturally tends to and abides in its proper place."

Cf. *Intermediate Physics* IV, i, 1, 6: "First, place surrounds the
object of which it is a place. Second, place does not exist in place
and is separable from the object and is no part thereof. Third,
first place is equal to the occupant, is neither greater nor smaller
than it. It is not smaller, because it surrounds the occupant. It
is not greater, because, by virtue of its being the first place of the
occupant, it cannot receive another body in addition to it."

הראשונה, שהמקום יקיף הדבר אשר הוא לו המקום. השנית, שהמקום בלתי
העומד במקום ושהוא נבדל לו ואינו חלק ממנו. השלישית, שהמקום הראשון שוה
לבעל המקום, אינו יותר גדול ממנו ולא יותר קטן. חה שאינו יותר קטן, לפי שהוא

יקיף לבעל המקום, ולא יותר גדול, לפי שאי״א שיקבל עמו גשם אחר מצד מה
שהוא מקום ראשון.

76. "First place" is defined by Aristotle in the following passages:
"With respect to place also one is common (κοινός) in which
all bodies are contained, but another proper (ἴδιος) in which
any thing primarily subsists" (*Physics* IV, 2, 209a, 32–33). "And
such is the first (πρῶτος) place in which a thing subsists" (*ibid.*
4, 211a, 28–29). Cf. above n. 69.

Aristotle's ἴδιος τόπος is reflected in Ibn Gabirol's המקום הידוע
(*Likkuṭim min Sefer Meḳor Ḥayyim* II, § 23, 24). Cf. *Fons
Vitae* II, § 14, p. 48: "locus cognitus;" p. 49: "loci noti."

77. Cf. *Physics* IV, 4, 211b, 6–9: "For there are nearly four things
of which it is necessary place should be one. For it is either form
or matter, or a certain interval between the extremes of a thing
(τῶν ἐσχάτων); or the extremes (ἔσχατα), if there is no interval beside the magnitude of the inherent body."

Cf. *Intermediate Physics* IV, i, 1. 8: "It is possible for us to
show that this definition of place, arrived at by way of a categori-
cal demonstration, can also be established by means of another
kind of syllogism, whose force is the force (כחו כח ההקש, cf. above
n. 42) of a hypothetical disjunctive syllogism. For it appears that
place must necessarily be one of the following four: form, matter,
the surrounding limit, or the interval between the limits of that
which surrounds, that which is called vacuum."

וכבר אפשר לנו שנודה על שזה הגדר שנתחדש בדרך המופת המשאיי הוא גדר
המקום בצד מה אחר מן ההקש, כהו כח ההקש התנאיי המתחלק. זה שכבר
יחשב שיחוייב בהכרח שיהיה המקום בהכרח אחד מארבעה: אם הצורה ואם
ההיולי, ואם התכלית המקיף, ואם הרוחק שבין תכליות המקיף, וזהו אשר יקרא
חללות.

78. Aristotle identified this with Plato's view of place (*Physics*
IV, 2, 209b, 11–12). Whether Aristotle understood Plato right
or not is a question raised by his commentators. (Cf. Simplicius'
commentary on the *Physics*, ed. Diels, p. 539, line 8 ff., and
Taylor, *Physics*, p. 185, n. 1; Zeller, *Plato*, p. 306, n. 39).

79. This view, which identifies space with vacuum, was held by
the Atomists and the Stoics, and it is considered by some to be

the view of Plato. Cf. Simplicius' commentary on the *Physics*, ed. Diels, p. 571, line 25, and Taylor, *Physics*, p. 197, n. 1. Averroes says of it here: "This view had.been maintained by many of the ancients," וכבר אמרו בו רבים מן הקדמונים. Cf. also *Intermediate Physics* IV, ii: "For they believe that place is extension, and place and extension in their opinion are one in subject, two in discourse."

כי אשר יראו שהמקום רוחק, והמקום והרוחק אצלם אחד בנושא שנים במאמר.

80. Hebrew והוא אשר יקרא חללות. This phrase is taken from the *Intermediate Physics*. It is Averroes' own explanation, in popular terms, of the more technical expression "the interval between the limits of that which surrounds," הרוחק אשר בין תכליות המקיף. The latter is the exact translation of the Greek διάστημά τι τὸ μεταξὺ τῶν ἐσχάτων (*Physics* IV, 4, 211b, 7–8). What he means to say is that according to the definition now proposed by Aristotle place is nothing but what people ordinarily call a void occupied by a body. Cf. *Physics* IV, 7, 214a, 19–20: τὸ γὰρ κενὸν οὐ σῶμα ἀλλὰ σώματος διάστημα βούλεται εἶναι.

Cf. also *Epitome of the Physics* IV, p. 13b: "And this makes it clear that place is not the void or the interval between the surrounding limits, which, in the opinion of some people, is capable of existing independently by itself, and which is designated by them by the term vacuum."

ומהנה יראה כי המקום אינו הפנוי והרוחק אשר בין התכליות המקיפות, אשר היה אפשר פרידתו אצל אנשים, והוא אשר יורו עליו בשם הרקות.

The terms פנוי, חללות, פנאי, רקות, are all translations of κενός خلا, فضا (cf. Prop. I, Part II, n. 31, p. 418).

81. "It is not, however, difficult to see that it is impossible for either of these to be place. For form and matter are not separated from the thing" (*Physics* IV, 2, 209b, 22–23). "For these things, viz., matter and form, are something belonging to that which is inherent" (*ibid.*, 3, 210b, 20–31).

There is nothing in the *Intermediate Physics* to correspond to this passage.

82. Cf. *Metaphysics* V, 17, 1022a, 4–6: "Limit (πέρας) . . . is applied to form, whatever it may be, of a spatial magnitude or of a thing that has magnitude."

83. Cf. *Physics* IV, 4, 211b, 12–14: "Both (i. e., place and form), therefore, are limits (πέρατα), yet not of the same thing; but form is the limit of the thing contained, but place of the containing body."

Cf. *Intermediate Physics* IV, i, 1, 8: "For form, though assumed by us to be a limit, is the limit of that which is surrounded, not the limit of that which surrounds." לפי שהצורה, אם הנחנו שהוא תכלית, הנה הוא תכלית המוקף, לא תכלית המקיף.

The term מקיף, *surrounding, circumambient, containing, enclosing,* is a translation of περιέχων, حاوى.

84. Hebrew תכלית שהוא למה תכלית אלא יאמר בו תכלית ולא שאינו והאמת להיולי ותגבילהו. Literally: "The truth is, it is not a limit, and it is said to be a limit only because it is the limit of matter and it bounds it." This statement is taken from Averroes but does not occur in the corresponding passage of Aristotle. The original statement in Averroes reads as follows: שהצורה אינה והאמת תכלית אבל היא הנותנת עצם הדבר, ואם נאמר בה תכלית לפי שהיא תתן תכלית הדבר ותגבילהו.

The meaning of these allusive affirmations about form not being a "limit" and being a "limit" and being a "limit" in a certain sense may be brought out by the following considerations.

The term limit (πέρας), according to Aristotle, means (1) the last point (ἔσχατον) of a thing, (2) the form (εἶδος = σχῆμα = μορφή) of a magnitude or of a thing having magnitude, (3) the end (τέλος) or final cause (οὗ ἕνεκα), and (4) the substance (οὐσία) and the essence (τί ἦν εἶναι) of a thing. See *Metaphysics* V, 17, and Schwegler's and Ross's commentaries *ad loc.*

Now in Hebrew the same word תכלית, reflecting here the Arabic نهاية or غاية or both, translates the Greek πέρας, ἔσχατον, τέλος, οὗ ἕνεκα. What Averroes is therefore trying to say here is that the term תכלית, or whatever Arabic term underlies it, has many shades of meaning, inasmuch as it reflects different Greek words, and while in one sense it may apply alike to both place and form, there are other senses in which it does not apply to them alike.

In so far as תכלית is a translation of πέρας it applies to both place and form. But there is the following difference. To place it applies in the sense of ἔσχατον. To form, however, it applies in the other senses enumerated by Aristotle. For form has many meanings and fulfills many functions. (1) Form (εἶδος) is the shape (μορφή) of a thing. *Metaphysics* V, 8, 1017b, 25–26: "And of this nature is the shape or form of each thing." (2) It is the substance (οὐσία) and essence (τί ἦν εἶναι) of a thing. *Ibid.* VII, 7, 1032b, 1–2: "By form I mean the essence of each thing and its primary substance." (3) Furthermore, it is an end (τέλος) and hence a final cause (οὗ ἕνεκα). *Ibid.* V, 4, 1015a, 10–11: "And form or essence, which is the end of the process of becoming." *Ibid.* II, 2, 994b, 9: "Further, the final cause is an end." (4) Finally, form is that which defines and circumscribes (ὁρισμόν), for matter is indefinite (ἀόριστον). *Ibid.* VII, 11, 1036a, 28–29: "For definition is of the universal and of the form." *Ibid.* 1037a, 27: "For there is no formula of it with matter, for this is indefinite."

With all these passages in mind, Averroes therefore argues here: (1) Form is not תכלית in the sense of ἔσχατον, והאמת שהצורה אינה תכלית. (2) Form is primarily the οὐσία and the τί ἦν εἶναι of a thing, אבל היא הנותנת עצם הדבר. (3) Still it is called πέρας, ואם נאמר בה תכלית, but only in the other senses mentioned by Aristotle, as follows: (a) οὐσία and τί ἦν εἶναι, אבל היא הנותנת עצם הדבר, (b) τέλος and οὗ ἕνεκα, לפי שהיא תתן תכלית הדבר, (c) εἶδος = μορφή, inasmuch as it is an ὁρισμός, ותגבילהו.

In accordance with this interpretation, the passage of Averroes is to be translated as follows: "The truth is that form is not a limit but it is rather that which constitutes the substance and essence of a thing. If we call form a limit it is because it furnishes the final cause of a thing and defines the thing." Crescas' restatement of this passage here is also translated accordingly.

85. This sudden reference to Aristotle would seem to be rather out of place in a passage which is entirely a paraphrase of Averroes' restatement of Aristotle. This reference to Aristotle occurs originally in the *Intermediate Physics* after a lengthy digression in which Averroes gives his own views on the impossibility of identifying space with the vacuum. In its original context, therefore, the expression "And Aristotle says" is the equivalent of saying, "Let us now resume our exposition of Aristotle." Here, Crescas

could have omitted it, inasmuch as he had not reproduced Averroes' digression. The retention of the phrase was simply due to
an oversight and to the mechanical copying of notes of which
this part of the *Or Adonai* is composed.

Cf. *Intermediate Physics* IV, i, 1, 8: "What remains for us to
explain is that place is not the three dimensions between the
limits of that which surrounds, i. e., length, breadth and depth.
The opinion that place is those three dimensions and that those
dimensions are separable from bodies is subject to formidable
doubts, even though it had been maintained by many of the
ancients. Indeed, there is a great plausibility in its favor, for at
first thought one would be inclined to believe that place must be
a certain emptiness and void which becomes the recipient of a
body, for, if place were a body itself, then two bodies would occupy one place at the same time. This kind of reasoning is almost
identical with that which leads to the belief in the existence of a
vacuum, as we shall explain hereafter. Furthermore, from the
fact that the empty space within a vessel is successively filled by
different bodies, they came to believe that emptiness itself is
something which has independent existence and is capable of receiving different objects in succession. But Aristotle says . . . "

הנה אשר נשאר עלינו לבאר שהמקום איננו הרחקים הג' אשר בין תכליות המקיף,
ר"ל רוחק האורך והרוחב והעמק, כי המאמר בשהמקום הוא אלו הרחקים השלשה
ושהם נבדלים הוא מאמר חזק הספקות, וכבר אמרו בו רבים מן הקדמונים.
ואמנם יחשב זה כן, לפי שהמקום יחשב בהתחלת המחשבה שמהכרחיותו שיהיה
המקום פנוי וריק ואז יקבל הגשם; ואם לא, היה המקום האחד בעצמו יקבל שני
גשמים יחד. וזאת המחשבה כמעט שתהיה המחשבה אשר תביא אל המאמר במציאות
הריקות, כמו שנבאר אחר זה. ועוד שהפנוי אשר בכלי, למה שהיה שיבואו הגשמים
עליו זה אחר זה, ידומה להם בו שהוא דבר אחד בעצמו קיים, יקבל הגשמים שיבואו
עליו זה אחר זה. ואריסטו יאמר...

86. Hebrew שיהיו המקומות מתנועעים ושיהיה המקום במקום. So also in
Averroes' *Intermediate Physics*. In Gersonides' supercommentary, however, the passage reads: שיהיו המקומות מתנועעים ויהיה
המקום במקום. "That the places would be movable, and *so* one
place would exist in another place."

Gersonides' reading reflects more closely the Greek, which is
as follows: "And at the same time, too, the place will be changed;

so (ὥστ') there will be another place of place" (*Physics* IV, 4, 211b, 23–24).

In *Iḳḳarim* II, 17, the reading is likewise ויהיה, as in Gersonides. Cf. Commentaries *Shorashim* and *Anafim, ad loc.*

87. Hebrew, יטרידהו, a literal translation of the Arabic شغل. Cf. Munk, *Guide des Egarés*, I, p. 185, n. 2; Mélanges, p. 102, n. 4; Kaufmann, *Attributenlehre*, p. 380, n. 30.

Averroes has here עם הרוחק אשר ייחדהו והוא ייחד לו, *the interval to which it particularly belongs, and which particularly belongs to it*, instead of Crescas' אשר יטרידהו, *which it occupies*. But the term יטריד occurs later in Averroes in the same passage. In Gersonides' supercommentary the term מחזיקים in the following passage יחוייב שיכנסו חלקי המים עם רחקיהם אשר הם מחזיקים בו בכלי ברחקים האחרים אשר יעתקו עליהם seems to be, like יטרידו, another Hebrew translation of شغل. Cf. מחזיק מקום in '*Olam Ḳaṭan* I, iii, p. 13.

88. I have rendered the expression אשר הם מקומות להם as if the pronoun להם referred both to חלקי המים and to עם הרחקים המיוחדים להם, thus proving at once the untenability of the two aformentioned conclusions.

In the original text of Averroes, this passage applies only to the first of the untenable conclusions, trying to show that one and the same thing would have many places at the same time. This is clear from the fact that later Averroes takes up the same illustration and uses it in refutation of the second untenable conclusion, introducing it with the following words: "From this, too, can be shown the impossibility of the second conclusion, namely, that the places would be movable and that they would exist in other places." ולוי מזה גם כן חיוב השקר השני, והוא שיעתקו המקומות ושינוחו במקומות. Crescas, however, has changed the phrasing of the last part of the passage so as to make it applicable at once to both the conclusions.

The original passage reads as follows: "So also would be affected the parts of the water, that is to say, they would be translated together with their intervals, which are their respective places, to other intervals, with the result that, beside and simultaneously with former places, those other intervals would also become places of the parts of the water."

362   CRESCAS' CRITIQUE OF ARISTOTLE   [157

כן יעשו חלקי המים, ר"ל שהם יעתקו עם מרחקיהם המיוחדים בם, אשר הם
מקומות להם, אל מרחקים אחרים, והיו גם כן מקומות להם עם המקומות הראשונים.

89. All the terms used here by Crescas in his definition of space
are to be found in Aristotle (see above n. 75). Still it is not an
exact translation of Aristotle's formal definition of space as given
in *Physics* IV, 4, 212a, 5–6: τὸ πέρας τοῦ περιέχοντος
σώματος. An exact translation of it is to be found in *Intermediate
Physics* IV, i, I, 8: המקום הוא תכלית הגשם המקיף. Crescas' version
of Aristotle's definition here occurs, however, in Narboni's commentary on the *Kawwanot ha-Pilosofim* III: הנה גדר המקום שהוא
תכלית מקיף שוה נבדל. (Similarly in his commentary on *Moreh* I,
73, Prop. 2). Narboni adds that according to Aristotle space is to
be further qualified by the statement that it is "immovable
essentially:" ואריסטו הוסיף עוד הבדל אחד בסוף ואמר בלתי מתנועע בעצם
Cf. *Physics* IV, 4, 212a, 18 ff.

In Crescas' paraphrases throughout these passages we may
note two variations from the original. (1) Crescas has substituted
here as well as elsewhere the term שטח, *surface*, for the term
תכלית, *limit*, which is used by Aristotle. (2) Without exception (but
see p. 176, l. 20), he uses the expression התכלית המקיף, *the
surrounding limit*, (similarly השטח המקיף, *the surrounding surface*), instead of תכלית המקיף, *the limit of that which surrounds*, as
the phrase runs in the original definition of Aristotle.

The substitution of the term "surface" for "limit" occurs also
in the reproduction of Aristotle's definition, quoted anonymously,
by the Iḥwan al-Safa: "It is also said that place is the *surface* of
the containing body which bounds that which is contained in it."
وقد قيل ان المكان هو سطح الجسم الحاوى الذى على المحوى فيه
(Dieterici, *Die Abhandlungen der Ichwân es-Safâ*, p. 30; German
translation in *Die Naturanschauung und Naturphilosophie der
Araber im X. Jahrhundert*, p. 9). It is also used in the definition
quoted by Algazali in the name of Aristotle: "It is a term signifying the *surface* of the containing body, I mean, the inner surface,
contiguous to that which is contained." وهو انه عبارة عن سطح
(*Makaṣid al-Falasifah* الجسم الحاوى أعنى السطح الباطن المس المحوى
III, p. 246). In one anonymous Hebrew translation of the
Makaṣid (MS. Adler 1500), the definition is rendered as follows:

שהוא מליצה משטח הגשם המקיף, ר"ל, השטח הפנימי המחזיק המוקף. In another anonymous translation (MS. Adler 978), the last part of the definition reads: רצוני השטח הפנימי שהוא מקום המוקף. Evidently neither of these translators had in the Arabic text the reading المس.

Narboni, in his commentary on the *Kawwanot ha-Pilosofim* points out that Algazali's definition tallies in every respect with that of Aristotle's: "Towards the end of his discussion, Algazali cites the definition of place, saying that it is the inner surface of the surrounding body. This is identical with the definition we have cited, for 'surface' means here 'limit.' The statement that it is the 'inner surface of the surrounding body' means to say that it is that which touches or that which is separate, inasmuch as it is the surface of the surrounding body. And it is equal, inasmuch as it is the inner part of the surrounding body. And it is that which surrounds. Hence place is a *surrounding, equal, separate limit.*"

ואבוחמאד יביא בסוף גדר המקום ויאמר שהוא השטח הפנימי מהגשם המקיף, והוא אחד עם הגדר אשר גדרנוהו [=תכלית מקיף שוה נבדל], כי שטח יורה על תכלית. ואמרנו הפנימי מהגשם המקיף יורה על הפוגש, שהוא הנבדל, אחר שהוא מהגשם המקיף. והוא שוה, אחר שהוא פנימי מהגשם המקיף. והוא המקיף. הנה שהוא תכלית מקיף שוה נבדל.

Two of the terms used by Aristotle in the definition of place, *surrounding* and *equal*, are implied in the following passage in *Cuzari* I, 89: "Moses is the rational, discriminating soul which is incorporeal, not bounded by place nor too large for place." ומשה נפש מדברת מכרת איננה גשם ואיננה נגבלת במקום ולא יצר ממנה מקום.

It will be noted that if we take out the parenthetical remark from Algazali's definition what is left is, with but a slight verbal difference, identical with the definition given by the Iḥwan al-Safa. Both these definitions have at the end, after the expression "the containing body," the additional statement "which bounds that which is contained in it" or "contiguous to that which is contained." That additional statement does not occur in Aristotle, but it does occur in Plutarch's version of Aristotle's definition *De Placitis Philosophorum* I, xix, 2: Ἀριστοτέλης, τὸ ἔσχατον τοῦ περιέχοντος συνάπτον τῷ περιεχομένῳ.

The term "surface" is also used in Ibn Gabirol's paraphrase of what seems to be Aristotle's definition of place. *Likkuṭim min Sefer Meḳor Ḥayyim* II, 21: המקום יחייב דבקות שטח גוף בשטח גוף אחר. Cf. *Fons Vitae* II, 14: "Locus autem non est nisi applicatio superficiei corporis ad superficiem corporis alterius." It occurs also in *Emunah Ramah* I, 4, p. 16: "For anything that is in place is enclosed by the surfaces of its place." לפי שכל מה שהוא במקום, שטחי מקומו כופים עליו cf. above n. 73.

It is also used by Averroes in the following reproduction of Aristotle's definition: وامّا سطوح الاجسام المحيطة به فهى له مكان. (M. J. Müller, *Philosophie und Theologie von Averroes*, Arabic text, p. 66).

A justification for the substitution of the term "surface" for "limit" may be found in Aristotle's own statement in *Physics* IV, 4, 212a, 28–29: καὶ διὰ τοῦτο δοκεῖ ἐπίπεδόν τι εἶναι.

A peculiar definition of place is given by Saadia in *Emunot we-Deot* I, 4 (Arabic, p. 51): "The true essence of place is not what our opponent thinks, but it is the meeting of two contiguous bodies and the locus of their contiguity is called place, or rather either one of the contiguous bodies becomes the place of the other."

כי אמתת המקום איננו כמו שחשב, אבל הוא פגישת שני הגשמים המתמששים ויקרא מקום משושם מקום, אבל ישוב כל אחד מהם מקום לחברו.

Similarly in II, 11 (Arabic, p. 102): "Furthermore, that which requires a place is a body, which occupies that which meets it and becomes contiguous to it, so that either one of the contiguous bodies is the place of the other."

ועוד כי הצריך אל מקום הוא הגשם אשר הוא ממלא מה שיפגשהו וממששו, ויהיה כל אחד מן המתמששים מקום לאחר.

That Saadia's definition is Aristotelian is quite obvious, for its purpose is to show that place implies the existence of one body in another. The expression "contiguous" is only another way of expressing Aristotle's περιέχων, as we have seen in the quotation from Algazali in this note above. But there would seems to be the following difference between Saadia's definition and the definition of Aristotle as generally understood. According to Aristotle, the body containing another body is the place of the contained body but not *vice versa*. According to Saadia, the two bodies, the containing and the contained, are each the place of the other. But we shall see that according to Themistius' interpretation of Aristotle the contained body is as much the place of the containing body as the containing body is of the contained body (see Prop. I, Part II, notes 54, 59, pp. 432, 443). Saadia's definition, therefore, reflects Themistius' interpretation of Aristotle. (But cf. discussion of this passage by the following authors: Kaufmann, *Attributenlehre*, p. 63, n. 117; Guttmann, *Die Religionsphilosophie des Saadia*, pp. 78–79; Efros, *The Problem of Space in Jewish Mediaeval Philosophy*, pp. 63–64.)

90. Cf.' *De Caelo* I, 5–7: Averroes, *Intermediate De Caelo et Mundo* I, vii (השמים והעולם האמצעי, מאמר א', כלל ז'). In the original the arguments from circular motion come first.

91. This argument does not agree with the first argument from rectilinear motion found in *De Caelo* I, 6, 273a, 7–21, and given in Averroes as the first part of the first argument.

It is in the main the second of the physical arguments found in the *Physics* III, 5, 205a, 8–205b, 1; *Metaphysics* XI, 10, 1067a, 7–25; and Averroes מה; המופת השני, ח"ב, פ"ד, כ"ג, מ"ג, שמע טבעי אמצעי and *Emunah Ramah* I, 4; which has been שאחר הטבע אמצעי, מ"י omitted by Crescas above (see above n. 65). Part of the original argument of *De Caelo* is reproduced later (see below n. 104 and 107).

This argument contains also an interpolation taken from Gersonides' supercommentary on the *Intermediate Physics* (see below n. 100).

92. Hebrew תיחדהו. The same term occurs also in the corresponding passage in Averroes. The term ordinarily would mean "individuates it," in which sense it is also used later, p. 200, l. 7. But here I prefer to take it in the sense of "properly belongs to it," as the equivalent of המיוחדים להם used above, p. 156, l. 4. The underlying Arabic term was probably خصّ which means both "to impart something as a property or peculiarity to something" and "to be the property or peculiarity of something." The Hebrew יחד may thus also have been used in these two senses.

Cf. the use of the word ייחד in the passages quoted above, n. 87, and below, n. 94.

93. I have added this, because in discrete bodies the part exists in the whole as in place, the place of the whole thus not being the place of the part. (See quotation from Aristotle below p. 444).

94. I. e., up or down. Averroes has here: "In the case of everything that has motion, i. e., rectilinear motion, and rest the place of the whole and of a part is the same in kind, for the place of one clod of earth is essentially the same as the place of the whole earth, namely, the lower region, and the place of one spark is essentially the same as the place of the whole fire, namely, the up, and it is to that place which is appropriate to the whole that the part is moved and in it does it rest."

וכל מה שיתנועע וינוח, ר"ל תנועה ישרה, מקום הכל והחלק אחד במין. חה
שמקום גוש וֹרגב [cod. 943: אחד בעצמוْ]הוא מקום כל הארץ אשר הוא המקום
השפל, ומקום הנצוץ האחד בעצמו הוא מקום כל האש אשר הוא המעלה. ואל
זה המקום אשר ייחד הכל יתנועע החלק ובו ינוח.

95. Hebrew מתדמה החלקים. Averroes has here: ומתדמה החלקים. ויהיה אחד במין או בלתי מתדמה החלקים ויותר מאחד במין. See quotation below, n. 96.

96. The Hebrew text here is obscure. In Averroes, the main outline of the argument is as follows:

(a) The fact that the place of the whole and the part of an homogeneous body is the same, would make every part of the homogeneous infinite be in its proper place wherever that part might happen to be.

(b) Again, the place of an infinite must be infinite. And so, the place of the infinite body cannot have the distinction of up and down.

(c) But for a body to have rectilinear motion implies two things: First, an ability to be within its proper place as well as without it. Second, a distinction of up and down in the medium through which it moves.

(d) Consequently, an infinite body cannot have rectilinear motion. It will have either to be permanently at rest or to move in a circle.

The text of the *Intermediate Physics* III, iii, 4, 2, Second Argument, is as follows: "Having laid down these two propositions as true, we resume our argument: The infinite body must inevitably be either of similar parts and one in species or of dissimilar parts and more than one in species. If it is simple and of similar parts, it is moved by nature either rectilinearly or circularly. But if it is moved rectilinearly, then the place of a part and of the whole of it will be essentially one and toward it the body will move. And if the place of a part and of the whole of it is one essentially and is infinite, the body occupying it will not be moved at all by nature. Thus the infinite will not be a natural body, for every natural body is movable. That it will not be moved at all is evident from this. Since it is assumed to be infinite, its place will be infinite, and if the place of the whole is to be infinite, there will be no place in which the repose of the part would be prior to [or "more proper than"] its motion and a place wherein its motion would be prior to [or "more proper than"] its repose, inasmuch as there would be no two places in one of which the object would move and another in which it would rest, as is the case of the simple bodies. And if we assumed that all its parts were at rest by nature, there would then be no natural rectilinear motion, inasmuch as the whole would have either to be at rest or to be moved circularly. But sense perception testifies as to the existence of rectilinear motion. Since rectilinear motion exists, the body endowed with that kind of motion must be finite, for the cause of rectilinear motion is the division of the ubiety of the movable body into a part that is natural to it and a part that is unnatural, and that division of the ubiety is made possible only by the fact that it is finite, and the finitude of the ubiety necessarily determines the boundary of the body which occupies a place in it. In the same manner it can be shown that rectilinear motion would not exist if we assumed the existence of an infinite having circular motion.

All this having been made clear, we may resume our argument, that if there is rectilinear motion there can be no simple infinite body, for if an infinite existed, it would have to be infinite in all its diameters, and thus it would either rest in its totality or be moved circularly in its parts. But rectilinear motion does exist. Hence there is no simple infinite body." (Latin, pp. 453 v b M— 454 r a A B).

הנה כאשר התאמתו אלינו שתי אלו ההקדמות, נשוב ונאמר שהגשם הבב״ת לא ימנע מאשר יהיה מתדמה ויהיה אחד במין, או בלתי מתדמה ויותר מאחד במין.

ואם יהיה פשוט מתדמה, אם שיהיה מתנועע בטבע תנועה ישרה או תנועה סבובית.
אבל אם היה מתנועע תנועה ישרה, היה מקום החלק והכל ממנו אחד בעצמו ואליו
יתנועע. ואם מקום החלק והכל ממנו אחד בעצמו, והוא בב"ת, חויב שלא יהיה
מתנועע כלל בטבע. הנה לא יהיה נשם טבעי, לפי שכל נשם טבעי מתנועע. ואמנם
יחוייב שלא יתנועע כלל, כי לפי מה שהיה אין תכלית לו, הנה מקומו אין תכלית
לו, ואם היה מקום הכל אין תכלית לו, לא יהיה בכאן מקום מנוחת החלק בו יותר
ראשון (ראוי) מתנועתו, ומקום תנועתו בו יותר ראשון (ראוי) ממנוחתו, לפי שלא
יהיה בכאן שני מקומות, מקום יתנועע בו הדבר ומקום ינוח בו, כענין בנשמים
הפשוטים. ואם הנחנו כל חלקיו נחים בטבע, חוייב שלא תהיה בכאן תנועה ישרה
בטבע, לפי שיחוייב אם שיהיה הכל נח ואם שיתנועע בסבוב, והחוש יעיד במציאות
התנועה הישרה. ולפי שהיתה התנועה הישרה נמצאת, הנה יחוייב שיהיה הנשם
המתנועע בה ב"ת, לפי שסבת התנועה הישרה אמנם הוא החלק האנה לנשם המתנועע
אל טבעי ובלתי טבעי. והחלק האנה אמנם הוא מצד היותו ב"ת, והיותו ב"ת יגזור
בהכרח תכלית הנשם הלוקח בו מקום. וכמו כן יחוייב העלות (הסתלקות:
(Cod. 943 התנועה הישרה ממציאות מה שאין תכלית לו שיתנועע בסבוב.
וכאשר התישב זה כלו, נשוב ונאמר שאם היתה הנה תנועה ישרה אין הנה נשם
פשוט בלתי ב"ת. וזה שאם היה בב"ת יהיה בב"ת בכל קטריו, ויהיה הכל אם נח
ואם מתנועע בחלקו בסבוב. אבל בכאן תנועה ישרה, הנה אין בכאן נשם פשוט
בלתי ב"ת.

97. Hebrew: ואם לא היה מתדמה החלקים, הנה החלקים אם שיהיו ב"ת
במספר ואם שיהיו בב"ת. Averroes has here: "But if the infinite
were of dissimilar parts and composite, then the dissimilar parts
of which it is composed would have to be either infinite in kind
or, if they were finite in kind, one or more than one of its parts
would have to be infinite in magnitude."

ואמנם אם היה בלתי ב"ת בלתי מתדמה החלקים ומורכב, יחוייב שיהיו החלקים
הבלתי מתדמים אשר הורכב מהם אם בב"ת במין ואם שיהיה אחד מהם או יותר
מאחד מהם בב"ת בגודל אם היה ב"ת במין.

But Gersonides in his supercommentary on the *Intermediate
Physics*, paraphrases this passage as follows: "But if we assumed
it to be composite and of dissimilar parts, then either those dissimilar parts of which the infinite whole is composed will be
infinite in kind, that is to say, *infinite in number*, in which case
we may assume each part to be finite in magnitude, or, if we say
that they are *finite in the number* of their kinds, one of those parts
or more than one will have to be infinite in magnitude, for otherwise an infinite magnitude could not arise from a finite number
of parts, as has been explained."

אבל אם הנחנוהו מורכב ובלתי מתדמה החלקים, הנה החלקים בלתי מתדמי
החלקים אשר הורכב מהם יהיו בהכרח אם בב"ת במין, ר"ל חלקים אין תכלית
למספרם, ובזה אפשר שנניח כל אחד מהחלקים בעלי תכלית בגודל, או אם נאמר
שהם ב"ת במספר מיניהם, יחוייב שיהיה אחד מהם או יותר מאחד מהם, אין תכלית
לו בגודל, כי בזולת זה לא יתחדש מהב"ת במספר בב"ת בגודל כמו שקדם.

From the use of the expressions of "finite in number" and
"infinite in number" by Crescas it is evident that in his restatement of the argument he had been following the text of Gersonides.

Crescas' paraphrase, however, is carelessly done. By using
Gersonides' term מספר, number, without the latter's qualifying
term מין, of kind, Crescas has exposed the text to a serious ambiguity. For taken by itself, the expression בב"ת במספר might
mean an infinite number of individuals belonging to a finite number of kinds (see below n. 100). This, however, is not what is
wanted here. We should expect Crescas to use some such expression as במספר במין, number with respect to kind, which is a common
expression and is opposed to במספר באיש, number with respect to
individual, as in the following quotations:

Epitome of the Physics III, p. 11a: ואולם אם הונח הגשם אשר אין
לו תכלית מורכב כמו שהיו רבים מן הקודמים חושבים אותו בכל התחייב שיהיה
מורכב אם מפשוטים שאין להם תכלית במספר במין וכל אחד מהם יש לו תכלית
בגודל או שאין להם תכלית בגודל אם כלם או אחד מהם ויהיו הם יש להם תכלית
במספר במין.

Ibid., p. 11b: ואולם שהוא בלתי אפשר שיונח זה הגשם אשר אין לו תכלית
מורכב מפשוטים שאין להם תכלית למספרם באיש ואם היו יש להם תכלית במין
במספר במין.

Happalat ha-Pilosofim I: ההתחלפות בין ב' מחוייבי המציאות הוא
במספר במין.

In the original argument of Aristotle the word "number" does
not occur. Physics III, 5, 205a, 21–22: ἔπειτα ἤτοι πεπερασμένα
ταῦτ' ἔσται ἢ ἄπειρα τῷ εἴδει.

98. The reason given here by Crescas for the impossibility of
one part of the heterogeneous infinite to be infinite in magnitude
does not agree with the reason given by Aristotle. Aristotle
argues that such an infinite part would be destruction to its contrary. Cf. Physics III, 5, 205a, 24–25; Metaphysics XI, 10,
1067a, 20.

In Averroes, however, there is a suggestion for the reason as
given here by Crescas.

Cf. *Intermediate Physics* III, iii, 4, 2, Second argument: "If one or more than one of the parts were infinite in magnitude, the whole would be destroyed. The same inevitable conclusion will follow whether we assume the infinite to be infinite in the number [of similar parts] or infinite in magnitude, for an infinite number of [similar] parts become by contiguity and conjuncture an infinite magnitude, and it has already been shown previously that an infinite body of similar parts cannot exist because, if it existed, there would be no rectilinear motion." (Latin, p. 454 r a—b).

ואם אחד מהם בב"ת בגודל או יותר מאחד נפסד הכל. ובין שיהיה בב"ת במספר או בגודל המתחייב אחד, לפי שהבב"ת החלקים יהיו ממנו במשוש והתדבקות אחד בב"ת בגודל, וכבר התבאר במאמר הקודם שאי אפשר שימצא גשם מתדמה החלקים בב"ת, לפי שאלו נמצא לא היה בכאן תנועה ישרה.

99. Hebrew ‏ואם היו בב"ת במספר חוייב שיהיו מיני האנה בב"ת במספר‏. Averroes has here ‏אבל אם היו בב"ת בצורה ובמין חוייב שיהיו המקומות‏ בב"ת. Gersonides paraphrases it as follows: ‏אבל אם היו החלקים‏ ‏המתחלפים במין בב"ת במספר, חוייב לפי מה שקדם כי יהיו מיני האנה אין‏ ‏תכלית להם‏. From the use of the expression ‏מיני האנה‏ instead of ‏המקומות‏ by Crescas it is evident that he has been following the text of Gersonides.

100. The entire passage from here to the end of the argument is based upon Gersonides' supercommentary on the *Intermediate Physics*. There is nothing in the *Intermediate Physics* itself to correspond to it.

The following is an outline of the text of Gersonides:

A. A restatement of the proof as it is given by Averroes and reproduced here by Crescas up to this point. See above n. 97, 99.

B. Gersonides' own additional argument that the places must be finite in kind, for (1) the existence of proper places is derived from the existence of rectilinear or circular motion, and (2) rectilinear motion is from and toward the centre. (3) Hence, the kinds of places must be limited, i. e., up and down.

C. Two arguments that each of the places must be finite in magnitude.

D. There cannot be an infinite number of proper places and elements one above the other, for (1) there would be no absolute height and lowness, as (2) their sum would make an infinite magnitude and an infinite has no centre, and as also (3) the places must be each finite in magnitude as shown in C.

Crescas, it should be noted, reproduces Gersonides' B(1) and B(2), but he adds to B(2) the expression והסבובית היא סביב האמצע and replaces B(3) by Gersonides' D(2). He omits Gersonides' C altogether. He then reproduces Gersonides' D(1) and proceeds with part of the original argument from the *Intermediate De Caelo* (see below n. 104).

The text of Gersonides reads as follows:

A. "But if we assumed it to be composite and of dissimilar parts, then either those dissimilar parts of which the infinite whole is composed, will be infinite in kind, that is to say, infinite in number, in which case we may assume each part to be finite in magnitude, or, if we say that they are finite in the number of their kinds, one of those parts or more than one will be infinite in magnitude, for otherwise an infinite magnitude cannot arise from a finite number of parts, as has been explained. But if those parts which differ in kind were infinite in number, it would follow, according to what has been said, that the kinds of ubiety would be infinite, inasmuch as each part would have a natural ubiety appropriate to it. But this will have been shown subsequently to be impossible. And if one of the [dis]similar parts were infinite in magnitude . . . .

B. Now we shall explain that the variety of kinds of natural ubiety cannot be infinite. The argument is as follows: The existence of natural ubiety is derived from either rectilinear or circular motion. But rectilinear motion is either from the centre or toward the centre. Hence the kinds of ubiety are limited in number.

C. That the natural localities must be finite in size, [literally, quantity], may be shown as follows: If any of them are infinite in size, there could not be more than one kind of ubiety. Furthermore, the existence of opposite motion, upward and downward, conclusively proves that the interval between up and down must be limited, for an infinite distance cannot be traversed.

D. We might, however, be tempted to say that the respective places of these simple natural elements are one above the other, and this to infinity, in the same manner as the place of fire is above the place of water, even though both fire and water are moved in an upward direction. But if this were the case, there would be no absolute up and no absolute down, inasmuch as the magnitude of their totality would have to be infinite, and that which is infinite has no centre. Furthermore, the distinction of kind within the ubiety, as has been explained, conclusively proves that the place of rest must be limited in size.''

A. אבל אם הנחנוהו מורכב ובלתי מתרמה החלקים, הנה החלקים בלתי מתדמי
החלקים אשר הורכב מהם יהיו בהכרח אם בלתי ב"ת במין ר"ל (?) חלקים אין
תכלית למספרם, ובזה אפשר שנניח כל אחד מהחלקים בעלי תכלית בגודל, או
אם נאמר שהם ב"ת במספר מיניהם, יחוייב שיהיה אחד מהם, או יותר מאחד מהם,
אין תכלית לו בגודל, כי בזולת זה לא יתחדש מהב"ת במספר בב"ת בגודל כמו
שקדם. אבל אם היו החלקים המתחלפים במין בב"ת במספר, חוייב לפי מה שקדם
כי יהיו מיני האנה אין תכלית להם, אחר שלכל אחד יהיה אנה טבעי תיחדהו.
וזה כבר התבאר אחר זה שהוא שקר. ואם היה אחד מהחלקים המתדמים בב"ת
בגודל...

B. ועתה נבאר שאי"א שיהיו מיני האנה אין תכלית להם. וזה שהאנה הטבעי
לקוח או מתנועעה הישרה אם מהסבוביית. אבל התנועה הישרה הנה אם מן האמצע
או אל האמצע. א"כ מיני האנה מונבלים במספר.

C. ואולם היותם מונבלי הכמות, שאם היה מהם בב"ת בכמות לא יהיה
בכאן מן אנה מינים. ועוד שהתנועה מהמטה אל המעלה או הפך, תגזור בהכרח
שיהיה מה שביניהם מונבל, כי לא ידרוך הדורך אל מה אין תכלית לו.

D. ואי אפשר ג"כ שנאמר שיהיה מקום אלו הפשוטים הטבעיים זה למעלה
מזה, וזה אל לא תכלית, על צד מה שמקום האש למעלה ממקום המים ושניהם
מתנועעים למעלה, שאם היה הדבר כן, לא יהיה הנה מעלה מוחלט ולא מטה,
כי יחוייב שיהיה גודל הכל אין תכלית לו, ואין במה שאין תכלית לו אמצע. ועוד
שהחלק האנה, כמו שקדם, יגזור שיהיה האנה מונבל בכמות.

101. Hebrew והסבובית היא סביב האמצע.

This expression is not found in Gersonides (see above n. 100B).
It seems that Crescas has added it in order to give the argument
a different turn.

102. Hebrew ואם היה בכאן גודל בב"ת בין חלקי הגשם לא יהיה בכאן אמצע.
This is based upon Gersonides' statement כי יחוייב שיהיה גודל הכל
אין תכלית לו ואין במה שאין תכלית לו אמצע. (See above n. 100D).

It certainly cannot be a repetition of Crescas' own previous
statement: ואם היו בב"ת במספר חויב שיהיה אחד מהם בב"ת בגודל. The
expression בין חלקי הגשם, I take in the sense of מכל חלקי הגשם.

103. The meaning of this passage is as follows: What has been shown so far is that there cannot be more than two kinds of motion, centrifugal and centripetal. But there still remains to be shown that these two kinds of motion cannot be infinite in number. For, why should we not conceive the universe to consist of an infinite number of concentric spheres? The motions in the universe would then be finite in kind, that is, centrifugal and centripetal; but there would be an infinite number of centrifugal motions, since there would be an infinite number of peripheries. These centrifugal motions, would indeed each be limited in extent, but they would be infinite in number. It will thus be possible to have an infinite number of different elements without having an infinite number of different kinds of places.

This argument is taken from Gersonides, quoted above in n. 100D. It is also found in an anonymous commentary on Averroes' *Epitome of the Physics* (MS. Bodleian 1387), where it is made still stronger by pointing out that the different proper places of the elements must not necessarily be different in kind. Fire and air, for instance, have each a proper place of its own, but their places are one in kind, that is, above.

"If one should raise an objection arguing that even if there were only two kinds of motion, namely, from the centre and toward the centre, we might still maintain that there could be an infinite number of simple elements one above the other in the same manner as the four elements are supposed to be arranged according to the Philosopher, even though we see that he has enumerated only two kinds of motion for these four elements—the answer is as follows: Inasmuch as reason conceives a kind of motion which is round the centre, from which it is deduced that there must be a simple element [i. e., the fifth element] which is endowed with that kind of motion, it must therefore follow that there exists an absolute up which is limited, namely, the periphery, and an absolute down, namely, the middle or centre. Hence the kinds of motion between these two, namely, the up and down, are limited and finite."

ואל שאלת השואל כי יאמר, כי אף שלא יהיו רק שני מיני תנועה והם מן האמצע ואל האמצע נוכל לאמר שיהיו גשמים פשוטים אין תכלית למספרם זה למעלה מזה על פי הדרך הפילוסוף שישים היסודות הארבעה, וראיתי כי לאותן הארבעה

רק שני מינים מתנועה. תשובה: כי אחר שהשכל יציור מין התנועה אשר סביב האמצע,
ויתחייב מזה גם פשוט למין התנועה ההוא, א"כ יהיה במציאות מעלה במוחלט שהוא
מוגבל, והוא עד המקיף ההוא, ומטה מוחלט והוא האמצע, ר"ל המרכז. אם כן
יהיו מיני התנועה אשר בין שני אלו, המעלה והמטה, מוגבלים ובעלי תכלית.

Cf. Averroes' *Epitome of the Physics*, III, p. 11b: "That it is
impossible to assume that that infinite body is composed of simple
elements which are numerically infinite in individual but finite
in kind will be explained in *De Caelo et Mundo*. For it will be
shown there that there can be no plurality of universes."

ואולם שהוא בלתי אפשר שיונח זה הגשם אשר אין לו תכלית מורכב מפשוטים
שאין להם תכלית למספרם באיש ואם היו להם תכלית במין, הנה יתבאר זה
בספר השמים והעולם, כי הוא ממה שיתבאר שם שהוא אי אפשר שימצא מחלקי
העולם שנים באיש.

See below p. 474, n. 128, 130.

104. This bracketed passage occurs in the printed editions and
in the MSS. as part of the succeeding argument, where, however,
it is entirely out of place. I have inserted it here, because it seems
to belong here. The passage is taken from Averroes' *Intermediate
De Caelo* I, 7, corresponding to *De Caelo* I, 6, 273a, 7–15. It is the
first part of the original first argument from rectilinear motion
(see above n. 91 and below n. 107).

The passage in *Intermediate De Caelo* I, vii, reads as follows:
"Of the four elements, one moves absolutely upward, and that
is fire, one moves absolutely downward, and that is earth, and two
move relatively upward, and these are air and water, for water
moves downward in relation to air and upward in relation to
earth, and similarly air moves upward in relation to water and
downward in relation to fire. Since the motions of those two
elements of which one moves absolutely upward and the other
absolutely downward are contraries, it follows that their places
must be absolutely contrary to each other, and that is absolutely
up and absolutely down. If one of these places is limited, then
the other place must be limited, inasmuch as it is a contrary, for
it is necessary that either one of them must be most distant from
the other and that their distance from each other must be the
same in either direction. As this opposition between these two
places is known to us from the fact that they are contraries

and as it is clear that the lower place is limited, it follows that the upper place must also be limited." (Latin, p. 279 v, b, K–L).

הגשמים הארבעה, מהם מה שיתנועע למעלה בחלטות, והוא האש, ומהם מה שיתנועע למטה בחלטות, והוא הארץ, ומהם מה שיתנועע למעלה בערך, והם האויר והמים, שהמים מתנועעים למטה בערך אל האויר ואל המעלה בערך לארץ. וכן האויר יתנועע למעלה בערך אל המים ואל מטה בערך לאש. ואחר שהיו תנועות השני גשמים אשר יתנועע אחד מהם למעלה בחלטות והאחד למטה בחלטות הפכיות, ראוי שיהיו מקומותיהם הפכיים בחלטות, והוא מעלה ומטה בחלטות. ואם יהיה אחד מאלה השני מקומות נגדר, ראוי שיהיה המקום השני נגדר, מצד מה שהוא הפך. חה שיתחייב שיהיה כל אחד מהם מחברו בתכלית הרחק, ושיהיה רחוקם רחוק אחד, וכשיהיה זה ההתנגדות מבואר מעין אלה השני מקומות מצד מה שהם הפכים, והיה נראה מעין המקום השפל שהוא נגדר, ראוי בהכרח שיהיה המקום העליון נגדר.

105. See *Categories* 6, 6a, 17–18: τὰ γὰρ πλεῖστον ἀλλήλων διεστηκότα τῶν ἐν τῷ αὐτῷ γένει ἐναντία ὁρίζονται. Cf. *Metaphysics* X, 4, 1055a, 5.

106. Cf. *De Caelo* I, 6, 273a, 21–274a, 18, and Averroes: השמים והעולם האמצעי, מאמר א' כלל ז'.

107. See above n. 104.

108. See above n. 105.

109. Hebrew ותניחהו עוד נבדל ממנו. In Averroes: ותבדיל מן הגשם הב"ת הנרשם עליו.

110. Hebrew שכל תנועה בזמן. In Averroes: "For every finite magnitude traverses a finite distance in a finite time, as has been shown in the sixth book of the *Physics*." Cf. *Physics* VI, 7. שכל בעל שעור בעל תכלית הוא מתנועע המרחק הב"ת בזמן ב"ת לפי מה שנתבאר במאמר הששי מספר השמע.

111. This last conclusion is not found in Averroes.

112. Cf. *De Caelo* I, 7, 274a, 30–274b, 32; and Averroes השמים והעולם, מאמר א', כלל ז'.

113. Hebrew שיתמששו. In the *Physics* V, 3, Aristotle defines the following terms:

| τὸ ἅμα | simul | at once | יחד. |
| χωρίς | separatim | separately | נפרדים. |

ἄπτεσθαι	tangere	to touch, to be contiguous	משש.
μεταξύ	interjectum	intermediate	במה שבין.
ἐφεξῆς	deinceps	successive	נלוים, or (והמשך) נמשכים.
ἐχόμενον	cohaerens	adhering	(כרוך) נכרכים.
συνεχές	continuum	continuous	(הדבקות) מתדבקים.

To be contiguous is defined by him as follows: "Those things are said to touch each other, the extremities of which are together." (*Physics* V, 3, 226b, 23).

Cf. also *Physics* VI, 1, 231b, 17–18: "The extreme of things continued is one, and touches."

See *Epitome of Physics* VI, p. 25b: והיו הדברים המתדבקים הם אשר
יקרה להם כאשר יתמששו שיהיו תכליותיהם אחדים.

Cf. also *Olam Ḳaṭan* III, ed. Horovitz, p. 49: וכן לא יעבור שיהיה
הגוף כי אם ממשש קצתו לקצתו או שיהיה מפורד חלק מחלק.

114. Crescas does not complete the reasoning. Aristotle has here: "For the first motion being finite, it is also necessary that the species of simple bodies should be finite, since motion of a simple body is simple, and simple motions are finite." (*De Caelo* I, 7, 274a, 34–274b, 4).

Cf. *Intermediate De Caelo* I, 7: "It is impossible that there should be bodies infinite in form, for it has already been shown that the simple forms are finite, inasmuch as the simple motions are finite, and for each simple body there is a simple motion." ומן השקר שיהיו הנה נשמים בב"ת בצורה מפני שכבר נתבאר שצורות הפשוטות
בעלות תכלית, כי התנועות הפשוטות בעלות תכלית, ולכל גשם פשוט תנועה פשוטה.

115. Hebrew זה אמנם מצד התנועה. This remark is not without significance. For the next argument, though included by Crescas among the arguments from motion, is treated by Averroes as a class by itself. I have therefore added within brackets the adjective "proper."

116. Cf. *De Caelo* I, 7, 274b, 33–275b, 8, and Averroes השמים
והעולם האמצעי, מאמר א', כלל ז'.

117. Hebrew ואמנם נרצה בהפעלות ההפעלות אשר בזמן. Based upon the following statement in the corresponding passage of Averroes: "By 'acting' and 'suffering action' he means to refer here to that whose motion comes to an end and whose action and suffering of action are completed. He does not mean to refer to that which is in motion perpetually, for it has already been shown that there is no perpetual motion except in locomotion."

ור"ל הנה בפועל ומתפעל מה שכלתה תנועתו ונשלמה פעולתו והתפעלותו פעולתו והתפעלותו לא מה שהוא בתנועה תמידית, שכבר התבאר שלא ימצא שנוי תמידי כי אם ברחק ובמקום.

Thus the term הפעלות here in Crescas stands for פעולה והתפעלות, ποίησις καὶ πάθος, action and passion, in Averroes.

The term הפעלות by itself may stand either for "action" or for "passion", the one being vocalized הַפְעָלוֹת and the other הִפָּעֲלוּת (but cf. Klatzkin's translation of Spinoza's Ethics, Torat ha-Middot, pp. 394–395). In the corresponding passage in the second part of this proposition (p. 204), Crescas uses the expression הפעל והפעלות. There it is clear that הפעלות stands for "passion."

What Averroes and, following him, Crescas mean to say is this. When Aristotle argues that there could be no action and reaction between an infinite and a finite or between two infinites, he means an action and reaction that has been completed and has come to an end, and not an action and reaction which come under the class of change or motion which, according to Aristotle, is an incomplete process of realization (cf. below Proposition IV). This qualification had to be made because, according to Aristotle himself, it is possible to have an eternal circular motion which is to continue in an infinite time (cf. below Proposition XIII). Such a continuous motion, always in a process of realization but never fully completed, would be possible between infinites, even though it implied an infinite time. What Aristotle is arguing here is that no action which is a completed motion and which must have taken place in a finite time would be possible between infinites or between an infinite and a finite.

The source of Averroes' remark seems to be following passages in Aristotle.

De Caelo I, 7, 275a, 22–24: "But neither will it move or be moved in an infinite time; for it has not an end; but action and passion have an end." Ibid. 275b, 2–4: "In no finite time therefore is it possible for the finite to be moved by the infinite. Hence it is moved by it in an infinite time. An infinite time, however, has no end; but that which has been moved has an end."

Cf. Themistius, *In Libros Aristotelis De Caelo Paraphrasis*, ed. Landauer.

Latin text, p. 40, l. 35—p. 41, l. 7: "At actio omnis affectioque tempore perficitur. in infinito autem tempore nec agere quicquam nec affici potest; motus enim qui infinito tempore instituitur, termino ac fine caret, actio vero omnis affectioque terminum ac finem habent, quorum uterque veluti forma ac perfectio existit. per actionem autem affectionemque hoc in loco minime eae intelleguntur, quae in motu, sed quae in eo, quod jam fuit, consistunt. quod enim in continua generatione consistit, esse non habet, atque eo *minus in* alia [*affectione?*] turpe est enim existimare eo quicquam moveri, quo nunquam pervenire potest."

Hebrew text, p. 27, ll. 10–17.

כי כל פעולה או נפעל הוא בזמן, ובזמן זולתי הבעל תכלית לא יפעל ולא
יתפעל, כי התנועה אשר תהיה בזמן בלתי בעל תכלית אין סוף לה ולא קץ, וכל
פעולה והפעלות אחרית ותכלית, כי כל אחד מהם כמו השלמות והצורה.

ולא ירצה לומר בפועל ובנפעל במקום זה אשר יהיה בו מנוחה ובתנועה
[read: אבל אשר יהיו במה שכבר היה, חה כי הדבר אשר יהיה במה שיתהוה
תמיד אין לו מציאות, כל שכן מזולתו, כי מה שאי אפשר שיגיע אליו דבר מן הדברים
לא יחשוב התנועה אליו דבר מן הדברים.

118. Hebrew המתפעל יחס המתפעלים בשני מתחלפים פועלים כשיפעלו ,והב'.
הפועל אל הפועל כיחס המתפעל אל . The text here is incomplete. Averroes has: "The second proposition is that when two agents act and complete their action in equal time, the relation of one agent to the other is like that of one object to the other."

וההקדמה השנית הוא כשיפעלו שני פועלים חלופים בזמן שוה ונשלמה פעלתם,
שיחס הפועל אל הפועל כיחס המתפעל אל המתפעל.

119. Hebrew בזמן יפעל שהפועל ,והג'. Averroes has here: "Third, every agent acts upon an object in finite time, i. e., it completes its action, for, as has been shown, there can be no finite action in infinite time."

והשלישית שכל פועל הוא פועל במתפעל בזמן ב"ת, ר"ל שתשלם פעלתו, שאי"א
שתהיה פעלת ב"ת בזמן בב"ת כמו שנתבאר.

120. Not found in Averroes.

121. Not found in Averroes.

122. Hebrew ואם נכפול עוד המתפעל, יתחייב שיתפעל הבב"ת מהבב"ת
בזמן מועט מהפעלותו מהבב"ת. This according to Adler MS. The
Munich, Jews' College, Paris, Vienna, Vatican, Parma, Oxford,
and Berlin MSS. read הבב"ת instead of הבב"ת מהבב"ת, which is
obviously a scribal error. Ferrara edition omits the first מהבב"ת
and reads מהפעלות instead of מהפעלותו. Undoubtedly, מהפעלות
was meant to be an abbreviation of מהפעלותו, but the abbreviation mark was erroneously omitted in the printing. Or, it is
possible, that in the MS. from which the Ferrara edition was
printed the reading was מהפעלות הב"ת מהבב"ת, but the הב"ת was left
out by mistake. Johannisberg edition attempted an unsuccessful
emendation of the text, as follows: יתחייב שיתפעל הבב"ת בזמן מועט
מהפעלות (מהב"ת) מהבב"ת. Vienna edition follows Ferrara reading
but spells out מהפעלותו. The reading here adopted is what is
required by the context. The pronominal suffix in מהפעלותו is to
be taken to refer to המתפעל in ואם נכפול עוד המתפעל.

123. Cf. below Proposition XIII.

124. Originally this argument was given by Averroes as class by
itself (cf. above n. 115).

125. Averroes has here: "He thought that it was fitting to start
his investigation with the simple elements. Of these he selected
the circular element and tried to show that it must be finite. In
this connection he has advanced six arguments" (Latin, p. 277vb,
35. The last two sentences are missing in the Latin). וראה שהראוי
בזה שישים התחלת החקירה על הגשמים הפשוטים, והתחיל בהם בגשם הסבובי,
ובאר מעינו שהוא ב"ת, והביא בזה ששה מופתים.

126. Cf. De Caelo I, 5, 271b, 27–272a, 7; and Averroes: השמים
והעולם האמצעי, מאמר א', כלל ז', המופת הא'. Averroes introduces
this proof by four preassumed propositions.

127. Hebrew חצי קטרו. In Averroes קו יוצא מן המרכז.

128. Averroes' fourth preassumed proposition: "Fourth proposition. If from the centre of the circular element more than one line
proceeds and these lines revolve until they return to the place
where they are assumed to have started their revolution, and if,
furthermore, one of these lines is assumed to be at rest and an-

other to revolve, then the revolving line may fall upon the line
at rest" (Latin, p. 278ra, A). וההקדמה הרביעית שהגשם הסבובי כשיצאו
ממרכזו יותר מקו אחד, והיה אפשר שיתנועעו הקוים ההם, עד שישובו אל המקום
אשר נחשבו מתנועעים ממנו, ואם נחשבו האחד מהם נח והאחר מתנועע, יהיה
אפשר שיתנועע המתנועע עד שיתדבק אל הנח.

129. Averroes' second preassumed proposition: "Second, if the
radii were infinite [in length], the distance between them would
inevitably have to be infinite, for the longer the radii the greater
the distance between them, that is to say, between their extreme
points. It necessarily follows that if the radii are infinite the
distance between them will be infinite, for having assumed that
the distance increases with the elongation of the radii, then if the
elongation is infinite, the distance must likewise be infinite"
(Latin, p. 277vb, M). והשנית שאם היו הקוים היוצאים ממרכזו בב״ת,
יתחייב בהכרח שיהיו הרחקים אשר ביניהם בב״ת, מפני שכל מה שיהיו הקוים
היוצאים מהמרכז יותר ארוכים, יהיה המרחק בינהם יותר גדול, ר״ל בין קצותם.
ומחוייב הוא, שאם הקוים בב״ת שיהיו הרחקים אשר ביניהם בב״ת, שאם נציע
שהרחק אשר ביניהם יתוסף בתוספת הקוים, והיתה התוספת בהם בב״ת, ראוי
שיהיה הרחקים ביניהם בב״ת.

130. Averroes' first preassumed proposition: "First, in an infinite
circular body the lines proceeding from the centre must inevitably
be infinite [in length]" (Latin, p. 277vb). אחת מהן שכל גשם סבובי
בב״ת ראוי בהכרח שיהיו הקוים היוצאים ממרכזו אין תכלית להם.

131. Averroes' third preassumed proposition: "Third proposition.
No moving object can traverse an infinite distance" (Latin, pp.
277vb–278ra). וההקדמה השלישית שא״א שיחתוך המתנועע מרחק בב״ת.

132. Averroes illustrated this proof by the following figure:

Let ACB be an infinite circle.
Let CA and CB be infinite radii.
Let CA revolve on its centre C and let CB be fixed.
If an infinite sphere could rotate upon itself, CA
would sometimes have to fall on CB.

But the distance AB is infinite, and an infinite distance cannot
be traversed.

Hence, CA could never fall on CB.

Hence, no infinite body could have circular motion.

133. The reference is to Altabrizi. The argument is designated by him as מופת הסולמי, i. e., "the proof of the scale."

Originally it is given as follows:

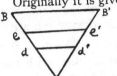

Let AB and AB′ be two infinite lines diverging from a common point A.

Let AB and AB′ be successively intersected by common lines at points dd′, ee′, etc. up to BB′.

Since AB and AB′ are infinite, BB′ must be infinite.

Again, the number of the intersecting lines between A and BB′ must likewise be infinite.

But BB′ is bounded by AB and AB′ and the total number of intersectors are bounded by A and BB′.

Thus infinites would be bounded, which is impossible.

Altabrizi's proof reads as follows:

(a) Isaac ben Nathan's translation:

ונתחיל עתה בבאור המופת הסולמי ונאמר לו היו המרחקים בב"ת אפשר לנו
שנניח שתי התפשטויות יוצאות מהתחלה אחת בתמונת משולש לא סר המרחק בין
שניהם נוסף בשעור אחד מהתוספת כמו שהמרחק הראשון אם היה אמה היה השני
אמה וחצי, ונוסיף השלישי על השני בזה השעור ג"כ, וכן הרביעי על השלישי, ויהיה
זה השעור נשמר בזה התוספות, וילכו אל בלתי תכלית עם התוספת הנזכר ויהיה
כל מרחק עליון בין שניהם מקיף על כל המרחקים אשר תחתיו על כל אופן, ואפשר
שימצא מרחק אחר יהיה מקיף על כל אותן המרחקים הבב"ת במרחק אחר, כי אם
לא יהיה אפשר שימצאו המרחקים הבב"ת במרחק אחד אשר היה אפשר שימצא
במרחק אחד, אמנם הוא מרחקים מוגבלים מהמרחקים הבב"ת, ואצל זה יחוייב
שיפסקו ההתפשטויות, כי נמצאו אחר זה אפשר שימצאו מרחקים נמצאים במרחק
אחד יותר ממה שאי"א יותר מזה והוא אותם המרחקים המוגבלים. זה חלוף. אבל
השאר שני התפשטויות מתרחקים על אותו המין מהתוספת אל בלתי תכלית אפשר
בהכרח, הנה מחויב שימצא אז מרחק אחד בין שניהם מקיף על כל אותם המרחקים
הבב"ת, ויהיה אותו המרחק אם כן בלתי תכלית עם היותו מוקף בין השני מקיפין.
זה שקר.

(b) Anonymous translation, which is much clearer:

ואמר עתה המופת הסולמי. מה שאם היו הרחוקים בעלי תכלית היה אפשר
שנחשוב ונניח שני המשכת מרחקים יוצאים מהתחלה אחת בצד שיתרחקו זה מזה
ברוחק על שעור מיוחד בהם, שאם עיינת עד"מ אליהם על רוחק אמה היתה מוצא
שביניהם ברחב רוחק אמה, וכאשר תהיינה שתי אמות תהיינה עוד שתים המרחקים.
כמו כן אל לא תכלית. ומן הידוע בהכרח שאין המנועות במרחקים על זה הצד,
ואם היה הרוחק ביניהם מתמיד כפי שעור המרחקים הנמשכים, ר"ל באורך כן

ברוחב, והמרחקים נמצאים מבלתי תכלית עם המרחק הנזכר, ר"ל אשר ברוחב,
יהיה א"כ ביניהם מרחק אין תכלית לו בהכרח, אחר אשר המרחק שביניהם אין
לו תכלית, מצורף אל היותו נאסר ונעצר בין שני קוים נמצאים אוסרים ועוצרים
אותו. וזהו השקר שיהיה נאסר בגבולים שהם הקוים ויהיה בב"ת, א"כ בהכרח
יתחייב היות לו תכלית, ואם יש לו תכלית, יהיה תכלית לקוים אשר הנחנו בב"ת.

It will be noted that Altabrizi's proof is reproduced only in the
last part of Crescas' proof and is introduced by him with the
words והנה יתחייב זה בכל שני קוים. Originally in Altabrizi there is
no indication of the connection between this proof and the Aristotelian proof reproduced by Crescas from Averroes. But Crescas
must have surmised that Altabrizi's proof was merely a modification of the Aristotelian, the difference between them being merely
that whereas the Aristotelian proof is connected with the rotation
of an infinite sphere, Altabrizi's proof argues from the existence
of any two infinite lines. Crescas has therefore reproduced it as
another version, more general in its application, of Aristotle's
proof.

On the margin of the Vatican MS. there is the following note:
"This argument is taken by the author from the commentary of
Altabrizi where certain doubts are raised against it and are
answered by him."

הנה המופת הזה לקחו המחבר מדברי התבריזי במופת ובמקומו יתבאר ספקות
מה עליו והתירם.

134. Hebrew כי המאמר בהיותו מוקף ובב"ת סותר נפשו.

In Isaac ben Nathan's translation of Altabrizi it reads: ויהיה
אותו המרחק ג"כ בב"ת עם היותו מוקף בין השני מקיפין. זה שקר.

In the anonymous translation it reads: וזהו השקר שיהיה נאסר
בגבולים שהם הקוים ויהיה בב"ת.

135. Hebrew היוצאים מהמרכז, *proceeding from the centre.*

Altabrizi: יוצאות מהתחלה אחת, *proceeding from one beginning.*

136. Cf. *De Caelo* I, 5, 272a, 7–20; and Averroes: השמים והעולם
האמצעי, מאמר א', כלל ז', המופת השני.

Averroes again introduces this proof by preassumed propositions.

In Averroes this proof is divided into two parts. The first corresponds to the last part in Aristotle (*De Caelo* 272a, 11–20). The

second corresponds to the first part in Aristotle (*De Caelo* 272a, 7–11).

Crescas reproduces now only the first part of Averroes' proof. (see below note 141).

137. By Averroes' first preassumed proposition, in which reference is given to the *Physics* (i. e. VI, 7): "First, every object that is moved in finite time is moved with a finite motion over a finite distance. This has been demonstrated in the *Physics*" (Latin, p. 278rb, E). האחת, שכל מתנועע שיתנועע בזמן ב״ת הוא מתנועע תנועה ב״ת ובמרחק ב״ת. וזה דבר כבר נתבאר בשמע הטבעי.

138. Averroes' fifth preassumed proposition: "Fifth, if from the centre of the infinite circular element we extend a line and cause it to pass through it, the line will be infinitely extended. Similarly, if we extend a chord through the infinite circular body, the chord will be infinite at both its ends" (Latin, p. 278rb, E). והחמישית, שהגשם הסבובי הבב״ת, כשנוציא ממרכזו קו, ונעבירהו בו, ילך אל בלתי תכלית. וכן כשנוציא בו מיתר ג״כ ילך ג״כ אל בלתי תכלית משתי קצותיו.

139. Averroes' fourth preassumed proposition: "Fourth, the circular body completes its revolution in finite time" (Latin, p. 278rb, E). והרביעית, שהגשם הסבובי ישלם סבובו בזמן ב״ת.

140. Averroes illustrates this proof by the following diagram:

Let C be an infinite circle.
Let CD be a radius infinite at D.
Let AB be a chord infinite at A and B.
Let CD revolve on its centre C.
CD will complete its evolution in a finite time, during a part of which it will intersect AB.

Therefore, CD will pass through AB in a finite time.

But an infinite distance cannot be passed through in a finite time.

141. This proof is of a composite nature. Its phraseology and construction are borrowed from Averroes' third proof, corresponding to *De Caelo* I, 5, 272a, 21–272b, 17. In substance, however, it is the second part of Averroes' second proof (see above n. 136). A similar proof is given by Avicenna in his *Al-Najah*, p. 33, which is also found in Algazali's *Maḳaṣid al-Falasifah* II, p. 126, and in Altabrizi, where it is called מופת הנכוחות (anonymous translation: מופת הנוכחיי), "the proof from parallel lines." It seems that Crescas' object in putting here this proof in place of the original third proof of Averroes was in order to be able afterwards to refute it by an objection raised against it by Altabrizi himself (see below p. 468, n. 117).

The following are the texts illustrating this note:

(a) Averroes third proof:

"Third argument. He introduces this argument by two propositions.

First, if two finite bodies are parallel to each other and are placed alongside each other, and each one of these bodies turns on a pivot (literally: is moved) in the opposite direction of the other, or one body is moved and the other remains at rest, both these bodies will cut through each other in finite time and then part from each other. There is no difference whether both bodies are moved or only one body is moved, except that in the former case their departure from each other will begin sooner.

Second, if of two magnitudes of this description, i. e., parallel to each other and alongside each other, one is infinite or both are infinite, and one is moved while the other is at rest or both are moved opposite to each other and then become parted, they will have to cut through each other in infinite time. For it has already been shown by a demonstration in the sixth book of the *Physics*, [ch. 7], that if an infinite distance is traversed it must be traversed with an infinite motion and in infinite time.

Having laid down these two propositions, if we now assume that the celestial sphere is infinite, it will follow that the celestial sphere will traverse a finite distance in a finite time, for we observe that it traverses a section of the earth in finite time. It will thus follow that two magnitudes, one infinite and the other finite, will traverse each other in finite time. But this is an impossible absurdity" (Latin, p. 278vb).

המופת השלישי, וזה המופת הוא מקדים לו שתי הקדמות.
האחת מהם הוא כשיהיו שני גשמים ב"ת, האחד מהם נכחי לאחר ומונח על צדו,
והתנועע כל אחד מהם לצד הנוכחיי לתנועת חברו, או שהתנועע האחד מהם
והאחר נח, שכל אחד מהם חותך חברו בזמן ב"ת ויפרד ממנו. ואין הפרש ביניהם
בזה, אלא כשיתנועע כל אחד מהן נוכח תנועת חברו יהיה הפרדם יותר מהרה.

וההקדמה השנית הוא כשיהיו שני בעלי שעור על זה התואר, ר"ל שהאחד מהם
מונח על צד חברו ונכחו, ויהיה אחד מהם בב"ת או שניהם בב"ת, והתנועע האחד
מהם ונח האחר, או התנועעו יחד על הנכה, שאם נודה שיפרד ממנו ויהתכהו יתחייב
מזה שיהיה חתכו לו בזמן בב"ת. וזה שכבר נהבאר שהמרחק הבב"ת אם יחתך
אמנם יחתך בתנועה בב"ת ובזמן בב"ת לפי מה שנהבאר במאמר הששי מספר השמע
הטבעי.

וכשנתייסבו אלו השתי הקדמות, ונציע שהגשם הרקיעי בב"ת, יתחייב שיחתך
המרחק הב"ת בזמן בעל תכלית שאנחנו נרגיש הגלגל חותך בכללו החתכה מן הארץ
בזמן ב"ת, יתחייב שיחתכו שני בעלי שעור, האחד מהם בב"ת והאחר ב"ת, כל אחד
מהם לחברו בזמן ב"ת, וזה שקר אי אפשר.

(The term נוכחי represents here the Arabic موازى, *parallel*,
which occurs in the quotation from Algazali given below in this
note. Cf. also below n. 142. The expression מונח על צדו, literally,
placed beside it, seems to me to mean also *parallel* and to be an
attempt to give a literal translation of the Greek term which
means *beside of one another*. The Latin translation renders נוכחי
by "obuius" and מונח על צדו by "iuxta positus.")

(b) The second part of Averroes' second proof:

"Furthermore, everything finite has a beginning. This being so,
then the intersection of the radius CD and the chord AB (see
diagram above in n. 140) must have a first point and that is the
point at which the two lines first meet and come in contact with
each other. But if we assume these two lines to be infinite, they
can have no first point of intersection. For when the two lines
described in the diagram meet, they cannot first meet at some
point in the middle. It is quite clear that they must first come in
contact with each other at a point at the extremity of one of the
lines or of both. But an infinite line has no extremity. Hence no
infinite line can come in contact with another line and can have no
first point of intersection. But the assumption is that the infinite
lines in the diagram meet at a first point of intersection. Hence an
impossible absurdity. Since it has been shown that in the circular
body under consideration the two lines must have a first point
of intersection by reason of the fact that the time of the intersection has a beginning, it has thus been demonstrated that a
circular body moving circularly cannot be infinite" (Latin, p.
278va—b).

ועוד שכל בעל תכלית יש לו התחלה, ואם הדבר כן, יש לחתכת קו ה"ג לקו
א"ב התחלה, והוא הנקודה הראשונה שיפגשו בה השני קוים, וידבק האחד מהם
באחר. אך כשנציע השני קוים בב"ת, לא יהיה אפשר שימצא להם נקודה ראשונה
שיחתכו עליה. מה שהשני קוים אשר יפגשו לפי התנועע על זאת ההצעה מפני
שהיה אי"א שיפגשו באחת הנקודות אשר באמצע, מה שמבואר הוא א"כ שראוי שידבק
האחד מהם באחר בנקודה אשר בקצתו או בשניהם, והקו הבב"ת אין לו קצה. על
כן לא ידבק בו דבר ולא ימצא לו התחלת החתוך, וכבר הוצע שהוא נמצא לו,
מה שקר אי אפשר. מפני שכבר נתבאר מענין הגשם הסבובי אשר בזה התאר שימצא
בו התחלת חתוך לאלה השני קוים מפני שימצא לזמן החתוך התחלה, הנה נתבאר
מזה המאמר שאי אפשר שימצא הגשם הסבובי המתנועע סביב האמצע בב"ת.

(c) Algazali's proof in *Kawwanot ha-Pilosofim* II (*Maḳaṣid
al-Falasifah* II, p. 126):

אולם שקרות סלוק התכלית מהמרחקים הנה נודע בשתי ראיות. אחת מהם,
שאנחנו לו הנחנו קו ג"ד בלי תכלית, והנענו קו א"ב בעגול ה' אל צד ג' מקו ד"נ
עד שב בנכחו, היה זה הנעה אפשרית בהכרח.
ולו הנענוהו מהנוכח אל צד הקורבה ממנו

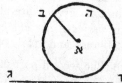

הנה אי"א מבלתי שירמה ממנו נקודה היא ראשית
הנקודות הרמחות. עוד אחר זה ירמה שאר
הנקודות עד שישוב מהרמחות בתכלית אל הנכח
מהצד האחר. מה שקר, לפי שאם שוער נטיה
אליו מהנכחיות מבלתי רמיזה, הנה הוא שקר, והרמיזה שקר, לפי שהרמיזה תפול
[ראשונה] על נקודה, ואין על הקו אשר לא יניע לתכלית נקודה היא ראשונה.
וכל נקודה הונחה לרמיזה ראשונה, הנה א"א מבלתי שתהיה כבר נרמזה מה שלפניה
קודם הרמיזה לה בהכרח. ולא תרמה כל העת שלא ירמז מה שאין תכלית
לו. עוד לא יהיה בה נקודה ראשונה היא נקודה הנרמזת, והוא שקר. מה מופת
חותך הנדסי בשקרות קיום מרחק בלי תכלית שוה הונח למלוי או לרקות.

(d) Altabrizi's version of the proof in Isaac ben Nathan's
translation:

ואולם מופת הנכוחות צורתו: שאנחנו נניח במרחק הבלתי בעל הכלית קו בב"ת
והוא קו א"ב. ונניח כדור יוצא ממרכזו קו ב"ת נכחי לקו הבלתי בעל תכלית,

והוא קו כמו זה. הנה כאשר התנועע הכדור עד סר קו ג"ד מנכחות
קו א"ב אל נכח ראשו, אי אפשר מבלתי שיחודש בקו א"ב נקודה
היא ראשית הנקודות אשר יפלו עליהם הפנישות נכח ראשו. אבל
זה בקו הבלתי ב"ת שקר, כי אין נקודה בו אם לא למעלה ממנה
נקודה אחת, והפגישה נכח ראשו עם הנקודות העליונות קודם
הפנישה עם הנקודות התחתונות, כי כאשר הגענו קו ג"ד אל קו
א"ב הנה הזוית אשר תחודש עם הנקודות העליונות תהיה יותר חדה מאשר תחודש
עם הנקודות התחתונות. והוא נגלה. ומן השקר שתהיה שם נקודה היא ראשית

הפגישה, אבל היא מחייבת אל סור הקו מהנכוחות אל הפגישה. וזה קבוץ בין שני
סותרים. אבל כל אשר הנחנוהו בהקדמות מהנחת הכדור ותנועתו וצאת קו בעל
תכלית ממרכזו נכחי לקו האחר אמתתו ידוע בהכרח, אם לא הנחת הקו הבב"ת,
הוא מחוייב לענין הבטל ויהיה בטל. הנה כל גודל ושעור יחוייב שיהיה בעל תכלית
והוא הדרוש.

In the light of these passages quoted the proof reproduced here
by Crescas is as follows:

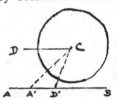

Let C be an infinite circle.
Let CD be a radius infinite at D.
Let AB be an infinite line parallel to DC.
Let CD revolve on C toward AB.
Let angle D' be the acutest angle formed
by the meeting of lines CD and AB.

D' will thus be first point of intersection of CD and AB.

But since D' is not the extreme of either CD or AB, it is possible to take any other point A' at which CD and AB would form
a more acute angle than at D'.

Hence angle D' is both the first point of intersection and not
the first point of intersection.

In restating the argument this way, I have drawn upon Altabrizi, whose refutation of this argument is made use of by Crescas
later in his criticism. Cf. below p. 468.

142. Hebrew קוים נוכחיים. The term נוכחי has several meanings.

(a) Here in the sense of *parallel* it is a translation of the Arabic
موازى which occurs in the corresponding argument in *Maḳaṣid
al-Falasifah* II, p. 126. See above n. 141.

(b) נוכחות as the equivalent of the Arabic جيب, *sine* in trigono-
metry, has been noted by Steinschneider, *Uebersetzungen*, p. 516.

(c) In the expression נוכח הראש, *zenith*, (see quotation from
Altabrizi above in n. 141 and *Sefer ha-Gedarim*, s.v.), the term
נוכח represents the Arabic سمت in سمت الراس. In the same sense.
is לעמת הראש used in *Cuzari* II, 20.

(d) In the following passage in *Milḥamot Adonai* VI, i, 11,
היה הזמן נפסד בהתהוותו על נכחותו, רצוני שכל אשר נתהווה ממנו חלק יפסד
ממנו חלק the phrase על נכחותו means *in a forward direction.*

143. Hebrew והאחר [ונח. The word נח does not occur in any of
the MSS. or printed editions. It is, however, required by the

context. In justification of its insertion here, compare the expression או שהתנועע האחד מהם והאחר נח in quotation (a) above in n. 141.

144. Cf. *De Caelo* I, 5, 272b, 17–24, and Averroes: השמים והעולם האמצעי, מאמר א', כלל ז', המופת הד'. Averroes again introduces this proof by a formal statement of preassumed propositions.

145. Cf. Averroes' proof for his third proposition: "As for the third proposition, it can be demonstrated by what has already been said, for it has already been shown that if there exists circular motion there must also exist a body circular in form, whence it follows that if circular motion is infinitely circular, the circular form implied by the circular motion must likewise be infinite" (Latin, p. 279ra—b).   ואמנם השלישית מבוארת נ"כ ממה שקדם,
וזה שאם כבר נתבאר שאם תמצא תנועה סבובית ראוי שימצא גשם סבובי בצורה,
מבואר הוא שאם תהיה התנועה הסבובית בב"ת בסבוב שהצורה הסבובית הנמצאת
לה תהיה בב"ת.
Cf. *De Caelo* II, 4, 287a, 4–5: "It follows that the body which revolves with a circular movement must be spherical."

146. Hebrew רשם, רשׁ, ὑπογραφή, *descriptio*, which is distinguished from גדר, حـد, ὁρισμός, *definitio*. Averroes uses חק, حق, *essentia*. (MS. Paris, Cod. Heb. 947.)

147. Hebrew מהנדס. Averroes has here תשברי" (MS. Paris, Cod. Heb. 947).

148. Averroes: "As for the first proposition, it is evident from the definition of figure, inasmuch as figure is defined by the geometrician as that which is contained by any boundary or boundaries" (Latin, p. 279ra).   אמנם ההקדמה הראשונה ו=שכל צורה ב"ת] תראה מחק
הצורה, וזה שאחר שהצורה היא אשר יאמר בה התשברי" בחקה שהיא אשר יקיף
בה גדר או גדרים.
Cf. Euclid, *Elements*, Book I, Definition XIV.

149. In Averroes: "In general, finitude exists in a thing only by reason of form and lack of finitude by reason of matter" (Latin, p. 279ra).   ובכלל התכלית אמנם ימצא לדבר מצד הצורה והעדר התכלית
מצד החומר.

150. Cf. *De Caelo*, I, 5, 272b, 25–28; and Averroes: השמים והעולם האמצעי, מאמר א', כלל ז', המופת הה'.

151. Hebrew עמוד על הקטר. אם הונח עמוד על הקטר. In Averroes: ונוציא ממנו קו על
זוית נצבה.

152. Hebrew ואי"א שיחתוך קו בב"ת בזמן ב"ת. The phrase בזמן ב"ת
is Crescas' own addition. In the original, this proof like the first
is based upon the general proposition that no infinite distance is
traversible, and not, like the second and sixth, upon the proposition that no infinite distance is traversible in finite time. That
this addition was not intentional may be inferred from the fact
that in his criticism he groups it together with the first proof
(See below p. 466, n. 113).

153. Averroes illustrates it by the following figure:

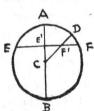

Let C be an infinite circle with C as its centre.
Let AB be its diameter infinite at both sides.
Take any point E in AB outside C and draw
through it infinite line EF at right angles
with AB.

Draw CD infinite at D intersecting EF at
any point F'.

Let AB and EF be stationary and let CD revolve on C.

CD could never pass through EF, for EF is infinite, and no
infinite distance is traversible.

Hence, no infinite could have circular motion.

The figure is given by Aristotle, who makes use of the line
AB. In Averroes' Paraphrase line AB in the figure serves no
purpose.

154. *De Caelo* I, 5, 272b, 28–273a, 6, and Averroes: השמים
והעולם האמצעי, מאמר א', כלל ז'.

The argument in the original has two parts. 1. If the heaven
were infinite, an infinite body would traverse an infinite distance
in a finite time. 2. Since the heaven is convolved in a finite time,
it must be a finite magnitude. Aristotle calls the second part
the converse of the first ἔστι δὲ καὶ ἀντεστραμμένως εἰπεῖν.
Averroes terms it "a more direct argument" ואפשר שיבא המופת הזה
על דרך הישר בזה.

Only the first part is reproduced here by Crescas.

155. Averroes refers here to the *Physics* [i. e. VI, 7]. לפי מה שנתבאר בספר השמע.

156. Hebrew: בבאור כולל. Aristotle has here λογικώτερον (*De Caelo* I, 7, 275b, 12). Cf. above n. 5.

157. Averroes has in this class four arguments, of which Crescas reproduces here only the first two.

158. *De Caelo* I, 7, 275b, 12–24 and Averroes: השמים והעולם האמצעי, מ"א, כ"ז, הבאור הג', האחד מהם.

159. Aristotle as well as Averroes introduces this by a statement that the infinite must consist of similar parts.

160. Cf. *De Caelo* I, 7, 275b, 25–29 and Averroes: השמים והעולם האמצעי, מ"א, כ"ז, הבאור הג', ובאור אחר.

161. Cf. *De Anima* II, 5, 417a, 2 ff.

162. This is not found in Averroes. What the author means by this additional argument may be restated as follows: If an infinite magnitude is possible, an *infinite number* of magnitudes must likewise be possible (cf. below Proposition II). Furthermore, if two infinite magnitudes are possible, there is no reason why an infinite number of infinite magnitudes should not be possible. But the assumption here is that the two infinite magnitudes are related to each other as *movens* and *motum*. Hence, it should also be possible that an infinite number of infinite magnitudes should be related to one another as *movens* and *motum* and thus forming an infinite series of causes and effects.

163. This refers to the two other arguments from gravity and levity which Averroes includes within this class of arguments.

164. Hebrew מהמקומות. I take מקומות here as well as below in the expression ומקומות ההמעדה as reflecting the Greek τόποι in its technical sense of *loci* or *sedes argumentorum*. Thus also is Aristotle's *Topics* called ספר המקומות, *Emunah Ramah* II, iv, 3, p. 65: חכרום אריסטו בספר הנצוח והפכם אלפראבי אל ספר המקומות. Cf. Steinschneider, *Uebersetzungen*, p. 47, n. 26, and p. 48: ספר הספסטניא לאבו נצר האלפרבי והוא ספר המקומות. In the same technical sense is

to be taken the expression מואצע אגלוטאאתהם, מקום טעותם, *the locus of their fallacy*, in *Cuzari* V, 2, and מואצע אלטען, מקום הטענה, *the locus of the argument*, in *Moreh* II, 16.

165. Hebrew המטעים, *causing error*, *misleading*, The Paris, Munich and Berlin MSS. read המנועים. This reading may be explained as a scribal error arising from the splitting of the ט in המטעים into נו. Still, if the reading of these three MSS. is correct, we have here a new meaning of the word מנועים, used in the sense of *subject to objections, refutable*. A similar use of the noun מניעות, in the sense of *objections, strictures*, is to be found in Isaac ben Nathan's translation of Altabrizi, Proposition I, in his discussion of the ידע שהחלק השני מזאת המליצה מליצה משותף בינה ובין: המופת הסולמי המליצה הראשונה ועליו מניעות חזקות.

166. Hebrew צורה. The term צורה is used here in the logical sense of the form of an argument as contrasted with its content. Cf. Crescas' reference to *material* and *formal* fallacies in the expression נפסד החומר והצורה, p. 192.

PART II

1. In order to understand the meaning of this passage, it is necessary for us to summarize the chief points in Aristotle's argument against which Crescas' criticism here is directed. Aristotle has laid down four premises: (1) There is no immaterial quantity, be it magnitude or number. (2) An infinite, by definition, must be divisible. (3) An infinite cannot be composed of infinites. (4) Everything immaterial is indivisible. By the first premise he disproves the existence of an infinite quantity. By the remaining three premises he shows that an infinite cannot be an immaterial substance, that is to say, a substance which is infinite in its essence, just as soul is said to be soul in its essence.

In his opposition to this, Crescas rejects outright the premise that there cannot be an immaterial magnitude. The vacuum, he says, if one admits its existence, is such a magnitude. He then proceeds to identify this immaterial magnitude, or vacuum, with the infinite. He furthermore argues, in effect, that the infinite vacuum has the following three characteristics: (1) It is infinite in its essence, as an immaterial infinite should be. (2) Still it is divisible, in conformity to the definition of infinity. (3) But though divisible, it is not composed of infinites.

This, however, would seem to be contradictory to Aristotle's premises which we have enumerated above. For, in the first place, according to Aristotle, nothing immaterial can be divisible. In the second place, if you say that the infinite vacuum is divisible, it would have to be composed of many infinites, or, to quote Aristotle, "the same thing cannot be many infinites, yet as a part of air is air, so a part of the infinite would be infinite, if the infinite is a substance and a principle" (*Metaphysics* XI, 10, 1066b, 15–17).

A way of reconciling these apparent contradictions is found by Crescas in appealing to the case of a mathematical line. Crescas, however, does not go beyond a mere allusion to the mathematical line, and so we must ourselves construct the argument by the aid of what we know about the definition and the nature of a line and their implications. The argument, we may state at the outset, rests upon a comparison of the terms "infinite" and "linear," and its purpose is to show that whatever is true of the latter, even according to Aristotle himself, can be true of the former.

(1) In the first place, a mathematical line is an immaterial magnitude (see definition of mathematics in *De Anima* I, 1, 403b, 12–15), and is linear in its essence, for a line, according to Aristotle, is a continuous quantity and does not consist of points (cf. *Physics* VI, 1, 231a, 24–26). The line must, therefore, be said to be linear in its essence.

(2) In the second place, a mathematical line, though immaterial, is still said to be divisible. Aristotle speaks of a line as being divisible into that which is always divisible (Cf. *Physics* VI, 1, 231b, 15–16). That is to say, it is always divisible into parts which are in themselves linear.

(3) Finally, a mathematical line, though divisible into linear parts, is not said to be composed of many lines. To prove this statement, it must be recalled that Arabic and Jewish philosophers usually quote Euclid's second definition of a line, namely, that "the extremities of a line are points." Cf. *Elements*, Book I, Definition III, and Averroes' *Epitome of Physics* III, p. 10b: והקו כמו שנאמר בגדרו הוא אשר תכליותיו שתי נקודות. Cf. also *Sefer Yesodot* II, ed. Fried, p. 45: לפי שהאורך הוא מרחק הגיע בין שתי הנקודות

חזו הוא הקו. Now, if a line must have points at its extremities, a mathematical line cannot be said to consist of lines, as that would make it contain points. Thus, while on the one hand, a mathematical line is said to be divisible into lines, on the other, it is maintained that it is not composed of lines.

The anomaly of this last statement, we may add in passing, is explained by Aristotle himself in the *Metaphysics* VII, 10. He tries to show there that to say that a certain whole is divisible into parts does not always mean that the whole is composed of those parts. The mutual implication of the terms "divisibility" and "composition" depends upon the circumstance as to whether the definition of the whole involves the definition of its parts or not. The definition of a syllable, for instance, involves the definition of the letters of speech. The letters, therefore, exist prior to the syllable. A syllable, consequently, is said to be divisible into letters and also composed of letters. The definition of a line, however, does not involve the definition of a point. The latter can be obtained only by dividing the line into parts. The point, therefore, does not exist prior to the line. Hence, though a line is divisible into parts, it is not composed of those parts. To quote Aristotle: "For even if the line when divided passes away into its halves, or the man into bones and muscles and flesh, it does not follow that they are composed of these as parts of their essence, but rather as matter; and these are parts of the concrete thing, but not of the form, i. e., of that to which the formula refers" (*Metaphysics* VII, 10, 1035a, 17–21). In other words, Aristotle's statement amounts to this: An actual line may be actually broken into parts and again be composed of those parts. An ideal, mathematical line, however, while it is thought to be infinitely divisible, it is thought to be so only in potentiality, and consequently it is not thought as being composed of parts.

The same holds true, according to Crescas, in the case of the infinite vacuum. As a mathematical line is linear in its essence, so is the infinite vacuum infinite in its essence. Again, the infinite is said to be divisible in the same sense as the mathematical line is said to be divisible, namely, into "parts of itself" מחלקיו, i. e., infinites in the case of the former, and lines in that of the latter. Finally, just as the mathematical line is not composed of the parts into which it is divisible, that is to say, its parts have no actual co-existence with the whole, so the infinite parts of the infinite have no actual co-existence with the whole infinite. Or to use Crescas' own words, the definition of infinity must not necessarily be applicable to its parts: ולא יתחייב שנדרר הבב״ח יצדק על חלקיו. The infinite no less than the line is simple and homogeneous, having no composition "except of parts of its own self," ולא יתחייב הרכבה בו כלל אלא מחלקיו, that is to say, of parts into which the whole is thought to be potentially divisible rather than of which the whole is actually composed.

As for the use made by Spinoza of Crescas' discussion of this argument, see my paper "Spinoza on the Infinity of Corporeal Substance," *Chronicon Spinozanum* IV (1924–26), pp. 85–97.

A criticism of Crescas' argument is found in Shem-ṭob Ibn Shem-ṭob's supercommentary on the *Intermediate Physics* III, iii, 4, 1:

"Rabbi Ḥasdai in the *Or Adonai* raises here an objection, arguing, that he who affirms the existence of an immaterial infinite will undoubtedly affirm also the existence of an immaterial number and magnitude, and so it is necessary first to establish that number and magnitude cannot be immaterial in order to prove afterwards that infinity, which is an accident of number and magnitude, cannot be immaterial.

To this we answer, that his contention is quite right, but Aristotle is addressing himself here to men of intelligence and understanding, who do not deny those true propositions, namely, that number and magnitude are undoubtedly inseparable from matter. This is Aristotle's method in most of the arguments he has advanced here.

It may also be said that Aristotle has anticipated this objection in his statement that 'the essence of number and magnitude is not identical with the essence of the infinite.' Aristotle seems to reason as follows: If the essence of the infinite were identical with that of number and magnitude, the opponent would be right in contending that, inasmuch as he maintains that the infinite is immaterial, he also believes that number and magnitude are immaterial, seeing that they are identical, and then, indeed, it would be necessary for us to establish by proof that number and magnitude are not separable from bodies. But inasmuch as thou, who art of sound mind, already knowest that the essence of number

and magnitude is not the essence of the infinite, and that they are
two accidents, as we have stated, there is no need for further discussion, and what we have said is quite enough."

והרב חסדאי באור י"י ספק כאן ואמר, שאין ספק שמי שיאמר שיש כמה נבדל
יאמר שיש מספר ושעור נבדל, וראוי שיאמר היות המספר והשעור בלתי נבדל
תחלה, ואחר יבאר שהבב"ת אשר הוא יקראו למספר ולשעור יהיה בלתי נבדל.
ונשוב לזה שהאמת כן הוא, אבל אריסטו ידבר עם אנשי השכל והתבונה אשר לא
יכחישו אלו ההקדמות האמתיות ההם, כי המספר והשעור הם בלתי נבדלים בלי
ספק. זה דרך ברוב המופתים אשר עשה בכאן. ואפשר שיאמר שאריסטו השיב
לזה הספק ג"כ באמרו כי מהות וכו'. וזה שאם היה מהות הבב"ת הוא בעצמו
המספר והשעור, היתה אמרו כי מאחר שהוא יסבור בב"ת הוא נבדל יסבור ג"כ
במספר והשעור שהוא נבדל, אחר שהם דבר אחד, ואז היה ראוי שנעשה המופת
במספר והשעור, ר"ל בהיותו בלתי נבדל. אבל אחר שאתה, הבריא השכל, כבר
ידעת שמהות המספר והשעור בלתי מהות וכו', ושהם שני מקרים כמו שאמרנו; ואין
צורך ליותר מזה, ודי.

An allusion to this argument is also found in Isaac ben Shem-
ṭob's *second* supercommentary on the *Intermediate Physics*, loc.
cit.:

"An opponent may contend that Aristotle's argument from the
fact that number and magnitude are inseparable from sensible
objects is a begging of the question, for he who believes that the
infinite is an immaterial substance does not admit that number
and magnitude are inseparable from sensible objects; but, quite
the contrary, he denies it absolutely. That this is so can be shown
from the fact that the Pythagoreans hold that the infinite is nothing but number itself and Plato similarly believes that it is the
universal, immaterial Great and Small. One may, therefore, question Aristotle as to what justification he has for taking it for granted
(מושלם, see below p. 426, n. 42) that number and magnitude are
inseparable from sensible objects, therefrom to argue against the
Metaphysicians, when as a matter of fact, the latter do not admit
it but rather maintain the contrary."

יש לאומר שיאמר שמה שאמר אריסטו הנה שהמספר והשעור בלתי נבדלים מן
המורגש שהוא מערכה על הדרוש, וזה שהאומר שמה שאין תכלית לו הוא עצם
נבדל שאינו מודה (שהעצם) [שהמספר] והשעור [אינם] נבדלים מן המורגש, אבל
ההפך, וזה שהם יכחישו זה תכלית ההכחשה. וזה מבואר מאשר סיעת פיתגורס
אינו אומר שהדבר שאין תכלית לו הוא דבר זולתי עצם המספר, ואפלטון אמר
ג"כ שהוא הגודל והקוטן הכללי הנבדל. וא"כ יש לשאול לאריסטוטליס איך לקח

הנה כמושלם שהמספר והשעור בלתי נבדלים מן המורגשים להלוק על זה באלהיים,
עם היותם אומרים בהפך ואינם מודים בו....

A similar allusion to this argument is also found in Isaac ben
Shem-ṭob's *first* supercommentary on the *Intermediate Physics,*
loc. cit.:

"The question may be raised, that those who admit the existence of an infinite deny that quantity cannot be immaterial, for
they maintain that the infinite is immaterial and identify it with
the number. In answer to this we may say that Aristotle has
assumed it here as something self-evident, inasmuch as it is generally acknowledged that number and magnitudes are accidents,
and accidents do not exist apart from their subject."

ו ב ט ל ש י ה י ה כ מ ה נ ב ד ל ו כ ו' ב ל ת י נ ב ד ל י ם ל מ ו ח ש. ו י ש
לשאול שהרי הם יכחשו זה, שהם אמרו שזאת ההתחלה נבדלת והוא המספר. ונכל
לומר שהניחו בכאן לדבר מבואר לפי שידוע הוא שהמספר והשעור מקרים, והמקרים
לא ימצאו נבדלים מנושא.

2. Hebrew מופת מספיק. The term מספיק reflects here the Arabic
اقناع, as in *Cuzari* V, 2: אקנאעאת, ראיות מספיקות (p. 297, l. 2, and
p. 296, l. 1). Both the Hebrew and the Arabic terms mean
"satisfying," but the Arabic means in addition to this also "per-
suading" and "convincing."

In Zeraḥiah ben Isaac's translation of Themistius' commentary
on *De Caelo* the Arabic term is Hebraized and taken over into the
Hebrew translation from which it is rendered into Latin by
persuasibilis. From the context it is clear that the term is applied
by him to an argument which, on the one hand, does not establish
the truth as it is, i. e., it is not a demonstrative argument, and, on
the other hand, is not an eristic argument. Cf. *Themistii in Libros
Aristotelis De Caelo Paraphrasis,* ed. Landauer. Hebrew text, p. 88,
l. 9: אמר כי זה הדעת אשר אמרתם אמנם הוא על צד ההקנעה בלתי היותו אמת.
Latin text, p. 131, ll. 23–24: "Haec autem vestra sententia
persuasibiliter (inquit Aristoteles) non autem vere dicitur."
Hebrew text, p. 91, l. 31: אבל המאמר האחר הורם אהב הנצחון מכל צד
ואע״פ שיהיה מקנע על נגלה העניין. Latin text, p. 136, ll. 33–34: "Alius
autem sermo est sermo sophisticus, tametsi prima fronte persuasibilis videatur." In this last passage of the Latin translation
the term *contentiosus* would be a more accurate translation of

נצחון than *sophisticus*. For מקנע the term מכניע (other readings: מכנים and מכריע) occurs on p. 8, l. 34.

The precise technical meaning of the term מקנע, מספיק, may be gathered from Algazali's *Mozene Ẓedek* (ed. Goldenthal, 1838; Arabic original *Mizan al-'Amal*, Cairo, A. H. 1328). Algazali enumerates first three classes of arguments: (1) contentious and litigious, הנצוח והמחלוקת, الجدل والمناظرة, ἀγωνιστικόν καὶ ἐριστικόν; (2) demonstrative, המופת البرهان (see above p. 326, n. 13); (3) rhetorical, האלכטאבה, خطابة=הלצה, cf. *Millot ha-Higgayon*, ch. 8. The last one is described by him as an argument the purpose of which is to persuade. Hebrew text, p. 170: ליֿשב הנפש, Arabic text, p. 159: الى اقناع النفس. Later he designates the rhetorical type of argument by the term "persuasion." Hebrew text, p. 172: וההחישב והוא אל אקנאע; Arabic text, p. 162: الاقناع. Hence the terms התישב, מופת מספיק, הקנעה, all mean *persuasion* and refer to the *rhetorical* argument which is known as הלצה. The connection between these two terms is to be found in Aristotle's definition of *rhetoric* as "a faculty of considering all possible means of *persuasion* (πιθανόν) on every subject." (*Rhetoric* I, 2, 1355b, 26–27). Thus מקנע, מספיק is πιθανόν; הקנעה and התישב are πίστις.

This contrast between a *demonstrative* and a *persuasive* argument underlies the following passages in the *Cuzari:* I, 13: "Because they are arguments of which some can be established by *demonstration* [להעמיד עליהם מופת, יברהנוא עליהא] and others can be made to appear plausible by *persuasion* [יספיקו בם דבר שתחישב או הדעת עליו, יקנעון פיהא]. I, 68: "Thus far I am satisfied with these *persuasive* [המספיקות, אלמקנעֿה] arguments on this subject, but should I continue to have the pleasure of your company, I will trouble you to adduce the *decisive* [החותכות=המפסיקות, אלקאטעֿה] arguments."

3. Hebrew לפי שבזה תועלת אינו מעט בחכמה הזאת. By a similar statement Aristotle introduces the problem of infinity in *De Caelo* I, 5, 271b, 4–6: "For the existence or non-existence of such a body is of no small but of the greatest consequence to the contemplation of truth." Cf. *Themistii in Libros Aristotelis De Caelo Paraphrasis*, ed. Landauer. Hebrew text, p. 14, ll. 19–21: ואמר שהוא ראוי לחקור

על זה, כי שיעורו גדול בידיעות האמת המבוקש בכל הענינים, כלומר אם העולם
בעל תכלית או הוא בלתי בעל תכלית. Latin text, p. 22, ll. 4–7:
"Necesse autem est, ut de eo inquiratur, videlicet utrum
mundus sit finitus an infinitus, quia magni est momenti ad
veritatis cognitionem, quam omnibus in rebus quaerimus."

The expression אינו מעט, *no small*, which is the reading here
according to all the MSS. instead of גדול, *great*, in the printed
editions, reflects the Greek οὔ τι μικρὸν in the corresponding
passage of Aristotle quoted above. The expression אינו מעט is
again used by Crescas in *Or Adonai* I, iii, 1: וכן יפול ספק אינו מעט.

4. An allusion to Crescas and his argument here is found in two
identical passages in Isaac ben Shem-ṭob's *first* and *third* supercommentaries on the *Intermediate Physics* IV, ii, 5.

"There is some one who raises here a question, saying that those
who admit the existence of a vacuum do not maintain its existence
on the ground of its being one of those enumerated causes of
motion but rather on the ground that it is necessary for motion,
even though not a cause thereof, just as there are many things
without which some other thing could not exist even though the
former are not the cause of the latter. Consequently, even though
he has demonstrated that the vacuum cannot be any one of the
causes, this does not make it impossible for it to be something
necessary for motion."

ויש מי שישאל ויאמר שהאומרים ברקות לא יאמרו שהוא מצוי בשביל שהוא
לתנועה סבה מאחת הסבות המכרות, אבל על שהוא מוכרח לה אע"פ שאינו סבה,
כמו שיש דברים רבים שהדבר אינו יכל להמצא זולתם, אעפ"י שאינם סבות. לכן,
אעפ"י שבאר שאינו אחד מהסבות, לא ימנע מפני זה שיהיה מוכרח לתנועה.

Pico Della Mirandola refers to this argument in *Examen Doctrinae Vanitatis Gentium* VI, 6: "Negat et eos qui vacuum astruxere
id ipsum causam motus asserivisse, praeterquam ex accidenti, ne
videlicet fieret corporum penetratio."

5. Hebrew ונעזרו בזה גם כן מהצמיחה וההתוך והספוגיות והמקשיות ומדמויים
אחרים. In *Physics* IV, 6 and 9, Aristotle reproduces a number
of alleged proofs for the existence of a vacuum, all based upon
various natural phenomena. Averroes has grouped them into five
classes. *Intermediate Physics* IV, ii, 2: "Those who affirm the
existence of a vacuum support their view by five examples . . . ."

locomotion . . . . motion of increase . . . . rareness and denseness . . . . weight and lightness . . . . augmentation and division."   ואמנם אשר יאמרו במציאות הריקות הנה בזה דמים חמשה.. מפני
תנועת ההעתק ... מפני תנועת הצמיחה... מהמקשיות והספוניות ...מהכבד
והחלק ... מהרבוי והקלות. In referring to these proofs, Crescas
quotes only the first three, and alludes to the others by the
phrase "and other illustrations."

The term וההתוך is not found in the original. Crescas has added
it apparently for no special reason, except out of the habit of
coupling the terms צמיחה and התוך together, as in the expression
..צמיחה והתכה.

As for the meaning and use of the terms מקשיות, ספוניות, התוך,
צמיחה, the following observations are in point:

צמיחה and its synonyms גדול and פריה are the Hebrew equivalents of
the Greek αὔξησις, Arabic نمو , used in the sense of natural growth
and increase, as in the following examples: *Intermediate Physics*
IV, ii, 2: מפני תנועת הצמיחה (Kalonymus' translation), מפני תנועת
הגדול (Zeraḥiah's translation). Altabrizi, Prop. IV: כי אותו
התוספת כבר יהיה בהתחבר נשם אחר אליו בכח טבעי ויקרא גדול (Isaac
ben Nathan's translation); והוא התוספת הנעשה בהתחברות גוף אחר אליו
בכח טבעי...זה נקרא צמיחה או גדול (Anonymous translation).
לתנועה בכמות שהוא פריה וחסרון (Hillel of Verona, Prop. XIV).

התוך or התכה is the Hebrew translation of (a) اضمحلال or ذبول ,
φθίσις, and (b) تحليل , ἀνάλυσις as opposed to σύνθεσις. In
the former sense it is opposed to צמיחה or גדול, as in the expression of אלנמו ואלאצמחלאל, הצמיחה וההתוך αὔξησις καὶ φθίσις, *increase
and diminution* (*Moreh* II, Introduction, Prop. IV). Its synonyms are חסרון, השחתה, כליון, as in the following passages:
Altabrizi, Prop. IV: וההתכה הוא התיבשות הצמיחה (anonymous
translation); והדבר יהיה בהחסר חלק ממנו ויקרא כליון והתכה, והוא
מקבל הגדול בצומח יחסר שעורו בכליון והתכה (Isaac ben Nathan's
translation). *Ibid.* Prop. XIV: לפי שאי"א שיותך דבר מחלקי הנתך עד
ילך חסר בכמותו אל ההתכה וההשחתה. Averroes, *Epitome of
Physics* V, p. 22a: יהיו סוני התנועות שלשה...והב' בכמות והיא הנקראת
צמיחה וחסרון. In the latter sense it is used as the antonym of
באלתרכיב ואלתחליל, הרכבה תركيب , as in the expression הרכבה
וההתכה "synthesis and analysis" (*Cuzari* V, 12).

ספוניות and מקשיות are translations of تَخَلْخُل, μανός, rarus, and تَكَاثُف, πυκνός, densus, respectively (see Maḳaṣid al-Falasifah III, p. 237). The synonyms of ספוניות are התחלחלות, הררפות, דקות; those of מקשיות are התעבות, התכנסות, as in the following passages: Altabrizi, Prop. IV: ויקרא התעבות, ר"ל דחיקת הגוף המתפשט ומתחלחל כדמות שקי צמר נפן הנדחקים בספינה ויעשו קטנים, והוא ענין ההתעבות (anonymous translation). Maimonides, Mish-neh Torah, Yesode ha-Torah IV, 5: וכן הרוח מקצתו הסמוך לאש משתנה ומתחלחל ונעשה אש. See also quotation from Albalag below in n. 23). Themistius on De Caelo IV, 2, Hebrew Text, p. 148, 11. 34-35: ואמנם גדרות האדם לקל ולכבד ברבוי ובקטנות, כלומר בדקות ובקושי (mollitie) או ברפיון (crassitie) החלקים ועביים (tenuitate) הוא כזב (duritie).

6. Hebrew דמוים, used here in the sense of משלים. Cf. Milḥamot Adonai VI, i, 3: מה שמי שיאמר בהויית העולם והפסדו פעמים אין תכלית להם, יש לו קצת דמיים יקיימו דעתו לפי מה שיחשב. מהם שכבר נמצא בכל הדברים אשר אצלנו שהם הווים נפסדים, ולזה יחשב מזה החפוש. Cf. also Ḥobot ha-Lebabot I, 10: וראיתי לקרב לך הענין בשני דמיונים (مثالين=משלים) קרובים.

7. In *Physics* IV, 6, Aristotle mentions two views with regard to the vacuum. (1) The Atomists' view, according to which the vacuum is an interval separate from bodies, having actual existence and pervading through every body, so that bodies are not continuous. (2) The Pythagorean view, according to which the vacuum exists outside the world, the world itself being continuous. (Cf. Plutarch, *De Placitis Philosophorum* I, 18).

Narboni, in his commentary on *Moreh* I, 73, Prop. II, describes these two views accurately and finds an allusion to them in the text of Maimonides: "Similarly those who believe in the existence of a vacuum are divided into two classes. Some believe that the vacuum is interspersed in bodies, diffused throughout them, and existing in actuality. Others believe that it is not interspersed in bodies after the manner of pores in porous objects but that it is rather something entirely unoccupied by a body, existing, as it were, outside the world and surrounding it. Having explained this, I say that these two views are summed up by Maimonides in his statement that 'the Radicals also believe that there is a

vacuum, i. e., one interval or several intervals which contain
nothing.' By the expression 'one interval or several intervals' he
refers to the two views of the vacuum, by the latter referring to
the kind that is interspersed in bodies and by the former to the
kind that is not interspersed in bodies but is existing separately
and unoccupied by anything."

וכמו כן מאמיני הרקות נחלקו לשתי כתות: מהם שאמרו שהוא מעורב בגשמים
ומסתבך בהם ונמצא בם בפועל. ומהם שאמרו שהוא בלתי מעורב בגשמים, כאלו
הרקות [read :ומקום] מקום יהיה גשם שם ואין הספוניים, בנקבים תאמר וכאלו
הוא חוץ כל העולם, מקיף בו. ואחר שההתבאר זה, אומר כי רבינו משה כלל זה
או רחק רחק באמרו רצה ...בו. יאמינו כן גם השרשיים כי באשמר,
רחקים שיכלל המסובכים והבלתי מסובך אבל נבדל בלתי מקומם.

See also Narboni on *Moreh* II, 14: "As we have said, the
existence of a vacuum is impossible, for the existence of separate
dimensions is impossible whether outside the natural bodies or
within them." כמו שאמרנו שמציאות הרקות נמנע, כי מציאות רחקים נבדלים
נמנע חוץ לגשמים הטבעיים ובתוכם.

8. Hebrew האותות. This term is the Hebrew translation of the
Arabic موافقة, *fitness, agreement, sympathy, analogy, resemblance,*
and is used synonymously with הסכמה (Moritz Löwy, *Drei
Abhandlungen von Josef B. Jehuda,* German text, p. 38, n. 2;
Steinschneider, *Uebersetzungen,* p. 369, n. 4). Hence it may be
translated here by *affinity, inclination, attraction.* It seems to
reflect the Greek ἐπιτηδειότης, *fitness, suitableness,* which is
used in a context similar to this in the following passage: τί δὲ
διοίσει πυρὸς ἐπιτηδειότης ἐπὶ τούτου ἤπερ ὕδατος. (Simplicius
in *Physica* IV, 8, ed. Diels, p. 665, lines 9–10). In the Latin
translations from the Hebrew, האותות is sometimes rendered
by *convenientia,* as in the following passage of Averroes' Intermediate commentary on the *Meteorology* (MS. Bibliothèque
Nationale, Cod. Heb. 947, f. 138v): ואמנם כפי דעת אלכסנדר הנה לא
כלל האותות אריסטו מאמר ובין האחרים [ומאמר] בין יהיה. "Sed secundum
opinionem Alexander nulla est *convenientia* inter dictum istorum
et dictum Aristotelis" (Averroes on *Meteorology* I, p. 409va–b).

For other meanings of האותות see Caspar Levias, *Oẓar Ḥokmat
ha-Lashon,* p. 29, under אות.

9. I take קרובו או רחוקו to refer to רקות which is used here through-out as masculine.

10. The argument may be restated fully as follows: The vacuum is not the producing cause of motion. It is called cause only in an accidental sense, that is to say, it makes motion possible in its midst. As for the producing cause of motion, argues he, it will remain the same when you assume the existence of a vacuum, through which the elements are to be dispersed, as when you deny it. It will always be due to the fact that each element has a place to which it is naturally adapted, toward which it moves by an inner momentum, and in consequence of which it tries to escape from any other place in which it happens to be. Now, you say that the elements could not try to escape from one part of the vacuum in order to be in another, since the parts of a vacuum cannot differ from one another. True enough. The parts of a vacuum cannot differ from each other in anything pertaining to their own constituent nature; but they can still differ from each other with reference to something external to their nature, namely, their respective distances from the lunar sphere (המקיף, *the periph-ery*) and the earth (המרכז, *the centre*). Thus, when fire moves from one part of the vacuum into another in upward direction, it is not because it tries to escape one part of a vacuum in order to be in another, but rather because in its endeavor to get nearer to its proper place, which is the concavity of the lunar sphere, it natu-rally has to leave those remote parts of the vacuum and occupy the parts which are nearer to its proper place.

It should be noted that this explanation of motion within a vacuum is advanced by Crescas only for the purpose of scoring a point against Aristotle. The real explanation of motion according to those who believe in a vacuum, is given by Crescas later. See below n. 22.

This argument is reproduced by Pico Della Mirandola: "Nunc ex Graecis expositoribus digressi, parumper videamus quid Hebraeus R. Hasdai de eodem vacuo senserit. Arbitratur nihil iuvare Aristotelem, eam quae dicitur loci ad collocatum corpus convenientiam, cum fieri queat ut elementa etiamsi sint inmixta, vacuo eam possideant, et diversos etiam habeant et suos terminos, quibus factum est nomen a quo, et ad quem, ex propinquitate videlicet distantia ad circumferentiam et centrum" (*Examen Doctrinae Vanitatis Gentium* VI, 6).

11. Reference to the Pythagoreans. See above n. 7.

12. According to Aristotle the circular motion of the spheres is performed within one place, and it is not from one place to another. Cf. Proposition XIII, p. 623, n. 18. See also *Moreh* II, 4: "For it moves toward the same point from which it moves away, and it moves away from the same point toward which it moves."
כי כל מה שאליו יתנועע ממנו יתנועע, וכל מה שממנו יתנועע אליו יתנועע
and *'Olam Ḳaṭan* I, 3, p. 10: "For circular motion has neither beginning nor end, for every part thereof is like any other part, and no one can say that the motion begins in one place and stops at another. Consequently, circular motion requires no place, for any one part thereof is a place for any other part."
כי תנועת ההקפה אין לה התחלה ולא סוף, שכל חלק מחלקיו דינה כדין חבירתה, ולא יוכל אדם לומר מכאן התחילה התנועה וינוח במקום אחר. ועל כן אינו צריך למקום, שכל חלק ממנו מקום.

Pico Della Mirandola restates this argument as follows: "Atque ut cetera obstarent vacuo, nihil tamen officere, quin orbiculare corpus in eo moveatur, cum in motu circulari, nec terminus a quo, nec terminus ad quem motus tendat, inveniatur: et secundum Aristotelem maxime qui motum nunquam voluit incepisse." (*Examen Doctrinae Vanitatis Gentium* VI, 6).

13. The passage following abounds in cryptic allusions to a lengthy discussion found in Averroes' *Intermediate Physics*, in Gersonides' supercommentary thereon, and in Narboni's commentary on Algazali's *Kawwanot, Physics*, On the Vacuum. From the general arrangement of this passage, and from the use of the illustration from a "fatigued person," which is found only in Gersonides, it is evident that Crescas has been following here Gersonides.

Following are the texts illustrating this passage:

A. *Intermediate Physics* IV, ii, 5:

§1. "From the following it will appear that a stone can have no motion in a vacuum, for the medium is a condition in the existence of this particular motion of the stone. It is, therefore, not to be thought of that the motion of a stone in air and in water is essentially of equal speed and that the medium in which it moves acts only as a resistance to that motion. Quite the contrary, its motion in the air is more rapid than that in water in the same sense as that in which we say that the keen edge of iron is more cutting than that of bronze. Accordingly, there can be no motion at all without a medium. The inquiry into the nature of this kind of motion and the explanation of the reason why it needs a medium in which it is to operate are out of place here, and it is not here where the discussion of these phases of motion belongs.

§2. The objection raised by Avempace in the seventh book of this work is based upon the assumption that the stone has something to impede its natural motion when it moves in water and in air, but has no impediment for its natural motion when it moves in a vacuum. For he contends that it is not the relation of one motion to another that equals the relation of one medium to another medium, but it is rather the relation of the retardation caused to one object in motion by its medium to that caused to another object by another medium that equals the relation of one medium to another. In a similar manner he maintains that if anything were moved in a vacuum it would be moved in time, for he believes that if the cause of the retardation were eliminated there would still remain its original motion.

§3. But this is all an impossible fiction. For when the rate of a motion is changed on account of a change in its medium, the relation between the earlier and the later motion does not equal the relation between the retarded part of one motion and that of the other motion but it rather equals the relation of one motion as a whole to the other motion as a whole. To assume that the retardation is a motion added to the original motion is an impossible fiction, for if there had been an original, natural motion, it would have already been destroyed by the retardation which accrues to it, so that the resultant motion would be entirely different, and there would be no relation between it and the original motion.

§4. Hence it is clear that if we assume the possibility of an object having motion in a vacuum, it will result that the same object will traverse an equal distance [in equal time] in the medium of a vacuum and in that of a plenum. For let a certain object traverse a certain distance in a certain time in a vacuum. Let the

same object traverse the same distance in air in a longer time.
Then, let the same object move in a medium (literally: body)
[more] attenuate [than air], whose receptivity for motion is related
to the receptivity of air as the relation between the time required
for the motion in air and in a vacuum. It will follow that the
same object will traverse the same distance in this attenuate
medium (literally: body) and in a vacuum in equal time. But this
is an impossible contradiction.

The suggestion put forward that when something moves in a
resistant medium there occurs some retardation to the natural
motion, so that it is not the relation between two such motions
that is equal to the relation of their respective impediments but,
as says Avempace, rather the relation between their respective
retardations, is pure fancy and utterly an impossible fiction. Our
argument is as follows: An object in motion has only one motion
and one time, and that motion as a whole and that time as a whole
are described by the terms slow and fast. Consequently, if two
such moving objects happen to be impeded in different degrees
by different media, it is the relation between their respective motions that is equal to the relation of one impediment to another.
This view is accepted in Book VII of this work."

§ 1 ומהנה יראה שאי אפשר שתהיה בריקות לאבן תנועה, לפי שהממוצע תנאי
במציאות זאת התנועה. ולזה לא יתכן שידומה שתנועת האבן באויר ובמים הוא
בהכרח שוה ולא שתהיה לה מעיק מפני מה שבו יתנועע, אבל אמנם תהיה תנועתו
אשר באויר יותר מהירה מאשר במים כמו שנאמר שהחדרות אשר בברזל יותר חותך
מאשר בנחשת, לא שיהיה אפשר בו תנועה בזולת אמצעי. והחקירה בזאת התנועה,
ומה הסבה בצרכה אל דבר שיתנועע בו אין זה מקומו, ואין אלה אופניו בזה המקום.

§ 2 והספק אשר חייבו אבובכר בשביעי מזה הספר אמנם הוא בני על שהאבן
יש לה מעיק מתנועתו הטבעית, כאשר תתנועע במים ובאויר, ואין לה מעיק מתנועתה
הטבעית, כאשר תתנועע בריקות. וזה שהוא יאמר שאין יחס התנועה אל התנועה
כיחס הממוצע אל הממוצע אבל יחס האיחור אל האיחור אשר יקרה למתנועע
מהממוצע הוא יחס הממוצע אל הממוצע. וכמו כן חשב שהמתנועע אלו יתנועע
בריקות יתנועע בזמן, לפי שהוא חשב שאם היה היה מסתלק ממנו]תנועת( ]סבת[
האיחור תשאר תנועתו השרשית.

§ 3 וזה כלו דמי בטל. כי התנועה כאשר יתחלף יחסה בהתחלפות הממוצע
אין אותו היחס יחס המאוחר אל המאוחר, אבל אותו היחס הוא יחס התנועה בכללה
אל התנועה בכללה. ולדמות שהאחור תנועה נוספת מהתנועה השרשית דמי בטל,
לפי שאלו היה בכאן תנועה שרשית טבעית היתה כבר נפסדת עם האיחור אשר
יקרה לה והיתה התנעה אחרת אין בינה ובין התנועה השרשית יחס.

§ 4 ומזה בעצמו יתבאר שכאשר הנחנו המתנועע האחד בעצמו יתנועע בריקות,
יתחייב ממנו שיהיה הדבר האחד מתנועע מהלך אחד בעצמו באמצעות הריקות
והמלוי. חה כשנייחהו פעם אחד יתנועע מהלך אחד בעצמו בזמן מה בריקות,
ויתנועע אותו בעצמו באויר בזמן יותר גדול, הנה כאשר הנחנו נשם דק, יהיה יחס
הקבול אשר בו לתנועה אל הקבול אשר באויר יחס הזמן אל הזמן. חוייב שיתנועע
זה המתנועע בזה הגשם הדק ובריקות אותו המהלך האחד בעצמו בזמן שוה. חה
חלוף בלתי אפשר.

ומה שאפשר שיצוייר הדבר כאשר יתנועע במונע שבכאן קרה איחור לתנועה
הטבעית, ולא יחוייב מזה שיהיה יחס התנועה אל התנועה הוא יחס המונע אל המונע,
אבל יחס האיחור אל האיחור, כמו שיאמר אבובכר, הוא ענין בעינו דמיון, והוא
ציור בטל: שלא ימצא במתנועע רק תנועה אחת חמן אחד, ואותה התנועה בכללה
ואותו הזמן בכללו יתוארו באיחור ומהירות. ולזה אפשר כאשר יתוארו בשנים
מונעים מתחלפים שיהיה יחס אחת משתי התנועות אל התנועה האחרת הוא יחס
המונע אל המונע. חה דבר קובל בשביעי מזה הספר.

B. Gersonides' Supercommentary on the *Intermediate Physics*,
loc. cit.:

§1. "From the following it will appear that a stone can have
no motion in a vacuum, for the medium is a condition in the existence of this particular motion of the stone, in view of the fact that
the medium has something of the nature of a terminus ad quem, that
is, we claim that the medium does not merely accelerate the
motion or retard it but rather it is a condition in its existence
· · · · · The motion of the stone in air is said to be faster
than that in water in the same sense in which we say that the
keenness of iron is more cutting than that of bronze, which does not
mean that there can exist a keenness without a subject. Similarly
here, the relation between one speed and another is said to be
equal to the relation between one medium and another without
implying that there can be motion without a medium, for it is the
possession on the part of the medium of the nature of an incomplete *terminus ad quem* that is the cause of the motion of the stone.

§2. Avempace, however, in his treatise argues in the manner
stated above, namely, that it is the relation between one kind of
retardation and another that is equal to the relation between one
medium and another, and that there exists an original time. To
illustrate by the example of two ships · · · · · ·

§3. But Averroes says that all this is an impossible fiction, for
the retardation is not a motion added to the original motion in

the manner illustrated above by the movement of the ship, so that by the elimination of the retarded motion there could still remain an original motion. Quite the contrary, if there had existed a natural, original motion, it would have already been destroyed by the retardation which accrues to it, for there is only one kind of motion in the movement of a stone in air and in water, and consequently, if an original motion is assumed, it will have to disappear completely, and an entirely new motion will take its place, and this new motion as a whole will be related to the medium; as we say, for instance, in the case of the motion of a *fatigued person* that his motion as a whole bears a certain relation to the fatigue rather than to the retardation. To illustrate: If Reuben's rate of motion is one mile per hour, but when he is slightly fatigued his rate of motion is one-eighth of a mile per hour, we then say that if he is twice as much fatigued his rate of motion will be one-half of an eighth of a mile per hour but not that the relation between one state of fatigue and the other will be equal to the relation between one degree of retardation and that of another, for that would not be so. But what we do say is that the relation between one rate of motion and that of another is equal to the relation between one impediment of the motion and that of another, as is accepted in Book VII of this work.

a. Says Levi . . . . . (Here follows an argument against Averroes' refutation of Avempace).

b. But the real refutation of Avempace's objection here is Averroes' contention that *the medium is a condition in the existence of the motion*. This is true and beyond any doubt. Consequently Aristotle's reasoning here is well established.

§4. Similarly Averroes' argument in refutation of Avempace, that if an original motion were assumed to exist in a vacuum, it would follow that the same object would traverse the same distance in equal time both in a plenum and in a vacuum, is subject to the following difficulty.

a. First . . . . . .

b. Second . . . . .

c. Hence Avempace's objection here is to be answered only by Averroes' contention that *the medium is a condition in the existence of motion*. Let us now return to where we were."

§ 1 והנה יראה שאי אפשר שתהיה לאבן בריקות תנועה, לפי שהממוצע תנאי
במציאות זאת התנועה ב מ ה ש ב ו מ ט ב ע מ ה ש א ל י ו, לא שנאמר שהממוצע
ממהר התנועה או מאחר, אבל הוא תנאי במציאותה.... אבל אמנם יאמר שתנועתו
אשר באויר יותר מהירה מאשר במים על צד מה שנאמר שהחדות אשר בברזל
יותר חותך מאשר בנחשת, לא שיהיה אפשר שיהיה שם חדות בולת נושא, כן יאמר
הנה שיחס המהירות אל המהירות הוא יחס הממוצע אל הממוצע לא שיהיה אפשר
התנועה בולת הממוצע, כי טבע מה שאליו הבלתי הגמור אשר בממוצע הוא סבת
התנועה בו.

§ 2 ואמנם אבובכר במאמר טוען כמו שקדם: יחס האחור אל האחור כיחס
הממוצע אל הממוצע, ויש זמן שרשי, משל משתי ספינות.....

§ 3 ויאמר בן רשד שזה כלו דמוי בטל, לפי שאין האחור תנועה נוספת על
התנועה השרשית על צד מה שקדם במשלנו בתנועת הספינה, עד שיהיה אפשר
שבהסתלק תנועת האחור תשאר התנועה השרשית, לפי שאלו היתה בכאן תנועה
שרשית טבעית היתה כבר נפסדה עם האחור אשר יקרה, כי אין בכאן בתנועת
האבן באויר או במים כי אם תנועה אחת, ולזה תסתלק התנועה השרשית בכללה,
אם היה שניחה, ותהיה זאת תנועה אחרת תיוחס בה בכללה אל הממוצע, כמו
שנאמר ב ת נ ו ע ת ה א י ש ה י נ ע תיוחס בכללה אל היגעות לא לאחורה.
והמשל בו שתהיה תנועת ראובן בשעה מיל אחד, וכאשר יהיה ינע יניעות מה תהיה
תנועתו בשעה שמינית מיל, הנה נאמר שאם היה ינע מזה הכפל תהיה תנועתו בשעה
חצי שמינית מיל, לא שנאמר שיהיה יחס היניעות אל היניעות יחס האחור אל האחור,
כי זה בלתי אפשר, אבל נאמר שיחס התנועה אל התנועה הוא יחס המונע אל המונע,
כמו שיקובל זה בז' מזה הספר.

a אמר לוי ...

b אמנם הבטול העצמי לספק אבובכר פה, הוא מה שיאמר שהממוצע תנאי
במציאות התנועה. והוא אמת אין ספק בה, ולזה יתאמת חיוב אריסטו פה.

§ 4 וגם כן הספק אשר חייב אבן רשד, אם הונחה בכאן תנועה שרשית בריקות,
שתהיה תנועת המתנועע האחד בעינו בריקות ובמקבל שוה, הנה בזה החיוב מן
הספק מה שאומר:

a אם תחלה ...

b ואולם הספק השני.....

c ולזה מה שנאמר כאשר יסתר ספק אבובכר פה הוא מה שיאמר אבן רשד,
שהממוצע תנאי במציאות התנועה. ונשוב אל אשר היינו בו.

14. Hebrew ידוע אצל הטבע, *known to nature*. According to some
readings ידוע אצל הטבעי, *known to the natural philosopher*. My
translation of this phrase, however, is based upon the following
consideration:

The existence of an "original time" of motion is explained by
Crescas later (p. 205) as being due either to the medium (אמצעי,

here: מקבל, *receptacle*) in which motion takes place or to the nature of motion itself (להכרח היות התנועה בזמן or מפאת התנועה). When, therefore, Crescas argues here that even by eliminating the medium or receptacle there will still be an original time on account of the fact שהתנועה תחייב זמן לעצמותה .... ידוע אצל הטבע, the alternative reason he offers here must correspond to the alternative reason he offers later. The expression ידוע אצל הטבע is thus equivalent to the expression ביחס ידוע אל הטבע which occurs in Prop. IX, Part II; cf. also Prop. XII, Part II, n. 6 (p. 612).

15. Hebrew במקצת. The qualifying term במקצת is rather misleading. Crescas has borrowed the theory of an "original time" of motion in its entirety from Averroes, who quotes it in the name of Avempace.

16. The reference is to Averroes' answer that has been refuted by Gersonides. See above n. 13, B, §3a, §4a, b. Thus relying upon Gersonides' refutation, Crescas dismisses Averroes in this summary fashion.

As for the expression והרבה דברים מרבים הבל, see Ecclesiastes 6.11.

17. The reference is to Gersonides rather than to Averroes, though Gersonides' answer is based upon Averroes. (See above n. 13, B, §3b, §4c. Cf. also Narboni on the *Kawwanot*, *Physics*, On the Vacuum: "The learned Averroes has solved this difficulty by explaining that the relation of one motion to another is equal to the relation of one medium to another, for the medium is not simply an impediment as was thought by Avempace." והחכם בן רשד התיר זה הספק בשבאר שיחס התנועה אל התנועה כיחס הממוצע אל הממוצע, כי אין הממוצע מונע כמו שחשב אבובכר. The expression הממוצע מונע, *the medium is . . . impediment*, reflects the Greek τὸ μὲν οὖν δι' οὗ φέρεται αἴτιον ὅτι ἐμποδίζει in *Physics* IV, 8, 215a, 29.

18. That is to say, the difference in the motion of the same object by the same agent in two *media*, in air and in water, for instance, is not due to the fact that water offers a greater resistance than air to a hypothetical original motion, but rather to the fact that motion in water is essentially different from motion in air, for the medium is an inseparable condition of motion. Averroes compares

motion to the keenness of the edge of a blade. The fact that the edge of an iron blade is keener than one made of bronze, he says, does not imply that there exists an original keenness, independently of the metal, which in varying degrees is dulled by the metal in which it inheres, and by bronze more than by iron, but what it means is that the keenness of the edge of an iron blade is essentially different from that of a bronze blade, the metal being an inseparable condition of the keenness, as there can be no keenness without metal. So also in the case of motion, there can be no motion without a medium, i. e., without space. See above n. 13, A.

19. Hebrew זה להאותות טבעו למה שאליו. This explanatory remark is not found in the corresponding passage in Averroes. It reflects the following statement of Gersonides quoted above in n. 13, B, §1: שהממוצע תנאי במציאות זאת התנועה במה שבו מטבע מה שאליו... כי טבע מה שאליו הבלתי גמור אשר במוצע הוא סבת התנועה בו.

What Crescas wants to say here is this: The medium is an essential condition of motion, because when an object moves toward its proper place, it is not the object alone irrespective of its medium that moves, but rather the object in so far as it is in a certain medium. Every point within the medium which the object has to pass in order to reach its goal is in itself a relative goal and acts upon the object as a *terminus ad quem*. The medium itself thus becomes charged, as it were, with a certain power to carry the object toward its objective. If that medium should be eliminated, the object would cease to move. Consequently, there can be no motion in a vacuum.

20. The purpose of this passage is to prove that the medium is not a necessary condition of motion and that motion is possible in a vacuum. Crescas, however, does not attack the problem directly. He starts rather with a flanking movement, arguing that weight and lightness need no medium, and seems to leave it to ourselves to supply the conclusion that whatever is proved to be true of weight and lightness must also be true of motion.

Such a conclusion may be properly supplied. For according to Aristotle, weight and lightness are only other terms for downward and upward motion. "But I call that simply light which is always naturally adapted to tend upward, and that simply

heavy which is always naturally adapted to tend downward unless something impedes" (De Caelo IV, 4, 311b, 14–16). We may
therefore infer that if it can be shown that weight and lightness
are independent of a medium so will also be upward and downward motion.

In showing that weight and lightness are independent of the
medium, Crescas advances a theory which dispenses with the
necessity of an inner striving of the elements towards their proper
places. This is not original with Crescas. It is reported by Aristotle as the view of the ancients, Plato and the Atomists. According to Plato, as reported by Aristotle, the difference in the weight
of bodies is due to the difference in the number of "triangles" of
which all things, he says, consist. According to the Atomists, the
difference in weight is due either to a difference in the number of
void interspaces a body contains or to a difference in the size and
density of the atoms of which bodies are composed. (Cf. De Caelo
IV, 2.)

According to these views, as may be inferred, the difference in
weight is due to a difference in the internal structure of bodies.
Crescas, therefore, characterizes them by saying "that the moveable bodies have weight and lightness by nature" (Compare the
account of the different theories of gravity and levity as given by
Plutarch in his De Placitis Philosophorum I, 12).

21. That is to say, the theories of weight and lightness just stated
might be said to deny altogether the existence of absolute lightness. There are according to these theories only different degrees
of weight. This interpretation suggested by Crescas agrees with
what Aristotle himself has said of those ancient views: "Of those,
therefore, who prior to us directed their attention to those things,
nearly most spoke only about things which are thus heavy and
light, of which both being heavy, one is lighter than the other.
But thus discussing the affair, they fancied the discussion was
about the simply light and heavy" (De Caelo IV, 2, 308a, 34–
308b, 2).

22. This correctly describes the explanation of upward motion as
given by Democritus and Plato. According to both of them, the
less heavy bodies move upward not on account of their own nature but by the pressure of the heavier bodies. (Cf. Zeller, Pre-

Socratic Philosophy, Vol. I, pp. 701, 713; Vol. II, p. 420; *Plato*, p. 376, n. 30). This view is also quoted by Avicenna and is attributed by him to some unnamed philosophers. *Al-Najah*, p. 41, quoted by Carra de Vaux in *Avicenne*, p. 193.

Pico Della Mirandola, in *Examen Doctrinae Vanitatis Gentium*, VI, 6, discusses this argument of Crescas as follows: "Et praeterea nihil efficere eas quae sunt excogitatae contra vacuum rationes, et fundatae super motu recto, quando intermedium nullum sit necessarium: et dici queat gravitatem et levitatem naturaliter corporibus inesse mobilibus, nec ea mediis indigere. Dici etiam possit omnibus corporibus inesse gravitatem, eaque vocari levia, quae videlicet gravia sint minus, eaque ipsa moveri sursum ex eorum, quae magis gravia sunt impetu et violentia. Ac memini etiam ex nostris theologis, qui causam quod ligna supernatent aquae, referant in gravitatem atque, quae minus gravibus sua parte natura non cedit. Sed quod attinet ad Hebraeum omnia corpora gravia non negat, et aerem descensurum, si terra loco moveretur affirmat, ob gravitatem verius, quam ne vacuum detur."

Cf. the following statement in *op. cit.* VI, 18: "Negaret alius fortasse etiam in ipsis corporeis authoritate Scoti, decernentis gravia et levia se ipsis moveri. Cui videtur assensus Hebraeus Hasdai."

23. This argument is not unanswerable. Aristotle has forestalled it by the theory that all elements, except fire, have gravity in their own place. "For all things, even air itself, have gravity in their own place except fire" (*De Caelo* IV, 4, 311b, 8–9). "But as earth, if the air were withdrawn, would not tend upward, so neither would fire tend downward; for it has not any gravity in its own place, as neither has earth levity. But the two other elements would tend downward, if that which is beneath were withdrawn; because that is simply heavy which is placed under all things; but that which is relatively heavy tends to its own place, or to the place of those things above which it emerges through a similarity of matter" (*op. cit.* IV, 5, 312b, 14–19).

Cf. Gersonides on the Epitome of *De Caelo* IV: "*This is an indication that air has some gravity in its own place.* Aristotle cites here another illustration for this from the fact that, when water or earth is withdrawn, air is easily attracted to the lower place, but the contrary does not happen, namely, when air is withdrawn, earth and water do not tend to move upward."

וזה ממה שיורה לאויר כבדות מה במקומו. ויאמר [אריסטו] גם כן מן הראיה על זה המשך האויר אל המקום השפל בקלות, כשהוסר ממנו המים או הארץ, ולא ימצא העניין בהפך, רצוני, שכשהוסר האויר, לא תמשך אחריהם כלל.

The same illustration with the inference that the descent of air is due to the impossibility of a vacuum is given by Gershon ben Solomon in *Sha'ar ha-Shamayim* I, 1:

"It may further be made clear to you by the following illustration. If a man makes a digging in the ground, the air will descend into that digging and fill it up. But how, then, is it possible for the air to move downward against its own nature, seeing that it does not ordinarily descend but rather ascend? The explanation is that its descent is due to the fact that no vacuum can exist, for which reason the vacuum attracts the air and causes it to move downward against its own nature, for there can be no vacuum at all."

עוד תוכל להבין אותו, שאם יחפור אדם חפירה בקרקע ירד האויר בחפירה ההיא ותמלא אותה. ואיך ירד האויר נגד טבעו, שהרי אין מדרכו הירידה אלא העלייה? אלא מפני שאין ריקות נמצא, מושכה הריקות האויר ומורידו אותו חוץ מטבעו, מפני שאין הריקות נמצא כלל.

This view that motion is due to nature's abhorrence of a vacuum is quoted in the name of Avicenna by Shem-ṭob in his commentary on *Moreh* II, Introduction, Prop. XVII: "It has been said by Avicenna that all motions, whether violent or natural, take place on account of [the impossibility of] a vacuum."

וכבר אמר בן סינא שכל התנועות, בין הכרחיות בין טבעיות, ימצאו מהכרח [והמנעות] הריקות.

Another explanation for the descent of air into a ditch is given by Albalag in his comments on Algazali's *Makaṣid al-Falasifah* III, On Place. According to him the descent of air under such circumstances is not locomotion but rather a form of expansion, that is to say, it is not local change but quantitative change:

"Says the translator: Inasmuch as the place of water is the inner surface of air and as the nature of each element is to tend toward its own place and not toward the opposite direction, would that I knew why it is that, when we withdraw, for instance, half of the water from a ditch, its place is taken by air? This evidently cannot be explained except on the ground that the air moves toward the water; but, if so, the air will then have a downward motion. One would rather expect the water to move upward toward the air, inasmuch as it is the object which moves toward its place rather than the place toward its object. The answer is that the motion of the air in this particular instance is not due to locomotion. It is rather due to the rarefaction and expansion of the parts of the air with the result that they spread over and occupy a larger area. It has already been explained by Algazali that this kind of motion belongs to motion in the category of quantity."

אמר המעתיק, אם מקום המים הוא שטח האויר הפנימי, וטבע היסוד להתנועע כלפי מקומו, לא כנגד, מי יתן ידעתי כשנוציא חצי המים אשר בתעלה, על דרך משל, איך ימלא האויר חסרונם, כי זה אי אפשר כי אם בהתנועעו כלפי המים, ונמצא האויר יורד למטה, ויותר היה ראוי שיתנועעו המים כלפי שטח האויר, כי מן הדין המתקומם מתנועע למקום, לא המקום למתקומם. התשובה, כי התנועה הזאת אשר לאויר אינה מקומית אלא התפשטות חלקיו והתרפותם עד שיטרידו גבול גדול. וכבר ביאר אבוחמד, כי התנועה הזאת היא ממיני התנועה אשר בכמות.

A similar illustration is cited by Bruno in his criticism of Aristotle's theory of light and heavy. His explanation of the descent of air is like that offered by Albalag, namely, that it is due to expansion. Cf. *De l'Infinito Universo et Mondi* III, p. 356, l. 18 ff. Cf. Prop. VI, n. 18, p. 539.

24. This is arguing for the Pythagorean view of a vacuum. See above notes 7, 11.

Pico Della Mirandola restates this argument as follows: "Nec impediri ex intermedio quin vacuum extra mundum reperiri queat" (*Examen Doctrinae Vanitatis Gentium VI*, 6).

25. This refers to the circular motion of the celestial spheres which does not involve change of place. See below Proposition XIII, n. 18.

26. Pico Della Mirandola reproduces this argument as follows: "Parvi facit etiam illam non penetratorum corporum, ob dimensiones rationem, cum dimensiones materiae iunctas id efficere posse dicendum sit, non seiunctas, et ab omni prorsus materia separatas" (*Examen Doctrinae Vanitatis Gentium VI*, 6).

27. Hebrew מורכב יצדקו הנה ... נפרדים יצדקו לא ואם. The terms מורכב, נפרד are borrowed from logic, where they are used in technical senses with reference to the fallacies of *compositio*, σύνθεσις, and *divisio*, διαίρεσις. Cf. *Epitome of Sophistic Elenchi*, p. 55a: מורכבים יצדקו נפרדים כאשר אשר בדברים יתאמת אמנם וזה. I have translated these terms freely, however, as required by the context.

28. This argument of Crescas contains many phrases which seem to be aimed at Aristotle's commentators, especially Averroes and Gersonides, who insist upon showing that the impenetrability of bodies is due exclusively to their pure, incorporeal tri-dimensionality.

Averroes' *Epitome of the Physics* IV, pp. 14b–15a: "We may also explain this in another way. Bodies exist in place through their dimensions and not through their accidents. The impossibility for two bodies to exist in one place at the same time is not due, for instance, to the fact that one is white and the other black, but rather to the impossibility of dimensions to penetrate each other . . . . . . Now, if place were identical with the vacuum, bodies would penetrate each other. But this is absurd."

במקום יחולו אמנם הגשמים כי זה אחרים. בפנים העניין זה שנבאר אפשר והנה לא אחד במקום יחד שיחולו גשמים בשני נמנע ואמנם במקריהם, לא ומרחקיהם]וב קצתם המרחקים הכנס המנעות מצד אבל משל, דרך שחור, זה לבן שזה מצד זה בגשמים הגשמים שיכנסו מתחייב היה הפנוי הוא המקום היה ואלו . . . . בקצת. שקר.

The same question is raised by Simplicius: "For why should these be prevented proceeding through each other, but a vacuum not? Shall we say that these are hot, or white, or heavy, or are replete with certain other passive qualities which happen to them, but that a vacuum is deprived of these? To assert this, however, would be absurd, for it has been shown before that bodies exist in place according to intervals alone" (Simplicius in *Physica* IV, 8, ed. Diels, p. 681, lines 21–26; Taylor's translation of the *Physics*, p. 228, n. 2).

Gersonides' Commentary on the *Epitome of the Physics*, *loc. cit.*, elaborates Averroes' statement as follows: "One cannot argue that while indeed it is impossible for corporeal dimensions to penetrate into other dimensions on account of the impenetrability

of bodies, it should still be possible for dimensions, which exist
apart from bodies, to penetrate into each other; for as against such
an argument, the following may be urged: It has already been
explained that corporeality is not the cause which makes the
interpenetration of bodies impossible, but the cause of that impossibility is rather the fact that a body possesses dimensions.
Consequently, if dimensions of any kind and under any conditions were capable of interpenetration, then the reason given for
the impenetrability of bodies would be no reason at all. Suppose,
for instance, we raise the question why man is incapable of flying.
If we answer that it is because he possesses life or because he is a
featherless animal, the reason given would not be a valid reason,
for the ability to fly is possessed by those who are animals and by
those who are featherless, though it is quite true that that
particular animal called man, or that particular featherless being
called man, does not happen to possess the ability to fly. But if
we answer that is because man is wingless, we have given the true
reason, for we do not find anything wingless that can fly. Similarly in this case if it were in any way at all possible for dimensions to penetrate into bodies, there would be no cause for the
impenetrability of bodies, for it is certain that mere corporeality
cannot be the cause."

ואין לאומר שיאמר, שזה אמנם נמנע ברחקי הגשם שיכנסו ברחקי הגשם להמנע
הכנס הגשמים, אבל הרחקים המופשטים מן הגשם יכנסו. זה שכבר קדם במאמר
שאין הסבה המונעת מהכנס גשמים טבע הגשמות, אבל הסבה המונעת הוא היותו
בעל רחקים. ואם היה אפשר שיכנסו הרחקים, איך ומה שהיה, הנה כבר נתנו
סבה שאינה סבה. כמו אם נשאלנו למה לא יהיה האדם מעופף, שאם השיבונו
לפי שהוא חי, או לפי שהוא חי בלתי בעל נוצה, הנה כבר נתנו סבה שאינה סבה,
מפני שכבר ימצא העופפות לחי או לבלתי בעל נוצה, אעפ"י שהחי שהוא אדם,
או הבלתי בעל נוצה שהוא אדם, אי"א שיהיה מעופף. ואמנם אם השיבונו לפי
שהוא בלתי בעל כנף, כבר נתנו הסבה האמתית, מפני שלא ימצא בלתי בעל כנף
מעופף. וכן הנה אם היה אפשר בשום פנים ברחקים שיכנסו בגשם לא תשאר
בכאן סבה ימנע בעבורה הכנס הגשמים, כי לא ימנע זה בהם מצד הגשמות".

Cf. Narboni on the *Moreh Nebukim* I, 73, Prop. 2: "The
impossibility of the interpenetration of bodies is due only to
the impossibility of the interpenetration of the dimensions."
והמנע הכנס גשם בגשם אינו אלא מפני הכנס הרחקים

29. Pico Della Mirandola refers to this argument as follows: "Negat praeterea dimensiones esse corporis extrema" (*Examen Doctrinae Vanitatis Gentium* VI, 6).

30. Hebrew מי יתן ואדע. Cf. Job 23, 3. The expression as here given by Crescas was frequently used by mediaeval Hebrew writers, as, e. g., Gersonides' *Milḥamot Adonai* III, 4.

According to Shem-ṭob Falaquera, it is a rendering of the Arabic phrase לית שערי, ليت شعري. He also quotes Avempace's explanation of the meaning of this phrase. Cf. *Moreh ha-Moreh* II, 15: במאמר האחרון מהשמע הטבעי, בחקרו התנועה, ח"ל: אמר מי יתן ואדע אם התחדשה התנועה הראשונה..... וכבר ידוע שזה המלה שהוא בערבי לית שערי ובלשוננו מי יתן ואדע, לא יאמר אותה האומר אלא בדבר שאינו יודע ידיעה אמתית ומתאוה לדעת אותו, וזו המלה לא אמר אריסטו בשאר הדברים שחבר אותם ואמר אותה בזה המקום .... וכתב אבובכר בן אלצאיג על זה המלה, אמר: חה המלה והוא לית שערי שמש אותה במה שאין דעת בו ולא אמתה, ולפעמים ישמשו אותה בשישתנו המחשבות בדבר מה אצל האומר, ומה שישתוו בו המחשבות הוא במדרגת הסכלות, שהוא על דרך השלילה, כי השני הסותרים אצלו שוים באפשרות הצדק.

Cf. also *Moreh ha-Moreh* I, 73, Prop. VII: פיאלית שער. העתיק אבן חבון, ואני תמיה. וכן בפי"ט מח"ב, ואני תמה. והנכון להעתיק, מי יתן ואדע.

31. The implication of this statement is that by defining place as a vacuum it does not mean that there is no difference in the use of these two terms. It rather means that what is called vacuum when it contains no body but is capable of receiving a body is called place when it does contain a body. This is in accord with the following statement of Aristotle; "For those who assert that there is a vacuum consider it as it were a certain place and vessel. And it appears to be full when it possesses the bulk which it is capable of receiving; but when it is deprived of this it is void; as if a vacuum, plenum, and place were the same, but their essence not the same" (*Physics* IV, 6, 213a, 15–19). A similar statement is found in Plutarch's *De Placitis Philosophorum* I, 20: "The Stoics and Epicureans make a vacuum, a place (τόπον) and a space (χώραν) to differ. A vacuum is that which is void of any thing that may be called body; place is that which is possessed by a body; a space that which is partly filled with a body, as a cask with wine." Similarly the Brethren of Purity explain that those who define place as a *vacuum* (الفضا, Dieterici: *Weite*) call it vacuum when considered apart from body but place when considered as possessing a body (Cf. Dieterici, Arabic text: *Die Abhandlungen der Ichwân Es-Safâ*, pp. 30–31; German translation: *Die Naturanschauung und Naturphilosophie der Araber*, p. 9).

32. Cf. below Second Speculation, Third Argument.

33. I. e., it is said to be "small and great" but not "much and few," because it is a continuous quantity. Cf. *Physics* IV, 12, 220a, 32–220b, 3: "It is also evident why time is not said to be swift and slow, but much and few, and long and short: for so far as it is continuous it is long and short, but so far as it is number it is much and few."

Pico Della Mirandola restates this argument of Crescas as follows: ". . . . . . quas explodi miratur cum magni et parvi nomine donentur, et per eius partes queamus illas dimetiri" (*Examen Doctrinae Vanitatis Gentium* VI, 6).

34. Hebrew והוא משוער בחלק ממנו.

Crescas evidently uses this expression here to prove that a vacuum must be a *continuous* quantity.

Abraham ibn Daud, however, uses it only as a definition of quantity in general and not necessarily of continuous quantity. *Emunah Ramah* I, 1: והחכמה הוא ענין ימצא בכל דבר שאפשר שישוער כלו בחלק ממנו, כמו הגשם הגדול אשר אפשר שיכרת ממנו חלק קטן וישוער בו כלו....‏ והחכמה שני מינים מתדבק ומתחלק.

Cf. Isaac ben Shem-ṭob's *first* supercommentary on *Intermediate Physics* IV, iii, 4: ונדר הכמה הוא הדבר שישוער בחלק ממנו.

Gersonides, on the other hand, uses it as a definition of continuous quantity. *Milḥamot Adonai* VI, i, 10: ונאמר שהוא מבואר בנפשו כי הזמן הוא מהכמה, זה שכבר יאמר בו שוה או בלתי שוה, שהם מסגולת הכמה. ואולם מאיזה חלק מהכמה הוא, הנה הוא מבואר שהוא מהכמה המתדבק, כי כבר יאמר בו ארוך וקצר. ועוד שהוא ישוער כלו במה שהוא חלק ממנו בהגנחה לא בטבע, זה מסגולות הכמה המתדבק. Crescas himself, in another place, uses this expression as the definition of quantity in general. Cf. *Or Adonai* III, i, 4, p. 67b: זה שאומר בב״ת אם היה שהוא ינדור הכמה שהוא אשר ישוער בחלק ממנו.

All these definitions of כמה are reproductions of Euclid's definition of the *multiple* of a *magnitude*, in *Elements*, Book V, Definition 2. "The greater is a multiple of the less when it is measured by the less."

It will be noted, however, that this Euclidian definition, which in Book V is applied to *magnitude*, i. e., a continuous quantity, is in Book VII, Definition 5, applied also to *number*, which, according to Aristotle, is a discrete quantity.

It is possible that in citing this definition Crescas merely meant to reason from the fact that a vacuum is *measured* (משוער) and not *numbered* (ספור), on which account it must be a continuous quantity. See *Metaphysics* V, 13, 1020a, 8–11: "A quantity (ποσόν) is a plurality (πλῆθος) if it is numerable (ἀριθμητόν), magnitude (μέγεθος) if it is measurable (μετρητόν). 'Plurality' means that which is divisible into non-continuous parts, 'magnitude' that which is divisible into continuous parts."

But here, too, it will be noted that Euclid uses the term *measured* (καταμετρῆται) with reference to both magnitude and number.

It is curious that in *Ḥobot ha-Lebabot* I, 5, Euclid's definition of *part* is reproduced from *Elements* V, Def. 1, and there the original term *measures* (καταμετρῇ) is replaced by the term *numbers* (סופר, ‎بـﻌد‎) though it is used with reference to *magnitude* (שיעור, ‎מقدار‎): כי השיעור הקטן סופר את הגדול כאשר זכר אקלידס בתחילת המאמר ‎ החמישי מספר השיעור.

Cf. Pico Della Mirandola's restatement of this argument in the passage quoted above in n. 33.

35. The implication of this statement is that a continuous quantity is either *time* or *magnitude*, גודל. However, inasmuch as a continuous quantity includes in addition to *time* also *line, surface, body* and *place*, it is evident that Crescas uses here the term *magnitude*, גודל, in a general sense to include all these four which are *magnitudes* as opposed to *multitudes*. Cf. above n. 34.

The following excursus on the various enumerations of quantity will be of interest.

Aristotle enumerates seven kinds of quantity, of which two are *discrete* (διωρισμένον), number and speech (λόγος), and five are *continuous* (συνεχές), line, surface, body, place and time (*Categories*, 6, 4b, 20–25). Cf. *Intermediate Categories* II, 2:

המתחלק שנים: המספר והדבור. והדבוק חמשה: הקו והשטח והגשם, ומה שיחזיק
בגשמים והיה בם, והם הזמן והמקום.

Algazali follows Aristotle in his general classification, but instead
of five *continuous* (متصلة, מתדבקת) quantities he speaks of four,
omitting place, and instead of two *discrete* (منفصلة, מתפרדת) quantities he mentions only one, number. (*Maḳaṣid al-Falasifah*
II, pp. 100–1).

Probably following Algazali, Abraham ibn Daud speaks of five
quantities, of which four are *continuous* and one *discrete* (מתחלק),
and concludes his discussion by saying that these five are the
only quantities "and he who made them more erred." ואלה
החמשה הם מיני נשאי הכמה, ומי ששם אותם יותר טעה. He was evidently
not aware that Aristotle himself made them more than five. He
must have had in mind Solomon ibn Gabirol who alludes to
seven kinds of quantity (*Meḳor Ḥayyim* III, 21: מיני השבעה
cf. *Fons Vitae* III, 27, p. 143, l. 22) and perhaps also Saadia who,
in *Emunot ve-Deot* II, 2, likewise speaks of seven kinds of quantity:
בשבעה מיני הכמה. These seven kinds of quantity are enumerated
by Saadia in his commentary on the *Sefer Yeẓirah* (*Commentaire
sur la Séfer Yesira*, ed. Lambert, Arabic text, p. 18; French
translation, p. 36).

The Hebrew translation of that passage in *Sefer Yeẓirah*
(quoted by Guttmann, *Die Religionsphilosophie des Saadia*, p. 97,
n. 4) contains several unusual terms. The passage reads as follows:
לפי שהכמיות שבעה מינים, חמשה מהם משותפים, והם הכתב, והנג, והעולם, והמקום,
והזמן, ושנים מהם זולתי משותפים, והמה, הספור והמנין. The term משותפים,
مشتر ك, in this passage is undoubtedly to be taken as synonymous
with מתדבקים متصلة, the latter being the usual translation of the
Greek συνεχής (see Proposition XV, Part II, p. 654, n. 23). כתב
is a literal translation of the Arabic خط which like the Greek
γραμμή means both *writing* and *line*. (Cf. Guttman, *ibid.*). נג is
a tolerable translation of the Arabic سطح, the latter of which
means both *roof* and *surface*. (Cf. Solomon Gandz, "On the
Origin of the Term Root," *American Mathematical Monthly*,
Vol. 33, 1926, p. 263, n. 2). It is in this sense of *surface* that
נג is used in the following passage: גוף שיש לו שורה וג ועומק (quoted
in Pinsker's *Liḳḳute Ḳadmoniyot, Nispaḥim*, p. 200). נלם for גשם

or גוף is quite simple. It is similarly used for חומר by Maimonides,
Sefer ha-Madda' I, ii, 3: כל מה שברא הקב"ה בעולמו נחלק לשלשה חלקים.
מהם ברואים שהם מחוברים מגולם וצורה. The term ספור, which Guttmann declares to be a mistranslation of the Greek λόγος should
be read סִפּוּר which is the equivalent of דבור, النطق, and a perfectly
good translation of λόγος. Cf. Cuzari IV, 25: ורצה בספור הדבור
והקול.

The Aristotelian classification of quantity is faithfully reproduced in the encyclopedia of the Brethren of Purity (Dieterici,
Arabic text: Die Abhandlungen der Ichwân Es-Safâ, pp. 343, 360;
German translation: Die Logik und Psychologie der Araber,
p. 7). Under discrete quantity they mention number and الحركة.
The latter term is translated by Dieterici as Bewegung. But this
makes no sense. It happens, however, that حركة means also
syllable (see Dozy, Supplément aux Dictionaires Arabes, s. v.), and
vowel, like the Hebrew תנועה, and is thus a well-enough translation
of λόγος. It will be recalled that in the passage of Metaphysics
VII, 10, quoted above in n. 1, Aristotle speaks of a syllable as of
a discrete quantity.

36. Crescas' argument that outside and beyond the world there
must be either a plenum or a vacuum had been answered by Gersonides who maintains that beyond the world there is neither a
plenum nor a vacuum but absolute privation or non-being. This
state of absolute nothingness, he continues, is one of the assumptions that are often made and are to be considered as true even
though it cannot be grasped by the imagination. Milḥamot Adonai
VI, i, 21, p. 386: "But there are things which, though true, man
cannot grasp with his imagination, as, for instance, the termina-
tion of the world at absolute privation which is neither a vacuum
nor a plenum." כמו אבל שם דברים צודקים לא יתכן שידמה אותם האדם, כמו
כלות העולם אל ההעדר המוחלט שאינו לא רקות ולא מלוי. That there are
things which reason compels us to assume even though the
imagination fails to grasp them is elsewhere also admitted by
Crescas and is equally insisted upon by Maimonides. See below
n. 112.

Similarly, prior to both Gersonides and Crescas, Averroes
argues, anticipating Crescas, that beyond the world there cannot

be a body, "for were it so, it would be necessary that beyond
that body there should be another body and so on to infinity."
Nor could there be a vacuum beyond the world, "for the impossibility of a vacuum has already been demonstrated in the speculative sciences." But unlike Crescas and like Gersonides he
concludes that beyond the world there is nothing but "privation"
(העדר, عدم, στέρησις)." Cf. M. J. Müller, *Philosophie und
Theologie von Averroes*, German text, p. 63; Arabic text, p. 66;
Mohammad Jamil-ur-Rehman, *The Philosophy and Theology of
Averroes*, pp. 176–177.

The difficulty raised here by Crescas is alluded to by Albo and
is answered by him. His answer is that while the expression
חוץ לעולם, *outside* or *beyond the world*, would ordinarily imply the
existence of something by which the world would have to be
bounded from without and that something would have to be
either a plenum or a vacuum, still the term חוץ may be used in
this connection in a figurative sense, in no way implying the existence of anything outside the world. '*Ikkarim* II, 18: כמו שיאמר
שאין חוץ לעולם לא ריקות ולא מלוי, ואם יש שם חוץ בהכרח יש שם ריקות או
מלוי, אלא שמלת חוץ נאמר בהעברה ובהקל מן הלשון. In making that
distinction in the use of the term חוץ, Albo must have drawn
upon Maimonides who, in describing God as an incorporeal
agent, says that in that case "it cannot be said that the agent is
outside the sphere; it can only be described as *separate* from it;
because an incorporeal object can only be said metaphorically to
reside outside a certain corporeal object." *Moreh* II, 1, First Proof:
ואם היה חוץ ממנו לא ימלט מהיותו גשם או שיהיה בלתי גשם, ולא יאמר בו או
שהוא חוץ (כארנא) ממנו, אבל יאמר בו נבדל (מפארקא) ממנו, כי מה שאינו
גשם לא יאמר שהוא חוץ לגשם אלא בהרחבה במאמר.

Pico Della Mirandola restates this argument as follows: "Imo
accersiri vacuum ab eis vel nolentibus, quibus asseritur non
inveniri corpus infinitum. Nam si nullum et extra mundum
corpus, nec plenum ibi esse convincitur, vacuum potius et seiuncta
dimensio" (*Examen Doctrinae Vanitatis Gentium* VI, 6).

Similarly Bruno argues that according to Aristotle himself the
nothingness outside the finite world must be a vacuum and that
the vacuum, since it cannot be limited by a body, must be infinite.
Cf. *De l'Infinito Universo et Mondi* I, p. 310, l. 7 ff.

37. Crescas draws here a distinction between the infinite in the sense of being incapable of measurement and the infinite in the sense of having no limits, and points to the possibility of an infinite in the sense of immeasurable which may not be without limits. Such, for instance, are the lines in Altabrizi's proof, which are infinite on one side but finite on the other. When two such immeasurable but limited infinites are given, then while indeed one of them cannot be conceived as greater than the other in the sense that the total number of its parts can be expressed by a number which is greater, still it can be conceived as greater than the other in the sense that it can extend beyond the other on the limited side. The reason why one immeasurable infinite cannot be greater than another, suggests Crescas, is that their parts cannot be expressed by any number and therefore the terms great and small are inapplicable to them. As he says elsewhere (*Or Adonai* III, i, 4): "But when the time or the number of rotations is infinite, neither of these can be described by the terms much and few, great and small, equal and unequal, for all these terms are determinations of measure, and measurability does not apply to an infinite."

אבל כשיהיה הזמן או המספר בב״ת, לא יאמרו בו רב ומעט וגדול וקטן ושוה ולא שוה, למה שהם גבולי השעור, והשעור נמנע בבלתי תכלית.

As for the use made by Spinoza of Crescas' discussion of this argument, see my paper "Spinoza on the Infinity of Corporeal Substance," *Chronicon Spinozanum* IV (1924–26), pp. 99–101.

In the last statement of this passage, I have followed the reading in MSS. מ, ל, ו, ר, ד, ק, ב, א, ג. In the editions and MS. ז, the reading is: ואם היה נוסף מהאחר היה מהצד שהוא בעל תכלית. "Thus indeed the former line is not greater than the latter, and if it extends beyond the latter, it is on the side which is finite."

38. If time be eternal, the following objection might be raised. Divide eternal, infinite time, at any point at the present, into past and future. Past and future time will each be infinite and so will the whole time be infinite. But the whole is greater than the part. Thus, one infinite will be greater than another.

The answer, as suggested here by Crescas, is as follows: The whole time is said to be greater than past or future time only in so far as the latter are each bounded at the dividing point. In so far, however, as they are all infinite in the sense of being immeasurable the whole time cannot be said to be greater than the past or future time.

Both the objection and an answer are given by Gersonides in *Milḥamot Adonai* VI, i, 27, p. 406.

39. According to Crescas' view, the belief in creation does not necessarily imply a belief in the future destruction of the world. The world, according to him, must have had a beginning in the past, but may be endless in the future (*Or Adonai* III, i, 5, cf. *Moreh* II, 27). This view, however, exposes itself to the same criticism that has been raised against eternity, namely, that one infinite will be greater than another. For, before creation there had been an infinite time of non-existence. After creation there will be an infinite time of existence. The sum of these two kinds of time will make infinite time, and thus one infinite will be greater than another. The answer, of course, is the same as given before in the case of eternity.

Both the objection and a similar answer are given by Gersonides in *Milḥamot Adonai* VI, i, 27, pp. 405-6. The objection is reproduced by Crescas in *Or Adonai* III, i, 1, p. 62b, lines 7-10, and the answer in III, i, 3, p. 66a, lines 15-20.

40. This objection has been anticipated by Narboni in his supercommentary on the *Intermediate Physics* III, iii, 4, 2: "Two objections may be raised here: First, against Aristotle's statement that there can be no infinite surface, we may argue that he who maintains the existence of an infinite body also believes in the existence of an infinite immaterial surface." בכאן יש שני קושיות: הא', שהוא אומר שלא ימצא שטח בב"ת, נאמר שלדעת שאומר שימצא גשם בב"ת שהוא סובר שימצא שטח בב"ת נבדל.

Likewise Gersonides in his supercommentary on the *Intermediate Physics*, *loc. cit.*, has a remark to the same effect: "The proposition that every body must be bounded by a surface or surfaces, is based upon the analogy of bodies which are perceived by our senses." ואולם מה שכל גשם יקיפו שטח או שטחים, הוא הקדמה לקוחה מהגשמים המוחשים אשר אצלנו.

Isaac ben Shem-ṭob refutes Crescas' objection in his *second* supercommentary on the *Intermediate Physics*, *loc. cit.*: "By a

proper understanding of the minor premise of this syllogism one
may solve the difficulty raised by Ibn Ḥasdai, viz., the opponent
may dispute the truth of the proposition laid down by Aristotle
here that every body is surrounded by a surface or surfaces, for,
believing as he does in the existence of an infinite body, he does
not admit that every body is surrounded by a surface or surfaces.
But the answer to this is as follows: We have already shown that
every body must be predicated as being either circular or not-
circular, inasmuch as these two predications, circularity and non-
circularity, are contradictory to each other after the manner of
the contradiction between a positive and a negative predication,
and in such cases, when the subject ordinarily may be either one
or the other of the predications, it must necessarily be either one
or the other. Consequently, since the mathematician has defined
a circular body as something which is surrounded by one surface
and a non-circular body as something which is surrounded by
many surfaces, the aforesaid difficulty disappears."

הנה בבאור ההקדמה הקטנה מזה ההקש יותר הספק שעשה ן' חסדאי, והוא שמה
שאמר אריסטוטוליס הנה בזאת ההקדמה שכל גשם מקיף בו שטח או שטחים, שיהיה
חולק על זה בעל הריב, מה שהוא אמר בגשם הבעל בלתי תכלית, לא יודה שכל
גשם מקיף בו שטח או שטחים. וזה כי כמו שאמרנו הנה שהוא מחוייב שתצדק כל
גשם שהוא עגול או בלתי עגול, אחר שהם חולקות חלוקת הקנין וההעדר, והקנין
וההעדר מחוייב שיצדק אחד מהם על הנושא שמדרכו שימצא בהם בעת שימצאו
אל זה שהלמודי גדר הגשם העגול בשהוא הדבר אשר יקיף בו שטח אחד, ושהגשם
הבלתי עגול הוא אשר יקיפו בו שטחים רבים, לא ישאר ספק כלל.

See also his *first* supercommentary on the *Intermediate Physics*,
loc. cit.: "Some one has raised an objection, arguing that this syllogism is a begging of the question, for he who admits the existence
of an infinite body claims also that there exists a body which has
no surface; and so, how could Aristotle refute the opinion of his
opponent with a premise which the latter does not admit? Our
answer to this objection is that this premise is self-evident and
the opponent could not help but admit it."

כל גשם הנה יקיף בו שטח אחד אם היה סבובי וכו'. יש
מי שהקשה ואמר שזה ההקש הוא דרוש על המערכה, שהאומר גשם בב"ת כותו
לאמר שיש גשם בלא שטח, וא"כ איך סתר דעתו עם ההקדמה שהוא מכחיש. ונוכל
להשיב שזאת ההקדמה הכרח היה להם שיקבלוהו והוא מבוארת בעצמה, לפי וכו'.

41. Cf. below Proposition II.

42. Hebrew מהקדמות בלתי מודות. One would naturally take מודות as the active participle מודות. But the expression "admissive premises" is as awkward in Hebrew as in English. While the passive participle מודות does not occur in Hebrew, as far as we know, still by taking it here as a passive participle, we get the right expression "inadmissible premises." The term מודות occurs in a Hebrew version of Algazali's *Maḳaṣid al-Falasifah* as the translation of the Arabic مسلمة and تسلم both of which, to judge from the context, are to be vocalized as the passive مُسَلَّمَة and تُسَلَّم. In two other versions the same Arabic terms are translated by the passives מקובלות and משלמות. Cf. *Maḳaṣid al-Falasifah* I, p. 68: (الرابع المبادى) ونعنى بها المقدمات المسلمة

فى ذلك العلم.......واما ان لا تكون اولية ولكن تسلم من المتعلم.
Anonymous translation, MS. Jewish Theological Seminary, Adler 398: [read: והחכם] ואם שלא יהיו ראשונות ואם תהיינה מודות מן החכמה.
Anonymous translation, MS. *ibid.* Adler 978: ואם שלא יהיו ראשונות.
Isaac Albalag's translation, MS. *ibid.* ואבל תהיינה מקובלות מחכם. See use of או יהיו בלתי ראשונות אלא שהן מושלמות מן הלומד :Adler 131 מושלם in quotation from Isaac ben Shem-ṭob's *second* supercommentary on the *Intermediate Physics* above n. 1, p. 395.

43. Cf. *Physics* I, 7.

44. This criticism has been anticipated by Narboni in his supercommentary on the *Intermediate Physics*, I, ii, 2, 2: "Shouldst thou say that our contention that principles must be known is true indeed according to him who maintains that the principles are finite, but according to him who believes that the principles are infinite, they need not necessarily be known; quite the contrary, they cannot be known, inasmuch as the infinite is not comprehended by knowledge—the answer is as follows: Aristotle's statement that the principles must be known, is based upon his belief that in order to know a thing perfectly it is necessary to know it according to its causes and principles, as we have stated at the beginning of this work."

וא"ת מה שאמרנו שההתחלות יחוייב בהכרח שתהיינה ידועות הי' הכרחית מי שאמר בהתחלות הם ב"ת, אבל מי שאמר שההתחלות בב"ת לא יחוייב שתהיינה ידועות, אבל יחוייב שלא תהיינה ידועות, כי מה שהוא בב"ת לא תקיף בו ידיעה.

וי"ל שמה שאמר שההתחלות יחוייב שתהיינה ידועות לפי שהדבר לשיודע בשלמות
ראוי שיודע לסבותיו והתחלותיו כמו שאמרנו בתחלת הספר.

The same question has also been raised and answered in an
anonymous supercommentary on the *Intermediate Physics* I, ii, 2,
2, fol. 99v (MS. Adler 1744): " 'But the principles must be
known.' Who has told you that the principles of being must be
known? We answer that the reason underlying this statement is
the view that nature does nothing in vain, for inasmuch as nature
has implanted in us a desire to comprehend all things and these
things cannot be comprehended by us except through their causes
and principles, it follows that the principles must be known."

אבל ההתחלות יחוייב וכ"ז. מי הגיד לך כי ההתחלות ההויה ידועות. נשיב כי
הסבה היא זאת, כדי שהטבע לא יעשה דבר לבטלה, כי הוא נתן בנו חשק להשיג
כל הדברים, והדברים לא נוכל להשיגם כי אם בסבותם והתחלותם, אם כן ההתחלות
יחוייב שתהיינה ידועות.

Shem-ṭob Ibn Shem-ṭob, in his supercommentary on the *Intermediate Physics*, *loc. cit.*, answers Crescas as follows: "It is for
this reason that Rabbi Ibn Ḥasdai raised here an objection, arguing that it is a begging of the question, for he who believes that
the principles are infinite claims that the principles are unknown.
Either one of two answers may be given. First, Aristotle is
addressing himself here to a man of good sense. Now, it has already been demonstrated in Book VI of this work that when we
are deprived of the knowledge of something, we have a longing
for it, and no sooner do we come into the possession of that knowledge than the longing disappears. Hence we do know that we
have a knowledge of the principles, inasmuch as that knowledge
causes our longing for it to disappear. [Second], or we may answer
it in this way, which indeed is something very subtle. Aristotle
will first force the ancients to admit that they possess a knowledge
of things, and then he will use their admission as an argument in
their own confutation. For they claim that, because the existent
objects are infinite, the principles must be infinite. Thus we
do know that the principles are infinite, and this, perforce, constitutes a kind of knowledge. But, then, if, as they claim, the
principles are infinite, they could not have that knowledge."

ולזה הקשה הרב ן' חסדאי ואמר שהוא מערכה על הדרוש, חה שמי שאומר
שההתחלות בב"ת, אומר שההתחלות אין ידועות. ולזה יצא לו אחד מב' תשובות:

הא', שאריסטו ידבר עם בעל שכל, ובאר בו' מזה הספר במופת, שאנחנו כאשר
נעלמה ממנו ידיעת דבר מה אמנם נשתוקק אליו, וכאשר השגנו הידיעה בו סרה
התשוקה, ואמנם ידענו שידענו באשר לא נשארה לנו אחר אותה הידיעה תשוקה כלל.
או נוכל לאמר, והוא דבר דק מאד, והוא בהכרח יביא לאמר לקדמונים שיודעים
הדברים, וא"כ יביא מאמרו בהפלתה, שהם אומרים שההתחלות בב"ת, לפי שהדברים
הנמצאים בב"ת, א"כ כבר ידענו שהם בב"ת, ואת היא הידיעה, ואם ההתחלות
הם בב"ת, א"א שידעו זאת הידיעה.

A veiled refutation of Crescas' criticism is also found in Isaac
ben Shem-ṭob's *second* supercommentary on the *Intermediate
Physics*, *loc. cit.*: "He who is inclined to be skeptical may raise
here a doubt and contend against the first argument, wherein
Aristotle states that principles must be known, that it is a begging of the question, inasmuch as the opponent disputes its truth,
for he who maintains that the principles are infinite claims that
they cannot be known."

יש למספק שיספק ויאמר כנגד הטענה הראשונה שמה שאמר אריסטוטליס בה,
אבל ההתחלות חויב שתהיינה ידועות, שהוא מערכה על הדרוש, מחולק ושחולק:
[Cambridge MS.: עליו בעל הריב, חה שהאומר שההתחלות הן בלתי בעל
תכלית, יאמר שהוא בלתי אפשר שתהיינה ידועות.

Two indirect answers to this criticism, one like the answer given
by Shem-ṭob Ibn Shem-ṭob, are found in Isaac ben Shem-ṭob's
first supercommentary on the *Intermediate Physics*, *loc. cit.*: "The
principles must be known, that is to say, inasmuch as the knowledge of anything becomes complete by a comprehension of its
causes and principles, and, furthermore, inasmuch as many of
the existent things are known to us, consequently we are bound
to admit that we have a knowledge of their principles. Or we may
say that any agent who performs a certain thing must have a
knowledge of all the principles out of which he has produced the
thing . . . . . . Gersonides, however, explains it in another way."

יחויב שתהיינה ידועות. ר"ל, שהידיעה בכל דבר נשלמת בידיעת סבותיו
והתחלותיו, וא"כ אחר שהרבה מן הנמצאות ידועות אצלנו יחויב שנדע ההתחלות.
או נאמר שהטעם בזה הוא לפי שכל פועל שיפעל דבר יחויב שידע כל ההתחלות
שמהם יעשה אותו הדבר. . . . . והר"ל [=הרלב"ג] הביא בזה טעם אחר.

45. This is an argument against the rejection of an infinite neutral
element. See above p. 348, n. 61. The reason given by Averroes is
that an element in so far as it is an element must possess qualities

different from those of other elements. Crescas' contention is that the unqualified and formless infinite element would be the substratum of the four elements into which they would never have to be resolved.

46. Cf. *De Caelo* I, 3.

47. I. e., the argument that sublunar substances would be destroyed by the infinite, does not obtain if an infinite existed outside the world of the four elements, which is the view held by the Pythagoreans. See above n. 7.

48. This question is discussed by Narboni in his supercommentary on the *Intermediate Physics* III, iii, 4, 2: 'We may object to this by arguing that the assumption of an infinite body does not necessarily require that the infinite should occupy all the room in all the three directions, for by assuming the infinite element to be a magnitude infinite only in length but not in breadth there will be room for the other elements, even if we say that such an infinite magnitude exists. To this we answer that such an assumption is untenable. For we observe that when a body increases by natural growth it increases in all its directions. By the same token, if we assume an infinite magnitude, it will have to be infinite in all its directions. Hence there will be no room for any other element."

וא"כ לא יהיה מקום לנשארים כלל וכו': ונאמר שהנחת גשם בב"ת לא יתחייב שיהיה ממלא כל הפאות השלשה, לפי שאם הנחנו שהוא גודל בב"ת, והוא אורך בלי רוחב, וא"כ אעפ"י שנאמר שימצא גודל בב"ת, כבר יהיה מקום לנשארים וכו'. נשיב להם, שזה בלתי אפשר, לפי שאנו נראה שהגשם אשר יצמח, יצמח בכל קוטריו, וא"כ כשנאמר גודל בב"ת מחוייב הוא שיהיה בב"ת בכל קוטריו. התחייב א"כ שלא יהיה מקום לשאר.

Cf. Averroes, *Epitome of the Physics* III, p. 10b: "That the infinite must be assumed to be infinite in all its directions is made clear by him by the following argument: Inasmuch as a body is that which extends in all the three dimensions, it must necessarily follow that if anything is assumed to be infinite *qua* body that it must be infinite in all its directions. For if one of its dimensions were supposed to be finite, then infinity will be only an accident of that body and not essentially necessary, for the same reasoning that makes it possible for that one dimension *qua* dimension to be either finite or infinite must equally apply to all the other dimensions. Hence the infinite must necessarily be infinite in all directions."

ואולם שהוא מחוייב שיונח בלתי בעל תכלית בכל קטריו, הנה הוא מבואר ממה שאמר. בעבור שהיה הגשם הוא הנמשך בכל המרחקים השלשה, יתחייב בהכרח, אם הונח שאין לו תכלית במה שהוא גשם, שיהיה בלתי בעל תכלית בכל קטריו, כי כאשר הונח שיש לו תכלית באחד מהם, היה העדר התכלית לו במקרה ובלתי הכרחי, כי הדין על מרחק אחד, מצד מה שהוא מרחק בתכלית או לא תכלית, דין על כל המרחקים. ולזה יתחייב בהכרח שימצא בלתי בעל תכלית מכל קטריו.

Gersonides paraphrases Averroes' passage in his commentary on the *Epitome of the Physics, loc. cit.*, as follows: "That a body assumed to be infinite must be infinite in all its three dimensions may be shown in this way. If a body is assumed to be infinite *qua* its being a body and it is a body *qua* its three dimensions, it follows that it must be infinite in every one of its dimensions. For if one of its dimensions were assumed to be finite, then infinity would be only an accident of the body and not essentially neces- sary, since to assume the contrary, i. e., that infinity were essen- tially necessary, would imply that the body is infinite *qua* its being a body, and hence it would necessarily have to be infinite in all its dimensions. Furthermore, the very same nature of the body which makes it necessary for it to be infinite in one of its dimensions will also make it necessary for it to be infinite in its other dimensions, for the same reasoning must hold true for all the dimensions. Conversely, the very same nature of the body which makes it necessary for it to be finite in one of its dimen- sions will also make it necessary for it to be finite in the other dimensions."

ואולם שהוא מחוייב, אם הונח הגשם בב״ת, שיהיה בב״ת בכל הרחקים השלשה, דבר זה מבואר מזה הצד. זה שאם היה הגשם בב״ת במה שהוא גשם, והוא גשם במה שהוא שלשה רחקים, הוא מבואר שהוא מחוייב שיונח בב״ת בכל אחד מהרחקים, וזה שאם הונח ב״ת באחד מן הרחקים, היה העדר התכלית לו במקרה ובלתי הכרחי, שאם היה העדר התכלית לו הכרחי, היה מחוייב שיהיה בב״ת במה שהוא גשם. ולזה יחוייב שיהיה בב״ת בכל רחקיו. ועוד כי הטבע אשר יחייב לו העדר התכלית באחד הרחקים הוא יחייב לו העדר התכלית ברחקים הנשארים, כי המשפט אחד. והפך זה נ״כ, רצוני, שהטבע שיחייב לו התכלית באחד מהם יחייב לו התכלית בנשארים.

Cf. also Isaac ben Shem-ṭob's *first* supercommentary on the *Intermediate Physics* III, iii, 4, 2: "An objection may be raised that his statement that an infinite body must be infinite in all its directions is not true of a natural body *qua* its being natural which is here the subject of our investigation, for in the case of a natural body *qua* its being natural one body may differ from another and in the same body one dimension may differ from another, and this indeed must be due to its being a natural body and not simply a body—for if the equality of dimensions were true also of a natural body, then all bodies would be equal in their dimensions and all those dimensions would be of equal size. In the same way we may argue here that this body under consideration *qua* its being natural will have its length infinite while its breadth may still be finite. To this we answer that even though what has been said is true and that in natural bodies *qua* their being natural the dimensions may differ from each other, that difference will be only relative, that is to say, even though in natural bodies *qua* their being natural one body may differ from another, still any given difference between them must be relative to the other difference between them."

בב״ת בכל מרחקיו; ויש להקשות ולאמר שזה לא יתאמת בגשם במה שהוא גשם טבעי, כמו שעיוננו הוא בכאן, ר״ל, שהוא בגשם טבעי, לא יתאמת זה, לפי שהגשם במה שהוא טבעי יתחלף גשם אחד לגשם האחר, בגשם האחד בעצמו יתחלף המרחק האחד למרחק האחר, חה יהיה מצד שהוא גשם טבעי, לא מצד שהוא גשם, שא״כ כל הגשמים יהיו שוים במרחקיהם. ונ״כ כבאן נאמר שמצד שזה הגשם גשם טבעי יתחייב שיהיה ארכו בב״ת ורחבו ב״ת. לזה נשיב ונאמר שאעפ״י שהאמת הוא כמו שאמרו ושהלוף מרחק הגשמים זה מזה הוא מצד שהם גשמים טבעיים, זה יהיה חלוף יחסיי, ר״ל, שאעפ״י שמצד הטבע ימשך חלוף גשם אחד מן האחר, עם כל זה אותו החלוף יהיה לו יחס עם החלוף האחר".

49. Cf. *De Caelo* I, 3.

Similarly Bruno argues against Aristotle that the infinite would have neither weight nor lightness. Cf. *De l'Infinito Universo et Mondi* II, p. 328, 1. 24; also p. 335, 1. 12; *De Immenso et Innumerabilibus* II, iv.

50. The printed editions as well as all the MSS. read here שמקומו הוא שטח קעריריתו, *its place is the surface of its concavity*. But this is impossible, for it does not agree with any of the views on this question reproduced below in n. 54. I have therefore ventured to emend the text by introducing the word מצד. It will be noted that מצד שטח קעךירותו is fittingly counterbalanced by ומצד גבינותו.

51. Hebrew קעךירותו. Above (p. 188, l. 6) Crescas uses the adjective קעךורי. We should therefore expect here the form קעךוךיותו. But קעךירות is used by him later (p. 196, l. 9) and the same form also occurs in *Emunah Ramah* I, vi, p. 28.

52. As for the special meaning of the term "centre" מרכז used in this connection, see below n. 70.

53. Hebrew גבינותו. By analogy of the Biblical גבן and the Post-Biblical גבון, we should expect here גבנינותו. But the MSS. read here גבינותו with which גבינתו in the Ferrara edition is practically in agreement. Similarly later (p. 196, l. 2) the form גבינות is used. Some MSS. read there נבינות.

54. The implication of this statement that according to Aristotle there is a difference between the outermost sphere and the other spheres as to their places needs some qualification, for it touches upon a controversial point. Aristotle himself has only the following general statements on the subject. "And some things indeed are in place essentially; as, for instance, every body which is moveable, either according to lation, or according to increase, is essentially somewhere. But heaven (οὐρανός) is not, as we have said, anywhere totally, nor in one certain place, since no body comprehends it; but so far as it is moved, so far its parts (μορίοις) are in place; for one part adheres to another. But other things are in place accidentally; as for instance, soul and the heaven (οὐρανός); for all the parts are in a certain respect in place; since in a circle one part comprehends another" (*Physics* IV, 5, 212b, 7–13). Aristotle's commentators are divided in their opinion as to the meaning of this passage. The cause of their disagreement seems to lie in the vagueness of the term οὐρανός which might refer (a) to the universe (τὸ πᾶν) as a whole, mentioned previously by Aristotle, or (b) to the outermost sphere, the parts thereof thus meaning the inner spheres, or (c) to all the spheres individually. The discussion is reproduced in the texts accompanying this note. It will be noted that it is only one interpretation, that of Themistius, which makes the distinction, implied here in Crescas' statement, between the outermost sphere and the inner spheres. According to Alexander Aphrodisiensis the outermost sphere, which he believes to be immovable, is not in place at all. According to Avempace and Averroes, all the spheres without distinction have the "centre" as their place, though the former calls it essential place and the latter calls it accidental place.

The following texts are illustrative of this note as well as of the succeeding notes.

Averroes, *Intermediate Physics* IV, i, 1, 9, in which only his own view and that of Avempace are given:

"As for the univocal applicability of this definition of place to all bodies that have locomotion, it is something which is not so clear. For if place is the limit of the surrounding body, then every body which has some other body external to itself is, as Aristotle maintains, in place. But as it is only the rectilinearly moving sublunar elements that require the existence of something external to themselves, would that I knew what is the place of those bodies which have by nature circular motion, [and hence do not require the existence of something external to themselves], as, e. g., the celestial bodies?

Aristotle, however, solves this difficulty by saying that a body which is endowed with circular motion, as, e. g., the celestial bodies, is moved only with reference to its parts, in consequence of which it is not necessary to look for a place for the whole of it but only for its parts. This is a rather plausible explanation. Still the following inquiry is rather pertinent: Those parts which are considered to be moved essentially in the circularly moving celestial spheres must inevitably have as their place either the convexity of a spherical body about which the sphere of which they are parts revolves or the concavity of a spherical body which encloses the sphere of which they are parts from without. If we assume that the place of the parts of the celestial sphere is the concavity of another surrounding sphere, then it will follow that every such sphere will have to be surrounded by another sphere, and this will go on *ad infinitum*. It is therefore necessary to assume one of the following alternatives, namely, either we must say that not every body that has locomotion is in place or we must say that the place of the circularly moving celestial spheres is the convexity of their respective internal spheres about which they revolve. But the first alternative must certainly be dismissed as false. Hence the second alternative must be accepted.

Evidence for this . . . (Rest of paragraph is quoted below in n. 70).

Hence it is generally true that place is the limit of that which surrounds, but in the case of the rectilinearly moving sublunar elements the surrounding body is from without and in the case of the circularly moving celestial spheres the surrounding body is from within.

That the centre must be something separate . . . (Rest of paragraph is quoted below in n. 70).

It cannot be contended . . . (Rest of paragraph quoted below in n. 72).

But the universe as a whole is not in place except in so far as its parts are in place. This is what Aristotle has meant by saying that it is in place accidentally. For a thing is said to be in place *potentially* or *actually*, *essentially* or *accidentally*. Now, the universe is not in place *actually*, inasmuch as there is nothing which surrounds it from without. Nor is it in place *potentially*, inasmuch as there is no possibility that such a body surrounding it from without will ever come into existence. Still less is it in place *essentially*. Hence it must be in place *accidentally*. But to say that something exists accidentally may mean two things: First, with reference to some accidental property, as when we say, for instance, that the white man is a physician, if the physician happens to be white. Second, with reference to a part of the thing, as when we say, for instance, that the man sees, when as a matter of fact only a part of him sees, namely, his eye. It is evident, then, that the universe is not in place accidentally in the sense that it happens to be a quality of a thing which is in place essentially. Hence, we are bound to say that it is in place because its parts are in place. Aristotle, however, uses terms rather loosely, sometimes applying the term *accidental* in a general sense and sometimes in a specific sense.

What we have just stated with regard to the place of the circularly moving celestial spheres represents the view held by Avempace and before him by Alfarabi, namely, that they exist in place essentially, their place being their [so-called] centre (see below

n. 70). Accordingly, the term place is used in an analogical sense with reference to the celestial spheres and with reference to the sublunar elements endowed with rectilinear dimensions.

It seems, however, that it would be truer to say that the celestial spheres, whose place is the [so-called] centre which they enclose, are only accidentally in place, for that which is in place essentially must be surrounded by its place and not *vice versa* surrounding it. The *surrounding* limit corresponds to the *surrounded* limit. But it is only accidentally that a surrounding body is said to exist in that which is surrounded by it; so that when a certain body, as, e. g., the celestial spheres, does not exist in a body that surrounds it, it is not in place essentially; it is in place only by virtue of its existing in that which is surrounded by it, but that means being in place accidentally. This is the view of Aristotle. Avempace, however, does not see the homonymy between the place of the circularly moving celestial spheres and the corresponding place of the rectilinearly moving sublunar elements.

Inasmuch as a thing is said to be in place accidentally on account of its existing in something which is in place essentially, this must be the case of the celestial spheres in their relation to their [so-called] centre (see below n. 70), the [so-called] centre itself being in place essentially. This, according to my opinion, is the meaning of Aristotle's statement that the heaven is in place accidentally, that is to say, it exists in the elements which are in place essentially, for when a thing is said to be in place on account of its parts it is not the same as when a thing is said to be in place accidentally.

This interpretation agrees with what appears to be the opinion of the author as well as with the truth itself."

ואמנם הסכמת זה הגדר לכל הגשמים אשר יתנועעו תנועת ההעתק הוא ממה שיקשה. חה שאם היה המקום הוא תכלית הגשם המקיף, הנה כל גשם חוץ ממנו דבר, כמו שיאמר אריסטו, הוא במקום, והגשמים אשר חוץ מהם דבר הוא אשר תנועתם תנועה ישרה, ומי יתן ואדע מה מקום הגשמים המתנועעים בטבע בסבוב כמו גרמי השמים?

אבל אריסטו ישיב מזה, בשהמתנועע בסבוב, כמו גרמי השמים, אמנם יתנועעו בחלקיו, ולכן לא יתכן לדרוש לו מקום לכללותו אבל לחלקיו. חה יותר ראוי. אבל שאלו החלקים, אשר ימצאו מתנועעים בגשם הסבובי בעצם, לא ימנע שיהיה

המקום להם גבנונית גשם כדורי עליו יסבוב, או קבוב נשם אחר כדורי יחופף מחוץ.
ואם הנחנו מקום חלקי הכדור הוא קבוב כדור אחר, חייב שיהיה לכל כדור כדור,
וילך זה הענין אל בלתי תכלית. ולזה מה שיחוייב אל זאת ההנחה אחד משני ענינים:
אם שנאמר שאין כל נעתק במקום, ואם שנאמר מקום הכדור הוא גבנונית הגשם
אשר עליו יסבוב, והראשון כבר יחשב שהוא בטל, הנה השני מחוייב.

וכבר יעיד לזה ...

הנה המקום בכלל הוא תכלית המקיף אם לגשמים הישרים מחוץ ואם לסבובים
בפנים.

ואמ׳ שהמרכז יחוייב ...

ואין לאומר שיאמר ...

ואמנם העולם בכללו אינו במקום, אם לא בשחלקיו במקום, וזהו אשר רצה
ארי׳סטו באמרו שהוא במקום במקרה. וזה שהדבר יאמר שהוא במקום אם בכח
ואם בפועל, אם בעצם ואם במקרה. והעולם אינו במקום בפועל, לפי שאין חוץ
ממנו דבר. ואיננו במקום בכח, לפי שאי״א שימצא בעתיד חוץ ממנו נשם. ואיננו
נ״כ במקום בעצם. הנה לא נשאר אלא שיהיה במקום במקרה. אלא שמה שבמקרה
שני מינים: אחד מהם מצד המשיג, כמו שנאמר, שהלבן רופא כאשר קרה לרופא
שיהיה לבן. והאחר מפני החלק, כמו שנאמר שהאדם רואה, והוא אמנם רואה
בחלק ממנו, והוא עינו. ומבואר שהעולם אינו במקום במקרה, מפני שהוא יקרה
לדבר והוא במקום בעצם. הנה לא נשאר שנאמר בו שהוא במקום אלא מפני
שחלקיו במקום. ואריסטו יקל בשמות, שפעם יעשה מה שבמקרה בכללות ופעם
ביחוד.

וזה אשר אמרנוהו במקום הכדור הוא אשר סבר אבובכר בן אלצ׳ג ואבונצר
לפניו, ר״ל שהוא במקום בעצם, ר״ל במרכזו. והמקום יאמר בספוק על מקום
הגשם הכדורי ועל מקום הגשם הישר המרחקים.

אבל ידמה שיהיה היותר אמתי שיאמר כי הכדור במרכזו אשר יקיף בו במקום
במקרה, מפני שאשר במקום בעצם הוא מוקף בו ולא מקיפו. והמקיף מקביל
למוקף בו, אבל יקרה למקיף שיאמר שהוא במוקף בו, כי כאשר היה נשם מה,
כמו השמים, איננו במקיף בו, הנה אינו במקום בעצם. ואמנם הוא במקום במוקף
בו, וזה במקרה. הנה א״כ האמת שהגשם השמימי אם נמצא במקום הוא במקרה.
וזהו דעת אריסטו. ואמנם אבובכר לא יפלינ השתוף אשר בין הגשם הסבובי והגשם
הישר שיהיה המקום באחד מהם מקביל באחר.

ולמה שהיה מה שיאמר בו שהוא במקום במקרה אמנם יאמר בו זה מפני שהוא
בדבר הוא במקום בעצם, חייב שיהיה זה ענין הכדור עם מרכזו אשר הוא במקום
בעצם. וזהו אצלי ענין מאמר אריסטו שהשמים במקום במקרה, ר״ל שהם ביסודות
אשר הם במקום בעצם, לפי שמה שיאמר עליו שהוא במקום בחלקיו בלתי מה
שיאמר עליו שהוא במקום במקרה.

וזה הפירוש מסכים למה שנראה מהאומר ולאמת בעצמו.

In his Long Commentary on the *Physics*, *loc. cit.*, in his exposition of the various interpretations of the Aristotelian passage, Averroes reproduces also the view of Themistius, which is of particular importance for us here, as we shall find allusions to it in Crescas. We quote parts of it here from the Latin translation: "Themistius vero dicit respondendo quod corpus coeleste non est in loco secundum totum sed secundum partes, scilicet secundum orbes, quos continet maximus orbis . . . . . . sed quia corpus altissimum, v. g., orbis stellarum fixarum, non continetur ab aliquo, concessit quod hoc corpus est in loco propter suas partes intrinsecas tantum, scilicet quae sunt in concavo eius." (p. 141rb-va.) Cf. Themistius in *Physica* (ed. Schenkl), p. 120.

"Et etiam secundum expositionem Themistii, cum Aristoteles dicit quod coelum est in loco per accidens, intendit quod alterum coelorum est in loco, s. orbium; et illud quod apud Aristotelem attribuitur alicui propter suam partem est aliud ab eo quod attribuitur alicui per accidens: et ideo omnibus expositoribus, ut dicit Themistius, displicet ut coelum sit in loco per accidens et dicunt ipsum esse in loco secundum partes." (p. 141vb.)

Narboni on the *Kawwanot ha-Pilosofim* III, Motion, probably based on Averroes' Long Commentary on the *Physics*, gives a complete account of all the views:

"Know that Averroes in the *Physics* has discussed five views with regard to relation of place to the heavens. We shall briefly restate their essential points.

First, the place of the outermost sphere is the potential vacuum [which exists outside the world]. This view is to be rejected with the rejection of a vacuum.

Second, the view of Alexander, according to which the outermost sphere has no motion and does not exist in place, for it does not change its place nor is it divisible, in consequence of which its parts cannot be described as having motion, and so it does not exist in place.

Third, the view of Themistius, according to which the outermost sphere has motion with reference to its parts but not with reference to its whole, that is to say, the celestial body as a whole [is in place] on account of the individual spheres, all of which are in place with the exception of the outermost sphere. As for the outermost sphere it is in place on account of its concave parts which are in place, for the convexity of the sphere which is within it, being enclosed by it, equal to it and separate from it, is in place essentially, and is the subject of the outermost sphere. Aristotle's statement that the heaven is in place accidentally is to be explained by the fact that that which is said to be in place on account of its parts is not in true place.

Fourth, the view of Avempace, namely, that the place of a sphere *qua* its being a sphere is the convexity of the object which occupies a place within it and about which it revolves, and that Aristotle's definition of place as a *surrounding, equal, separate limit* must be understood with reference to the rectilinearly moving sublunar elements to mean an *external* limit but with reference to the celestial sphere an *internal* limit. If some of the celestial spheres happen to be also [externally] surrounded [by other spheres], it is to be considered only as an accident. According to this view, the outermost sphere is moved essentially and is in place essentially.

The fifth view is that of Averroes, and it is composed of the views of Themistius and Avempace. From Avempace he borrows the view that the fact that most of the circularly moving celestial spheres happen to be [externally] surrounded by other spheres should be considered only as an accident. From Themistius he borrows the view with regard to the outermost sphere, namely, that the convexity of the [so-called] centre (cf. below n. 70) should be considered as the place only of the concave surface of the sphere which surrounds it, for it is only that concave surface which the centre equals and not the surrounding sphere in its entirety . . . . . .

Thus, according to Averroes' interpretation, the natural bodies are in the opinion of Aristotle of three kinds: First, those which exist in place *per se*, namely, the rectilinearly moving sublunar elements. Second, those which are in place *per accidens*, namely, circularly moving celestial spheres. Third, those which are in place on account of their parts, namely, the universe as a whole.

Themistius, however, considers the case of the [outermost] celestial sphere as similar to that of the universe as a whole."

ודע כי בן רשד באר בשמע חמש דעות בענין יחס המקום אל השמים, ונקצר
הנה עניינם ונאמר.

הראשון, שהמקום הקיצון הוא הפנוי בכח. והוא בטל בבטול הרקות,

השני, הוא דעת אלכסנדר, שהגרם הקיצון אינו מתנועע ואינו במקום, כי איננו
ממיר מקומו ולא יתחלק, ולזה לא יתוארו החלקים גם כן בתנועה, ולזה איננו במקום.

השלישי, הוא דעת תמסטיוס, שהוא מתנועע בחלקיו לא בכלל, ר"ל הגרם השמימי
בכלל]והוא במקום[במה שחלקיו במקום, מלבד הקיצון. ואם הקיצון מפני שחלקיו
הקבוביים במקום, כי גבנונית הגלגל אשר בתוכו מחופף בו ושוה ונבדל, והוא הנושא,
והוא במקום בעצם. ואריסטו אמר שהשמים במקום במקרה, ואין אשר יתואר מפני
חלקיו הוא במקום אמתי.

הרביעי, הוא דעת אבובכר, והוא שמקום הכדור במה הוא כדור, הוא גבנוני
המקומם בו, אשר עליו יסבוב, ושגדר אריסטו במקום בשהוא תכלית מקיף שוה
נבדל, ראוי שיובן בנשם הישר בשהוא מחוץ, ובכדורי מבפנים, ושאם היו קצת
הגרמים השמימיים מוקפים, זה מקרה קרה להם. הנה הגלגל הקצון מתנועע בעצם
ולו מקום בעצם.

והדעת החמישי, הוא דעת בן רשד, והוא מורכב מדעת תמסטיוס ואבו בכר.
כי הוא יקח מאבובכר שהמתנועע בסבוב מקרה הוא מקרה לו היותו מוקף, ויקח
מתמסטיוס מה שאמרו בקיצון, והוא שגבנוני המרכז אינו מקום רק לשטח הקבובי
מהמקיף עליו, כי הוא שוה לו לבד, לא לכלל המקיף. . . . .
הנה א"כ דעת בן רשד, שהגשמים הטבעיים אצל אריסטו שלשה מינים: מין במקום
בעצם והם הישרים, ומין במקום במקרה והם הסבוביים, ומין במקום מפני חלקיו,
וזהו כל העולם.

ותמסטיוס ישוה משפט הגרם השמיי לכלל העולם.

In the *Epitome of the Physics* IV, p. 16b, Averroes mentions
still another view, that of Avicenna: "Avicenna's statement with
reference to circular motion that it is not in place at all but only
in position is past my understanding. I surmise that he meant
thereby that circular motion is translation from one position to
another without changing places as a whole. If this is what he
meant, it is true enough. But if he meant to say that circular
motion is in position itself, that is to say, in the category of position, then it is not true, for position has no existence but in place.
Furthermore, we shall show that there can be no motion at all in
position."

ומאמר אבן סינא בתנועה הסבובית אשר היא אינה במקום כלל ואמנם היא במצב
הנה לא אבין אותו ואחשוב בו שירצה בזה שהיא תעתק ממצב אל מצב מבלתי
שיחליף המקום בכללה. ואם היה זה, הנה הוא אמת. ואם רצה לומר כי תנועתה
במצב נפשו, אשר הוא המאמר, הנה אינו אמת, כי אחד ממה שיתקיים בו המצב
הוא המקום, וגם כן הנה נבאר כי המצב אין בו תנועה כלל.

Cf. Proposition VI, p. 504, n. 6.

Gersonides' supercommentary on the *Intermediate Physics, loc. cit.:* "Says Levi: It seems that Aristotle's statement reads only that the sphere is in place accidentally. This term 'sphere' was taken by Avempace to refer to the universe as a whole, and the reason for his taking it in that sense is because he believes that [every individual] celestial sphere is in place essentially. Averroes, on the other hand, according to my understanding of his discussion before us, took the word 'sphere' in Aristotle to mean that [every individual] celestial sphere is in place accidentally. For were Aristotle's own statement explicit on this point, Avempace would not have understood from it that every [individual] celestial sphere is in place essentially."

אמר לוי, ידמה שמאמר אריסטו שהכדור במקום במקרה, והבין ממנו אבובכר שיהיה זה הכדור כולל העולם בכללו, ויהיה זה ממנו לפי שהוא יחשוב שהגרם השמימי במקום בעצם. ואולם ב״ר הבין, לפי מה שאחשוב מזה המאמר, שהגרם השמימי במקום במקרה, שאם היה מאמר אריסטו זה מבואר, לא היה מבין אבובכר ממאמר אריסטו שיהיה הכדור השמימי במקום בעצם.

Isaac ben Shem-ṭob's *first* supercommentary on the *Intermediate Physics, loc. cit.:* "Averroes says: 'The meaning of Aristotle's statement that the sphere is in place accidentally is as we shall set forth.' All the commentators, however, agree that Aristotle did not say explicitly that the universe as a whole is in place accidentally, for were it so there would have been no room for the disagreement between Avempace and Averroes, as will appear in this chapter. What seems to be the case is that Aristotle said that the sphere is in place accidentally, which term 'sphere' is taken by Avempace to mean the universe whereas according to Averroes it means the individual celestial spheres."

ואמר: שזהו אשר כוון אריסטו באמרו שהוא במקום במקרה כמו שנפרש אח״ז. אבל המפרשים הסכימו שאריסטו לא אמר בפירוש שהעולם בכללו הוא במקום במקרה, שא״כ לא היו חולקים בזה אבובכר ון' רשד, כמו שיראה בזה הפרק. אבל מה שיראה שאריסטו אמר שהכדור הוא במקום במקרה, ואבובכר אמר שרצונו לומר העולם, ון' רשד אמר שכוונתו לומר הגלגל.

The following statements seems to reflect the view of Alexander:

Joseph Albo in *'Iḳḳarim* II, 17: "For the uppermost sphere is the absolute above, and it has been shown that it is not in place, inasmuch as there is no other body outside of it to surround it . . . . . . but this is based upon the view of Aristotle, who says

that the universe as a whole is not in place, inasmuch as there is
nothing outside of it to surround it."

שהרי הגלגל העליון הוא המעלה במוחלט, ונתבאר שאינו במקום, כי אין חוצה
לו גשם אחר יקיף בו... אלא שזה הוא בני על דעת אריסטו האומר .... כי
כלל העולם אינו במקום לפי שאין חוצה לו דבר אחר יקיף בו.

Cuzari II, 6: "The uppermost sphere carries the whole and
has no place." ‏והגלגל העליון נושא הכל ואין מקום לו.

55. This, as may be recalled, is one of the tentative definitions
of place advanced by Aristotle. See above p. 155, n. 80. According to Crescas' interpretation, following that of Averroes,
this definition identifies place with the vacuum (חללות; see above
p. 357, n. 80). And so subsequently in the course of his discussion
Crescas keeps on referring to place under this definition as being
identical with the vacuum (הפנוי).

56. Refers to Aristotle's argument that if place were the interval
of the body, an object would have an infinite number of places,
and place would be movable and exist in other places. See above
p. 155.

57. That is to say, there is no reason to assume that the interval
of the body would have to move together with the body. If the
interval was place it would remain unmoved just as the place of
Aristotle's definition.

This argument has been refuted by Shem-ṭob Ibn Shem-ṭob in
his supercommentary on the *Intermediate Physics* IV, i, 8: "By
this we may answer the objection raised by Rabbi Ibn Ḥasdai
who argues as follows: What makes it impossible to argue that
just as you, who define place as the limit of the surrounding body,
say that when a body is withdrawn from its place that place is
left behind it intact while the body is translated to another place,
so also would say those who identify place with the dimensions
that when a body is withdrawn from its place those dimensions,
which constituted its former place, are left behind it, and the
object assumes new dimensions which become its new place. And
the same will happen to any of its parts. Furthermore we observe
that even when a body is removed from a vessel, the dimensions
between the extremities of the vessel are left behind. When the

expression *occupying a place*, however, is well understood, the difficulty disappears of itself. We may state the answer as follows: When a body, [e. g., water], is lodged in dimensions and fills them up, those dimensions must of necessity be occupied and absorbed by that body [of water] and by all the parts of the water in the vessel, for were it not so, would that I knew where they go! Similarly, the contention that the dimensions are observed to remain in the original place of the vessel after the vessel has been removed to another place, will be rejected by them as inconsistent with their view, for they will contend that the dimensions do not remain behind but must rather be removed with the vessel by which they have been occupied and absorbed."

ובזה נשוב על ספק הרב ן' חסדאי, אשר ספק על זה ואמר, ומה המונע שכמו
שאתם אומרים שהגשם כאשר היה מקומו תכלית הגשם המקיף, כי אתם אומרים
כי כאשר נעתק הגשם ממקומו הניח המקום ההוא קיים ונשמר שם, והוא נעתק אל
מקום אחר, כן יאמרו בעלי המרחקים, כי כאשר נעתק הגשם ממקומו הניח המרחקים
אשר הם מקומו בתהלה ולבש מרחקים אחרים, והיו לו מקום, וכן כל אחד מהחלקים.
ועוד כי אנו נראה כי אעפ"י שכבר נעתק הגשם נשארו [ועם] המרחקים בין קצות
הכלי, ואבל כאשר תובן זאת הטרדה בטל הספק מעקרו. זה לפי שכאשר נח
הגשם במרחקים ומלא אותם, חוייב בהכרח הגמור שיהיו המרחקים נטרדים ונבלעים
בדבר ובכל חלקי המים אשר בכלי, שאם לא כן, מי יתן ואדע איפא הם. ומה שאנו
רואים במקום הכלי שנשארו שם מרחקים אחר העתק הכלי, זה ג"כ מהבטול
לסברתם, שאיננו חוייב שישארו שם מרחקים, וחוייב נ"כ שנעתקו עמם הכלי בהכרח
אחרי שכבר נטרדו ונבלעו בו.

It has been forestalled by Gersonides in his supercommentary on the *Intermediate Physics, loc. cit.*: "This objection cannot be raised against our view, for we maintain that it is the vessel, i. e., the place of the water, that is translated and that the water is only accidentally translated with it. Essentially the water always remains at rest within the vessel, never leaving its place, which place, as defined, is the limit of the body that surrounds it. The water and its parts thus never move essentially, for they are always in a place which is part of the place of the occupied vessel."

ואמנם אנחנו לא יתחייב לנו זה הספק, זה שאנחנו נאמר שהכלי, אשר הוא מקום
המים, הוא נעתק ונעתקו עמו המים במקרה, והמים נחים בעצם בכלי, אחר שלא
ימירו מקומם, והוא תכלית הגשם המקיף בם, לא שהמים וחלקיו יתנועעו בעצם,
לפי שהם במקום הוא חלק מהמקום הכלי המלא.

It has been adopted by Joseph Albo in 'Ikkarim II, 17: "This impossibility will indeed follow if the dimensions were capable of motion, but if we say that they are incapable of motion, and that it is only the body and its parts that are moved from one set of dimensions to another, this impossibility will not follow at all."

הנה מתחייב זה אם אם היו הרחקים מתנועעים, אבל אם נאמר שאינם מתנועעים,
ושהגשם והלקיו הם המתנועעים ממרחקים אל מרחקים, לא יתחייב מזה בטול כלל.

58. Similarly Bruno argues that Aristotle's definition of place does not apply to the place of the outermost sphere. Cf. *De l'Infinito Universo et Mondi* I, p. 309, 1. 16 ff.; *De Immenso et Innumerabilibus* I, vi, p. 221 ff.

59. Here again Crescas argues from Themistius' interpretation, according to which the places of the inner spheres are the concave surfaces of the spheres which respectively surround them, whereas the place of the outermost sphere is the "centre" round which it rotates. He therefore calls the places of the inner spheres essential whereas that of the outermost sphere accidental. No such distinction exists according to the other interpretations of Aristotle. See above n. 54.

60. In this argument Crescas will try to show that even the places of the sublunar elements cannot meet all the three conditions which are considered by Aristotle as essential of place, namely, *surrounding* (מקיף, περιέχων) the object, *equal* (שוה, ἴσος) to it, and *separate* (נבדל, χωριστός) from it. Cf. *Physics* IV, 4, 210b, 34 ff. and 211a, 24 ff.

61. Hebrew בעצם. The term בעצם is used here advisedly. For some parts are moved *essentially* with the whole, while others are moved only *accidentally*. The former is true of homogeneous bodies, the latter of heterogeneous bodies, as for instance, to use Aristotle's own illustration, the parts of the body and the nail in a ship. (Cf. *Physics* IV, 4). Speaking here of the simple elements, Crescas emphasizes the *essentiality* of the motion of its parts.

In order to understand the argument Crescas is about to advance, we must quote here the particular passage in Aristotle against which it seems to be directed. "And that which is continued is not indeed moved *in*, but together *with* it; but that which is divided is moved *with* it. And whether that which contains is moved, or whether it is not, it is not the less moved. Further still; when it is not divided, it is said to be as a part in the whole; as for instance, sight in the eye, or the hand in the body; but when it is divided, or touches, it is said to be as in place; as for instance, water in a wine vessel, or wine in an earthen vessel. For the hand is moved together with the body, and the water in the wine vessel" (*Physics*, IV, 4, 211a, 34–211b, 5).

The implication of this passage is that every part of air, for instance, by virtue of its being part of something continuous and homogeneous, is moved essentially *with* the whole and exists in the whole not as in place but as part in the whole. Crescas will hence investigate as to what is to be the place of that part.

62. Hebrew ערבות ודמיון. Cf. *De Caelo* IV, 3, 310b, 10–12: "It is to its like (ὅμοιον) that a body moves when it moves to its own place. For the successive members of the series are like one another; water, I mean, is like air and air like fire." Cf. also Averroes' *Epitome of the Physics* IV, p. 14a: "For place is that toward which the bodies move according to a desire, when they are out of it, and, having attained it, rest in it according to an agreeableness and likeness."

כי המקום הוא אשר יעתקו הגשמים אליו על צד התשוקה, כאשר היו חוץ ממנו, וינוחו בו, כאשר השיגוהו, על צד הערבות והדמיון. See below n. 69.

As for the meaning of ערבות throughout this passage, judged by its usage in the passage וכל שכן שיסוד האש ידרוש המעלה אשר מזה הצד יש לו ערבות ודמיון במקיף, it is to be taken in the sense of *agreeableness, fitness, suitability*, and seems to be used by Crescas as synonymous with האותות. Cf. above n. 8.

Were it not for that particular passage, one would be tempted to take it in the sense of *mixture*, i. e., the "mutual transformation" of the elements into each other. Cf. εἰς ἄλληλα μεταβολή in *De Generatione et Corruptione* II, 4, 331a, 11. It is in this sense that the term ערוב is used in the following passage of Averroes' *Epitome of the Meteorology* I (MS. Bibliothèque Nationale, Cod. Heb. 918, fol. 74r–v; Latin, fol. 404r–v): "It is also manifest in the *De Generatione et Corruptione* that the elements exist one within another according to *mixture* and proximity . . . . . . But

as for fire, it seems that in its own place it is simpler than all the
other elements, for the other elements have a certain weight in
their own place, as has been shown in De Caelo (cf. above n. 23),
and consequently are mixed with one another; but as they have
no lightness, their mixture with fire is difficult."

ונראה גם כן בספר ההויה וההפסד שהם [=היסודות] ימצאו קצתם בקצת על
צד הערוב ועל צד השכנות. . . ואולם האש, הנה ידמה שתהיה במקומה יותר פשוטה
מכלם, כי מה שזולתה מן היסודות להם כבדות מה במקומותם, כמו שהתבאר
בשמים והעולם, ולכן יתערב קצתם בקצת, ואין להם קלות, ויקשה התערבם באש.

63. That is to say, Aristotle's definition of place as something
surrounding the object, *separate* from it, and *equal* to it is inconsistent with his view that the elements have an affinity to their
proper places.

64. As to what are the proper places of the four elements, the
following statement is made by Algazali. "The place of fire is
the internal surface of the moon, the place of air is the internal
surface of fire, and the place of water is the internal surface of
air." *Kawwanot*, Physics, On Place (*Maḳaṣid* III, pp. 246–247):

ואולם האש מקומו מקיף הגלגל הירח מתוך; ומקום האויר השטח הפנימי מאאש;
ומקום המים השטח הפנימי מהאויר.

As for the place of earth, which Algazali does not mention,
there seems to be some confusion.

Aristotle himself speaks of earth as moving toward the centre
and of its resting there (*De Caelo* II, 13, 295b, 20 ff). But he does
not explicitly state what the place of the earth is. Simplicius
raises the question and argues that it cannot be the centre, inasmuch as it comprehends nothing. On the basis of a passage in
Physics IV, 4, 212a, 26–28, Simplicius concludes that the place
of earth is the boundary of the body which contains the earth,
which body partly consists of water and partly of earth. (Cf.
Simplicius in *Physica*, ed. Diels, p. 585, 1. 34 ff., and Taylor's
translation of the *Physics*, p. 204, n).

Averroes evidently follows this interpretation and makes the
explicit statement that the place of earth is the inner limit of
water. He goes even further to say that earth moves toward that
limit and rests in it. *Epitome of the Physics* IV, p. 15a-b: "In
accordance with what is established by evidence, we may assume

that the lower limits are the limit of water and the limit of air, for we observe that earth is at rest at the limit of water and moves toward water, and water similarly is at rest at the limit of air and moves toward air by nature. In like manner we may propose here that the upper limits are the limit of the celestial body and the limit of fire, the former being [the place] of fire and the latter [the place] of air, as has been shown from their nature in *De Caelo et Mundo*, so that fire moves toward the limit of heaven and rests there, and similarly water moves toward the limit of fire and rests there."

וניח לפי מה שהוא נודע בעדות, כי התכליות השפלות הם תכלית המים ותכלית האויר, כי נראה כי הארץ נחה בתכלית המים ומתנועעת אליהם בטבע, והמים גם כן נחים בתכלית האויר ומתנועעים אליו בטבע. וכן נציע בכאן כי התכליות העליונות הם [ותכלית] הגשם השמימיי ו[תכלית האש, אמנם תכלית הגשם השמימיי הוא לאש, ואמנם תכלית האש לאויר, כפי מה שהתבאר בספר השמים והעולם מענין אלו הדברים, ושהאש מתנועעת אל תכלית השמים ונחה בה, והאויר מתנועע אל תכלית האש ונח בה.

The same view is given by Albo in *Ikkarim* II, 17: "And if the place of the element earth is the surface of the element water which surrounds it from without . . . . . ." ואם מקום יסוד הארץ
הוא שטח יסוד המים המקיף בה מחוץ...

As against this, Joseph ibn Ẓaddiḳ takes the centre to be the place of earth. *'Olam Kaṭan* I, 3, p. 15: "Having observed and studied the nature of the elements, we find that the earth is in the centre of the universe . . . . . . We know therefore that its proper place (מקומה הידוע, cf. above p. 356, n. 76) is the centre, which is a point in the middle of a circle, and that it is therefore in the middle of the universe." ולפי שידענו וחקרנו על היסודות מצאנו
הארץ בטבור העולם... ידענו מזה שמקומה הידועה לה היא הטבור, והיא הנקודה שבאמצע העגול, ושהיא באמצע העולם. Cf. below n. 77.

65. Hebrew ההאותות. In the printed editions and most of the MSS. the reading here as well as later in the expression האותות is לא יתכן בו ההאותות אשר אמרו במקומם בכלל.

If the reading האותות, without the definite article, ה, is correct, then האותות here as well as in the later expression cited is not to be read הָאוֹתוֹת but rather הָאֳחוֹת, that is, אֳחוֹת with the definite article ה. The term אֳחוֹת will then refer to the *distinguishing* or *characteristic marks* of place from which Aristotle arrives at its definition (see above p. 153). The term אות, عَلامْ *sign*, *mark*, *earmark*, is used in this sense with reference to place in the following passage of the *Kawwanot ha-Pilosofim* III, On Place, (*Maḳaṣid al-Falasifah* III, p. 246): ואם נאמר ומה אמתת המקום, נאמר מה שישב עליו דעת אריסטו, והוא אשר ישוב אליו הכל, והיא שהוא מליצה משטח הגשם, רצוני, השטח הפנימי, אשר הוא מקום המוקף, לפי שהאותות (العلامات) הארבעה הנזכרות נמצאות בו, וכל מה שנמצא בו אותם האותות (العلامات) הנה הוא מקום.

66. The text here is uncertain.

MSS. ק, ד, ר, ל read: ואמנם החלק האמצעי מן האויר אם שאינו במקומו הטבעי אשר לו האותות אשר אמרו, ואם הוא...

MSS. ב, א, ג read: ואמנם החלק האמצעי מן האויר אם שאינו במקומו הטבעי אשר יש לו ההאותות שאמרו, ואם הוא...

MS. ו reads: ואמנם החלק האמצעי מן האויר לא נמלט אם שהוא במקומו הטבעי אשר לו האותות אשר אמרו, ואם הוא...

MS. צ reads: ואמנם החלק האמצעי מן האויר לא נמלט אם שהוא במקומו הטבעי למה אם שאינו במקומו הטבעי למה אשר לו האותות אשר אמרו, ואם הוא...

Printed editions and MSS. ז, מ read: ואמנם החלק האמצעי מן האויר לא נמלט אם שהוא במקומו הטבעי אם שאינו במקומו הטבעי אשר לו האותות אשר אמרו, ואם הוא...

I have adopted the last reading, with the exception of ההאותות, and understand the passage to argue as follows:

Take the element air, for instance. Its place as a whole is the concave surface of fire. This place indeed meets all the conditions. It is *surrounding, equal,* and *separate.* Furthermore, it is the proper and natural place of air, for there is a likeness between them. But then take any part of air from anywhere in the middle. That part of air will never move *in* the whole air; but will always move *with* it (see above n. 61). Consequently, that part of air will never reach the concave surface of fire; it will always be surrounded by air in which it will exist as part in the whole (see above n. 61).

Crescas now raises the following question: According to Aristotle's definition of place, where does the part of an element, say the part of fire, exist? Does it exist in a place which is natural to it or does it exist in an unnatural place and out of its own natural place? He seems to think that neither of these alternatives is possible. He does not tell us, however, why it cannot be assumed to exist out of its natural place. He tells us only that it cannot be assumed to exist in its natural place, and for this, too, he states the reason rather briefly, asserting only that, under this assumption, the place of the part will differ from the place of the whole, without telling us how they would differ. We must therefore try to reason the matter out for ourselves. The argument in full may be restated as follows:

A. The part of air cannot be assumed to exist outside of its natural place. For if it existed outside its natural place, it would move *in* the whole as in place and not *with* the whole as part of it, for when elements are out of their natural place they tend to move toward it. But according to Aristotle the elements are homogeneous substances and any part of the elements moves *with* the whole as part of the whole and not *in* the whole as an object in place (see above n. 61). Hence the part of air cannot be assumed to exist outside its natural place.

B. Nor can the part of air be assumed to exist in its natural place. For what would be its natural place? Two alternatives are possible. (1) The parts of air adjacent to it and surrounding it. (2) The concave surface of fire which is also the natural place of the whole air. But in case (1), the place of the part will be totally different from the place of the whole. Furthermore, the place will not be *separate* from the object of which it is place. In case (2), while indeed the place of the part will be identical with the place of the whole, the place will not be *equal* to the object of which it is place, and thus the place of the part will differ in definition from the place of the whole. Thus in either case, the place of the part will differ in some respect from the place of the whole.

This argument seems to be underlying the following passage in *'Iḳḳarim* II, 17: "This view is obviously false, for as a consequent of it he will be compelled to say that the place of the part and that of the whole are different. Take, for instance, the parts of fire. They are not surrounded from without by a limit but are rather surrounded by parts of fire and air, and as the natural place of the element fire is the concavity of the lunar sphere, the place of the whole of fire will thus be different from the place of the part of fire. The same reasoning may be applied also to the other elements. Furthermore, he will be compelled to say that the elements abide in their respective places by compulsion, for the natural place of the element fire is the concavity of the lunar sphere which is above, and thus all the parts of fire, except those in the proximity of the surface of the [lunar] sphere, will be in their place by compulsion. The same reasoning may be applied also to the other elements."

וזה הדעת מבואר ההפסד, כי יתחייב אליו לומר שמקום החלק והכל מתחלפים
כי חלקי האש אין להם תכלית מקיף מחוץ אלא חלקים אחרים אשיים או אוירים,
והמקום הטבעי ליסוד האש הוא מקוער גלגל הירח, והוא מתהלף למקום חלקי
האש, וכן בשאר היסודות, ועוד יתחייב לו לומר שהיסודות הם עומדים מוכרחים
במקומם, כי המקום הטבעי ליסוד האש הוא מקוער גלגל הירח שהוא למעלה,
ויהיו לפי זה כל חלקי האש עומדים מוכרחים זולתי העומדים אצל שטח הגלגל,
וכן יתחייב זה בשאר היסודות.

The argument is also reproduced by Pico Della Mirandola in *Examen Doctrinae Vanitatis Gentium* VI, 4: "Hebraeus quoque Hasdai asserit multa contra loci definitionem, inter quae illa, vitium non fuisse antiquis permultis, loci definitionem ab Aristotele traditam corporibus, quae motu recto perferuntur convenire: quoniam proprius partium locus, quae ad totius motum agitantur, non est superficies circundans aequalis adeo, ut seorsum habeat cum partibus loci convenientiam. Nam si (causa exempli) suprema pars aeris conveniet imae continentis et circum vallantis ignis, media tamen pars ei non ita conveniet, nec in suo naturali reponetur loco, qui si assereretur parti ipsi suapte natura congruere, tamen diversus habebitur a loco totius et integri corporis collocati."

67. Here Crescas has departed from Themistius and is arguing now from the points of view of Avempace and Averroes. According to both of these the places of all the spheres is the "centre" round which they rotate. But whereas Avempace calls it essential place, Averroes calls it accidental place. According to Themistius, the places of the inner spheres are the concave surfaces of the spheres which respectively surround them. See above n. 54.

68. An allusion to this argument is to be found in the following passage of Pico Della Mirandola; "Praeterea omnia quae collocantur corpora, suis congruere locis falsum esse aperiri, et ex supremi coeli circunferentia . . . " (*Examen Doctrinae Vanitatis Gentium VI, 4*).

69. According to Aristotle, the elements air and water are each similar to the elements which are both above them and below them. Fire, however, has no similarity to the element below it, and its motion, therefore, is absolutely upward. Cf. *De Caelo* IV, 3, 310b, 11–13: "For the successive members of the series are like one another: water, I mean, is like air and air like fire, and between intermediates, i. e., water and air, the relation may be converted, though not between them and the extremes, i. e., earth and fire."

Still, though fire is not like air, the transformation of fire into air is possible, according to Aristotle. Cf. *De Generatione et Corruptione* II, 4, 331a, 13 ff. Hence the following statement by Maimonides in *Mishneh Torah, Yesode ha-Torah* IV, 5: "Similarly in the case of fire, that part of it which borders upon air is transformed and condensed and becomes air." וכן האש מקצתה הסמוך לרוח משתנה ומתכנס ונעשה רוח.

Cf. also *Intermediate Physics* IV i, 1, 10: "It is further clear that by introducing this element into the definition of place he is enabled to explain why each of the natural bodies tends to its proper place and rests there, that is to say, why heavy bodies move downward and light bodies move upward. The reason for their moving toward the limits of each other is to be found in the likeness existing between them, that is to say, between the element that moves and the limit of the body in which it comes to rest, as, for instance, the likeness of the limit of the [lunar] sphere to fire, the likeness of the limit of fire to air, of the limit of air to water, and of the limit of water to earth. For in all these cases, the element surrounding is like a form and entelechy to the element surrounded, and the element surrounded is like matter. The discussion of this subject will be taken up in a whole book in *De Caelo et Mundo*."

ומבואר עוד, שמזה הצד אשר הושם לנדר המקום יוכל להביא הסבה אשר
בעבורה היה כל אחד מהגשמים הטבעיים יעתק אל מקומו המיוחד וינוח בו, ר"ל
הגשמים הכבדים המתנועעים למטה והקלים המתנועעים למעלה, ושהם אמנם יעתקו
קצתם אל תכלית קצת להדמות אשר ביניהם, ר"ל בין הנעתק ותכלית הגשם אשר
בו ינוח, כמו הדמות תכלית הגלגל לאש, והדמות תכלית האש לאויר, ותכלית
האויר למים, ותכלית המים לארץ. וזה שהמקיף בכל אלו במדרגת הצורה והשלמות

למוקף, והמוקף בו במדרגת ההיולי. ויתבאר זה בספר השמים והעולם במאמר שלם.

Cf. above n. 62.

70. The reference is to Aristotle's theory according to which the circular motion of a sphere implies the existence of another spherical body round which the circular motion of the former sphere is performed and it further implies that the other spherical body must be itself fixed and separate from the revolving sphere. It is by this theory that Aristotle proves that the earth must be spherical in form and at rest, existing in the middle of the universe (cf. *De Caelo* II, 3, 286a, 12–22, and II, 14). This separate, spherical and fixed body, round which the sphere moves, is called by Aristotle "centre" in a special sense, not to be confused with the term "centre" in the mathematical sense, which is only a point (cf. *De Motu Animalium* 1, 698a, 15–698b, 1).

Intermediate Physics IV, i, 1, 9: "Evidence for this may be found in the fact observed concerning the celestial sphere that by virtue of its sphericity it must have a figure and also a convex stationary body about which it is to revolve, that body being called centre. This is something which has been demonstrated by Aristotle in *De Caelo et Mundo*, namely, that the circular motion of the celestial sphere would be impossible without a stationary body about which the circular motion is to be performed, which body is called centre and constitutes the place of the circularly moving sphere, and because it constitutes a place of the sphere, it must be stationary, for it has been shown that the place of a thing must be essentially at rest. Furthermore, that centre must be something separate from the sphere, that is to say, it must not be a part of the sphere, and being thus separate it must be a body [i. e., it cannot be a mere point], for that which is indivisible [i. e., a point] cannot exist as something separate and by itself. Since every celestial sphere must have such a separate, stationary centre, which centre is its place, it follows that [the place of the spheres] is the convexity of that [so-called] centre which is the limit of that which surrounds the celestial spheres from within."

וכבר יעיד לזה מה שיראה מענין הכדור שהוא יצטרך בטבע במה שהוא כדור אל תמונה ואל גשם גבוני נח עליו יסבוב, והוא הנקרא מרכז. חה דבר באדו אריסטו בספר השמים והעולם, ר״ל שהתנועה הסבובית אי״א לה מבלתי גשם נח עליו יסבוב,

והוא המרכז, אשר הוא מקום המתנועע בזאת התנועה, ולזה היה נח, לפי שכבר
התבאר שהמקום ראוי שיהיה נח בעצם. ועוד כי המרכז יחוייב שיהיה נבדל לכדור,
ר"ל שיהיה אינו חלק ממנו, והנבדל גשם בהכרח, לפי שמה שלא יחלק לא יובדל
וכאשר היה כל כדור לו מרכז נח, וזאת היא סגולת המקום, הנה נבונית המרכז
הוא תכלית המקיף מבפנים בכדור.

Cf. *'Olam Ḳaṭan* I, 3, p. 11: "We say that the sphere has circular motion, and everything that is moved with such motion must
perform its motion round something stationary . . . . . . Furthermore, a circumference cannot be without a centre . . . . . .
Hence the moving circumference is the celestial sphere and the
stationary centre is the earth." הואיל והתבאר זה נאמר שהגלגל מתנועע
תנועת ההקפה, וכל מתנועע תנועה כזאת הוא מתנועע סביב לשוקט. . . . אם כן
יהיה המתנועע המקיף הוא הגלגל והנקודה השוקטת היא הארץ.

Cf. also *Moreh Nebukim* II, 24: "Again, according to what
Aristotle explains in natural science, there must be something
fixed round which the motion takes place; this is the reason why
the earth remains stationary." ועוד שהצעות אריסטו בחכמה הטבעית
שאי אפשר בהכרח מבלתי דבר קיים סביבו תהיה התנועה, ולזה התחייב שתהיה
הארץ קימת.

It is because the earth is the stationary and separate centre of
the spheres that Avempace and Averroes consider the surface of
the earth to be the place of those spheres. See above n. 54.

The special text against which Crescas' criticism here is directed
is the passage quoted below in this note.

In this passage Averroes tries to prove that the centre round
which a sphere rotates must be a stationary body. The language
of the passage is rather misleading, as Averroes uses there mathematical terms which, however, as has been pointed out by Gersonides, he could not have meant to be taken in their purely
mathematical sense. The argument may be restated as follows:

Let C. be a sphere rotating on C.
Draw a radius from C to A in the periphery.
Let CA revolve on C.
Any point taken in the radius CA will describe cir-
cles concentric with the periphery of the sphere.

The last point C in CA, therefore, will likewise describe a circle
concentric with the others.

That circle will have to be somewhere, that somewhere being
either a plenum or a vacuum.

But a vacuum does not exist.

Hence, it must be a plenum.

Now, that plenum must be at rest, for if it rotated the same reasoning might be repeated and the thing would thus go on *ad infinitum*.

Hence, C is a magnitude and at rest.

It is against this proof of Averroes that Crescas raises his objections. He argues thus: If the last material point on the bar at C must describe a circle on a stationary magnitude, then the radius CA at C must be implanted in a stationary body. But that is absurd.

Intermediate Physics IV, i, 1, 9: "That the centre must be something separate and stationary may be demonstrated as follows. If we draw a line from the centre to the periphery [of the sphere] and imagine that line to move on its centre until it returns to its original position, then every point assumed in that line will in the course of its motion describe an arc similar to that great arc described by the further end of the line upon the periphery of the sphere itself. This being so, then all the parts of the line must of necessity perform movements all of which are related to the movement of the whole line in exactly the same way, so that the point at the end of the line [at the centre] must inevitably describe a circle similar to the circles described by all the other points in the line. Now, that circle must inevitably exist either in a spherical body or in a vacuum. But the existence of a vacuum will be shown to be impossible. Hence it must exist in another spherical body. But that other spherical body, again, must either be at rest or move in a circle. In the latter case, if that other spherical body were assumed to move in a circle, then by the same reasoning applied in the case of the former sphere, there will have to be still another spherical body [and that would go on *ad infinitum*]. Hence the celestial spheres must needs have a stationary body round which they are to perform their circular motion."

ואמנם שהמרכז יחוייב שיהיה נבדל [נח] זה מבואר מאשר אנו כאשר הוצאנו
קו מהמרכז אל המקיף, ודמינוהו מתנועע עד שישוב אשר התחיל, הנה כל נקודה
הניח על זה הקו, הנה היא תחדש בתנועתה קשת דומה לקשת הגדול אשר יחדשהו
קצה הקו במקיף הכדור עצמו, וכאשר היה זה כן, הנה הקו כלו מתנועע בכללו,
וכל חלקיו יתנועעו על יחס אחד, והנקודה אשר היא תכלית הקו תחדש הקו בהכרח

עגולה דומה לשאר העגולות. הנה אותה העגולה לא תמנע מאשר תמצא בגשם
כדוריי או רקות. ומציאות רקות יתבאר שהוא שקר, הנה בהכרח אם שימצא גשם
כדוריי נח ואם מתנועע. ואם היה מתנועע בסבוב חוייב בו במה שיחוייב בראשון,
הנה בהכרח שיהיה לגשם הכדורי גשם נח עליו יסבוב.

In his supercommentary on the *Intermediate Physics, loc. cit.*,
Gersonides argues that Averroes could not have used his term
centre in a strictly mathematical sense, for the mathematical
centre of a moving radius does not describe a circle, contrary to
what is implied in Averroes' discussion. He suggests that Averroes must have used the term centre in the sense of the convexity
of the enclosed sphere. "Says Levi: His conclusion is inconsequent, for while that line as a whole will indeed move on its centre,
its extremity at the centre, which is the centre, will not be moved
at all . . . . . . But if by centre here he does not mean a centre
in the true sense of the term, but rather the convexity of another
sphere enclosed within it, then he is justified in arguing as he does."

ואמנם שהמרכז יחוייב שהיה נבדל לכדור... זה מבואר מאשר אנו כאשר הוצאנו
קו מהמרכז.... אמר לוי, והנה זה החיוב בלתי צודק, לפי שזה הקו כבר יהיה
מתנועע ותכליתו האחד אשר הוא המרכז בלתי מתנועע.... ואם אמר שאין המכוון
באמרו מרכז הנה מרכז על דרך האמת אבל קבוב הכדור מבפנים, ואם יצדק
באמרו...

See above in this note on Aristotle's use of the term "centre."

71. The expression ויתפוצצו אם כן חלקיו, used here by Crescas,
is suggestive of the identical expression used by Maimonides in
describing the Mutakallimun's explanation of the revolution of a
millstone in accordance with their atomistic theory of motion.
See *Moreh* I, 73, Prop. 3: והיתה תשובתם כי יתפוצצו חלקיו עם הסבוב.
The Mutakallimun, in order to defend their theory of atomistic
motion, were forced to assume that during the circular motion of
a millstone the parts of the millstone separate from each other.
Crescas, therefore, challenges here Aristotle, or rather Averroes,
as follows: If you say that the place of the world is a stationary
centre of a certain magnitude, and on this centre the spheres
perform their revolution, then like the Mutakallimun you will be
forced to assume that during the rotation of the spheres the
centre will fall apart.

72. The meaning of this passage is as follows: In Averroes' proof, C is nothing but a mathematical point and is thus the ideal centre of the sphere and likewise the ideal extremity of the radius. As such it is neither in motion nor at rest by itself and does not therefore describe any circle that would have to be "somewhere." It is on this ideal point that the sphere is in rotation. Thus the earth itself rests on the ideal centre of the universe, which is a point, as in place. But an ideal point cannot be place.

This objection has been suggested by Averroes himself, in *Intermediate Physics* IV, i, 1, 9: "It cannot be contended that the centre is only a point, for a point cannot be described as being either at rest or in motion except accidentally and in so far only as it is the extremity of something at rest or in motion, as will be shown in Book VI of this work. Avempace has already refuted this view in his work on the *Physics*, where you may find his discussion on the subject."

ואין לאומר שיאמר שהמרכז נקודה בלבד, כי הנקודה לא תתואר במנוחה ולא בתנועה כי אם במקרה ומצד מה שהוא תכלית נמצא בנח או במתנועע כפי מה שיתבאר בששי מזה הספר. וכבר סתר אבובכר בן אלצינ המאמר הזה בספרו בשמע, ושם אמרו.

Simplicius, too, has raised the same question and answered it. Cf. Simplicius in *De Caelo* II, 3, ed. Heiberg, p. 398, ll. 20–24; Taylor's translation of *De Caelo*, p. 176, n. 2.

That the centre is only a point is also asserted by Gersonides in his commentary on Job, ch. 27, in באור דברי המענה: כי השם מטה הצפון אשר שם היישוב על תוהו, כי הוא נשען על מרכז הארץ אשר אינו כי אם נקודה, ותולה הארץ על בלימה, ר"ל שהיא נשענת ונסמכת על הנקודה שהיא מרכזה, לא בדבר חוץ ממנה, כמו שהתבאר בטבעיות.

73. Cf. *Physics* VI, 10, 240b, 8 ff.

74. See above n. 55.

75. Similarly Albo concludes his arguments against Aristotle's definition of place by setting up against it a definition which identifies place with the vacuum. *'Ikkarim.* II, 17: "But if place is identified with the void or vacuum into which the body is entered, none of these impossibilities will arise." אבל אם המקום הוא הפנוי והרקות שיכנס בו הגשם, לא יתחייב דבר מאלו הבטולים.

76. I. e., if place is the intervals of a body, and wherever a body happens to be that is its proper place, natural motion can no longer be explained by the alleged tendency toward the proper place. What the cause of motion would according to the present theory be is expounded by Crescas above, p. 410, n. 20.

77. Hebrew כאשר בקשנו ליסוד הארץ מקום. The phrasing suggests the passage from 'Olam Ḳaṭan quoted above in n. 64.

78. This would seem to argue from the assumption that the place of the earth is the centre, thus reflecting the view of Joseph ibn Zaddiḳ in 'Olam Ḳaṭan quoted above in n. 64, with which the phrasing of this passage has some resemblance. See preceding note.

However, it is possible that the argument is here incompletely stated and is to be carried out in full somewhat as follows: If we were to determine the place of the earth by the same reasoning as in the case of the other elements, namely, by the consideration of its absolutely downward motion, it would have to·be the absolute below, that is, the centre. But since the centre is only a point and cannot therefore be place, Aristotle will have to make the adjacent surface of water as its place. But then the place of the earth will not be what it should be by reason of its downward motion. This interpretation of the argument will make it correspond to the following passage in 'Iḳḳarim II, 17: "And if the place of the element earth is the surface of the element water which surrounds it from without, the place of the earth will not be the absolute below, as has been assumed by him, for the absolute below is the centre." ואם מקום יסוד הארץ הוא שטח יסוד המים המקיף בה מחוץ, לא יהיה מקום הארץ המטה במוחלט, כמו שהניח הוא, לפי שהמטה במוחלט הוא המרכז.

Pico Della Mirandola reproduces this argument as follows: "Praeterea omnia quae collocantur corpora, suis congruere locis falsum esse aperiri, et ex supremi coeli circunferentia et etiam ex terra, cui locus assignatur non superficies sed punctus imus, cui loci nomen iure non congruit" (*Examen Doctrinae Vanitatis Gentium* VI, 4).

79. Hebrew ולוה היה האמת עד לעצמו ומסכים מכל צד. Cf. *Analytica Priora* I, 32, 47a, 8: δεῖ γὰρ πᾶν τὸ ἀληθὲς αὐτὸ ἑαυτῷ ὁμολογούμενον εἶναι πάντῃ. This Aristotelian formula has many different Hebrew translations and paraphrases, a collection of which was made by Steinschneider. (Cf. *Monatsschrift für Geschichte und Wissenschaft des Judenthums*, Vol. 37, (1893), p. 81; *Uebersetzungen*, Endenote 11; *ibid.*, p. 56, n. 75b).

80. That is to say, the place of a thing taken as one whole must be *equal* (שוה ἴσος) to the place of the same thing when broken into parts. But if you accept Aristotle's definition that place is the boundary of that which surrounds, the place of a two-foot cubic block for instance, will be twenty-four square feet, whereas the place of the same block cut into eight one-foot cubic blocks will be forty-eight square feet.

This argument is thus the nucleus of the following passage in *'Ikkarim* II, 17: "Similarly he will be compelled to say that one thing will have many places differing according to great and small, for if a body is broken up into parts, its parts will require a greater place than that required formerly by the whole, and the same will happen if those parts are broken up again into other parts, and the other parts into still other parts. But this is contrary to what has been laid down by Euclid in his work on *Weight and Lightness* [a pseudo-Euclidian work; see Steinschneider, *Uebersetzungen*, p. 503, n. 20] wherein he says that things which are equal occupy equal places." וכן יתחייב לו לומר שהגשם האחד יהיה לו מקומות רבים מתחלפים בגודל וקוטן, כי הגשם האחד כשיתחלק, יצטרכו חלקיו אל מקום יותר גדול מאשר בתחלה, וכן כשיתחלקו חלקיו לחלקים אחרים וחלקים לחלקים, זה הפך מה שהניחו אקלידוס בספרו הכבדות והקלות, שאמר שם כי הגשמים השוים ימלאו מקומות שוים.

The commentary *Shorashim* on the *'Ikkarim* has failed to notice this similarity, and describes it as one of the original arguments of Albo which was not borrowed by him from his teacher: אבל ב' קושיות שהקשה המחבר אח"כ והם שיהיה מקום החלק וכו', הוא הוסיף מדיליה ואינו מקושית רבו.

81. Hebrew אינו נותן האמת בדרוש. The term דרוש or מבוקש is the technical Hebrew word for the thesis, or that which is to be proved (مطلوب *quaesitum, probandum*) as contrasted with תולדה, نتيجة, which is the conclusion already proved. See *Makaṣid al-Falasifah* I, p. 30.

82. Crescas is indirectly alluding here to some implied difference between his definition of place and that of Aristotle. According to Aristotle, place is different from form (see above p. 155). Again, according to Aristotle, there is a difference between *general space* and *proper place* (see above Part I, n. 76, p. 356). Furthermore, according to Aristotle, Crescas has already tried to show, there must be a difference between the place of the whole and that of the part (see above p. 197). But if the place of a thing is identical with the vacuum occupied by the thing, it is like the form of the thing. There is no distinction between *general space* and *proper place*. Nor is there any distinction between the place of the whole of the thing and that of the part, except that the latter is part of the former.

83. Cf. *Shebu'ot*, 7b.

84. Cf. *Mekilta*, Ki Tissa, I (ed. Friedmann, p. 103b). For this reference I am indebted to Prof. Louis Ginzberg. Cf. W. Bacher, *Die Exeg. Terminologie der jüdischen Traditionsliteratur* I, p. 8.

85. Cf. *Horayot*, 11b.

86. This is an allusion to Maimonides' explanation of the term "place" as meaning "degree" or "position." Cf. *Moreh* I, 8.

87. Cf. *'Abodah Zarah*, 40b.

88. Hebrew: לא על דעתך אנו משביעים אותך כי אם על דעתנו ועל דעת המקום. This is evidently a composite quotation made up from phrases in the following passages: (a) *Shebu'ot* 29a: הוי יודע שלא על דעתך (b) *Shebu'ot* 39a: אנו משביעין אותך אלא על דעתנו ועל דעת בית דין הוי יודע שלא על דעתך אלא על דעת המקום ועל דעת בית דין. (c) *Nedarim* 25a: הוו יודעים שלא על דעתכם אני משביע אתכם אלא על דעתי ועל דעת המקום.

89. Genesis Rabbah 68, 9, and elsewhere.

90. Isaiah 6, 3.

91. Referring to the three times that the word "holy" occurs in the verse.

92. In David Ḳimḥi's commentary on Isaiah 6, 3, the threefold repetition of the word "holy" is said to refer to God's separation from the three worlds, which are named as follows: (1) The world of angels and souls. (2) The world of spheres and stars. (3) The terrestial world: חכר שלש פעמים קדוש כנגד שלש עולמות, עולם העליון והוא עולם המלאכים והנשמות, ועולם התיכון והוא עולם הגלגלים והכוכבים, ועולם השפל והוא זה העולם. A similar interpretation of the verse is given in Solomon ben Immanuel Dapiera's *Batte ha-Nefesh* (Hebrew translation of Abu 'Imran Moses Tobi's *Al-Saba'niyyah* with commentary, ed. Hirschfeld in the *Report of the Judith Montefiore College*, 1894), p. 45: ונעלה משלש ישולש ותקדש בשלש משלש ותקדש בסוד הנפרדים, ר"ל שהשם שהוא למעלה מהג' עולמות שהם עולם המלאכים ועולם הגלגלים ועולם היסודות השפל והוא נקרא ג"כ עולם.

From the entire tenor of Crescas' discussion here, however, it would seem that he has reference to the Cabalistic Sefirot and their threefold division. As preliminary to the understanding of this passage the following remarks are pertinent.

The term כבוד in the Biblical expression כבוד ה', *the glory of the Lord* (Ex. 24, 16), was taken from earliest times by Jewish philosophers to refer either to the essence of God or to something emanating from His essence (see next note). In the Cabala the term כבוד was appropriated as a designation for the Sefirot. Cf. Azriel, *Perush 'Eser Sefirot*, p. 5a: דע כי כל הספירות נקראות כבוד. The ten Sefirot were divided into three worlds, as follows: (1) The world of mind, עולם השכל, (2) The world of soul, עולם הנפש, (3) The world of body, עולם הגוף (*op. cit.* p. 3b). All the Sefirot, with the exception of the last, have both an active and passive quality, i. e., they are both emanating and receiving. In the language of Cabala these two qualities are designated as the masculine and the feminine qualities. Cf. *'Iḳḳarim* II, 11:

לפי שחכמי הקבלה ייחסו כל יום מימי בראשית אל עלול אחד מן הז' שכלים האחרונים, ויקראו העלולים ספירות... ויאמרו שהשכל האחרון שהוא השכל העשירי והוא השכל הפועל והוא הספירה העשירית שיקראו שבת נתרעמה לפני ה' ית' למה שבת בה מההשתלשלות ולא היה לה בן זוג, כלומר נמצא אחר שיהיה שכל עומד בעצמו וכמו שהוא בשאר השכלים שיהיה מושפע ממנה עד שנשארה היא כנקבה ומושפעת ולא משפעת.

In view of these considerations, Crescas uses the expression יסוד העבור, *the element of impregnation*, as a designation of the emanative process whereby the Divine influence is extended to the terrestial world. Ordinarily, it may be remarked in passing, the term עבור refers to metempsychosis, as in the expression סוד העבור in Baḥya ben Asher's commentary on the Bible, Ex. 34, 7: פוקד עון אבות על בנים זהו סוד העבור. Deut. 3, 26: כי 'ויתעבר ה רמז לסוד העבור.

Crescas' interpretation of the verse, therefore, is as follows: Though God is exalted above the three worlds into which the Sefirot are divided, still through the emanative quality of His Glory, i. e., the Sefirot, He is present in the terrestial world.

It may also be remarked here, that the term יסוד in Cabala is the name of the ninth Sefirah which in the figure of the Adam Kadmon, πρωτόγονος, represents the genital organs. Cf. Azriel, *Perush 'Eser Sefirot*, p. 3b: יסוד עולם בכח הגיד. It is not impossible to find in the expression יסוד העבור here an allusion to this.

Similar uses made of this verse to prove the presence of the Divine influence in the terrestial world is to be found in many places, as, for instance, in *Sefer ha-Bahir* 48: 'ומאי הוי קדוש קדוש ה צבאות מלא כל הארץ כבודו, אלא קדוש כתר עליון, קדוש שירת האילן, קדוש דבוק מיוחד בכלן, ה' צבאות שמו מלא כל הארץ כבודו, and *Ma'amar Yikkawu ha-Mayyim*, ch. 8, pp. 31–32.

93. In the following passage Crescas alludes to an old question as to whether the Biblical expression "the Glory of the Lord" refers to the essence of God or to something emanated from His essence.

The question is raised by Philo in his attempt to explain away the implication of spatial motion in Exodus 24, 16: "And the Glory of the Lord came down," *came down* being here the Septuagint reading for the masoretic וישכן *did abide*. According to Philo, the term "Glory" in this Biblical verse refers either to (a) the presence of His powers by which God manifests Himself in the world, or to (b) the subjective manner in which the human mind apprehends God. Cf. J. Rendel Harris, *Fragments of Philo Judaeus*, p. 60; Wendland, *Neu Endeckte Fragmente Philos*, p. 101: Philo Judaeus, *Opera Omnia*, ed Richter, Vol. VII, p. 310.

Maimonides discusses the same question in the *Moreh Nebukim*. According to him, the expression "the Glory of the Lord," as used in different places in the Bible, has three meanings: (a) An emanation from God designated by him as "the created light," and

in this connection he quotes Exodus 24, 16, which is also quoted
by Philo. (b) The essence of God itself. (c) Human glorification
or conception of God. "The same is the case with 'the Glory of
the Lord.' The phrase sometimes signifies the created light which
God caused on a certain place to show the distinction of that place
. . . . . . Sometimes the essence and the reality of God is meant
by that expression . . . . . . Sometimes the term Glory denotes
the glorification of the Lord by man or by any other being"
(*Moreh Nebukim* I, 64). The similarity between Philo's two explanations and Maimonides' first and third explanations is striking. It has been definitely shown, on other grounds, that Philo's
writings were not altogether unknown to mediaeval Jews. See
Harkavy's additions to Rabinovitch's Hebrew translation of
Graetz's *Geschichte der Juden*, Vol. III, pp. 497–8.

The first interpretation of Glory is referred to by Maimonides
also in *Moreh* I, 10; I, 76; III, 7.

The term כבוד as an emanated Divine Light identical with
Shekinah occurs also in the works of other Jewish philosophers.

Saadia, *Emunot ve-Deot* II, 11; II, 12; II, 10: ועם זה כבר הראה
בו אורו הנברא אשר הקדמנו זכרו הנקרא שכינה וכבוד. Cf. commentary on
Sefer Yezirah, ch. 4 (ed. Lambert, Arabic text, p. 72, French
text, p. 94), Malter, *Life and Works of Saadia Gaon*, p. 189.

Jehuda ha-Levi, *Cuzari* II, 8: אמר החבר, כן הכבוד נצוץ אור אלהי
והכל שב בהשתלשלות אל האלהים. אך אשר V, 20: מועיל אצל עמו ובארצו
יהיה במגרה גרידה הוא הכבוד והאותות, וזה אין מצריך אל סבות אמצעיות.
Cf. also II, 4.

Pseudo-Baḥya, *Ma'ani al-Nafs*, ch. 16, ed. Goldziher, p. 54;
Broyde, *Torat ha-Nefesh*, p. 71. Cf. Harkavy's additions to
Rabinovitch's Hebrew translation of Graetz's *Geschichte d.
Juden*, Vol. V, p. 18.

In accordance with these interpretations of the term Glory,
Maimonides interprets Isaiah 6, 3 in two ways, one taking the
term כבוד to mean the essence of God and the other to mean an
emanation (*Moreh* I, 19).

Now, just as כבוד has these two meanings so the Sefirot which
are identified by the Cabalists with כבוד have two meanings with
reference to their relation to God. According to some Cabalists,
the Sefirot are identical with God's essence while according to

others they are emanations of God's essence. Abraham Shalom compares this cabalistic controversy to the philosophic controversy as to whether the Prime Mover is identical with God or is something emanated from Him. *Neveh Shalom* V, 11, p. 81b:

והנה המקובלים נחלקו בענין זה לשתי כתות: יש מהם שהאמין שמי' ספירות הוא
הסבה הראשונה ית', ומהם מי שהאמין שהסבה הראשונה לא יודבר ממנה לא ברמז
ולא בפירוש ויקראוהו אין סוף.... ושהוא מצוי ושהוא סבה ועלה למה שזולתו
ואומר ששפע ממנו י' שכלים יקראו אותם שכלים. וקרה זה למקובלים כמו שקרה
לחכמי המחקר אם המניע הראשון עלול או הוא הש"י.

What Crescas is trying to do in this passage is to transfer Maimonides' discussion of the term כבוד as he understood it to the term כבוד as it was understood by the Cabalists in the sense of the Sefirot.

Assuming first that כבוד, or the Sefirot, is identical with God, Crescas interprets the verse to mean as follows: "The blessedness (ברוך) of the Glory of God (כבוד ה')," i. e., of the Sefirot, "is from Glory's place (ממקומו)," i. e., from the essence of God, inasmuch as Glory or the Sefirot are identical with God's essence.

He takes ברוך not as a passive participle but as a substantive.

94. Referring now to the other Cabalistic view, that the Sefirot are intermediaries and tools of God, Crescas interprets the verse as follows: "Blessed is (ברוך) the glory of God (כבוד ה')," i. e., the Sefirot, "from His place (ממקומו)," i. e., from God's essence.

The entire passage, as will have been observed, is a Cabalistic version of Maimonides' discussion in *Moreh* I, 19.

95. Cf. *Moreh* I, 8.

96. Hebrew ולזה יהיה החיוב חלקי. I. e., ἀπόδειξις κατὰ μέρος, *particular demonstration*, as opposed to ἐπὶ τοῦ καθόλου, באור כולל, *universal demonstration*. Cf. *Anal. Post.* I, 24, 85a, 13 ff., *De Caelo* I, 6, 274a, 20.

97. That is to say, there may exist an infinite number of concentric spheres, so that while all the motions toward the circumference are one in kind they are infinite in number terminating as they do at each of the infinite number of circumferences. The argument is taken from Gersonides' commentary on *Intermediate Physics*. Cf. above p. 373, n. 103.

98. Crescas refers here indirectly to the answer given by Gersonides himself to his own argument for an infinite number of upper places. Gersonides' answer is as follows: If there were an infinite number of upper places there would be no absolute above, and without an absolute above, there would be no absolute below. Crescas does not explicitly state here his reasons for rejecting this answer. He summarily dismisses it as inconclusive. His reason for that may be supplied as follows: The centre of the earth is called the absolute below only in relation to the periphery of its surrounding sphere. But if those peripheries are infinite, the centre of the earth can no longer be called the absolute below. In fact, the very idea of an above and a below in the universe is based upon its finitude. Anaximander and Democritus who deny the finitude of the world likewise deny the distinction of an above and a below within it. So also Plato denies the distinction of above and below (Cf. *De Caelo* IV, 1).

99. Crescas argues here, in the first alternative, that the hypothesis of an original time of motion might be tenable even if we admit the impossibility of motion within a vacuum. For even according to Averroes' contention that the medium is a necessary condition of motion and that within a vacuum motion cannot take place, we may still maintain that within the medium of any plenum there is a common original time of motion which can never disappear, no matter what the agent or the magnitude may happen to be, for that original time is due to the very medium itself in which the motion takes place.

100. In this second alternative Crescas rejects Averroes' contention that the medium is a necessary condition of motion, but, following Avempace, he argues that the original time of motion may be due to the nature of motion itself and must thus exist even in a vacuum. See above n. 19.

101. Crescas refers here to the difference between "motion" and "change." Motion is always in time. Change is without time. Change in place is "motion," whereas change in quality is "alteration" (cf. Propositions IV and V).

That locomotion is gradual, i. e., in time, whereas qualitative change may be instantaneous, i. e., in no-time, is the view of

Aristotle in *De Sensu*, ch. 6, 446b, 29–447a, 2: "Local movements, of course, arrive first at a point midway before reaching their goal. . . but we cannot go on to assert this in like manner of things which undergo qualitative change. For this kind of change may conceivably take place in a thing all at once." Cf. also *Kawwanot ha-Pilosofim* III (*Maḳaṣid al-Falasifah* III, p. 236): "As for quality, a sudden translation is possible in it, as, e. g., a sudden blackening." כמו פתאום, העתקה בו אפשר הנה האיכות ואולם פתאום ההשתחר. Cf. Prop. IV, notes 3 and 4.

102. Similarly Bruno dismisses all of Aristotle's arguments that an infinite would be incapable of circular motion by contending that those who believe the world to be infinite believe it to be immovable. Cf. *De l'Infinito Universo et Mondi* II, p. 326, l. 29; *De Immenso et Innumerabilibus* II, ii.

103. While number and magnitude must be actually finite, still, says Aristotle, they are both infinite in capacity, but with the following distinction. Number is infinitely addible, and magnitude is infinitely divisible. It is in this sense that an infinite is possible, "for the infinite is not that beyond which there is nothing, but it is that of which there is always something beyond" (*Physics* III, 6, 207a, 1–2). Number, however, being a discrete quantity, cannot be infinitely divisible, nor can magnitude, which is by its nature limited, be infinitely addible (*ibid.*, III, 7).

Cf. *Epitome of the Physics* III, pp. 12–13: "Aristotle believes that magnitude is not infinitely addible . . . . . . But that magnitude is infinitely divisible will be shown in Book VI . . . . . . Number is infinitely addible but not infinitely divisible."

ואריסטו סובר שאי אפשר בשעור שיתוסף אל לא תכלית. . . ואולם החלק השעור
אל לא תכלית הנה יתבאר במאמר השש. . . ולזה היה אפשר במספר שיתוסף אל
לא תכלית, אבל שיחלק אל לא תכלית לא.

Cf. also *Milḥamot Adonai* VI, i, 11, p. 334: "The case here is analogous to the case of number, that is to say, it is like number which, though infinitely addible, is always potentially some finite number."

והנה הענין בזה כמו כמו הענין במספר, רצוני שכמו שהמספר יתוסף אל
מה שיתוסף תמיד מולת שיהיה בכח אלא מספר בעל תכלית.

104. Cf. *Metaphysics* XI, 3, 1061a, 19: Ἐπεὶ δ' ἐστὶ τὰ ἐναντία πάντα τῆς αὐτῆς καὶ μιᾶς ἐπιστήμης θεωρῆσαι.

105. Hebrew ספר החרוטים, كتاب الخروطات κωνικὰ στοιχεῖα of Apollonius (Book II, Theorem 13). Cf. Munk, *Guide* I, 73, p. 410, n. 2.

Crescas seems to have quoted the problem referred to from *Moreh* I, 73, Prop. X. The entire passage here is full of expressions taken from Maimonides. See below n. 112.

106. Hebrew יצא. MSS. ב and ג read יצא. MS. א reads יוצאים. In the corresponding passage of the *Moreh* our texts read יצאו, and so also in the reproduction of this passage in Isaac ibn Laṭif's *Rab Pe'alim* 63. But the Arabic אכרנא in the *Moreh* would suggest a passive form like יוצא or more likely the new form יֻצָּא.

107. Hebrew שיש שם Similarly later the negative שאין שם (p. 216, l. 1). The word שם in these expressions is not the adverbial "there" but rather the pronominal "there," reflecting the Arabic ثَمّ which, like the English "there," is used as an indefinite grammatical subject of a verb. Cf. Bacher, *Über den sprachlichen Charakter des Maimûni'schen Mischne-Torah* in *Aus dem Wörterbuche Tanchum Jerusalmi's*, p. 121; I. Friedlaender, *Der Sprachgebrauch des Maimonides*, p. 15; S. Rawidowitz, *Sefer ha-Madda'*, p. 73, n. 20.

108. Cf. Euclid, *Elements* I, Def. 23.

109. Hebrew השרשים which stands here for השרשים המונחים. We should naturally expect here ושאר הגדרים, *and the other Definitions*, for in our present editions of Euclid the First Principles are called Definitions, Postulates and Axioms, but not Hypotheses. But the use of Hypotheses here instead of Definitions may be explained on the ground that in Crescas' copy of Euclid's *Elements* the term Hypotheses was used instead of Definitions. The confusion of these two terms are traced to Proclus. (Cf. T. L. Heath, *The Thirteen Books of Euclid's Elements*, Vol. I, p. 122). Similarly Algazali in his *Maḳaṣid al-Falasifah* I, p. 68, quoting Euclid, leaves out Definitions and divides the First Principles (ἀρχαί, المبادى, ההתחלות) into the following three classes: (1) Axioms (أولية, ראשונות) or Common Notions (κοιναὶ ἔννοιαι, κοιναὶ δόξαι, متعارفة علوما, ידיעות נודעות, In Albalag's translation:

(ידיעה). (2) Hypotheses (اصول موضوعة, שורש מתנח, Albalag: עיקר מתנח).
(3) Postulates (αἰτήματα, מصادراة, מערכה, Albalag: הקדמה).

The force of Crescas' reasoning here may become clearer in the light of Aristotle's statement that a hypothesis, unlike a definition, assumes the existence of the thing defined and reasons from that assumption. Cf. *Anal. Post.* I, 10, 76b, 35 ff.

110. Hebrew שהוא מן הידיעות הראשונות, literally, *one of the axioms*. But see preceding note. Cf. Euclid, *Elements*, Book I, Postulate I.

111. Similarly Bruno contends in connection with another of Aristotle's arguments that when an infinite acts upon another infinite or upon a finite the action itself will be finite. Cf. *De l'Infinito Universo et Mondi* II, p. 340, l. 32 ff.; *De Immenso et Innumerabilibus* II, vii.

112. Hebrew ואם היה רחוק מן הציור השכל מחייבו. By ציור here is meant ציור הדמיון.. Cf. Averroes, *Intermediate De Anima* III: כי הציור בשכל ממנו דמיון וממנו סברא.

The statement here is based upon the discussion in *Moreh* I, 73, Proposition X, where the problem from the *Conic Sections* referred to above by Crescas is also mentioned. Maimonides discusses there the difference between imagination and reason: "And the action of the imagination is not the same as the action of the intellect," ואין פעל הדמיון פעל השכל, and concludes: "It has consequently been proved that things which cannot be perceived or imagined, and which would be found impossible if tested solely by imagination, are nevertheless in real existence." הנה כבר התבאר מציאות מה שלא ידומה ולא ישימהו הדמיון אבל הוא נמנע אצלו. Cf. *Phys.* III, 2, 202a, 2–3: χαλεπὴν μὲν ἰδεῖν, ἐνδεχομένην δ' εἶναι.

As for the use made by Spinoza of Crescas' discussion of this argument, see my paper "Spinoza on the Infinity of Corporeal Substance," *Chronicon Spinozanum* IV (1924–26), p. 101–3.

113. Originally "sixth," הו', in all the texts. But the sixth proof is based upon the impossibility of an infinite to be passed through in finite time and not upon the general proposition that no infinite can be passed through at all, and should thus be grouped together

with the second proof which is taken up next by Crescas. The fifth proof, however, is originally in Averroes based on the proposition that no infinite can be passed through at all. See above p. 389, n. 152.

114. Originally "fourth," 'הד, in all the texts.

115. I. e., as in the *third* argument from *circular* motion in the Third Class of Arguments (above p. 173).

116. I. e., as in the *second* and *sixth* arguments from *circular* motion in the Third Class of Arguments (above pp. 171, 175).

117. In order to understand the meaning of this passage, it is necessary to summarize here part of Aristotle's discussion in the sixth book of the *Physics*.

He shows there how in motion three things are to be considered: that which changes, i. e., the magnitude; that in which it changes, i. e., the time; and that according to which it changes, i. e., the category of the motion, as, for instance, quality, quantity, place. (Cf. *Physics* VI, 5, 236b, 2–4).

He also shows that in none of these three respects can motion have an absolutely fixed beginning. He puts it as follows:

(1) "That there is not a beginning of mutation, nor a first time in which a thing is changed" (*Physics* VI, 5, 236a, 14–15).

(2) "Neither in that which is changed, is there any first part which is changed" (*ibid.*, 27–28).

(3) Nor is there any first with reference to motion of place or quantity (cf. *ibid.*, 236b, 9 ff.).

He then concludes with the following statement: "Everything which is moved must have been previously moved" (*Physics* VI, 6, 236b, 32–34; *Metaphysics* IX, 8, 1049b,35 ff.).

The upshot of all this is that there is no absolute beginning of motion. No beginning which we may assume of motion, either with reference to its time, its magnitude or its place, can be definitely designated by a fixed, irreducible quantity, since motion is infinitely divisible in all these respects. Whatever quantity we may assume to designate the first part of motion, we can always conceive of a smaller quantity which would have to be prior to that alleged first part.

With this in mind, Crescas now endeavors to answer the *second*, *third* and *sixth* arguments from *circular* motion in the Third Class of Arguments (above pp. 171, 173, 175).

He first tackles the *third* argument. His answer may be paraphrased as follows:

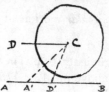

You say that CD cannot meet AB at D′ without having met it first at some point A′. This indeed would be true if D′ were a definitely fixed point on AB. But D′ is a point in infinity. The argument therefore falls down.

This refutation of Averroes' proof is taken from a tentative objection raised by Altabrizi against the corresponding proof by himself (see above p. 384, n. 141). The final answer by which Altabrizi justifies his own proof does not apply to the Averroesean proof adopted by Crescas.

The refutation as given by Altabrizi is as follows: "Against this proof many objections may be raised, of which the recent philosophers had no inkling. It may be argued as follows: Why do you say that the sphere in the course of its rotation, when its radius ceases to be parallel to the other line and is about to meet it at the vertex, that the former would undoubtedly have to meet the latter at a point which is the first point of the points of intersection? Why should it have to do so? Their meeting at the vertex cannot come about except as a result of motion, but, inasmuch as motion is potentially infinitely divisible, a first meeting at the vertex with the infinite line will be impossible, seeing that the extremity of the finite line which is moved along with the motion of the sphere is potentially infinitely divisible so that we cannot assume any point of the points of intersection without the possibility of assuming another point before it . . . . . . The result is that the meeting of the two lines at the vertex cannot be effected but by motion, which is potentially infinitely divisible, and similarly any parts of the lines that meet must be infinitely divisible. Consequently we cannot assume that any point is the first of the points at which the lines meet."

ועליו ספקות חזקות לא ישיגום המתאחרים, והוא שיאמר: למה אמרתם שהכדור
כאשר התנועע עד סר מהנכח אל הפנישה נכח הראש אין ספק שיחודש בקו הבב״ת

נקודה היא ראשית הנקודות הנפגשות? וזה שהפגישה נכח הראש בין שניהם אמנם
תתחדש בתנועה, והתנועה מתחלקת לעולם בכח, הנה יהיה שקר הפגישה נכח
הראש בקו הבב״ת, עם שקצה הקו הב״ת המתנועע בתנועת הכדור לעולם מתחלק
בכח, ואי אפשר הנחת נקודה מנקודות הפגישה אם לא שאפשר הנחת נקודה אחרת
לפניה... הנה המניע שהפגישה נכח הראש אמנם תגיע בתנועה, והיא מקבלת לחלוקה
בבלתי תכלית בכח, וכן מה שתפגשהו מן הקו, הנה אי אפשר הנחת נקודה היא ראשית
הנקודות הנפגשות בו נכח הראש.

118. Hebrew בשעור ב״ת, *in a finite magnitude.*

119. In this part of the passage he means to answer the *second* and
sixth arguments. These two arguments are based upon the impossibility of the infinite chord AB to be passed through by the revolving line CD in finite time.

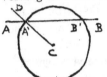

Crescas' answer may be paraphrased as
follows:

Point A′, at which CD first meets AB, is
indeed a point in infinity. But A′B′ which
is part of AB forming a chord in the circle
generated by CD is finite. It is therefore, only a finite distance
that is traversed by CD in finite time.

120. Hebrew וזה להכרח קצה התחלת התנועה בזולת זמן. This passage
is misplaced. Logically it is an explanation of the previous statement הנה לא יתחייב מציאות נקודה ראשונה מהפגישה. One is tempted
to emend the text here as follows: והנה למה שהתבאר המנעות חלק
ראשון בתנועה, למה שחויב שכל מתנועע כבר התנועע, הנה לא יתחייב מציאות
נקודה ראשונה מהפגישה, וזה להכרח קצה התחלת התנועה בזולת זמן. ולזה איננו
רחוק שיפגוש הקו בשעור ב״ת בתנועה ב״ת.
"Since, however, it has been shown that there can be no first part
of motion, because every object that is moved must have already
been moved, it does not follow that there would have to be a first
point of meeting, and this indeed because of the fact that the
extreme beginning of motion must take place in no-time. It is not,
inconceivable, therefore, that the infinite line [in question] should
meet the other line in a finite distance with a finite motion."

The meaning of this statement is as follows: The reason why
there can be no absolutely first part of motion is that an absolutely first part of motion would have to take place in an indivisible instant. But motion is infinitely divisible and cannot take

place in an instant, except qualitative motion in a certain aspect (see above n. 101). To quote Aristotle's original statement upon which this statement of Crescas seems to be based: "But that, in which that which is changed is first changed, is necessarily an indivisible" (*Physics* VI, 5, 235b, 32–33).

Cf. *Epitome of the Physics* VI, p. 32a: "No part of motion can be called first, inasmuch as motion is infinitely divisible. But the same is not true of the end of motion, for that is called end which refers to something that has already come into existence and is completed, so that a certain definite time can be assigned to it, and of such a nature is the entelechy which is the end of motion. But as for the beginning of motion, it exists in an instant rather than in time, on account of which it cannot be definitely designated in the same way as the entelechy, for the latter is the limit of [a completed] motion and not, as in the case of the former, the limit of something that does not yet exist."

והתנועה אי אפשר שימצא חלק ממנה ראשון, כי היא מתחלקת אל מה שיתחלק
תמיד. ואולם תכלית התנועה, הנה אין ענין בו כן, כי הוא אמנם נלקח תכלית מה
שכבר נמצא ונשלם, והיה אפשר שידמה אליו זמן, כי זה דרך השלמות אשר הוא
תכלית התנועה. ואולם התחלת התנועה, הנה מציאותה בעתה ולא בזמן. ולזה
אי אפשר שירמז אליו, כמו שאפשר זה בשלמות אשר הוא תכלית התנועה, לא
תכלית מה שלא ימצא עדיין כעניין בהתחלה.

121. All the MSS. and the printed editions read here "fifth," הה'.

122. Similarly Bruno argues against Aristotle that the infinite would be without figure. Cf. *De l'Infinito Universo et Mondi* II, p. 326, l. 29; *De Immenso et Innumerabilibus* II, x.

123. This argument has been anticipated by Averroes in his *Intermediate De Caelo* I, 4: "It cannot be argued that the existence of circular motion implies only the existence of a body that is capable of circular motion but not necessarily the existence of a spherical body, seeing that fire and air, for instance, are by their nature capable of circular motion. The answer may be stated as follows . . . . . . " (Latin, p. 273vb, L). ואין לאדם לומר שלא יתחייב

ממציאות התנועה הסבובית כי אם נשם מתנועע בסבוב לא נשם כדורי, כמו האש
והאויר אשר יראה מעינינם שהם מתנועעים להם בסבוב. וזה...

124. A suggestion of this argument may be discerned in Isaac ibn Laṭif's *Rab Pe'alim*, 60.

He first makes the following statement: "The rays furnish an argument for the non-existence of a vacuum and so does also the visibility of the stars, for the sun's ray coalesces with them gradually until they reach the sense of vision." הנצוצות בם מופת לבטול הרקות, וכן ראיית הכככים, כי הנצוץ השמשי מתלכד בם ראשון ראשון עד שמגיעין להרגש הראות.

(The term מתלכד here seems to reflect the Greek συμφύεσθαι in *De Sensu*, ch. 2, 438a, 27).

As far as one can make out the meaning of this argument, it seems to rest on Aristotle's theory that the perception of vision requires some medium and that "if the intermediate space became a void. . . an object could not be visible at all." (*De Anima* II, 7, 419a, 15–21). But see the interpretation of this passage by Efros, *The Problem of Space in Jewish Mediaeval Philosophy*, p. 73.

Then he proceeds to say: "This proof for the impossibility of a vacuum is itself a proof for its existence. Consider this, for it is a sealed mystery." והמופת לבטל הריקות הוא בעצמו מופת למציאותו. והבן זה כי חתום הוא.

This mystery may perhaps be unsealed for us with the aid of Crescas. What Isaac ibn Laṭif may have wished to say is that the same argument from the sun's rays, or the rays of any luminous object, which proves the non-existence of a vacuum *within the world* must prove its existence *outside the world*, as is maintained by the Pythagoreans (see above n. 7). For by an argument from the rays of a luminous object we may prove, as shown here by Crescas, the possibility of the existence of something infinite outside the world. But that something infinite outside the world, again as argued above by Crescas (see p. 189), must be either a plenum or a vacuum. As it cannot be a plenum, it must of necessity be a vacuum (see *ibid.*). Hence the argument from the rays of a luminous object proves the existence of a vacuum outside the world.

The reference in Isaac ibn Laṭif, however, may be to some such argument for the existence of a vacuum from the transmission of light as is reported by Simplicius in the name of Straton Lampsacenus. "Straton Lampsacenus endeavored to show that there is a vacuum which intercepts every body so as to prevent its continuity, for he says that light would not be able to pervade through water or air or any other body. . . unless there were such a vacuum: for how could the rays of the sun penetrate the bottom of a vessel." (Simplicius in *Physica* IV, 9, ed. Diels, p. 693, l. 11 ff.; Taylor's translation of the *Physics*, p. 237, n. 9).

125. Similarly Bruno argues against Aristotle that the infinite would have neither an end nor a middle. Cf. *De l'Infinito Universo et Mondi* II, p. 328, l. 22.

126. *Analytica Priora* II, 18, 66a, 16: ὁ δὲ ψευδὴς λόγος γίνεται παρὰ τὸ πρῶτον ψεῦδος. Cf. *De Caelo* I, 5, 271b, 8–9: εἴπερ καὶ τὸ μικρὸν παραβῆναι τῆς ἀληθείας ἀφισταμένοις γίνεται πόρρω μυριοπλάσιον. Of this last quotation there are the following Hebrew versions: *Intermediate De Caelo* I, 7: הטעות אשר יפול בהתחלת הדרך יביא האדם אל טעות גדול. Themistius, *In Libros Aristotelis De Caelo Paraphrasis*, ed. Landauer. Hebrew text, p. 14, ll. 24–26: כי אנו כשנטעה בהתחלה ואפילו בדבר מועט נתרחק במה שכחנו בו מתחלת העניין פי שניים ממה שנפל בו ממנו הטעות. Latin text, p. 22, ll. 13–15: "Entenim si initio vel in re minima a veritate deflexerimus, longe plurimum deinde ab ∞ scopo errabimus, quem ab initio intendebamus."

It is interesting to note that this statement, with which Crescas introduces here his discussion of the existence of many worlds, is also quoted by Bruno in the middle of his discussion of the same subject (*De l'Infinito Universo et Mondi* IV, p. 369, lines 39–40). As we shall see, Crescas' argument against Aristotle's denial of many worlds has something corresponding to it in Bruno. See below n. 130. The statement, however, occurs in *De Caelo* which is the principal source of the problem of many worlds.

127. The discussion of the problem of the existence of many worlds would seem to be quite irrelevant in this place. Crescas, however, has introduced it here because Aristotle happens to take it up immediately after his disposing of the problem of infinity (cf. *De Caelo* I, 8). Then also Crescas needed it for his criticism of Maimonides' proofs of the existence of God. The problem is again taken up by Crescas in Book IV, 2. Cf. *Milḥamot Adonai* VI, i, 19, and *Emunot we-Deot* I, 1, First Argument.

128. The passage as it stands would seem to contain one single argument of which the first part (שהוא חייב תחילה) is the premise and the second part (וחייב שאלו היה שם) is the conclusion. I take it, however, to contain two distinct arguments. The first is suggestive of one of the arguments against the existence of many worlds used by Crescas later in Book IV, 2. The second is taken from Aristotle's discussion of the same problem in *De Caelo* I, 8.

The first argument is incompletely stated here. Only the premise is given. In its full form, as given in Book IV, 2, the argument reads as follows:

"If there existed many worlds at the same time, the following disjunctive reasoning would be inevitable, namely, that between those worlds there would have to be either a vacuum or a plenum. But the existence of a vacuum outside the world is impossible, according to the opinion of the ancients. Hence there would have to be a body between those worlds. Now, that body would inevitably be either transparent or not. If it were transparent, it would follow that we would be able to see numerous suns and moons on such occasions as when the suns and the moons of the various worlds happened to be together on the horizon. And if it were opaque, then, inasmuch as the dark celestial bodies receive light from other bodies, as the moon, for instance, receives light from the sun and as do also certain stars in the opinion of some people, it would follow that the opaque body between the worlds would receive light from the suns and it would be possible for us to see many stars from one or more of the other worlds."

שאם היו בכאן עולמות יחד, לא ימלט הענין מהלוקה, אם שיהיה במה שבין העולמות רקות או גשם. והיות שם רקות נמנע אצל הקודמים, יחוייב אם כן שיהיה ביניהם גשם. והגשם אם שיהיה ספיריי אם לא. ואם הוא ספיריי, יחוייב שנראה בקצת הזמנים שמשים וירחים יותר מאחד, כשיהיו שנידם על האופק. ואם הוא גוף חשוך, הנה לפי מה שנמצא בגרמים השמימיים החשוכים שיקבלו האורה מזולתם, כמו הירח שיקבל אורה מהשמש, וקצת הכבבים, לדעת מי שיראה כן, הנה יתחייב שיקבל אורה מה שבין העולמות מהשמשים, ויתכן שנראה כוכבים רבים מעולם או מעולמות אחרים.

Similarly the refutation given by Crescas of this argument in Book IV, 2, is the same as here, namely, that the impossibility of a vacuum outside the world has not been conclusively demonstrated.

The second argument against the existence of many worlds is somewhat as follows: If there were other worlds, they would all have to possess the same nature as this world of ours. The elements of those other worlds would, therefore, have to possess upward and downward, i. e., centrifugal and centripetal, motions, the same as the elements in our world. Furthermore, the centre from and toward which all those elements would move would have to be one in all the worlds, that is, it would have to be identical with the centre of our own world. Consequently, if there were other worlds, the earths in those worlds would all tend toward the centre of our world and the fires in those worlds would move toward the periphery of our world. But that is impossible, since in that case the earth and fire in those worlds would move away from their own respective centre and periphery. Cf. *De Caelo* I, 8.

129. Ecclesiastes 6, 11.

130. The meaning of this argument may be stated as follows: It is true that the elements in all the other worlds would have to have two kinds of motion, upward and downward. It is not true, however, that their motions would all have to be from and toward the same centre. For our knowledge that those elements would have to possess two kinds of motion is based only upon the assumption that they would have to be of the same nature as our elements. But what does that assumption mean? Certainly it does not mean that those elements would have to be a continuation of our elements. It only means that, while they were distinct from our elements, they would have to present the same characteristics, namely, some being light and some heavy, some warm and some cold, etc. Or, in other words, those elements would be the same as ours *in kind* but not *in number*. By the same token, when we say that those elements would have to move upward and downward like ours, it does by no means imply the same upward and downward, from and toward the same centre. It is therefore possible to conceive of many worlds, each with a centre of its own, from and toward which their own respective elements have their motion. The motions of the elements in all those worlds would thus be one in kind, i. e., centrifugal and centripetal, but many in number, i. e., with reference to different centres.

This criticism is found in Gersonides' commentary on the
Epitome of De Caelo I: "One may argue that if many worlds existed, the elements in those worlds would exist in their respective
natural places and their movements would follow the order of the
movements of their respective worlds, without necessarily giving
rise to the conclusion that the natural place of the parts of the
same element would not be one. The only conclusion given rise
to by such an assumption would be that the below would constitute the place of the heavy elements, that is to say, the heavy
elements would sink beneath all the other elements that exist
together with them. Nor will it follow from the principle that
contraries are those things which are most distant from each other
that the places of the parts of an element must be one in number.
That this is not to follow can be illustrated by the following example. Take a certain black object that is undergoing a gradual
change from blackness to whiteness. Then take other black
objects which are likewise being in the process of changing to
whiteness. This does not mean that the whiteness into which all
these black objects are being changed and which constitute the
opposite of the *terminus a quo* in their changing process is one
and the same in number. What it implies is only that they are
all changed to colors which are one and the same in kind. Similarly if there were many worlds, it might be said that the element
earth in every one of those worlds would move away from the
above and downward toward the below, but this would not mean
that the above from which the different terrestial elements moved
would be one in number; it would rather mean that they would
be one in kind, that is to say, it would be the concavity of the
circularly moving celestial sphere."

הנה לאומר שיאמר, שאם נמצאו עולמות רבים, היו היסודות בהם במקומם
הטבעי, ותנועותיהם מסודרות על צד סדור התנועות]וב[זה העולם, ולא יחוייב
מפני זה בחלקי היסוד האחד שלא יהיה מקומם אחד, אבל מה שיחוייב שיהיה
מקום הכבדים המטה, ר"ל שישקעו תחת שאר כל הגשמים הנמצאים עמהם יחד.
ולא יחוייב גם כן, מפני חיוב היות ההפכים בתכלית המרחק, שיהיו אחדים באיש.
ומשל זה, כי השחור יתנועע מהשחרות אשר הוא בו אל הלובן, וכאשר יתנועעו
שאר דברים שחורים אל הלובן, לא יחוייב שיהיה אל הלובן, ההפך אשר ממנו
התנועה, אחד במספר, אבל מה שיחוייב שיהיה אחד במין. כן גם כן יאמר התנועה
בכל ארץ הוא מהמעלה אל המטה, לא שיהיה המעלה אשר יתנועעו מהם הארצות

אחד במספר, אבל הוא אחד במין, והוא קבוב המתנועע בסבוב במה שהוא מתנועע
בסבוב.

A similar refutation of this argument of Aristotle against the existence of many worlds is found in Bruno. Cf. *De l'Infinito Universo et Mondi* IV, p. 365, l. 31 ff.

131. Ecclesiastes 1, 14.

132. Ḥagigah 11b.

PROPOSITION II

PART I

1. The Hebrew version of this proposition is taken from Isaac ben Nathan's translation of Altabrizi.

2. This entire proof is a paraphrase of Altabrizi.

Aristotle proves the impossibility of infinite number by the following argument. *Physics* III, 5, 204b, 7–10: "But neither will there be number, so as to be separate and infinite; for number or that which possesses number is numerable. If, therefore, that which is numerable can be numbered, it will be possible for the infinite to be passed through." (Cf. *Metaphysics* XI, 10, 1066b, 24–26).

This Aristotelian proof is faithfully reproduced by Abraham ibn Daud in *Emunah Ramah* I, 4, p. 16: "For when you say that things which have number exist in actuality, it means that their number is an actually known number. But when you say they are infinite, it means that you cannot arrive at the end of their number. Consequently, he who says that an infinite number exists in actuality is as if he has said: I have completely enumerated that which is infinite and I have come to the end of it, despite its being endless."

כי אמרך דברים נמים נמצאים בפועל יורה שמספרם מספר ידוע בפועל,
ואמרך בלתי בעל תכלית יורה על שאתה לא תוכל להגיע אל מספרם. והאומר
זה, כאלו אמר כבר מניתי מה שאין תכלית לו וכבר באתי עד קצו, והוא בלתי
בעל תכלית.

PART II

3. This proof, taken directly from Altabrizi, is to be found in the following sources.

Algazali, *Happalat ha-Pilosofim* I (*Tahafut al-Falasifah* I, p.9, ll. 23–24; *Destructio Destructionum* I, p. 19va): "We say number is divided into even and odd, and it is impossible that anything should be outside of this distinction whether it be existent and permanent or non-existent."

אמרנו המספר יחלק אל זוג ונפרד, ושקר הוא שיצא מזו החלוקה, בין שהיה הדבר נמצא נשאר או כלה.

Averroes, *Intermediate Physics* III, iii, 4, 2 (Latin, p. 453rb, E): "It can likewise be demonstrated that every actual number is actually numbered and everything numbered is either even or odd. Consequently everything numbered is finite."

וכן יתבאר שכל מספר בפועל הנה הוא ספור בפועל וכל ספור הנה הוא זוג או נפרד. הנה כל ספור בעל תכלית. *Epitome of the Physics* III, p. 10b: "Again, every number is even or odd. Either one of these two is finite. Consequently every number is finite."

וגם כן כל מספר אם הוא זוג ואם נפרד, וכל אחד מאלו השנים בעל תכלית. אם כן כל מספר בעל תכלית.

Gersonides, *Milḥamot Adonai* VI, i, 11; "We may also say that number is finite, because every number is either even or odd, and this constitutes its finitude."

וכן נאמר שהמספר הוא בעל תכלית, לפי שכל מספר הוא אם זוג אם נפרד, זהו תכליתו.

Cf. Proposition III.

4. The reference is here to the view held by Maimonides and Avicenna that infinite number is impossible only with reference to things that exist in space but that immaterial beings, such as disembodied souls, can be infinite. From this Crescas infers that they do not admit that infinite number must be subject to the division of odd and even. Cf. Proposition III, Part I.

5. The reference is to the passages of the *Intermediate Physics* and the *Epitome of the Physics* quoted above in n. 3. The argument does not occur in the corresponding passage of Averroes' Long Commentary on the *Physics*.

6. Crescas' argument is especially directed against the passage in *Physics* III, 5, 204b, 7–10, quoted in Prop. II, Part I, p. 476, n. 2. Aristotle, it will be recalled, argues that "number" (ἀριθμός, מספר) is the same as "that which possesses number" (τὸ ἔχον ἀριθμόν, בעלי המספר) and that both are "numerable" (ἀριθμητόν, מדרכם שיספרו) and that both "can be numbered" (ἐνδέχεται ἀριθμῆσαι, ספור בפועל), and consequently neither of them can be infinite. Crescas is attacking here the original assumption that "that which possesses number" is the same as "number," arguing that while the latter cannot be infinite the former may be so.

7. The implication of this argument is that the fact that number must be divided into odd and even does not by itself prove the impossibility of infinite number, for unless it is established independently that number cannot be infinite, it is possible to assume the existence of an infinite number of dyads no less than of monads. This argument must have been suggested to Crescas by the following passage in *Milḥamot Adonai* VI, i, 11: "The same can be demonstrated with regard to number, in the following manner. Seeing that every number must be finite, it follows that every even number must be finite; and the same must be true with regard to the even-times even number and the even-times odd number." (Cf. ἀρτιάκις ἄρτιος and ἀρτιάκις περισσός in Euclid *Elements* VII, Definitions 8 and 9).

חה יתבאר במספר ממה שאומר: והוא כי מפני שהיה כל מספר בעל תכלית,
הנה יתבאר שכל זוג הוא בעל תכלית, וכן הענין בזוג הזוג ובזוג הנפרד.

8. For a full discussion of the sources of this distinction, see Prop. III, Part I, notes 8–9.

Crescas' use of this distinction as a criticism of the proposition denying the possibility of an infinite number is not novel. It is to be found in the following works.

Algazali, *Tahafut al-Falasifah* I, p. 9, ll. 19–20: "Should one say that only the finite is described by even and odd but that the infinite is not to be described by them, we answer etc."

Narboni, Supercommentary on the *Intermediate Physics* III, iii, 4, 2: "Second, how can it be proved that there is no infinite number on the ground that number is divided into even and odd, when those who affirm the existence of an infinite number may also claim that such a number is not divisible into even and odd but into an infinite number of parts, etc. To this we answer that Aristotle is arguing here in accordance with the truth, namely, that there is no infinite actual body [that is to say, Aristotle is not arguing here from the premises of his opponents]."

והשני, איך שלא ימצא מספר בב״ת לפי שהמספר יחלק אל זוג ונפרד שג״כ לפי דעת שאומר שימצא מספר בב״ת סובר שלא יחלק אל זוג ונפרד אלא אל חלקים בב״ת וכו'. נשיב שאריסטו לא דובר אלא על צד האמת, וזה שלא ימצא נשם בפועל בב״ת.

An answer to Crescas' criticism is given by Isaac ben Shem-ṭob in his *second* supercommentary on the *Intermediate Physics* III, iii, 4, 2: "By what we have said in explanation of this proposition may be solved the difficulty raised by Ibn Ḥasdai, namely, that the argument is a begging of the question, for he who affirms the existence of an infinite number does not admit that everything actually numbered must be either even or odd but; quite the contrary, he will deny this. In view, however, of what we have said, namely, that the relation of even and odd to number is like that of priority and posteriority to time, the objection disappears. For just as there can be no time without the prior and the posterior (cf. definition of time in Proposition XV), so there can be no number without even or odd. Hence the proposition is absolutely true."

ובמה שאמרנו בביאור זאת ההקדמה יותר הספק שעשה ן' חסדאי והוא שזה מערכה על הדרוש, וזה שהאומר במספר שהוא בלתי בעל תכלית לא יודה שכל מסופר בפועל הוא או זוג או נפרד אבל יכחיש זה. אבל במה שכבר אמרנו, שהערך ייש לזוג והנפרד עם המספר הוא כמו הערך שיש לקודם ולמתאחר עם הזמן, הנה לא נשאר ספק כלל. וזה שכמו שאי אפשר שימצא זמן ולא ימצא הקודם והמתאחר כמו כן אי אפשר שימצא מספר ולא יהיה או זוג או נפרד, ואם כן יתחייב שיהיה זאת ההקדמה צודקת בההלט.

PROPOSITION III

PART I

1. The Hebrew version of this proposition is taken from Isaac ben Nathan's translation of Altabrizi, with the following exception: Altabrizi reads לא תכלית for בלתי תכלית.

The term מבואר in מבואר הבטול is to be taken here in the sense of "demonstrably" rather than "evidently" (Munk: *évidemment*), for in *Moreh* I, 73, Eleventh Proposition (quoted in the next note) Maimonides speaks of the impossibility of an infinite series of causes and effects as having been demonstrated by proof, התבאר במופת.

2. This introductory comment is based upon Altabrizi: "The verification of the first and second propositions is not sufficient in establishing the truth of this proposition, for what has been ascertained by the first two propositions is only the fact that things which have position and place, i. e., bodies, must be finite. Causes and effects, however, may sometimes be not bodies but rather beings free of matter and body and independent of them, called Intelligences . . . ... Hence Maimonides has made of this inquiry a separate proposition."

ואמתת ההקדמה הראשונה והשנית לא תהיה מספקת באמתת זאת ההקדמה,
כי הידוע מאותם השתי הקדמות אמנם הוא הגעת תכלית ענינים להם הנחה ומקום
והם הגשמים, והעלות והעלולים פעמים לא יהיו נשמים, אבל יהיו נמצאים מופשטים
מהחומר והנשמות, בלתי נתלה בהם, ויקראו שכלים. ... ולזה שם זאת החקירה
הקדמה נפרדת בעצמות.

The same distinction between magnitudes and causes is made by Maimonides himself. *Moreh* I, 73, Eleventh Proposition: "It has been already shown that it is impossible that there should exist an infinite magnitude, or that there should exist magnitudes of which the number is infinite, even though each one of them is a finite magnitude, provided, however, that these infinite magnitudes exist at the same time. Equally impossible is the existence of an infinite series of causes, namely that a certain thing should be the cause of another thing, but itself the effect of another cause, which again is the result of another cause, and so on to infinity, so that there would be an infinite number of things existing in actuality. It makes no difference whether they are *bodies* or *beings free of bodies*, provided they are in causal relation to each other. This causal relation constitutes [what is known as] the *essential, natural* order, concerning which it has been demonstrated that an infinite is impossible."

כי כבר התבאר המנע מציאות נשם אחד אין תכלית לו, או מציאות נשמים אין
תכלית למספרם ואף על פי שכל אחד מהם נשמי בעל תכלית, ובתנאי שיהיו אלו

שאין להם תכלית נמצאים יחד בזמן. וכן מציאות עלות אין להם תכלית שקר, ר"ל
שיהיה דבר עלה לענין אחר, ולדבר ההוא עלה אחרת, ולעלה עלה, וכן אל לא
תכלית, עד שיהיו מניים אין תכלית להם נמצאים בפועל יהיו גשמים או נבדלים,
אלא שקצתם עלה לקצתם, וזהו הסדור הטבעי העצמי אשר התבאר במופת המנע
מה שאין תכלית לו בו.

In the foregoing passage we have Maimonides' own commentary on his first three propositions and the source of the statements here by Altabrizi and Crescas. Maimonides first divides
the infinite into infinite *magnitude* and infinite *number*. The latter is subdivided by him into the number of *co-existent magnitudes*
and the number of *causes and effects*. Then, again, he describes
the relation between the causes and effects as an *essential, natural
order*. The term *essential* is used by him as the opposite of *accidental* which he proceeds to explain and which is taken up by
Crescas later (see p. 494, n. 19). The term *natural* is meant to be
the opposite of what Altabrizi and Crescas call here *order in
position*.

The expression בעלי סדור, without any qualifying term, occurs
in *Emunah Ramah* I, 4, p. 16: "It is also impossible that there
should be an infinite number of actually existing things having
order." וגם כן אי אפשר שימצאו דברים נמנים נמצאים בפועל בעלי סדור
בלתי בעלי תכלית. Judged from the context, however, the expression "having order" here may mean both "order in position"
and "order in nature," for the author seems to deal both with co-
existent magnitudes and with causes and effects. When he argues,
for instance, that "the things which have order are those things
which have a first, an intermediate or intermediates, and a last,"
כי הבעלי סדור הם אותם אשר להם התחלה ואמצע או אמצעיים וסוף, he
seems to be quoting phrases from Aristotle's proof for the impossibility of an infinite series of causes, quoted below in n. 4.

Equivalent expressions for סדר במצב are הדרגה בהנחה (Altabrizi) and סדר תשומיי (*Mif'alot Elohim* IX, 4, p. 62).

3. This last statement contains Crescas' own explanation of the
expression "order in nature." A similar explanation of the expression is found in *Kawwanot ha-Pilosofim* II (*Maḳaṣid al-Falasifah*
II, p. 125): "For the order between cause and effect is *necessary*
and *natural*, and should that order between them be eliminated
the cause will cease to be a cause." לפי שהסדור מהעלה והעלול

הכרחי טבעי, אם סולק בטל היותו עלה. It is on the basis of this interpretation of the passage that I have connected it with the statement preceding it rather than with the statement following it.

4. The proof for the impossibility of an infinite series of causes and effects reproduced here by Crescas is based directly upon the proof given in Altabrizi, which in turn is based upon a proof found in Avicenna, which in its turn may be considered as a free version of Aristotle's proof in *Metaphysics* II, 2, 994a, 1 ff. Crescas himself refers later to Altabrizi as his immediate source and describes the proof as having been suggested "in the eighth book of the *Physics* and in the *Metaphysics*" (see Prop. III, Part II, p. 225). Again, later, after refuting this Altabrizian proof of Aristotelian origin, Crescas quotes what he supposes to be another proof in the name of "one of the commentators." That proof, too, we shall show (p. 492, n. 16), is based upon the same proof of Aristotle, though Crescas unwarily advances it as something new.

The original proof of Aristotle, as interpreted by Averroes, may be analyzed as follows (cf. *Epitome of the Metaphysics* III, Arabic, p. 118, §64; Latin, p. 383va; Quirós Rodríques, p. 187; Horten, p. 140; Van den Bergh, p. 98):

I. In a series of causes and effects, consisting of three or more members, that is called *cause* proper which is the *first* in the series and is not preceded by any prior cause. That is called *effect* proper which is the *last* in the series and is not followed by another effect. The *intermediates* are both causes and effects. They are causes only in relation to what follows from them; in themselves they are effects, requiring thus a first uncaused cause for their existence. Cf. *Metaphysics* II, 2, 994a, 11–15: "For in the case of an intermediate, which has a last term and a prior term outside it, the prior must be the cause of the later terms. For if we had to say which of the three is the cause, we should say the first; surely not the last, for the final term is the cause of none; nor even the intermediate, for it is the cause only of one."

II. Intermediates will always be effects and thus require a first cause even if they were infinite in number. Cf. *ibidem*, 15–16: "It makes no difference whether there is one intermediate or more; nor whether they are infinite or finite in number."

III. Hence, there can be no infinite number of causes. For in an infinite number of causes all the causes would be intermediates, and intermediates, being also effects, could not exist without a cause which is not an effect. Otherwise, things would exist without a cause. Cf. *ibidem*, 16–19: "But of series which are infinite in this way, and of the infinite in general, all the parts down to that now present are alike intermediates; so that if there is no first there is no cause at all."

Avicenna's version of this proof, in its fullest and most elaborate form, is to be found in his *Al-Najah*, p. 62, quoted by Carra de Vaux in *Avicenne*, pp. 269–271. It is to be found also in the following places: Algazali, *Maḳaṣid al-Falasifah* II, p. 127, *Tahafut al-Falasifah* IV, p. 34, l. 12 ff. (*Destructio Destructionum* IV, p. 71va, I; *Muséon* 1900, pp. 376–377), *Teshubot She'elot*, pp. LI–LII; Moses ha-Lavi, *Ma'amar Elohi;* Altabrizi, Prop. III.

Though Crescas has taken his proof from Altabrizi, he does not follow him closely. Altabrizi's proof is more elaborate and is more like the original argument of Avicenna. It runs as follows:

I. In an aggregate (Altabrizi: מקובץ *Maḳaṣid al-Falasifah* II, p. 127: جملة, כלל), of causes and effects, let each member be conditioned by a preceding cause.

II. The aggregate itself will be conditioned.

III. Now, the cause of that aggregate will have to be one of these three:

    (a) The aggregate itself.
    (b) Something included within the aggregate.
    (c) Something outside that aggregate.

The first two, (a) and (b), being impossible, the third, (c), must be true.

IV. But that external cause must be causeless.

Crescas' statement of the proof, as may have been observed, is much shorter. It runs as follows:

I. Within the aggregate (כללם) of the infinite series of cause and effect, either all the members are conditioned or some of them are not.

II. If they are all conditioned, there must be a determining cause. "Outside the series" is to be understood here.

III. If any of the members is unconditioned, the series is no longer infinite.

The text of Altabrizi's proof reads as follows:

והראיה על אמתת זאת ההקדמה: שהנמצא אשר יהיה אפשר לעצמותו עלול,
ועלתו אם היתה בזה התאר גם כן, וכן עלת עלתו אל בלתי תכלית, או יהיה כבר
הגיע מקובץ עלות ועלולים בב"ת, כל אחד איפשר עלול. חה המקובץ מצד הוא
מקובץ היה ג"כ איפשר עלולי, שהמקויים האיפשר העלול יותר ראשון שיהיה איפשר
עלולי, ועלת אותו המקובץ, אם שיהיה עצמו, או דבר נכנס בו, או דבר חוץ ממנו.
והחלק הראשון בטל, כי העלה קודמת על העלול, והדבר לא יקדם על עצמו.
והחלק השני ג"כ בטל, כי אשר הוא נכנס באותו המקובץ לא יהיה עלה לעצמו
ולא לעלתו, ואם הוא הוא יהיה קודם על עצמו ועל זולתו חה שקרי, ולא יהיה עלה
למקובץ, כי עלת המקובץ תהיה ראשונה עלת חלקיו אחר באמצעי' חלקיו יהיו
עלה למקובץ. ואולם החלק השלישי, והוא שיהיה עלת המקובץ דבר חוץ מאותו
המקובץ, הנה אותו שהיא חוץ לא יהיה אפשר עלול, לפי שאנחנו כבר קבצנו כל מה
שהוא איפשר עלול באותו ההשתלשלות, הנה אשר הוא חוץ מהם לא יהיה אפשר
עלול, ואם היה היה יהיה נכנס בו. והנמצא אשר לא יהיה אפשר עלול יהיה מחוייב
לעצמותו, ויהיה ההשתלשלות העלות כלו אצלו, ויהיה הוא קצה להם, ולא יהיו אותם
העלות בלתי בעלי תכלית, אבל יהיו בעלי תכלית אל עלה ראשונה, הוא עלה
למה שאחריו מן העלות. חהו הדרוש.

5. Hebrew בשכלים או בנפשות. See at the end of the next note.

6. The question as to whether the infinity of disembodied souls is to be included within the rule of this proposition has been also raised by Altabrizi, who, though inclined to answer it in the negative, ends with the remark that God alone can solve such intricate difficulties. אבל הענין בו עומד על ראיה נפרדת בחלוק ובקיום, והאלהים יודע. This is expressed in simpler language by the anonymous translator: והענין בזה נעלם, והש"י יודע נסתרות. Unlike Altabrizi, however, Crescas, instead of relegating the problem to divine omniscience, tries to solve it with whatever help he could get from Avicenna, Algazali and Averroes.

Algazali's view as to the infinity of disembodied souls is to be found in the following places:

Kawwanot ha-Pilosofim II, i (*Maḳaṣid al-Falasifah* II, p. 125): "Similarly the human souls which are parted from the bodies at death can be infinite in number, even though they exist simultaneously, for there is not between them that order of nature the elimination of which would cause the souls to cease to be souls, for those souls are not causes of each other, but exist simultaneously without any distinction of priority and posteriority either in position or in nature. If they seem to have a distinction of priority and posteriority it is only with reference to the time of their creation, but their essences *qua* essences and souls have no order between them at all. They are rather all alike in existence, in contradistinction to distances and bodies, causes and effects."

וכן הנפשות האנושיות הנפרדות מהגופות במות אפשר סלוק התכלית למספרם, ואם היו נמצאים יחד, כי אין בם סדור הטבע בצד לו שוער סלוקו בוטל היותם נפשות, כי אין קצתם עלה לקצת, ואבל הם נמצאות יחד, מבלתי קדימה ואיחור בהנחה והטבע, ואמנם, ידומה הקדימה והאיחור בזמן חדושם. אולם עצמיותיהם מצד שהם עצמיות ונפשות, הנה אין סדור בם כלל אבל הם שוים במציאות, בחלוף המרחקים והגשמים, והעלה והעלול.

Happalat ha-Pilosofim I (*Tahafut al-Falasifah* I, p. 9, l. 26 ff.; *Destructio Destructionum* I, p. 20ra, l. 8 ff.; Horten, p. 29; *Muséon* 1899, pp. 281–282): "Furthermore, we argue against the philosophers thus: Even according to your own principles, it is not impossible to assume that at the present moment there exist things which are units [احاد, אחדים; but Latin: *eadem in esse*] qualitatively different from each other and still are infinite in number, namely, the souls of men which have become separated from the bodies at death [بالموت, במות, *hora mortis*], and these are things which are not described as either even or odd . . . . This view concerning the infinity of disembodied souls is one which Avicenna has adopted, and perhaps it is the view of Aristotle."

ועוד שאנחנו נאמר להם הנה זה לפי שרשכם אינט מן השקר שימצאו נמצאים הווים שהם אחדים משתנים בתואר ואין תכלית להם, והם נפשות האנשים הנבדלות מהגופים במות; והנה הם נמצאים שאינם מתוארים בזוג ונפרד...... וסברא זו בנפשות היא שבחר בה ן' סינא, ואולי שהיא סברת אריסטו.

Cf. the parallel discussion in *Happalat ha-Pilosofim IV* (*Tahafut al-Falasifah* IV, p. 33, l. 29 ff.; *Destructio Destructionum* IV, p. 71r; *Muséon* 1900, pp. 375–376).

Maimonides refers to this view of Avicenna in *Moreh* I, 74, Seventh Argument: "Some of the later philosophers solve this difficulty by maintaining that the surviving souls are not bodies requiring a place and a position on account of which infinity is incompatible with their manner of existence."

אמנם קצת אחרוני הפלוסופים התירו זה הספק בשאמרו הנפשות הנשארות אינם
נשמים שיהיה להם מקום והנחה שימנע במציאותם האין תכלית.

The original view of Avicenna is to be found in his *Al-Najah*,
p. 34, partly quoted by Carra de Vaux in his *Avicenne*, p. 203.
Cf. Shahrastani, pp. 403–404 (ed. Cureton).

It must, however, be noted that personally Algazali does not
admit the infinity of disembodied souls. He advances it merely
as an argument *ad hominem*. Crescas is following the general
method of quoting in the name of Algazali views contained in his
Kawwanot ha-Pilosofim, which Algazali himself later rejected.

The expression בשכלים או בנפשות "souls or intellects" call for
some comment. The term "intellect" does not occur in any of
the sources which we have reason to believe to have been drawn
upon by Crescas for his information. Altabrizi has here only the
term "souls," בנפשות בני אדם הנפרדות. So does also Algazali in
the *Kawwanot ha-Pilosofim:* ויאמר לנפשות האנושיות הנפרדות מהגופות
אין תכלית להם and in the *Happalat ha-Pilosofim:* נפשות הנבדלים
מהגופות במות.

It is quite obvious that by שכלים here Crescas does not mean
the "Intelligences" of the spheres, in which sense the term שכל
is used by Maimonides in the proposition. Such a rendering could
not be construed with the context.

It occurs to me that these two terms are used by Crescas for a
special purpose. He wants indirectly to call attention to his controversy with other philosophers as to the nature of the immortal
soul. According to Avicenna and others, it is only the "acquired
intellect," השכל הנקנה, that survives. But according to Crescas,
the soul as such is immortal in its essence (cf. *Or Adonai* II, vi, 1).

Accordingly what Crescas means to say here is as follows: It is
possible to have an infinite number of disembodied souls, whether
these disembodied, immortal souls be acquired *intellects* (בשכלים), as
is the view of Avicenna, or *soul essences* (בנפשות), as is my own view.

A similar indirect allusion to his controversy with the philosophers on the nature of the immortal soul occurs also in Prop. XVI,
Part II.

7. *Happalat ha-Happalah* I (*Tahafut al-Tahafut* I, p. 10, l. 6 ff.;
Destructio Destructionum I, p. 20rb, l. 26 ff.; Horten, p. 31): "I
do not know of any one who makes a distinction between that

which has position and that which has no position with reference
to infinity except Avicenna. As for all the other philosophers I do
not know of any one who maintains such a view. Nor is it in
harmony with their principles. It is rather a tale out of fairy land,
for the philosophers reject an actually infinite number of forms
whether it be corporeal or incorporeal, inasmuch as that would
imply that one infinite can be greater than another infinite.
Avicenna only meant to ingratiate himself with the multitude by
advancing a view concerning the soul which they had been accustomed to hear. This view, however, carries but little conviction or
persuasion. For if an infinite number of things existed in actuality
then the part would be equal to the whole."

ולא אדע אחד יפריש בין מה שיש לו מצב ומה שאין לו מצב בזה הענין אלא
בן סיני בלבד, ואולם שאר בני אדם כלם לא אדע אחד מהם אמר זה המאמר,
ולא יאות לשרש משרשיהם, והוא מהבלי הטפלות, כי הפלוסופים ירחיקו מצורות
מה שאין תכלית לו בפועל בין שהיה נשם או בלתי נשם לפי שיתחייב ממנו שיהיה
מה שאין תכלית לו יותר ממה שאין תכלית לו. ואולם בן סיני כון בו לפייס ההמון
במה שהורגלו לשמעו מענין הנפש, אבל הוא מאמר מעט הספקה והפיוס, כי אלו
נמצאו דברים בפועל אין תכלית להם היה החלק כמו הכל.

(Cf. a similar refutation by Averroes in *Happalat ha-Happalah*
IV; *Tahafut al-Tahafut* IV, p. 71, l. 23; *Destructio Destructionum*
IV, p. 71va, G).

It is evidently this passage of Averroes that is restated by
Narboni in his commentary on *Moreh*, I, 74, Seventh Argument:
"Averroes objects to it, and argues . . . .
Furthermore, it is a well recognized principle that that which
exists in actuality cannot be infinite whether it be material or immaterial, and there is no difference in this respect between that
which has position and that which has no position, as was thought
by Avicenna. For if actually existent things were infinite, the
part would be equal to the whole."

ובן רשד חלק ואמר. . . . . וגם כן שורש ידוע שמה שאין תכלית על מה שהוא
נמצא בפועל הוא נמנע, שוה היו נשמים או בלתי נשמים. ואין הבדל בזה בין מה שלו
הנחה ובין מה שאין לו הנחה כמו שחשב אבן סיני, כי לו נמצא דברים בפועל אין
תכלית להם היה החלק כמו הכל, ר"ל בלתי בעל תכלית בפועל.

According to Narboni (Commentary on the *Kawwanot, loc. cit.*)
Averroes' denial of the infinity of disembodied souls follows as a
result of his denial of individual immortality.

"It behooves you to know that this philosopher [i. e., Averroes] ...... objects............. to Algazali's statement that disembodied souls are infinite .... He says that this view is refutable .... It is not in agreement with Aristotle's view as to the immortality of the soul, for Aristotle does not believe that every man has an individual soul which is individual in its essence .... And consequently we do not have to adopt the view which Algazali was compelled to adopt. Ponder upon this. We further say that Algazali's statement here indicates that he has been following Alexander's view, who believes that the soul is only a predisposition .... and that it is created."

וצריך שתדע שזה שזה החכם... חלק... בשאמר אבוחמד שהנפשות הנפרדות מהגופות אין תכלית להם... זה בטל... אין זה אמת לפי דעת אריסטו בנפש הנשארת, כי הוא לא יסבור שלכל אדם נפש נבדלת בעצמו... ולא נתחייב להאמין מה שהאמינו אבוחמד. ודע זה. ונאמר כי ממה שכתבו הנה יראה כי אבוחמד מדעת אלכסנדר שהאמין כי הנפש היא הכנה לבד... ושהיא מחודשת.

8. Crescas is misrepresenting Averroes' view in attributing to him the distinction of odd and even as an argument against the infinity of disembodied souls. It is true that Averroes denied the possibility of an nfinite number of disembodied souls, but his reason for it is not that attributed to him here by Crescas. He rejects it on the following two grounds: (1) No infinite number is possible, whether material or immaterial. (2) There cannot be an infinite number of disembodied souls because the individual souls do not persist after death (cf. above n. 7 and below n. 9).

Crescas himself mentions Averroes' commentary on the *Physics* as his only source for the argument from odd and even (see Prop. II, Part II), and there is no indication there that the argument was directly applied by Averroes to the infinity of disembodied souls.

9. Crescas argument that the infinite by virtue of its being unlimited should likewise be indivisible into odd and even has been raised and refuted by Algazali. It is introduced in the following connection.

Algazali raises an objection against the eternity of motion on the ground that every number must be divisible into odd and even, whereas eternal motion would imply an infinite number of motions which could not be divided into odd and even. He then suggests himself that the eternalists might say that it is only a finite number that must be divisible into odd and even but not an infinite number (quoted above Prop. II, Part II, p. 478, n. 8). But he rejects this distinction and affirms that an infinite as well as a finite number must always be divisible into odd and even.

Happalat ha-Pilosofim I (*Tahafut al-Falasifah* I, p. 9, l. 23 ff.; *Destructio Destructionum* I, p. 19va, l. 11 ff.; Horten p. 27; *Muséon* 1899, p. 281): "We say number is divided into even and odd, and it is impossible that anything should be outside this distinction whether it be existent and permanent or non-existent. For when we assume a certain number we must believe that it must inevitably be even or odd, irrespective of whether we consider the things numbered as existent or as non-existent, for even if they cease to exist after having existed, this [disjunctive] judgment does not disappear nor does it change."

אמרנו המספר יחלק אל זוג ונפרד, ושקר הוא שיצא מזו החלוקה, בין שיהיה הדבר הנמנה נמצא נשאר או כלה. זה כשהנחנו מספר, מהחוייב עלינו שנאמין בשלא ימלט מהיותו זוג או נפרד, בין שנשער הספורים נמצאים או נפרדים, כי אם נעדרו אחר המציאות, לא תעדר זאת הגזרה ולא תשתנה.

Averroes, on the other hand, insists that it is only by virtue of its finitude that a number must be divisible into odd and even, be that finitude conceptual or real. Conceptual finites, however, as, e. g., future time, are only conceptually divisible into odd and even. The infinite, therefore, is not necessarily divisible into odd and even, inasmuch as the infinite has neither conceptual nor real existence, for it exists only in potentiality, and existence in potentiality is like non-existence.

Happalat ha-Happalah I (*Tahafut al-Tahafut*, p. 9, l. 3 ff.; *Destructio Destructionum* I, p. 19va, l. 24 ff.; Horten, p. 27): "This proposition is only true of that which has a beginning and an end outside the soul or in the soul, that is to say, it is only then that we are intellectually bound to think that it must be either even or odd irrespective of the circumstance whether it has actual existence or it has no actual existence. But that which exists only in potentiality, that is to say, a thing which has neither a beginning nor an end, cannot be described as either even or odd . . . . . .

for that which is in potentiality is like that which is non-existent." ....

חה המאמר אם יצדק במה שלו התחלה ותכלית חוץ לנפש או בנפש, ר"ל שמשפט השכל עליו בזוג והנפרד בעת העדרו ובעת מציאותו. ואולם מה שהוא נמצא בכח, ר"ל שאין לו התחלה ולא תכלית, לא יצדק עליו לא שהוא זוג ולא שהוא נפרד...

PART II

10. *Physics* VIII, 5; *Intermediate Physics* VIII, ii, 2. Cf. below n. 19.

11. *Metaphysics* II, 2. Cf. Prop. III, Part I, p. 482, n. 4.

12. See *Moreh* II, 22.

13. Crescas' argument here may be restated as follows: Suppose we have an eternal uncaused cause capable of producing more than one effect. Suppose again that these effects co-exist with the eternal cause and have order neither in space nor in nature. Under these circumstances, according to Maimonides' own admission, these effects may be infinite in number. Crescas now raises the following question: Why could not these effects be infinite in number even if we assume them to be arranged among themselves in a series of causes and effects? In other words, Crescas' contention is this. Assuming an uncaused eternal cause, with which its effects are co-existent, these effects should be possible to be infinite in number even if they form a series of causes and effects. As for the possibility of one simple cause to produce more than one effect, it is denied if the cause acts by necessity but is admitted if it acts by will and design (cf. *Moreh* II, 22).

The point of Crescas' reasoning will become all the more effective when taken as being especially directed against section II of Aristotle's proof in the *Metaphysics* as reproduced above in Prop. III, Part I, p. 482, n. 4. It will be recalled that Aristotle makes the statement that intermediates would require a first cause even if they were infinite. Now Crescas seems to turn on him and argue: Why not assume an infinite number of intermediates having a first cause and affirm the existence of an infinite series of intermediate causes and effects?

14. Hebrew קצת מהמפרשים "*one* of the commentators" and not as the expression would ordinarily mean "*some* of the commentators," for the reference is here to Narboni. The term קצת is used here in the sense of the Arabic بعض which means both *some* and *some one*. Thus in *Cuzari* I, 115, בעץ אלמלוך is translated by אחד מהמלכים "*one* of the kings," whereas in *Moreh* I, 74, Seventh Argument, בעץ מתאכרי אלפלאספה פחלוא הדא is translated by קצת אחרוני הפלוסופים התירו זה "*some* of the later philosophers have explained this." It was the ordinary understanding of the Hebrew קצת as "some" that caused here the corruption of חתר into חתרו in the printed editions and some MSS.

15. Hebrew יגיע. The term יגיע throughout this passage and elsewhere is used in an additional sense which it had acquired from its Arabic equivalent تناهى of which it was used as a translation. Both the Hebrew and the Arabic terms mean *reach, arrive, extend to, attain*. But the Arabic means also *be brought to an end, be accomplished, be limited*. Thus in *Ḥobot ha-Lebabot* I, 9; العلل متناهية فى الابتدا, העלות מגיעות בתחלתן מתנאהיה فى الابتدا, "the causes are limited *a parte ante*." Here I have translated it in each instance according to the requirements of the context but always in conformity with its original and acquired meanings.

Shem-ṭob ben Joseph Falaquera evidently was conscious of the new use of the term יגיע in philosophic texts but, unable to account for it, ascribes it to the intransitive meaning of the verb, which indeed is a good explanation as far as it goes. *Reshit Ḥokmah* III, 1, p. 62: וצריך לדעת כי מלת מגיע ברוב מקומות זה הספר הוא פועל עומד, כמו ואיוב כ י] וראשו לעב יגיע, ובא ממנו פועל יוצא וישעיה ה' ח'] מגיעי בית בבית, חכרתי זה לבל ישתבש הקורא ויחשוב היוצא במקום עומד והעומד במקום יוצא.

The influence of the Arabic تناهى *reach one's aim*, is also to be discerned in Samuel ha-Nagid's use of הגיע in the following verse in *Ben Kohelet:* אנש יחם בלבו מאויים, להגיעם ביום מחר יקוה. See Yellin, "Ben Kohelet of Samuel Ha-Nagid," *Jewish Quarterly Review*, n. s., XVI (1926), 275 [6], and Yellin's comment on p. 273.

For הגיע as a translation of بلغ, see quotations from Saadia and Baḥya in the next note.

16. This passage is a verbatim quotation from Narboni's commentary on *Moreh* II, Introduction, Prop. III.

This statement, however, is not original with Narboni. It is only a paraphrase of Aristotle's own words with which he clinches his arguments against an infinite series of causes upward, in *Metaphysics* I, 2, 994a, 18–19: "So that if there is no first there is no cause at all," and of the statement in *Physics* VIII, 5, 256a, 11–12: "And without the first mover, indeed, the last will not move." What Crescas, therefore, really does here after having refuted the Aristotelian proof of Altabrizi, is to quote again, this time *via* Narboni, another part of the same Aristotelian proof (see above p. 482, n. 4).

Other paraphrases of this statement of Aristotle are as follows:

Themistius in *De Caelo* I, 1, ed. Landauer, Hebrew text, p. 27, l. 15: חה כי הדבר אשר יהיה במה שיתהוה תמיד אין לו מציאות כל שכן מולתו. כי מה שאי אפשר שיגיע אליו דבר מן הדברים לא יחשוב התנועה אליו דבר מן הדברים. Latin text, p. 41, l. 4: "Quod enim in continua generatione consistit, esse non habet, atque eo *minus in* alia <*affectione?*> turpe est enim existimare eo quicquam moveri, quo nunquam pervenire potest."

Saadia, *Emunot we-Deot* I, 1, Fourth Demonstration: "For the mind cannot think backward infinitely and comprehend the infinite. By the same token, existence cannot proceed forward infinitely and complete an infinite process so as to reach us. And if existence could not reach us, we would not exist."

ומה שאין לו תכלית לא תעלה בו המחשבה למעלה ותעבור בו. העלה עצמה תמנע שתלך בו ההויה למטה ותעבור בו עד שתגיע (يَبلغ) אצלנו. ואם לא תגיע ההויה לא נהיה.

Baḥya ibn Pakuda, *Ḥobot ha-Lebabot* I, 5, Second Proposition: "It has already been shown that that which has no beginning has no end, for it is impossible in that which has no beginning to reach at a limit at which one can stop."

כי כבר נתברר שכל מה שאין לו תחלה אין לו תכלה, מפני שאי אפשר להגיע (يَبلغ) בדבר שאין לו תחלה אל נבול שיעמוד האדם אצלו.

Judah ha-Levi, *Cuzari* V, 18: "For that which is infinite cannot become actually realized." ומה שאין לו תכלית לא יצא אל הפועל.

Averroes' *Epitome of the Physics* VIII, p. 43b: "For if the intermediate causes go on to infinity, there will be no first, and if there

is no first, there will be no last. But the last exists. Hence the
first exists, and that is the self-mover."

כי אלו הלכו האמצעיים אל לא תכלית, לא יהיה שם ראשון, וכאשר לא יהיה
שם ראשון לא יהיה שם אחרון. אבל האחרון נמצא, הנה הראשון נמצא, והוא
המתנועע מצדו.

17. The line of reasoning employed by Crescas in the arguments
following bears some resemblance to Algazali's reasoning against
the impossibility of an infinite series of causes and effects, in
Happalat ha-Pilosofim IV (*Tahafut al-Falasifah* IV, p. 33, l. 24
ff.; *Destruction Destructionum* IV, p. 71r; *Muséon* 1900, pp. 375–
376).

Algazali's arguments may be outlined as follows:

I. According to the philosophers' belief in the eternity of the
universe it should be possible to have a series of causes and effects
which is infinite in the upward direction but finite in the downward direction, for of such a nature is time according to their own
view. (Cf. Refutation of Altabrizi's proof in Prop. I, Part II,
p. 423, n. 38).

II. If you say that time constitutes a successive series whereas
natural causes and effects are all co-subsistent, the answer is that
disembodied souls are admitted to be infinite even though they
are not in a successive line.

III. If you say that disembodied souls have no order at all,
neither that of *nature* nor that of *position*, whereas causes and
effects have order in nature, the answer is:

a. By admitting the infinity of disembodied souls, the philosophers have admitted the possibility of an infinite number at large.
If they are now to deny any particular kind of infinite number,
such as the infinite number of causes and effects, they must prove
that by a special argument.

b. It is not true that disembodied souls have no order. They
have order in time.

18. That is to say, Narboni's statement might hold true only in
case the causes are prior to their effects in time in addition to
their being prior to them in nature. In fact, in the original application of this argument to the problem of eternity, as we have
seen, there is the assumption of priority in time. The argument,

therefore, is insufficient to prove the contention of this proposition, namely, the impossibility of an infinite series of causes and effects where the priority involved is only that of nature.

The reasoning in this argument, it will be noticed, is just the opposite of that employed by Algazali. Cf. above n. 17, II.

19. The distinction between *essential* and *accidental* causes with respect to infinity is described by Maimonides in the following passage: "Equally impossible is the existence of an infinite series of causes . . . This causal relation constitutes [what is known as] the *essential natural order*, concerning which it has been demonstrated that an infinite is impossible. . . . . In other cases it is still an open question, as, e. g., the existence of the infinite in succession, which is called the *accidental* infinite, i. e., a series of things in which one thing comes forth when the other is gone, and this again in its turn succeeded a thing which had ceased to exist, and so on *ad infinitum*" (*Moreh* I, 73, Eleventh Proposition). Cf. above Prop. III, Part I, n. 2 (p. 481).

Similarly in Algazali's *Maḳaṣid al-Falasifah* II, pp. 124–5, the impossibility of an infinite series of causes is confined only to that which Maimonides describes as *essential*. "It follows that any number assumed to consist of units existing together and having order in nature and priority and posteriority cannot be infinite, and this is what is meant by infinite causes."

והמחוייב שכל מספר הונח אחדים נמצאים יחד ולו סדר בטבע וקדימה ואיחור, הנה מציאות מה שאין תכלית לו ממנו שקר, וזה בעלות אין תכלית להם.

This distinction is likewise discussed by Averroes in the following places:

Happalat ha-Happalah I (*Tahafut al-Tahafut* I, p. 7, l. 30 ff.; *Destructio Destructionum* I, p. 18vb, l. 7 ff.; Horten p. 21, l. 29–p. 23, l. 5): "This [impossibility of an infinite regress] is true and is conceded by the philosophers if the prior motions are assumed to be a necessary condition for the existence of the posterior motions . . . . Accordingly, in their opinion, the existence of an accidental infinite is possible but not of an essential infinite."

זה אמת ומקובל הוא אצל הפלוסופים אם הונחו התנועות הקודמות תנאי במציאות המתאחרות... והיה אפשר אצלם מציאות מה שאין תכלית לו במקרה לא בעצם.

Happalat ha-Happalah IV (*Tahafut al-Tahafut* IV, p. 70, l. 4
ff.; *Destructio Destructionum* IV, p. 70ra, l. 8 ff.; Horten, p. 187):
"According to the philosophers a series of infinite causes is in one
respect false and impossible but in another respect necessary.
They consider it impossible when the causes are essential and in
a straight direction, if, e. g., every preceding cause is a condition
in the existence of every succeeding one. But they do not consider it impossible if the causes are accidental and in a circular
direction."

הפלוסופים אומרים שעלות בב"ת נמצע מצד ומחוייב מצד. זה שהוא נמצע
אצלם כשהיו בעצם ועל היושר, אם היה קודם מהם תנאי במציאות המתאחר, ובלתי
נמצע אצלם כשהיו במקרה ובסבוב.

Intermediate Physics VIII, ii, 2: "As for the existence of an
infinite number of bodies one being the cause of the other, it is
impossible both essentially and accidentally if they all are assumed to be at the same time; it is impossible essentially but
possible accidentally if they are assumed to be not at the same
time."

ומציאות נרמים בלתי בעלי תכלית קצתם סבות לקצת, אם שיונחו יחד, זה שקר
בעצם ומקרה. ואם שיונחו, אבל לא יחד, הוא מהשקר בעצם אבל אפשר במקרה.

Throughout all these passages, it will have been noticed, in
addition to the distinction between *essential* and *accidental*
causes, a distinction is also made between *successive* causes and
co-existent causes, the former being described in one place as being "in a straight direction" על היושר. This distinction can be
traced to *Metaphysics* II, 2, 994a, 1 ff. Aristotle states there that
causes cannot be infinite either "in a straight direction," εἰς
εὐθυωρίαν or "according to kind," κατ᾽ εἶδος. Averroes offers
two interpretations of these Aristotelian phrases: "By *in a straight
direction* he means that the causes are coexistent, as if they were
in a straight line, and by *according to kind* he means that the
causes are one after the other and not together, after the manner
of things which belong to the same kind, that is to say, that one
individual exists after another individual and one group after another group, so that when the later comes into existence the earlier
passes away. It is possible, however, that by *in a straight direction* he means that the causes belong to the same kind as, e. g.,
man from man, and by *according to kind* he means that the causes

belong to different kinds under one genus, as, e. g., fire arising
from air, air from water, water from earth, for all these are causes
alike in genus." (Quoted by Abrabanel in *Mif'alot Elohim* IX,
4, p. 62b).

ירצה בדרך היושר שיהיו העלות נמצאות יחד, כאלו הם על קו ישר, וירצה
בדרך המין שיהיו העלות אחת אחר האחרת, לא יחד, על דרך הדברים המיוחסים
אל המין האחד, רצוני שימצא מהם אחד אחר אחר וכלל אחר כלל, על שהמתאחר
כאשר נמצא נפסד הקודם. ויסבול שירצה ביושר מה שהיה מהם ממין אחד, כמו
היות אדם מאדם, ובדרך המין מה שהיה מהם ממינים מתחלפים נכנסים תחת סוג
אחד, כמו שיהיה האש מהאויר והאויר מהמים והמים מהארץ, כי אלה כלם הם
עלות מסכימות בסוג.

Averroes' first interpretation is reflected in the following passage of Gersonides' Commentary on Averroes' *Epitome of the
Physics* III: "Another difficulty has been raised against this view,
which difficulty is based upon the proposition that an infinite
number of causes and effects is impossible, whether those causes
and effects exist together or not. This proposition has already
been demonstrated in the first book of the *Metaphysics*, [i. e.,
Book Alpha Minor]."

ועוד היו מסופקים בזה ספק אחר, וזה בנוי על שמציאות עלות ועלולים אין
תכלית למספרם שקר, היה שימצאו יחד או שלא ימצאו יחד, וזאת ההקדמה כבר
התבארה במאמר הראשון ממה שאחר הטבע.

A similar interpretation of that statement of Aristotle may
also be discerned in the following passage of Algazali, *Teshubot
She'elot*, p. xxxix: "Those causes must inevitably be in a straight
direction, i. e., existing together, or in coming one after the other."

לא ימנעו אותם הסבות והעלות עם שיהיו על השווי נמצאות יחד ואם בבוא זו
אחר זו.

20. The Hebrew text is rather vague. I take it as Crescas' own
criticism of the foregoing distinction. He now argues to the effect
that if an infinite series of *accidental* causes is possible, it will be
necessary to advance a special argument to prove that an infinite
series of *essential* is not equally possible.

The reasoning here is suggestive of the reasoning employed by
Algazali as reproduced above in n. 17, III, b.

21. As we have seen, the main point of Crescas' argument was,
that, assuming an uncaused eternal cause, it is not impossible to
have an infinite series of causes and effects coexisting with eternal

cause. And so he now concludes, quite logically, that while it is true that this proposition does not prove the impossibility of an infinite series of causes and effects, and hence does not prove the creation of the world in time, still it proves that the world is not its own cause but presupposes the existence of an uncaused cause.

There is in Crescas' conclusion the ring of a veiled challenge to Altabrizi's statement that the object of the proposition is to prove both (a) that the series of causes and effects cannot be infinite and (b) that they must culminate in an uncaused cause: "Now that you know this, you may understand that the purpose of this proposition is to prove that there must be an end to the series of causes and effects and that they must terminate at a cause which is entirely uncaused but has necessary existence by its own nature."

וכאשר ידעת זה, דע שהמכוון מזאת ההקדמה הוא באור תכלית השתלשלות העלות והעלוליות והגעתם אל עלה לא תהיה עלולה כלל אבל תהיה מחוייבת המציאות לעצמו.

PROPOSITION IV

1. The Hebrew text of this proposition is taken from Isaac ben Nathan's translation of Altabrizi.

2. Hebrew סתמי בשלוח. The term משולח is a literal translation of the Arabic مطلق. Both these terms are derived from a root originally meaning set free. They thus reflect the Greek ἀπόλυτος, which, from its original meaning loosed, free, came to be used in the sense of absolute. A still closer analogue of the Hebrew משולח is the Arabic مرسل, which, literally meaning sent, is used in the sense of absolute in the spurious Theology of Aristotle (cf. Dieterici, Die sogenannte Theologie des Aristoteles, Arabic text, p. 108, l. 3). The term מוחלט in the sense of absolute, which occurs often in Crescas (p. 152, l. 13) and elsewhere, is of Mishnaic origin and is to be considered as the equivalent of the Arabic and the Greek terms rather than a translation thereof. For the opposite of משולח and מוחלט there are several terms each of which designates a different shade of meaning of the term relative. (a) צרופי in the various senses of the category of relation, מצטרף, مضاف, πρός τι, (Prop. VI, p. 238, l. 9). (b) נמשך تابع, ἀκόλουθος, consequent upon or incident to Prop. XIV, Part II, n. 9, p. 631; Prop. XV, p. 282, l.

14; below n. 14). (c) מקושר, مقيّد, *restricted*, from a root meaning *bind*, as אם מקושרת ואם מוחלטת in Narboni quoted below n. 8.

The expressions סתם ביחוד and סתם במוחלט are used by Hillel of Verona in his discussion of this proposition.

3. Crescas endeavors to explain here why Maimonides has included substance among the categories of change, for, as we shall see in the course of this note, there had been two kinds of classifications, one which included substance and the other which did not. The distinction drawn here by Crescas between timeless change and change in time corresponds to the distinction he draws later, in Proposition V, between change proper and motion. The latter is always change in time. (Cf. Prop. I, Part II, n. 101, p. 463). What Crescas is therefore trying to say here is that Maimonides has used the term change in this proposition advisedly to include timeless change. This implied difference between change and motion and the further implication that the former includes substance and the latter does not has a history behind it, which I am going to trace here with some detail.

Aristotle himself seems to make a distinction between change, μεταβολή, and motion, κίνησις. While in one place he says: "for the present we do not have to make any difference between the terms motion and change" (*Physics* IV, 10, 218b, 19–20), in another place he states explicitly that "change differs from motion" (*Physics* V, 5, 229a, 31). The difference between motion and change is expressed by him as follows: Motion is the change from a certain subject to a certain subject (*Physics* V, 1, 225b, 2, and V, 5, 229a, 31–32), whereas change may be from a subject to a non-subject or from a non-subject to a subject (*Physics* V, 1, 225a, 3 ff.). Accordingly, Aristotle denies that "there is motion in the category of substance" (*Physics* V, 2, 225b, 10–11), inasmuch as generation and corruption, he says, which constitute the changes in substance, are changes from a non-subject to a subject and from a subject to a non-subject (*Physics* V, 1, 225a, 26 and 32).

Following out this distinction, Aristotle seems to be on the whole very careful in the use of the terms change and motion. When he uses the term *change* as the subject of his classification, he enumerates four categories, including substance. But when he uses the term *motion*, he enumerates only three categories, excluding substance. The following references to his writings will illustrate this point.

I. Passages in which the term *change* is used and the category of substance is included:

Physics III, 1, 200b, 33–34.

Metaphysics VIII, 1, 1042a, 32—b, 3; XII, 2, 1069b, 9 ff.

De Gen. et Corr. I, 4, 319b, 31 ff.

The category of substance is also included in the classification given in *Physics* I, 7, 190a, 31 ff. and *Metaphysics* VII, 7, 1032a, 13–15, where instead of *change* the term *generation*, γένεσις, is implied. In the first of these passages the categories of relation and time are also mentioned.

II. Passages in which the term *motion* is used and the category of substance is excluded:

Physics V, 1, 225b, 7–9; 2, 226a, 24–25; VII, 2, 243a, 6–7; VIII, 7, 260a, 26–28.

De Caelo IV, 3, 310a, 23–24.

De Anima I, 3, 406a, 12 ff. Here Aristotle speaks of four kinds of motion, but he gets the four not by including substance but by resolving the term *quality* into *diminution* and *growth*.

Topics IV, 1, 121a, 30 ff.: "If, then, *motion* be assumed as the genus of pleasure, we must see whether pleasure be not locomotion (φορά), nor alteration, nor any of the other assigned *motions*." By mentioning here under *motion* the categories of *place* and *quality* and by referring to the remaining kinds of motion by the plural 'other motions', by the 'other motions' Aristotle undoubtedly means here the categories of *substance* and *quantity*. Thus, by implication, substance is included under motion, contrary to Aristotle's general usage. This contradiction to his general usage will appear all the more forceful if we accept the reading φθορά in this passage instead of φορά. Then, indeed, substance will be explicitly mentioned under motion. It is, however, possible that by 'other motions' Aristotle means here 'growth' and 'diminution,' which terms are often used by him in place of 'quantity.'

Categories, ch. 14, 15a, 13 ff.: "Of motion there are six species, generation, corruption, augmentation, diminution, alteration, and change of place." It will be noticed that these six species of *motion* fall under the four categories, including substance.

This sixfold classification of motion given by Aristotle in the *Categories* seems to have been adopted by many Arabic and Jewish philosophers from the earliest times. Traces of this classification are found in the works of the following authors:

Al-Kindi, "Liber de quinque essentiis," in *Die philosophischen Abhandlungen des Ja'qub ben Isḥaq Al-Kindi*, by Albino Nagy, p. 35: "Motus autem diuiditur in sex species. quarum una est generatio, et secunda corruptio, tertia alteratio, quarta augmentum, quinta diminutio et sexta permutatio de loco ad locum."

Iḥwan al-Safa. See Dieterici, *Die Naturanschauung und Naturphilosophie der Araber*, p. 11; *Die Lehre von der Weltseele bei den Arabern*, p. 117.

Isaac Israeli, *Sefer Yesodot* III, pp. 62–63 (and cf. p. 71):

"For motion must inevitably be either essential or accidental. As for essential, it is, e. g., the motion of generation and destruction. As for accidental, it is of two kinds, either motion of quantity, as, e. g., motion of increase [and decrease], or motion of quality, as, e. g., alteration, and translation from one place to another."

לפי שהתנועה לא תמנע מהיותו אם בעל עצם או בעל מקרה. אולם העצמות הוא כגון תנועת ההוה והפסד. ואולם המקרית תהיה על שתי פנים, אם תנועת הכמות, כגון תנועת הגדול [וההתוך], או תנועת האיכות, כגון השנוי, וההעתק ממקום למקום.

Saadia, *Emunot we-Deot* II, 2: "And thus of the six species of motion." וכן בששת מיני התנועה.

Pseudo-Baḥya's *Kitab Ma'ani al-Nafs*, ch. 2. ed. Goldziher, p. 6: "And the species of corporeal motions are six: motion of generation, motion of corruption, motion of augmentation and motion of diminution, motion_of place and motion of alteration." ואנואס אלחרכאת אלגסמיّה סתّה: חרכّה כון וחרכّה פסאד, חרכّה נמוّ וחרכّה דבול. חרכّה נקלّה וחרכّה אסתחאלّה. The term אסתחאלّה is translated in Broydé's *Torat ha-Nefesh*, p. 7, by the Hebrew מנוחה, *rest*, which is obviously wrong. The term אסתחאלّה reflects the Greek ἀλλοίωσις (cf. Munk, *Guide* II, p. 7) which is specifically used by Aristotle as a designation for qualitative change which is otherwise described by him as κατὰ ποιόν (*Physics* III, 1, 200b, 34), κατ' εἶδος (*De Caelo* IV, 3, 310a, 24) and κατὰ πάθος (*De Gen. et Corr.* I, 4, 319b, 33). Narboni distinguishes between μεταβολή, تغیّر, שנוי and ἀλλοίωσις by using for the latter השתנות ביחוד (see quotation below n. 8). Hillel of Verona uses for it חלוף in *Tagmule ha-Nefesh* (see quotation below) and חלול in Propositions XIII and XIV. The term חלול, however, may be a corruption of חלוף. In *Sefer ha-Yesodot* it is simply שנוי (see quotation above).

Hillel of Verona, *Tagmule ha-Nefesh* I, 3, pp. 3b–4a: "Shouldst thou be inclined to say that the soul is moved essentially by the motion of the body, [you will find that] it cannot be moved by any of the six kinds of motion which are found in four out of the ten categories, namely, substance, quantity, quality, and place. Substance includes two opposite motions, i. e., generation and destruction. Quantity includes increase and decrease. Quality includes only one kind of motion, and that is the alteration from one property to another, as, e. g., from hot to cold, from black to white, and their like. Alteration occurs when a new property is generated, contrary to the one which exists in the subject now, while the subject itself remains the same. Place, too, includes only one kind of motion which in its turn is divided into other kinds. This kind of motion is prior in nature to all the other motions, that is to say, locomotion, which is the motion whereby the heavenly bodies are moved."

אם תאמר שהנפש מתנועעת בעצמה בתנועת הגוף, אי אפשר לה להתנועע משֵׁשת התנועות הבאות בד' מאמרות מן העשרה, ר"ל העצם, הכמות, האיכות, והאנה. בעצם נכנסות שתי תנועות מתנגדות, הם ההויה וההפסד; בכמות נכנסות הצמיחה והחסרון; באיכות נכנסת תנועה אחת והוא החילוף מדבר לדבר, כלומר מחום לקור, מלובן לשחרות, ודומה לזה. והחלוף הוא בהעשות דבר אחר, הפך הנמצא בו בהוה, עם השאר הנושא קיים. בה ונֹאבֹה [read ונאנה] נכנסת גם כן מין אחד מתנועה, ותחתיו יש עוד מינים אחרים, חה המין מן התנועה הוא הקודם בטבע לכל התנועות יותר, כלומר תנועת המקומית, שבה יתנועעו גופי השמים.

Al-Saba'niyyah by Abu 'Imran Moses Ṭobi with Hebrew translation and commentary *Batte ha-Nefesh* by Solomon ben Immanuel Dapiera (published by Hartwig Hirschfeld in the *Report of the Judith Montefiore College*, 1894), p. 46, speaks also of six kinds of motion. But these six motions all belong to the three categories of place, quantity and quality. The number six is obtained by counting upward, downward and circular motions as three kinds of motions under place, and augmentation and diminution as two kinds of motions under quantity. "The motions of animal beings are six . . . Motion includes the three in place, [and those] in quantity [and] in quality. The three [in place] have been explained above [see p. 45: upward, downward, circular]. Motion in quantity is twofold, towards augmentation and towards diminution. This makes it five. Motion in quality makes it six."

תנועות חי שש הם... כלל התנועות שלש באנה, בכמות, באיכות. שלש, והם הנזכרות למעלה. התנועה בכמות שתים: אל התוספת ואל החסרון, הרי חמש. ותנועה באיכות הרי שש.

Still among the Arabic and Jewish philosophers who were acquainted with the other writings of Aristotle the classification of *motion* does not include substance. Thus Algazali in *Maḳaṣid al-Falasifah* III, p. 236: "And the term motion does not apply to all the categories but only to four; motion of place, and translation in the categories of quantity, position and quality."

Algazali's fourfold classification, with its inclusion of the category of position and exclusion of the category of substance, is adopted by Abraham ibn Daud in *Emunah Ramah* I, 3, p. 13. In Shahrastani it is definitely stated that there is no motion in the category of substance (ed. Cureton, p. 397).

In view of all this, it is strange that Maimonides himself, in his own explanation of this proposition, should maintain that the term change as used by him here is identical with motion and is in time, though he includes under it the category of substance. It is stranger still that Crescas should not have known of Maimonides' own explanation and offer here an explanation which is diametrically opposed to it. See *Ḳobeẓ Teshubot ha-Rambam we-Iggerotaw* II, (Letter to Samuel Ibn Tibbon), p. 27b:

"With regard to your question concerning the phrasing of the fourth proposition, there is nothing wrong with it. You may recall the general statement we have made in the introduction of the book that I have written it for him who has read much in the sciences and that it is not intended for him who has never studied any of these profound and difficult subjects. It is one of the generally known principles, about which there can be no doubt, that every change is necessarily a motion, for every change is in time and time is the measure of motion according to the

prior and the posterior in motion, as we have explained it in its
proper place [see Prop. XV]."

מה שזכרת מסדר ההקדמה הד' אין בה חסרון. וכבר ידעת מה שכללנו בפתיחת
הספר, כי חברתיו למי שקרא הרבה מן החכמות, ולא חברתיו למי שלא קדם לו
לעולם עיון בדבר מענינים העמוקים האלה הקשים להבין. ומן הידוע אשר אין
ספק בו, כי כל השתנות תנועה על כל פנים, לפי שכל השתנות תהיה בזמן, והזמן
הוא שעור התנועה בקודם ומתאחר בתנועה, כאשר נבאר במקומו.

The difference between Maimonides and Aristotle as to the
use of the term *motion* is correctly set forth in *Ruaḥ Ḥen*, ch. 11:
"Know that all these kinds of changes are called *motion* according
to the Master's view, as is set forth by him at the beginning of
the second part of his noble work the *Guide of the Perplexed*. But
according to Aristotle, there is no motion in the category of
substance."

ודע שכל אלו השנויים נקראים תנועה, לפי דעת הרב ז"ל, כמו שזכר בראש
החלק השני במאמר הנכבד ספר מורה הנבוכים, אך לפי דעת אריסטוטלו אין
תנועה במאמר העצם.

4. The reference here is to *De Gen. et Corr.* I, 4, 319b, 31 ff., where
a distinction is drawn between change in the categories of quantity, place and quality and the change of generation and corruption, i. e., change in the category of substance. The difference,
however, is not expressed by Aristotle in the terms used here by
Crescas, i. e., between temporal and instantaneous change. As
Aristotle puts it, change in the first three categories implies a
substratum which is perceptible and persists throughout the
change (319b, 10–11), whereas in change of substance there is
nothing perceptible which persists in its identity as a substratum
(319b, 14–21). The view that change of substance is in no-time
is reported in the name of Avicenna by Shahrastani (ed. Cureton,
p. 397). It is also found in the comments on this proposition by
Altabrizi, Narboni, the *Moreh ha-Moreh* and the *Ruaḥ Ḥen*, ch.
11. But this view was a matter of controversy, as we shall see in
Prop. VII. Maimonides in his letter to Samuel ibn Tibbon, quoted
above in n. 3, is of the opinion that all changes, including that of
substance, is in time. A similar statement is found in *Physics*
IV, 14, 222b, 31. There seems to be, however, according to
Maimonides, one exception to this generalization, and that is
the generation and destruction of forms. See *Moreh Nebukim*

II, 12: "Every combination of the elements is subject to increase and decrease, and this comes-to-be gradually. It is different with forms; they do not come-to-be gradually, and have therefore no motion; they come-to-be or pass-away without time."

כל מזג מקבל התוספת והחסרון, והוא יתחדש ראשון ראשון, והצורות אינם כן, שהם לא יתחדשו ראשון ראשון, ולזה אין תנועה בהם, ואמנם יתחדשו או יפסדו בלא זמן.

No mention is made of the distinction between change in time and change in no-time in the passage in the *Intermediate De Gen. et Corr.* I, i, 4 (Latin, p. 354rb–va) corresponding to *De Gen. et Corr.* I, 4, 319b, 31 ff., quoted above.

5. This question has been raised by Altabrizi: "Know that against the author's statements many objections can be raised, viz., what does he mean by the term change in his statement that 'change exists in four categories'? Does he mean sudden change, or gradual change, or change in general, whether sudden or gradual? He could not mean sudden change, for change in quantity, quality and place are not sudden but rather gradual . . . . He could not mean gradual change, for change in substance is not gradual but rather sudden . . . . Nor could he mean change in general, inclusive of all the kinds of change he mentions, be they sudden or gradual, for change in this general sense is not confined to those four categories mentioned, for every one of the categories is generated in the subject in which it inheres, and thus every one of them has some change either sudden or gradual. Why then did he single out these four categories to the exclusion of the others?"

ודע שעל דבור המחבר ספקות חזקות: והוא שיאמר מה הנרצה מהשנוי במאמרו שהשנוי ימצא בד' מאמרות. אם הנרצה בו השנוי פתאום, או לא פתאום, או השנוי משולח, שזה היה פתאום או לא פתאום. ואם היה רצונו בו השנוי פתאום, הנה השנוי בכמה והאיך והאנה לא יהיה פתאום אבל על ההדרגה. . . . . . ואם היה רצונו בו השנוי על ההדרגה, הנה השנוי בעצם לא יהיה על ההדרגה, אבל יהיה פתאום. . . . ואם היה רצונו בזה השנוי הוא השנוי משולח, עד יקיף כל אשר זכר, היה פתאום או לא היה פתאום, הנה השנוי משולח לא ייוחד במאמרות הארבעה אשר זכרם, כי כל מאמר מן המאמרות הנה הם יחודשו במשכנם, ויהיה לכל מאמר שנוי מה אם פתאום או לא פתאום, ולמה זה ייוחד לזכר המאמרות הד' בלתי שאריתם.

6. The category of position is included by Algazali among the categories of motion. *Makaṣid al-Falasifah* III, p. 236: "The term motion does not apply to all the categories, but only to four, namely, motion in place, and translation in quantity, in position and in quality." Upon this there is the following comment by Albalag: " 'The term motion does not apply to all the categories, but only to four, namely, motion in place, and translation in quantity, in position and in quality.' . . . . Says the translator: . . . . This is the view of Avicenna with regard to the celestial sphere, namely, that its motion is not in place, inasmuch as it has no place. Moreover, its motion is circular, and circular motion is not in place . . . . Aristotle's view, however, is that motion is in three categories, in quantity, quality and place, and that the motion of the [celestial] sphere is in place."

ולא תפול התנועה מכלם אלא בארבע: התנועה המקומית, וההעתק בכמה ובמצב ובאיכות... אמר המעתיק. .. זהו דעת אבן סינא בגלגל העליון שאין תנועתו מקומית, לפי שאין לו מקום, ולא עוד אלא שתנועתו סבובית, והתנועה הסבובית אינה במקום. .. ודעת אריסטו כי התנועה בג' מאמרות בכמות ובאיכות ובאנה, וכי תנועת הגלגל מקומית.

A similar comment occurs in Narboni's commentary on the *Maḳaṣid:* "Avicenna calls the motion of the celestial sphere motion in position, not motion in place, because of the fact that the body of the sphere as a whole does not change its place. But Averroes has already caught him up on this, for the celestial sphere does change its place as a whole in form if not in substance."

ואבן סיני יקרא תנועת הגלגל תנועה במצב, לא תנועת האנה, למה שכלל הגשם לא ימיר מקומו בכללו, וכבר תפס בן רשד עליו, כי הוא ימיר מקומו בכללו בצורה לא בנושא.

So is 'position' also mentioned by Shahrastani in the name of Avicenna (ed. Cureton, p. 398).

The same view is followed by Abraham ibn Daud in *Emunah Ramah* I, 3, p. 13: "Motion is a term applied primarily to the translation of a body from one place to another or to the translation of its position."

התנועה שם נאמר ראשונה על העתק הגשם ממקום למקום או על העתק מצבו.

Similarly Altabrizi is for the inclusion of position: "Then the philosophers proceed to say that motion exists only in four categories, three of which are mentioned here by the author, namely, the categories of quantity, quality and place, and a fourth one which is not mentioned by him, namely, position."

אחר כן אמרו התנועה אמנם תמצא בד' מאמרות, שלש מהם זכרם המחבר,
והוא מאמר הכמה והאיך והאנה, והנה באחד מהם לא דבר, והוא מאמר המצב.

He explains, however, the omission of the category of position
by Maimonides on the ground that motion of position is identical
with circular motion, and the latter is to be included, according
to Maimonides, under locomotion.

והתנועה במצב היא כמו התנועה הסבובית... ולזה התנועה הסבובית אצלו
נכנסת בתנועה במאמר האנה.

Cf. Judah Messer Leon's commentary on *Categories* III, 2,
On Motion: "It would seem that there is motion in the category
of position, even though Aristotle does not mention it, as, e. g.,
the motion of things that remain in the same place, and of such
a description is the motion of the celestial bodies. If one should
try to forestall this objection by saying that the spheres have motion only with reference to their parts and those parts do change
their place by motion, the answer is that it is not so, for the parts
of the spheres have motion only accidentally, by virtue of the
motion of the whole, whereas the motion of the whole is essential, and consequently the motion of the spheres ought to be
identified with the motion of the whole which is essential. It is
for this reason that [Avicenna] has said that the motion of the
celestial bodies is in the category of position. Averroes, however,
rejects this view. But we shall discuss this problem in the *Physics*."

וכבר יחשב שתהיה במאמר המצב, ואם לא זכרו אריסטו, כמו תנועת שאנה להם
אחד תמיד, וכן תנועת הגרמים השמיים. אלא אם יאמר אומר שהתנועה באלו
הוא לחלקיהם, והם מתחלפים מקומם באנה, חה שקר, כי התנועה לחלקיהם היה
במקרה, מצד תנועת הכל, והתנועה לכל היה בעצמותו ולזה היה מחוייב שתיוחס
התנועה למתנועע בעצמות. ולזה אמר בתנועת הגרמים השמיים שהוא במאמר
המצב. ון' רשד ירחיק זה. וכבר נחקור בזה בספר השמע.

7. Whether Aristotle himself included the categories of action
and passion under motion is not clear. On the one hand, in
Physics V, 2, 225b, 11–14 and 226a, 23–24, he definitely states
that there is no motion in the categories of relation, action and
passion. But, on the other hand, in *Topics* IV, 1, 120b, 26–27,
Aristotle seems to state that there is motion in the categories of
action and passion (cf. Zeller, *Aristotle*, Vol. I, p. 277, n. 1).
According to the Stoics action and passion are included under
motion, and this view was later introduced into the Aristotelian

doctrine (cf. Zeller, *Stoics, Epicureans, and Sceptics*, p. 185, n. 3). Shahrastani in the name of Avicenna enumerates only four categories of motion, namely, place, quantity, quality and position, and explains in great length how in all the other categories motion is to be found only indirectly and accidentally (p. 398, ed. Cureton). In the *Intermediate Physics* V, ii, 4, Averroes enumerates only the three categories of motion and tries to show that there can be no motion in any of the other categories. A similar discussion occurs also in *Ruaḥ Ḥen*, ch. 11. As against all this, Altabrizi states that change in the general sense of the term, if no distinction is made between temporal and instantaneous change, is to be found in all the ten categories (text quoted above n. 5).

8. In raising the question, as we have seen above (n. 5), Crescas has been following Altabrizi. In trying now to answer it, however, he disregards Altabrizi and follows other sources.

As preliminary to our understanding of Crescas' answer, I shall reproduce here first certain texts from Narboni which are the underlying sources of Crescas' statements here, then I shall try to show how the distinctions made by Narboni can be traced to Aristotle, and finally I shall point out that while Crescas is following Narboni on the whole he departs from him in certain details.

The immediate source of Crescas' answer is the following passage in Narboni's commentary on this proposition in the *Moreh:*

A. "Change has two subjects, a sustaining subject, i. e., the body underlying the change, as e. g., water, and a material subject, i. e., the quality that passes from potentiality into actuality, as, e. g., heat or cold, or blackness and whiteness in a body that is becoming black or white. With reference to the change itself, i. e., the transition [of the sustaining subject] from one state to another without reference to the state, change belongs to the category of passion, that is to say, it is the process of suffering action and of being affected and the realization of a state of being which previously did not exist. But with reference to the material subject, i. e., the state of being itself with reference to which the body in question is undergoing a change in passing from that state to another, change belongs to the category to which that state belongs (see below n. 12), for when a potentiality with reference to any of the categories falls in some way under any given category, then the motion or change, which is a certain entelechy of that potential state of being, seeing that is a sort of realization whether relative or absolute, must be included under that category to which belongs the state of being that is passing from potentiality to actuality.

This is what is meant by this proposition wherein it is stated that change exists in certain categories. What is meant is that inasmuch as the material subject of change exists in four categories the change itself exists in those very same categories, for change is of the nature of the state that comes-to-be (see below n. 12) and, as such a state exists in four categories, change itself exists in them. These categories are then specified as follows:

'The category of substance,' and this change which occurs in substance is 'generation and corruption.' By this is meant the non-being and the coming-into-being of the form. With reference to the form which comes-to-be after it has-not-been, it is called generation, and this is a change from non-being to being. With reference to the form that passes-away, it is called corruption, and this is a change from being to non-being. But with reference to translation from one form to another form, it is called change from being to being. In the last mentioned case, there is only one change, but in the first two cases there are two changes.

And it exists 'in the category of quantity, which is growth and diminution,' thus again two opposite motions.

And it exists 'in the category of quality, which is alteration' in the proper sense of the term, as, for instance, when cold water becomes hot.

And it exists 'in the category of place, which is the motion of translation, and to this change of place the term motion proper is applied but of the other kinds of changes it is used in a general sense.' Truly speaking there is no motion in the category of substance, for substantial change takes place suddenly."

והשנוי לו שני נושאים, אם נושא מעמיד, והוא הגשם המשתנה, כמים, ואם נושא
חמרי, והוא הדבר היוצא מן הכח אל הפועל בעצמו, כחום או הקור או השחרות
והלובן בגשם המשתחר או המתלבן. ומצד עצמות השנוי, שהוא העתק מתואר אל
תואר בלי בחינת התואר, היה השנוי במאמר ההתפעלות לבד, והוא ההתפעלות
והתרשמות והגעת תואר לא היה. ובבחינת הנושא החמרי, והוא התואר אשר יעתק

הגשם בו והלך מן התואר אל התואר, היה השני במאמר אשר בו התואר ההוא,
כי כאשר היה הכח על מאמר נכנס באופן מה במאמר ההוא שהתנועה או השני,
אשר הוא שלמות מה לדבר, כי הוא הגעה אם מקושרת ואם מוחלטת, שראוי שנכנסו
במאמר אשר בו הדבר ההוא היוצא מן הכח אל הפועל.

חזו הנרצה בזאת ההקדמה, שאמר בה שהשני ימצא ממאמרות. ירצה שהנושא
החמרי לשני הוא בד' מאמרות, השני ההוא גם כן ימצא בהם בעצמם, כי השני
הוא מטבע התאר המתחדש, והוא נמצא בד' מאמרות, הנה השני ימצא גם כן בהם.
והנה מפרש זה ואומר.

מאמר העצם. חה השני הוא בעצם והוא ההויה וההפסד, ירצה העדר הצורה
ותתחדש הצורה. ובבחינת הצורה המתחדשת אחר שלא היתה יקרא הויה, והוא
שני מלא מציאות אל מציאות. ובבחינת הצורה הנפסדת יקרא הפסד, והוא שני
ממציאות אל לא מציאות. ובבחינת העתק מצורה אל צורה יקרא שני ממציאות
אל מציאות. ובזאת הבחינה יהיה השני אחד, ובשתי הבחינת הראשונות יהיה שני
שניים.

וימצא במאמר הכמה, והוא הצמיחה והחסרון. וכן שתי תנועות מקבילות.
וימצא במאמר האיכות, והוא ההשתנות ביחוד, כשוב המים הקרים חמים.
וימצא במאמר האנה, והוא תנועת ההעתקה, ועל זה השינוי באמר תאמר התנועה
בפרט ועל שאר השנויים בכלל. ואם במאמר העצם אין תנועה באמת למה שהוא
פתאומי.

B. A similar use of the terms "material subject" and "sustaining subject" is found in Narboni's commentary on the *Moreh* I,
73, The Third Proposition: "Know that motion is the entelechy
of that which is in potentiality, in so far as it is in potentiality,
while it has that entelechy. Therefore the entelechy which is
motion is an intermediate entelechy, that is to say, the *material
subject*, i. e., the thing itself which passes from potentiality into
actuality, is neither completely potential nor completely actual,
but its realization is taking place slowly and gradually so that
the potentiality cannot be distinguished from the actuality. If
the motion, for instance, is that of place, it is the gradual con-
sumption of distance. This is the *material subject* of motion, for
the *sustaining subject* refers to the thing that is being moved."

דע כי התנועה היא שלמות מה שבכח מצד מה הוא בכח עם היות לו זה השלמות.
ולכן היה זה השלמות אשר הוא התנועה הוא שלמות ממוצע, ר"ל שאין הנושא החמרי
בכח גמור ולא בפועל גמור, ר"ל הדבר ההוא היוצא מהכח אל הפועל בעצמו,
אבל הוא מניע מעט מעט מעט ראשון בלתי נכר הכח מן הפועל. ואם היא תנועת
האנה, הנה היא הגעת הדרך ראשון ראשון, והוא הנושא החמרי, כי המתנועע הוא
הנושא המעמיד.

C. Cf. also Narboni on *Moreh* II, Introduction, Prop. XXIV: "From this you may gather that the term 'possible' may be applied in general to two kinds of things: First, to that which receives, which may be named the *sustaining subject*, and an example of this is prime matter, which is potential with reference to form, and likewise body, which is potential with reference to accidents. Second, to that which is received, which may be named the *material subject*, and an example of this is the form [with reference to prime matter] or the accidents [with reference to body]."

ונראה לך מזה כי האפשרי יאמר על שני מינים, על המקבל, והוא הנושא המעמיד, והוא החומר הראשון, אשר הוא בכח אל הצורה, וכמו כן הגשם, אשר הוא בכח אל המקרים, ויאמר על המקובל, והוא הנושא החמרי, והוא הצורה או המקרים.

D. In his commentary on Algazali's *Kawwanot ha-Philosofim* III, on motion, Narboni quotes this distinction in the name of Averroes. "Said Averroes in the fifth book of the *Physics* . . . . that motion has two aspects, first, with reference to its matter, and, second, with reference to its form. The meaning of this is as follows: Motion has two subjects, (a) A subject in which it exists, and this is identical with that which is movable. It is with reference to this subject that motion is defined as the entelechy of that which is movable *qua* movable. (b) A material subject, and this is identical with that which is realizable in place or in quality or in quantity or in substance, if there be motion in the category of substance. It is with reference to this subject that motion is defined as the entelechy of that which is in potentiality (see about the two definitions of motion in Proposition V, p. 523, n. 5) . . . . Motion, then, when viewed with reference to its matter . . . . is to be included under the four categories . . . . But in general, when we consider motion only with reference to its form . . . . it is to be included under the category of passion, for it is the transition of a thing from state to another."

אמר בן רשד בחמישי משמע טבעי. . . . . . כי התנועה לה שתי בחינות, האחת מצד חמרה, והשנית מצד צורתה. ובאור זה, כי התנועה לה שני נושאים, נושא בו תעמוד, והוא המתנועע, ולזה כבר יאמר בגדר שהוא שלמות המתנועע במה הוא מתנועע, ונושא חמרי, והוא המגיע באנה או איך או כמה או עצם, אם היה במאמר העצם תנועה, והוא אשר יאמר עליו שלמות מה שבכח. . . הנה התנועה כאשר נבחנה מצד חמרה. . . היתה התנועה נכנסת בארבע מאמרות. . . ובכלל שנקח

מהתנועה צורתה לבד... הנה היא נכנסת במאמר שיתפעל, כי היא תמורת הדבר
מתאר אל תאר.

E. This distinction is made, without mentioning Averroes, in
an anonymous supercommentary on the *Intermediate Physics*
(MS. Adler 1744.2) V, ii, 4: " 'The contraries between which
there is an intermediate etc.' If the question is raised that motion
is known to exist in a category in which there is no intermediate
between the contraries, as, e. g., the categories of action and passion, our answer is that motion has two subjects, a material subject and a sustaining subject, and that the motion which exists
in the categories of action and passion is that with reference to
the sustaining subject which we have mentioned. But in three
categories, i. e., quantity, quality and place, there is motion, for
these categories there is an intermediate between the contraries."

ההפכים אשר ביניהם אמצעי וכו'. ואם נאמר וכבר נמצא תנועה במאמר שאין
ביניהם אמצעי, כמו מאמר שיפעל ושיתפעל, נשיבם התנועה לה שני נושאים, נושא
חומרי ונושא מעמיד, והתנועה שיש בהם הוא מפני הנושא המעמיד כאשר אמרנו.
ובשלשה, ר"ל בכמה ובאיך ובאנה, תמצא התנועה, כי ביניהם אמצעי, ר"ל בין
שני ההפכים.

F. The original statement of Averroes is not found either in his
Intermediate Commentary or in his *Epitome*. It is found only in
his Long Commentary on the *Physics* V, i, 3, of which the following passage is quoted from the Latin translation (p. 215ra, B):
"Motus igitur habet duplicem consyderationem. quoniam secundum suam materiam est in genere eius, ad quod est motus, secundum autem formam, idest secundum quod est transmutatio
coniuncta cum tempre, est in praedicamento passionis."

There is no single passage in Aristotle to which this distinction
of the two kinds of subjects in motion can be traced. But it can
be shown that on the whole it reflects the main trend of his views:

First, as pointed out by Narboni himself (quotations B and
D), it is based upon Aristotle's two definitions of motion, which
we shall discuss later in Prop. V, n. 5.

Second, it reflects Aristotle's discussion in *Physics* V, 1, 224a,
34–224b, 16. Aristotle names five things which are present with
motion, namely, the mover, that which is moved, time, that from
which the motion proceeds, and that to which it tends. He then
raises the question as to in which of these five things motion exists.

He eliminates outright the mover, time, and that from which motion proceeds. He takes up the remaining two and concludes that motion is in that which is moved (τὸ κινούμενον, המתנועע.) As for the *into which* (εἰς ὅ, מה שאליו), he draws a distinction. Taking the change of a thing in its process of becoming white as an example, he says that whiteness (λευκότης, לבן) is not motion, but becoming white (λεύκανσις, לבון) is motion (*Physics* V, 1, 224b, 15–16).

Now, taking this last example of Aristotle, the change undergone by a thing in its becoming white, Averroes would call the thing underlying the change (τὸ κινούμενον) the *sustaining subject* whereas the color that is becoming white (λεύκανσις) he would call the *material subject*.

Third, it may be traced to the following passage in *Metaphysics* VII, 7, 1033a, 7–12: "But though what becomes healthy is a man, 'a man' is not what the healthy product is said to come from. The reason is that though a thing comes both from its privation and from its substratum, which we call its matter (e. g., what becomes healthy is both a man and an invalid), it is said to come rather from its privation (e. g., it is from an invalid rather than from a man that a healthy subject is produced)." Now, in this illustration, Averroes would call "man" the *sustaining subject* and "invalid" the *material subject*.

Fourth, it reflects a lengthy discussion of Aristotle which occurs in *De Generatione et Corruptione* I, 4, 319b, 8 ff., and in *Physics* V, 1, 224b, 35 ff. I shall start with an analysis of the passages in the *De Generatione et Corruptione* and then correlate with them the passages in the *Physics*.

In the *De Generatione et Corruptione* Aristotle enumerates the four species of change, belonging to the four categories of *quantity*, *place*, *quality* and *substance* (319b, 31–320a, 2). Each of these changes is from contrary to contrary, as, e. g., growth and diminution in *quantity;* front and rear in *place;* hot and cold in *quality;* generation and corruption in *substance*. In each of these changes, furthermore, there is a subject or substratum (ὑποκεί-μενον) which is receptive of both the contraries. There is, however, the following difference between the subject in the changes of *quantity*, *place* and *quality* and that of *substance*. In the first three, the subject is perceptible (319b, 11) and the contraries are each "an accident in the general sense of the term" (320a, 1). In the change of *substance*, the subject is imperceptible (319b, 15), being "matter in the most proper sense of the term" (320a, 2), and the contraries generation and corruption do not exist in it as accidents. Cf. Joachim, *Aristotle on Coming-to-be and Passing-away*, p. 105 ff.

Aristotle goes further to say that in the categories of quantity, quality and place, the changes may be considered with reference to three things. First, with reference to the subject. Second, with reference to the categories to which the contraries, considered independently of their subject, happen to belong. Third, with reference to the contraries considered together with their subject, not as accidents but as forms of the subject. If we take, for instance, the qualitative change expressed in the statement "that the *musical man* passed-away and an *unmusical man* came-to-be, and that the *man* persists as something identical" (319b, 25–26), in that change three things are to be considered. First, *man* as the perceptible, persistent subject of the contrary properties musicalness and unmusicalness. Second, *musicalness-and-un-musicalness* as constituting a property or quality inhering in man. Third, the *musical man* and the *unmusical man* considered as two men. Now, says Aristotle, the changes will have different designations in accordance to each of these three aspects.

First, "as regards *man*, these changes are πάθη" (319b, 29). The meaning of πάθη here is uncertain. Joachim takes it, with some hesitation, in the sense of ἀλλοιώσεις. But from Narboni's and Averroes' statements in quotations A and F, it is clear that in the Arabic and Hebrew translations of the *De Generatione et Corruptione* the term πάθη here was taken in the sense of πάσχειν, i. e., the category of passion. Thus, according to this interpretation of the text, the changes with regard to the *subject* belong to the category of passion.

Second, with reference to *musicalness-and-unmusicalness* constituting "a property essentially inhering in man" (319b, 27), the change belongs to the category of *quality* and is therefore called alteration (cf. 319b, 33 and 30).

Third, "as regards *musical man* and *unmusical man*, they are generation and corruption" (319b, 29); i. e., they belong to the category of *substance*.

By the same token, we have reason to infer, if instead of "musical" and "unmusical," we take the predicates "great and small" or "front and rear," with reference to *man* the changes belong to the category of *passion;* with reference to *great* and *small* or *front* and *rear* they belong to the categories of *quantity* and *place* respectively; but with reference to *great man* and *small man* or *front man* and *rear man*, the changes belong to the category of *substance*.

But still, according to Aristotle, there is a difference between substantial change in this last illustration, which is only involved in the other three kinds of change, and substantial change which is a complete coming-to-be and a complete passing-away, as, e. g., the birth and death of a musical man. The former kind of substantial change may be called *relative* substantial change, or, to use Aristotle's own expression, it is "a certain" (τις: *Physics* V, 1, 225a, 14) change. The latter kind may be called *absolute* substantial change, or, to use again Aristotle's own expression, it is change "simply" (ἁπλῶς, *ibid.*). We may express this distinction between the relative and the absolute kind of substantial change in still another way, also suggested by Aristotle. Relative substantial change is from a subject to a subject, by which terms is meant a perceptible subject. Absolute substantial change is either from a subject to a non-subject or from a non-subject to a subject, i. e., either from a perceptible subject to an imperceptible subject or from an imperceptible subject to a perceptible subject.

Cf. *Intermediate Physics* V, ii, 3: "After it has been shown that motion is of two kinds, either from a subject to a subject, i. e., from a contrary to a contrary, or from a subject to a non-subject and from a non-subject to a subject, i. e., from being to non-being and from non-being to being, meaning by non-being here not absolute negation but rather privation which is inherent in matter, I say that motion cannot exist in change from a non-subject to a subject and from a subject to a non-subject. It exists only in the change from a subject to a subject. Although it is true that of both these kinds of change we say that it is from a non-subject to a subject, the meaning of the term 'non-subject' is like that of the term 'non-being' in the phrase from 'non-being to being' when applied to the same two kinds of change. For the prefix 'non' is used in both these cases equivocally. Its proper meaning, however, is evident. In the first kind of change we mean by 'being' and 'non-being' that absolute being is generated from absolute non-being, as, e. g., man is generated from non-man. This is absolute generation, and its opposite is absolute corruption. But in the second kind of change we mean by 'being' and 'non-being' that being is generated from non-being which is a certain being, i. e., white is generated from non-white which is black. This is not absolute generation; it is only a certain generation, and in the same way its opposite is not absolute corruption but only a certain corruption. In general, these two kinds of change are differentiated from each other in two ways. First, the change from a subject to a subject contains something actual which constitutes the subject of the change, whereas generation and corruption contains nothing actual to constitute the subject of the change. The latter is therefore called absolute generation and corruption, whereas the change in the former case is called a certain generation and corruption. The second differentia is that the change from a subject to a subject is from an existent contrary to an existent contrary and from an affirmation to an affirmation, whereas the change from a non-subject to a subject is from privation to existence and from negation to affirmation."

ואחר שכבר התבאר שהשנוי בו מינים, אם מהנושא אל נושא, ר"ל מהפך אל
הפך, ואם מנושא אל בלתי נושא או מבלתי נושא אל נושא, ר"ל ממציאות אל העדר
ומהעדר אל מציאות, ולא רצה בהעדר בכאן השוללת המוחלט אבל ההעדר הנמצא
בהיולי, אומר שהתנועה אי אפשר שתהיה בהשתנות אשר יהיה מבלתי נושא אל נושא
ומנושא אל בלתי נושא, ואמנם תהיה כאשר יהיה מנושא אל נושא. חה שהוא ואם
אמרנו בשני מיני השנוי שאשר יהיה מבלתי נושא אל נושא, אמנם ירצה לומר שכבר
יהיה מהבלתי אם כמו שנאמר שיהיה מבלתי נמצא נמצא, כי מלת בלתי נאמרת
בם בשתוף השם, וכי מהם הפירוש מבואר. כי אנו נאמר בראשון שהוא מבלתי
נמצא במוחלט יהיה נמצא במוחלט, כמו שנאמר מבלתי אדם יהיה אדם, וזאת היא
ההויה המוחלטת והפכה ההפסד המוחלט. ואמנם המין האחר מהשנוי אמנם נאמר
בו שהוא מבלתי נמצא מה שהיה נמצא, ר"ל מהבלתי לבן אשר הוא נמצא מה,
ר"ל השחור, יהיה לבן, חה לא יאמר בו שהוא ההתהוות במוחלט, אמנם יאמר בו
שהוא ההתהוות מה, כמו שיאמר בהעדרו הפסד מה לא הפסד מוחלט. ובכלל הנה
ב' השנויים מובדלים בב' ענינים. הא', כי השנוי אשר יהיה מנושא אל נושא דבר
מה בפועל אשר הוא נושא השנוי. והמין הב' מההויה וההפסד אין בו דבר בפועל

נושא השנוי, ולזה יאמר בו שהוא הויה מוחלטת והפסד מוחלט, ויאמר בשנוי שהוא
הויה מה והפסד מה. וההפרש השני שהשנוי אשר יהיה מנושא אל נושא אמנם יהיה
מהפך נמצא אל הפך נמצא ומחיוב אל חיוב, והשנוי הב' היה מהעדר אל מציאות
ומשוללות אל חיוב.

In the foregoing analysis of Aristotle I have purposely restated
his views in such a manner as to form a background of Narboni.
In Narboni's language, the ὑποκείμενον of Aristotle is called
נושא מעמיד, which he himself explains as נושא בו תעמוד, "the
subject in which the motion exists (or by which the motion is
sustained)." We may therefore translate נושא מעמיד by *sustaining subject*. The accidents of quantity, place and quality which
are predicated of the *sustaining subject* are called by Narboni
נושא חמרי, literally, *material subject*, but preferably, *subject matter*. This *subject matter* is identified by him, quite properly, with
"form and accidents" (see quotation C). It should be noticed
that throughout his discussion Narboni applies the expression
sustaining subject to primary matter, i. e., to the imperceptible
subject. He thus finds the distinction between the *sustaining
subject* and the *subject matter* in all the four categories, including
the category of substance.

On this last point Crescas seems to depart from Narboni. It
will be impossible to explain fully all of Crescas' statements unless we assume that he uses the expression *sustaining subject* with
reference to a perceptible or, as Averroes calls it, actual subject,
and the expression *subject matter* with reference only to accidents
of quantity, place and quality existing in the perceptible subject.
He does not seem to apply this distinction to absolute substantial
change where there is but an imperceptible sustaining subject.

9. Hebrew העתק המשתנה מתאר אל תאר. The term תאר here
reflects the Greek πάθος in *De Gen. et Corr.* I, 4, 319b, 8. But
the Hebrew cannot be translated here by *property*, for that would
apply only to the category of quality (cf. *Ibid.* 319, 33), whereas
Crescas uses it, as he proceeds to specify, with reference to the
three categories of quantity, quality and place. The term תאר
is therefore to be understood here in the sense of *accident* in
general. Cf. *ibid.* 320a, 1: πάθος ἢ συμβεβηκὸς ὅλως.

In Narboni (quotation A) the same term תאר is used also with reference to the category of substance. Accordingly I have rendered it there by *state* and *state of being*.

10. We have seen above in n. 7 that while some authorities did include the categories of position, action and passion in their classifications of motion, none of them included all the ten categories, with the exception of Altabrizi who makes a general statement to that effect. Furthermore, Narboni, who is the immediate source of Crescas here, says definitely that change with reference to the *sustaining subject* exists in the category of passion, which, as we have shown, is based upon a dubious statement in *De Gen. et Corr.* I, 4, 319b, 28 (see above n. 8). Consequently, this statement of Crescas here is to be rendered either "and the other categories," thus reflecting the statement of Altabrizi, or "and the other categories [mentioned above]." Crescas himself later in Prop. V says that change with reference to the *sustaining subject* belongs to the categories of action and passion.

Crescas' statement here, however, may perhaps reflect the following passage in *Kawwanot ha-Pilosofim* III (*Maḳaṣid al-Falasifah* III, pp. 235–236): "As for its true meaning, it is well-known that motion applies only to translation from one place to another, but by the common consent of the philosophers it has come to be used in a more general sense, signifying the transition from one descriptive quality to another . . . This transition from one state to another undoubtedly applies to all the ten categories, but motion does not apply to all the categories but only to four."

ואולם האמתות, הנה המפורסם שהתנועה תשולח על ההעתק ממקום למקום לבד, ואבל היתה בהסכמת האנשים מליצה מעניין יותר כולל ממנה, והוא ההליכה מתאר אל תאר . . . וההעתק מעניין (חאל) לעניין אמנם יפול במאמרות העשר בלי ספק, ולא תפול התנועה מכלם, אלא בארבעה.

11. The omission of substance is significant. Using the expression *sustaining subject*, as we have suggested (above n. 8), only with reference to a perceptible subject, Crescas similarly uses the expression *subject matter* only with reference to accidents which exist in a perceptible subject. Consequently, change with reference to the *subject matter* cannot exist in the category of substance.

12. Hebrew ‏ובבחינה הזאת הוא במאמר אשר בו חומר השנוי‏. Verbally this passage is undoubtedly a paraphrase of the following passage in Narboni (above n. 8, quotation A): ‏ובבחינת הנושא‏ ‏החמרי, והוא התואר אשר יעתק הנשם בו והלך מתואר אל תואר, היה השנוי‏ ‏במאמר אשר בו התואר ההוא‏. But it is used by Crescas in a different sense. Narboni's original statement means that change is named after the *terminus ad quem*. Cf. *Physics* V, 1, 224b, 7–8: "For change is more denominated from that into which, than from that from which, it is moved." Crescas' statement here means this: Change with reference to the accidents which exist in a perceptible substratum is to be found only in the three categories of quantity, quality and place. For it is only in these three categories that you have a perceptible subject receptive of contrary accidents, such as 'augmentation and diminution in *quantity*, blackness and whiteness in *quality*, front and rear in *place*. In *substance*, to be sure, there is generation and corruption, but these are not changes between accidental qualities but rather absolute substantial changes between being and non-being and there is no perceptible substratum there.

Cf. *Intermediate Physics* V, ii, 3: "It is evident that there is no motion in the category of substance, inasmuch as motion is defined as the entelechy of that which is movable, but there is nothing actual that is movable in this substantial kind of change." ‏והוא גלוי שאין בו תנועה אחר שהיתה התנועה כמו שיאמר בגדרה שהיא שלמות‏ ‏המתנועע, ואין מתנועע בזה המין מהשנוי נמצא בפועל.‏

Intermediate Physics V, ii, 4: "It is evident that there is no motion in substance, inasmuch as there is no contrary in it. Furthermore, substantial change, as we have said, has no actual subject, its subject being only potential." ‏והוא גלוי שאין בעצם תנועה, אחר שאין בו הפך. ועוד כי השנוי אשר בעצם,‏ ‏כמו שאמרנו, אין בו נושא בפועל לשנוי, ואמנם הנושא בו לשנוי הוא בכח.‏

13. That is to say, the proposition deals with change in which a *perceptible substratum* passes from one *accident* to a contrary *accident*, as, e. g., from one size to another, from one color to another, or from one place to another, and then, too, with reference only to the size, the color and the place involved, i. e., the *matter* of the change, but not with reference to the *substratum* underlying the change.

14. It will have been noticed that Narboni, by taking the *sustaining subject* to include an imperceptible subject, i. e., matter, and by taking also the *subject matter* to include forms in addition to accidents (see above n. 8), had no need of explaining the inclusion of the category of substance by Maimonides in this proposition. Crescas, however, by using the terms *sustaining subject* and *subject matter* with reference only to a perceptible subject and accidents, has to look now for an explanation for the inclusion of the category of substance in the proposition.

Crescas' explanation is expressed in the following statement: והיה השיני אשר בעצם נמשך לתנועה אשר באלו המאמרות, ייחד הרב אלו הארבעה מאמרות. In the English text I have given a literal translation of it. But what does it mean?

It would seem that the statement lends itself to three possible explanations:

(a) Change of substance, according to Maimonides, is always preceded by changes of place and quantity and always precedes change of quality (see Prop. XIV, p. 281). Hence, argues Crescas, since Maimonides has enumerated here the changes of quantity, quality and place, he also had to mention substance, inasmuch as it is involved in all these three.

(b) As we have seen above (n. 8), in every quantitative, qualitative and spatial change there is a relative substantial change. What Crescas, therefore, means to say here is this: Whenever there is a change of quantity, or of quality or of place there is always a relative change of substance. To take Aristotle's own example, when a musical man becomes an unmusical man, the change with reference to *musical man* and *unmusical man* and not with reference to *man* or to *musical* and *unmusical* is a relative change of substance. Now, argues Crescas, while indeed in absolute substantial change there is no distinction between *sustaining subject* and *subject matter* in the specific sense used by Maimonides, still he includes relative substantial change in the proposition because of its being concomitant with the other three changes. Similarly in Prop. XIV (Part II) Crescas points out that Maimonides deals only with relative generation and the term used by him there is the same as here, הויה נמשכת (see p. 282).

(c) The statement may reflect the following passage in *Metaphysics* VIII, 1, 1042b, 3–5: καὶ ἀκολουθοῦσι δὴ ταύτῃ αἱ ἄλλαι μεταβολαί, τῶν δ' ἄλλων ἢ μιᾷ ἢ δυοῖν αὕτη οὐκ ἀκολουθεῖ. The meaning of this passage is explained by Averroes in his Long Commentary (Latin, p. 211rb) as follows: That which has change of substance has also all the other three changes, but that which has change of place may not have change of substance, as, e. g., the celestial spheres. If this be the source of Crescas' statement here, then it does not mean, as it would literally suggest, that change of substance is incident to the motion of the other categories, but it is rather to be understood to mean that change of substance involves the motion of the other categories.

15. If the third interpretation given in the preceding note is right, then the reference here is clearly to the quotation from *Metaphysics* VIII, 1, 1042b, 3–5. Accordingly what Crescas means to say here is that the reason for Maimonides' inclusion of substance among the categories of change is Aristotle's statement in the *Metaphysics* that the change of substance involves all the other changes. Otherwise, the reference is to *Metaphysics* VIII, 1, 1042a, 32—1042b, 3, which is one of the places where Aristotle enumerates all the four categories of change. Accordingly what Crescas means to say here is that Maimonides' enumeration of the four categories of change in this proposition follows the enumeration given by Aristotle in the *Metaphysics*.

16. The emphasis is here on the word "right." It is an indirect allusion to his preference of Narboni's answer of the difficulty to that of Altabrizi's and also to his slight modification of Narboni's answer (see above n. 8).

17. Cf. *De Gen et Corr.* I, 4, 320a, 17–19: "Since it is evident that, whereas neither what is altering nor what is coming-to-be necessarily changes its place, what is growing or diminishing changes its spatial position of necessity."

Physics VIII, 7, 260b, 13–15: "The magnitude likewise of that which is increased or diminished, changes according to place."

Kawwanot ha-Pilosophim III (*Maḳasid al-Falasifah* III, p. 236): "Quantitative motion likewise cannot be without locomotion." והכמה לא ימנע גם כן מהתנועה המקומית.

The same question is also raised by Hillel of Verona: "From Aristotle's and Averroes' statements in *De Caelo et Mundo* and in *De Generatione et Corruptione* it is evident that growth-and-diminution is motion in place."

ומתוך דברי ארסטו ודברי אבן רשד בספר השמים העולם ובספר ההוה וההפסד נראה שהצמיחה וההתוך, ר"ל הפריה והחסרון, היא תנועה באנה.

18. Altabrizi: "As for change in the category of quantity, as growth and diminution, it almost deserves to be called motion; it is not called so, because the motion therein is imperceptible."

ואולם השנוי במאמר הכמה בצמיחה והתכה, הנה קרוב משיקרא תנועה בלשון, אבל לא יכנס בחוש.

A similar answer is given by Hillel of Verona: "The reason why the Master has ascribed growth and diminution to quantitative motion and not to locomotion is to be found in the fact that objects moved by locomotion are moved either both from within and from without, as in the case of animals and the motion of the heavens, or only from without, as in the case of the motion of artificial things. These motions are more known to the senses, whereas the motion of growth and diminution is more known by reason and nature, for nature is the principle of motion to that in which it is inherent essentially" (cf. Prop. XVII, n. 7).

שהטעם מדוע יחסה הרב אל התנועה בכמות ולא אל האנה הוא בעבור כי המנענע תנועת הצמיחה וצריך לתקן: האנה] הוא מבית ומחוץ, כמו הבעלי חיים ותנועת השמים, ובקצת ענינים הוא מחוץ לבד, כמו תנועת המלאכותי. ואלו הן תנועות מפורסמות יותר אצל החוש, ואותה היא יותר מפורסמת אצל השכל ואצל הטבע, שהטבע הוא מה שהתחלת תנועתו בו בעצם.

19. This seems to reflect the following passage in *De Gen. et Corr.* I, 4, 320a, 19–22: "For that which is being moved changes its place as a whole: but the growing thing changes its place like a metal that is being beaten, retaining its position as a whole while its parts change their places."

PROPOSITION V

1. The Hebrew text of this proposition reads alike in Ibn Tibbon's translation of the *Moreh* and in Isaac ben Nathan's translation of Altabrizi.

2. This statement is based upon Altabrizi: "But it is inconvertible, for generation is also a transition from potentiality to actuality and still is not motion." ולא יתהפך, כי ההויה גם כן יוצאת מהכח אל הפעל ואינו תנועה. Narboni similarly remarks: "It is evident from this that every motion is change but that not every change is motion, for motion does not take place suddenly but is rather a gradual transition from potentiality to actuality, whereas the transition from potentiality to actuality which is change may be either sudden or gradual."

הנה מבואר מזה שכל תנועה שנוי ואין כל שנוי תנועה, בשלא יהיה פתאומי כי אם ראשון ראשון בהדרגה ויצאה מן הכח אל הפעל, כי היציאה מן הכח אל הפעל הוא השנוי יהיה פתאומי או מעט מעט.

A similar remark is also made by Hillel of Verona: "While it is true that every motion is change, this is not an altogether convertible definition, for not every change is motion, that is, motion in the ordinary sense of the term."

ואף על פי שכל תנועה שנוי, אין זה גדר מתהפך לגמרי מכל צד, שהרי אין כל שנוי תנועה, כלומר, תנועה סתם.

Cf. above Prop. IV, p. 503, n. 4.

3. Cf. Prop. IV, p. 517, n. 10.

4. Taken literally the text contains the following argument: (a) The proposition is inconvertible. (b) It is inconvertible because change means both timeless and temporal change, and of these only the latter can be called motion. But if this is what was meant by Crescas, then his conclusive remark that none of the philosophers has been aware of this distinction is puzzling, to say the least, for we have seen that the incovertibility of this proposition has been asserted by both Altabrizi and Narboni (see above n. 2) and similarly the distinction between timeless and temporal change is not original with Crescas (see above Prop. IV, p. 503, n. 4).

What the text perhaps means to say, but says it imperfectly, may be stated as follows: (a) It is asserted that the proposition is inconvertible on the ground that change includes timeless change. (b) But inasmuch as Prop. IV has been explained to deal with change only in its respect to the "subject matter," in which respect change is temporal and is motion, Prop. V similarly uses the term change in that restricted sense. (c) The proposition is thus convertible, contrary to the assertions of Altabrizi and Narboni who failed to note this distinction. Here I have therefore retained the reading of the printed editions אבל השנוי ממנו אשר הוא בבחינת חומר השנוי אשר בו יצדק שם התנועה לבד as against most of the MSS. which omit ממנו and אשר and have translated the text according to the interpretation suggested above.

Cf. discussion on this point in Flensberg's commentary *Oẓar Ḥayyim* on *Or Adonai, ad loc.*

5. The following preliminary remarks will help toward an understanding of the rest of the chapter.

Aristotle phrases his definition of motion in two ways: (a) "Motion is the actuality of that which is in potentiality in so far as it is in potentiality." ἡ τοῦ δυνάμει ὄντος ἐντελέχεια, ᾗ τοιοῦτον, κίνησίς ἐστιν. (*Physics* III, 1, 201a, 10–11; cf. *Metaphysics* XI, 9, 1065b, 16). (b) "Motion is the actuality of that which is movable in so far as it is movable." ἡ κίνησις ἐντελέχεια τοῦ κινητοῦ, ᾗ κινητόν. (*Physics* III, 2, 202a, 7–8; cf. *Metaphysics* XI, 9, 1065b, 22–23).

The difference between these two definitions, it will be observed, is in the use of the term "potentiality" in the one and of the term "movable" in the other. Averroes discusses the relative merits of these two definitions. Bearing in mind that a definition, according to Aristotle, must not include the thing which is to be defined nor such terms as are derived from the definiend (*Topics* VI, 4, 142a, 34 ff.), that the terms it uses must be especially appropriate and applicable to the subject (*Topics* VI, 1, 139a, 31), and that these terms must not be equivocal (*Topics* VI, 2, 139b, 19 ff.), he finds certain defects in both of these definitions. The first definition is, according to him, equivocal and not especially appropriate to motion in the strict sense of the term. In the second definition he finds that the differentia is derived from the term which forms the subject of the definition. His discussion is contained in the following texts:

Intermediate Physics III, ii, 2 (Latin, p. 450rb, D): "It is evident that this [the first] definition applies to all the genera of motion, for motion in substance is the entelechy of that which is in potentiality with reference to substance, in so far as it is in potentiality. The same may be said of motion in quality and of every one of the four categories. This is a definition of motion derived from things which are applicable [to the term defined]."

ומבואר שזה הגדר ידבק על כל סוגי התנועה, כי התנועה בעצם היא שלמות מה שבכח העצם מצד מה שבכח, וכמו כן התנועה באיך ובאחד ובאחד מהמאמרות הארבעה, וזהו גדר התנועה הנלקח מן העניינים המיוחסים.

(In the Latin translation the last part of the sentence reads: "sumpta ex rebus proprijs, (seu proportionalibus)." The translator evidently had before him two readings, המיוחדים and המיוחסים, the former of which he translated properly by "proprijs" and the latter he translated quite justifiably, but erroneously, by "proportionalibus." Both of these terms are used in the anonymous supercommentary quoted later in this note.)

Ibid. III, ii, 3 (Latin, p. 450 rb, F—va): "This differentia, used in the present [the second] definition, though not the same as the differentia used in the first definition, being a differentia derived from the subject of motion, is still superior to the differentia used in the first definition, for it does not contain that equivocation which is contained in the term potentiality. For potentiality may be found in all the ten categories, whereas the potentiality used in the definition of motion is the potentiality which is to be found only in the four categories."

זה ההבדל הלקוח בזה הגדר, ואם היה בלתי ההבדל הלקוח בגדר הראשון, לפי שהוא הבדל מצד הנושא, הנה הוא גם כן כבר יעדיף זה ההבדל על ההבדל הלקוח בגדר הראשון, לפי שאין בו השתוף אשר בשם הכח. זה שהכח נמצא במאמרות העשרה, והכח אשר לוקח בגדר התנועה אמנם הוא הכח הנמצא בארבעה מאמרות.

The first part of this passage is elucidated by a paraphrase in an anonymous supercommentary (MS. Adler 1744.1): "This differentia, even though not as good as that used in the first definition, being a differentia derived from the subject of motion, whereas that of the first definition is derived from things which are only appropriate and applicable to motion, is still superior to the differentia used in the first definition . . ."

והנה זה ההבדל, גם אם היה שיש בחסרון מהראשון, כי הוא לקח בזה ההבדל מצד הנושא ובראשון לקח כי אם הדברים שהם מיוחדים ומיוחסים אל התנועה, הנה עם כל זה יעדיף זה ההבדל מזה הגדר על ההבדל הלקוח בגדר הראשון.

These two passages of Averroes are summed up in the aforementioned anonymous supercommentary as follows: "The first
definition is superior to this one, because it is made up of terms
that are appropriate and applicable to motion, which is not the
case with this definition. But, on the other hand, this definition
is superior to the first, because it cannot be applied to any other
category outside the four genera of motion, namely, substance,
quantity, quality and place, whereas the first definition may be
applied to all the ten categories, for in all the ten categories there
are a potential and an actual."

והנה יעדיף הגדר הראשון לזה הגדר, כי הגדר הראשון לקוח מעניינים מיוחדים
ומיוחסים אל התנועה, מה שאין כן בזה הגדר. ויעדיף זה הגדר לראשון, שזה הגדר
לא יצדק במאמר אחר כי אם בד' סוני התנועה, ר"ל העצם והכמה והאיך והאנה,
והגדר הראשון יצדק על כל המאמרות העשרה, כי בכל המאמרות העשרה יצדק
בהם מה שבכח ומה שבפועל.

The relation between Maimonides' definition of motion and
the first definition of Aristotle is described by Altabrizi as
follows: "They have already mentioned two ways of formulating
the definition of motion. The first we have already reproduced
[i. e., the transition from potentiality to actuality]. The other
is mentioned by the First Master who says that motion is a first
entelechy of that which is in potentiality in so far as it is in
potentiality."

וכבר זכרו בהודעת התנועה אופנים: אחד מהם מה שזכרונוהו. והאחד מה שזכרו
המלמד הראשון, כי אמר התנועה שלימות ראשון למה שבכח מצד מה הוא בכח.

As for the significance of the expression "first entelechy," used
by Altabrizi, see *De Anima* II, 1, 212a, 22–27.

Unlike Crescas, however, Shem-ṭob Falaquera, after quoting
"a certain learned man," probably Altabrizi, finds that Aristotle's
definition is not the same as that of Maimonides, and points out
the superiority of the former definition to the latter: *Moreh ha-
Moreh*, II, Introduction, Prop. 5, p. 66: "A certain learned man
said: 'motion is a first entelechy [of that which is] in potentiality
in so far as it is in potentiality, and if you prefer you may say
that it is a transition from potentiality to actuality.' The first
definition explains more accurately the nature of motion than the
second, for motion must exist potentially, being something inter-

mediate between potentiality and actuality . . . . . It must combine both potentiality and actuality."

ואמר חכם, והתנועה שלמות ראשון בכח מצד מה שהוא בכח, ואם תרצה תאמר
כי היא יציאה מהכח אל הפעל. עד כאן. והענין הראשון מבאר התנועה יותר
מהענין השני, כי התנועה בכח, והוא דבר אמצעי בין הכח והפעל... אם כן לא
נשאר אלא שתהיה מורכבת ממה שבכח והפעל.

6. Hebrew שלמות, كَمَال, ἐντελέχεια, completeness or actuality, as distinguished from פועל فعل, ἐνέργεια, which, strictly speaking, means activity or actualization. Aristotle, however, commonly uses these terms without distinction (cf. Zeller, Aristotle I, p. 348, n. 2). Both these terms are used by Aristotle in defining motion (cf. Physics III, 2, 201b, 31; 202a, 7; Metaphysics XI, 9, 1065b, 22–23), and they are both likewise used by Crescas in this chapter. I have translated both these terms here by "actuality," except in two places, where Crescas used both of them together, when I have translated them by "entelecheia" and "energeia." The Latin translation of Averroes renders שלמות by "actus (seu perfectio)."

A discussion as to the meaning of the terms "energy" and "entelechy" as used by Aristotle in the definition of motion is to be found in Simplicius on Physics III, 1, 201a, 9 (ed. Diels, p. 414, 1. 15 ff., and Taylor's translation of the Physics, p. 141, note).

7. Cf. above n. 5.

8. Cf. Physics III, 2, 201b, 27 ff.

9. Cf. Posterior Analytics II, 4, 91a, 16: "Now it is necessary that these [i. e., the definition and the thing defined] should be convertible." ταῦτα δ' ἀνάγκη ἀντιστρέφειν.

The Hebrew term המופת (Arabic ברהאן, cf. Steinschneider's Uebersetzungen, p. 54) corresponds to the Greek ἀποδεικτική and περὶ ἀποδείξεως by which the Posterior Analytics is called by Alexander and Galen respectively (cf. Zeller, Aristotle I, p. 68, note).

10. According to Maimonides' definition, motion is the transition from potentiality to actuality. As the definition must be convertible, it follows that every transition from potentiality to

actuality is likewise motion. Now, in the motivity of any motive agent there is also a transition from potentiality to actuality, in so far as it is first a potential motive agent and then becomes an actual motive agent. If every transition from potentiality to actuality is motion, then every motivity is motion. But every motion requires a motive agent (see Prop. XVII). Consequently, every motivity would require a motive agent, thus subverting Aristotle's contention as to the existence of an immovable mover.

This argument, as will have been observed, contains two elements. First, the convertibility of definitions. Second, the impossibility that everything which moves should be moved. These two elements occur in the following discussions of the definition of motion:

A. *Physics* III, 2, 201b, 20–22: "By some motion is said to be difference, inequality and non-being; though it is not necessary that any of these should be moved, neither if they be different, nor if they be unequal, nor if they be non-beings."

This passage is paraphrased in *Intermediate Physics* III, ii, 5 (Latin, 450vb, L) as follows: "Among them there were some who said that motion is difference and inequality and others who said that it is non-being. However, if motion is difference, as they say, it will follow that whenever a thing becomes different it is moved. But while all things are changed into one another, they are not all moved."

מהם מי שאמר שהתנועה שנוית ויציאה מן השווי ומהם מי שאמר שהיא
בלתי נמצאת. ולו היתה התנועה שנוית, כמו שאמרו, התחייב כל מקום הוא שנוי
שיהיה מתנועע, וכל הנמצאות משתנות קצתם בקצתם ואין כלם מתנועעות.

Upon this paraphrase of the *Intermediate Physics* there is the following comment in Gersonides' supercommentary: "Says Levi: Everything is clear until the end of the chapter, except the statement 'If motion is difference, as they say, it will follow that whenever a thing becomes different it is moved.' The explanation of this reasoning is to be found in the fact that *a definition is convertible into the definiendum*. Accordingly, since they say that motion is difference, this definition can be converted so as to read that difference is motion."

אמר לוי. זה כלו מבואר עד סוף הפרק. אלא מה שאמר: ולו היתה התנועה
זולתיות כמו שאמרו, חוייב שכל מה שיהיה זולת שיהיה מתנועע. ואולם הסבה בזה

החיוב הוא לפי שהגדר יתהפך אל הנגדר, והם אמרו שגדר התנועה הוא שהוא אל
זולתיות, ולזה יתהפך שכל מה שהוא זולתיות הוא תנועה.

(In the foregoing Hebrew quotations, it will have been noticed,
the second passage uses זולתיות for שניות of the first passage. Both
represent the Greek ἑτερότης. The Latin translator evidently
had before him the reading זולתיות, and being uncertain as to its
exact meaning translated it according to the various meaning of
the Hebrew word by the following Latin terms: "alietatem (seu
non ens, seu nihil, seu aliud)."

B. *Physics* III, 3, 202a, 21–31; restated in *Intermediate Physics*
III, ii, 6 (Latin, p. 451r, B ff): "There is, however, a logical
doubt . . . . . If the motive agent is different from the movable
object and their actions constitute together motion, I wish I knew
whether their actions are one or two . . . . . If their actions are
one and the same, it will follow . . . . . , but this is absurd. And
if their actions are different, ... the question is whether
motivity is in the agent and movability in the object or whether
both exist together either in the agent or in the object . . . . .
And if we say that movability is in the object and motivity in the
agent, seeing that they are two different things, i. e., two different
motions, it will give rise to these alternative conclusions, namely,
either *everything which moves will be moved* or that which possesses
motion will not be moved."

והיה משיג בזה ספק מה... מי יתן ואדע, אם היה המניע דבר בלתי המתנועע
ופעולותיהם יחד תנועה, אם פעולותיהם אחת או שתים... אם אחת... זה מגונה.
ואם היו פעולותיהם מתחלפות... האם הנעה בפועל והתנועעות במתפעל, או שניהם
ימצאו יחד אם בפועל ואם במתפעל... ואם אמרנו שההתנועעות במתפעל וההנעה
בפועל, על שהם ב' דברים, ר"ל ב' תנועות, יתחייב אחד מב' עניינים, אם שיהיה
כל מניע מתנועע אם שתהיה התנועה נמצאת בדבר בלתי מתנועע.

This last passage is made use of by Gersonides in *Milḥamot
Adonai* VI, i, 24: "For while indeed it is true that every change is
a transition from potentiality to actuality, as may be gathered
from its definition in the *Physics*, it does not follow that every
transition from potentiality to actuality is change. The reason
for this is as follows: Change is a transition from potentiality to
actuality only with reference to a passive object in its process of
suffering action, but it is not a transition from potentiality to
actuality with reference to an active agent in its process of carry-

ing out its action. This becomes self-evident from the definition
of motion, which reads: 'Motion is the entelechy of that which
is movable *qua* movable.' And in general, change exists in that
which is moved and not in that which moves. Were it not so,
the agent would be moved by the work it performs. Furthermore,
if the transition from potentiality to actuality in the agent is
change, we will have to say that *every mover undergoes change*,
in so far as it is a mover."

זה, כי כל שנוי הוא יציאה מכח אל הפעל, כמו שהתבאר מגדרו בספר השמע.
אבל לא יחייב מפני זה שתהיה כל יציאה מהכח אל הפעל שנוי. זה שהשנוי הוא
יציאה מהכח אל הפעל אשר במתפעל להתפעל, לא היציאה מהכח אל הפעל
אשר בפועל לעשות פעולתו. זה מבואר בנפשו ממה שנאמר בזה הגדר. זה שכבר
נאמר בתנועה שהיא שלמות המתנועע במה שהיא מתנועע. ובכלל הנה השנוי הוא
במתנועע לא במניע, ולולא זה היה הפועל מתנועע מהמלאכה. ועוד שאם היציאה
מהכח אל הפועל בפועל שנוי, הנה נאמר שיחוייב שיהיה כל מניע משתנה, מצד
מה שהוא מניע.

It can be readily seen how these passages with their references
to the convertibility of definitions and to the impossibility that
every mover should be moved could have suggested to Crescas
his argument here.

There is also a suggestion made by Aristote himself that from
his first definition of motion it might be inferred that every mover
is movable. *Physics* III, 1, 201a, 23–27: "Hence that which naturally moves is also movable; for every thing of this kind moves,
while being itself moved. To some, therefore, it appears that
every thing which moves is moved. Whether, however, this be
true or not, will be manifest from some other of our writings; for
there is something which moves and is itself immovable."

11. See above n. 5. Cf. Averroes' *Intermediate Physics* III, ii, 3
(Latin, p. 450rb, E F): "Aristotle says also that motion is the
entelechy of that which is movable *qua* movable. This definition
becomes evident by reasoning inductively from similars and particulars. For building is the entelechy of that which is buildable
qua buildable. Rolling is the entelechy of that which is rollable
qua rollable. Heating is the entelechy of that which is heatable
qua heatable. The act of building does not exist when the house
is already completed nor does it exist when the house exists only
in potentiality. The act of building is rather the passage from the

non-being of the house to its becoming a house in actuality and in complete reality. This being so, it is thus proved by this inductive method of reasoning that motion is the entelechy of that which is movable *qua* movable. The justification for including the term 'movable' in the definition of motion is evident from what we have already stated, namely, that the genus of motion is relation. We have therefore taken the term 'movable' in the definition of motion, because it is more known than motion. This differentia, used in the present definition, though not the same as the differentia used in the first definition, being a differentia derived from the subject of motion, is still better than the differentia used in the first definition, for it does not contain that equivocation which is contained in the term potentiality. For potentiality may be found in all the ten categories, whereas the potentiality used in the definition of motion is the potentiality to be found in the four categories."

ויאמר גם כן שהתנועה שלמות המתנועע במה שהוא מתנועע. זה הגדר גלוי
מחפוש הדומים והחלקים. זה כי הבנין שלמות הנבנה במה שהוא נבנה, והגלגול
שלמות המתגלגל במה שהוא מתגלגל, והחמום שלמות המתחמם במה שהוא מתחמם,
כי הבנייה לא תהיה עם שלמות הבית ולא תהיה גם כן בהיות הבית נמצא בכח.
ואמנם הבנייה הוא דרך מהעדר הבית אל מציאות בית בפועל ועל השלמות.
וכאשר היה זה כן, מבואר מזה החפוש שהתנועה שלמות המתנועע במה שהוא
מתנועע. ולקחנו המתנועע בגדר התנועה גלוי למה שהקדמנו, שהתנועה סוגה
ההצטרף. ואמנם לקחנו המתנועע בגדר התנועה, לפי שהוא יותר ידוע·מהתנועה.
זה ההבדל הלקוח בזה הגדר, ואם היה בלתי ההבדל הלקוח בגדר הראשון,
לפי שהוא הבדל מצד הנושא, הנה הוא גם כן כבר יעדיף זה ההבדל על
ההבדל הלקוח בגדר הראשון, לפי שאין בו השתוף אשר בשם הכח. זה שהכח
נמצא במאמרות העשרה, והכח אשר לוקח בגדר התנועה אמנם הוא הכח הנמצא
בארבעה מאמרות.

12. See above n. 6.

PROPOSITION VI

1. In the Arabic original of the *Moreh* and in its Hebrew translations there follows here the statement: "The latter kind of motion is a species of motion according to accident." והוא מין ממה שבמקרה. (cf. below n. 3). It is, however, omitted in Isaac ben Nathan's translation of Altabrizi, from which source the Hebrew version of this proposition is taken. Similarly, toward the end of the proposition, Altabrizi and most of the MSS. read וכל whereas Ibn Tibbon and the editions read וכן כל.

2. Hebrew מסמר בספינה, Arabic אלמסמאר פי אלספינה, a literal translation of the Greek ἐν τῷ πλοίῳ ἧλος (*Physics* IV, 4, 211a, 20–21).

3. Aristotle has several classifications of motion or change.

A. *Physics* IV, 4, 211a, 17 ff.: (1) According to itself or its own essence, καθ' αὐτό (2) According to accident, κατὰ συμβεβηκός. This accidental motion is subdivided into (a) what he elsewhere calls 'according to part,' illustrated by the motion of the parts of the body and of the nail of a ship and (b) what he elsewhere describes as 'inherent in the mover,' illustrated by the motion of whiteness and of knowledge (see B, C, E).

B. *Physics* V, 1, 224a, 21 ff.: (1) According to accident. (2) According to part, κατὰ μέρος. (3) According to itself.

C. *Physics* V, 2, 226a, 19 ff.: (1) According to accident. (2) According to part. (3) According to itself.

D. *Physics* V, 6, 231a, 10–11: (1) According to nature, κατὰ φύσιν. (2) Contrary to nature, παρὰ φύσιν.

E. *Physics* VIII, 4, 254b, 7 ff.: (1) According to accident, subdivided into (a) such as are inherent in movers and (b) such as are according to part. (2) According to itself, καθ' αὐτό, subdivided into (a) By itself, ὑφ' αὐτοῦ. (b) By something else, ὑπ' ἄλλου. (c) By nature. (d) By violence, βίᾳ, and contrary to nature.

F. *De Anima* I, 3, 406a, 4 ff.: (1) According to itself. . (b) According to something else, καθ' ἕτερον, or according to accident. Here, again, Aristotle identifies 'according to accident' with what he elsewhere calls 'according to part.'

In the foregoing classifications, it will have been noted, Aristotle draws no sharp line of distinction between 'according to accident' and 'according to part.' Both are sometimes treated as one class and contrasted with 'according to itself.' Similarly Algazali uses the term accidental in the sense of 'according to part.' *Kawwanot ha-Pilosofim* III (*Maḳaṣid al-Falasifah* III,

p. 238): "As for accidental, it is so called when a body is in another body and the enclosing body is moved and thereby motion is produced in the enclosed body."

ואשר במקרה הוא שיהיה הגשם בגשם אחר, ויתנועע הגשם המקיף וינוע בו הגשם המוקף.

It will also have been noted that Aristotle makes a distinction between καθ' αὐτό, בעצמו, and ὑφ' αὐτοῦ, מצדו, מפאת עצמו. The former means being moved independently of anything else, as opposed to accidental motion, whereas the latter means having the cause of motion in itself, as opposed to being moved by something external to itself. (Cf. Prop. XVII, n. 7). Similarly there is a difference between καθ' ἕτερον and ὑπ' ἄλλου. The former ,means being moved as a part of something else, whereas the latter means being moved by a cause which is external to oneself.

A very elaborate classification is given by Altabrizi in his commentary on this proposition. But stripped of its numerous and cumbersome subdivisions, Altabrizi's classification is in its main outline based upon Aristotle's classification E. It is as follows:

I. According to its essence, שתהיה אותה התנועה קיימת ועומדת בו. This is subdivided into two parts:

a. By something else, סבת אותה התנועה אם שיהיה דבר חוץ מן הגשם. This is also designated as motion "by violence", יאמר לו המתנועע בהכרח, and Altabrizi gives here an eightfold classification of violent motion.

b. By itself, אם היתה סבת אותה התנועה דבר בנפש אותו הגשם הנה יאמר לו שהוא מתנועע בעצמות. Under this Altabrizi includes "voluntary motion" and "natural motion."

והוא אם שתהיה מסודרת ממנו בכונה ובחירה והוא התנועה הרצונית, או מבלתי כונה ובחירה והיא התנועה הנמשכת... והרביעי שיהיה המניע מניע בהמשך אל צד אחד והוא הטבע.

II. According to accident, המתנועע במקרה. This is subdivided by him, as in Aristotle, into two parts:

a. According to part, כי המתנועע במקרה אם שיהיה חלק למה שהוא מתנועע האמת.

b. Not according to part, but existing as a quality in a subject, illustrated by the motion of "whiteness." או לא יהיה... משל הלובן בגשם.

What Crescas is, therefore, trying to say here is that Maimonides' classification of motion was not meant by him to be final. All that Maimonides wanted to establish in this proposition is the fact that motion can be classified in a general way under the headings of essential, accidental, violent, and according to part. Crescas then proceeds to show how Maimonides' classification can be reduced to the Aristotelian and Altabrizian pattern. In the succeeding notes we shall see how he does it.

4. I take the expression כהעתק הגשם ממקום אל מקום to be an explanation of עצמותית and not of הרצונית. This reclassification corresponds to sections I *a b* in Altabrizi's scheme. Cf. *Physics* VIII, 4, 254b, 12–20: "Of those things, however, which are moved essentially, some are moved . . . . . by nature, but others by violence and contrary to nature: for that which is moved by itself is moved by nature, as, for instance, every animal; since an animal is moved by itself. But of such things as contain in themselves the principle of motion, of these we say, that they are moved by nature. Hence, the whole animal, indeed, itself moves itself by nature; but the body happens to be moved by, and contrary to, nature: for it is of consequence with what kind of motion it may happen to be moved, and from what element it consists."

5. Corresponds to section II *b* in Altabrizi's scheme: "Second, when it is no part of that which is moved essentially nor is it capable of having motion indpendently, as, e. g., whiteness in a body, for when the body is moved, the whiteness is said to be moved accidentally." (Hebrew quoted below n. 8).

Cf. *Physics* VIII, 4, 254b, 8–10: "Accoding to accident, indeed, such as are inherent in movers or the things moved."

6. In Altabrizi there is no such subdivision under section I *a*. But in Aristotle there is mention of two kinds of "violent motion," one "according to its essence" and the other "according to accident," i. e., "according to part." *Physics* VIII, 4, 254b, 22–24: "Contrary to nature, indeed, as terrestrial things when moved upward, and fire downward. Again, the parts of animals are frequently moved contrary to nature, on account of positions and modes of motion."

The term "accidental," then, is used here by Crescas in the sense of "according to part." See below n. 13.

7. For instance, the parts of an animal, which are moved with the whole, may sometimes move by nature and sometimes contrary to nature. Cf. *Physics* VIII, 4, 254b, 17–20: "Hence, the whole animal, indeed, itself moves itself by nature; but the body happens to be moved by, and contrary to nature: for it is of consequence with what kind of motion it may happen to be moved, and from what element it consists."

8. This statement reflects the following passages:
Narboni: "The difference between 'accidental' and 'according to part' is that in the case of the latter it is possible for the nail to become separated from the boat and be moved essentially."

ההבדל ביניהם כי אשר בחלק כבר יהיה כי המסמר כבר יבדל מהספינה
ויתנועע בעצם.

Altabrizi: "Second, when it is no part of that which is moved essentially nor is it capable of having motion independently, as, e. g., whiteness in a body, for when the body is moved, the whiteness is said to be moved accidentally. Third, when it is part of that which is moved essentially and is capable of being moved independently, as, e. g., a body composed of other bodies, as the boards of which the boat is built and as the nails which are driven in them."

והשני מהם שלא יהיה חלק לו ולא מדרכו קבלת התנועה נפרד, המשל הלובן
בגשם, כי כאשר התנועע הגשם יאמר ללובן שהוא מתנועע במקרה. והשלישי מהם
מה שיהיה חלק לו ומדרכו שיקבלה נפרד, משלו הגשם המחובר מהגשמים, כנסרים
המסודרים בספינה והמסמרים התקועים בם.

Physics IV, 4, 211a, 18–20: "And those which are according to accident, some can be moved essentially, as, for instance, the parts of the body and the nail in the ship; but others cannot be so moved, but are always moved accidentally, as, for instance, whiteness and science: for these thus change their place, because that changes in which they subsist."

9. Hebrew אמנם. This is one of the many instances in this book, especially in the texts quoted in the notes, in which אמנם is used in the sense of "only," after the Arabic انما, of which it is com-

monly used as a translation, as, e. g., toward the end of the Introduction to *Moreh Nebukim* I (Arabic, p. 11a, last line): ‫ונחן אגמא כאן נרׂצנא. ואנחנו היה אמנם היה דעתנו.‬

10. Regarding the motion of the celestial spheres, there is a difference of opinion between Avicenna and Averroes. According to the former, the circular motion of the spheres is not locomotion (‫תנועה באנה‬ or ‫תנועה במקום‬), since the totality of the body does not change place at all. He therefore calls it "motion in position" (‫תנועה במצב‬ or ‫תנועה בתשומה‬). Averroes, however, maintains that it is locomotion. Cf. Prop. IV, p. 504, n. 6.

Hence, Crescas argues as follows: If Maimonides' definition of essential motion were true, namely, that it is the translation of a body from one place to another, the celestial spheres could not have essential motion.

11. Continuing his argument, Crescas proceeds to prove that the circular motion of the spheres must be essential. The crux of his argument is this: Essential motion, the καθ' αὐτό of Aristotle, must not be defined, as is done by Maimonides, as motion by which a body is translated from one place to another, but rather as motion by which a body is moved in virtue of itself whether from one place to another or within one place.

In the course of his argument, Crescas refers to the question as to the nature of the motion of the spheres. According to the view which he ascribes to Aristotle, the celestial spheres are animate and intelligent beings, endowed with souls and intellects. Their motion is, therefore, voluntary, as is the motion of animals. A statement of this view is given in Avicenna's *al-Najah*, p. 71 (see Carra de Vaux, *Avicenne*, pp. 249–250), in *Emunah Ramah* I, 8, p. 41, and in *Moreh* II, 4–5. Crescas discusses it in Book IV, 3. As to the antiquity of this view among the Jews, see Ginzberg's *The Legends of the Jews* V, p. 40, n. 112.

The opposite view, that the circular motion of the spheres is natural, is discussed by Crescas also in Prop. XII, Part II; in Book I, ii, 15; and in Book IV, 3. Here he describes it as our own view (‫לפי מה שיראה לנו‬).

As a matter of fact, this view is not original with Crescas, as is claimed by him, unless he means here by ‫לפי מה שיראה לנו‬ the view which he prefers to follow. Algazali devotes to it an entire chapter in his *Happalat ha-Happalah*: "Disputation XIV. Of their failure to establish a proof that the heavens are animate beings, worshipping God by their circular motion, and that they are moved voluntarily." השאלה הי"ד. בלאותם מהעמוד הראיה על שהשמים חיים עובדים לאל ית' ויתי בתנועתם הסבובית ושהם מתנועעים ברצון. His argument is contained in the following passage (*Tahafut al-Falasifah* XIV, p. 58, l. 25–p. 59, l. 2; *Destructio Destructionum* XIV, p. 118rb):

"The third [possibility for the motion of the spheres] is that the heavens are endowed with a particular property which property is the principle of their motion, analogous to the principle assumed by the philosophers in their explanation of the movement of a stone downward, and, again, like the stone, the heavens are unconscious of that principle. Their contention that the object which is sought after by nature cannot be the same as that which is fled from by nature is erroneous, for the celestial spheres have no numerical difference, being one in the corporeality of their substance and one in the circularity of their motion, and their corporeal substance is not actually divisible into parts [nor is their circular motion actually divisible into parts]; they are divisible only in the imagination. Furthermore, that motion of theirs is not due to a quest for a place nor to a flight from a place. It is quite possible for a body to be created with such a nature as to contain in itself something which determines circular motion. Thus it is motion itself that determines its own direction, and it is not the quest for a place that determines the particular kind of motion so that motion would be only an effort to reach that place. When you say that motion is due to the quest for a certain place or, if it is violent, to the flight from a certain place, you speak as if you consider nature as that which determines the quest for the place and regard motion not as an action purposeful in itself but as a means of approaching that place. But we say it is not impossible that motion itself, and not the quest for a place, determines its own direction. What is there to deny this view?"

והשלישית, הוא שיקובל שהשמים נתיחדו בתאר, והתאר ההוא התחלה לתנועה, כמו שהאמינוהו בירידת האבן למטה, אלא שלא ישערהו בה כמו האבן. ואמרם שהדרוש בטבע לא יהיה במה [שיברח] ממנו בטבע, הוא שבוש, לפי שאין שם הבדל

במספר אצלם, אבל הגשם אחד, והתנועה הסבובית אחת, ואין לגשם חלק בפועל, ואמנם יתחלקו בדמיון. ואין אותה התנועה לדרישת מקום ולא לברוח מהמקום. ואפשר שיברא גשם ובעצמותו יגזר עניין תנועה סבובית, ותהיה התנועה עצמה גזרת זה העניין, לא שתגזור התנועה ההיא דרישת המקום, עוד תהיה התנועה להגיע אליו, ואמרכם שכל תנועה היא לדרישת מקום או לברוח ממנו, בשיהיה הכרחי, כאלו תשימו דרישת המקום יגזור הטבע, ותשימו התנועה בלתי מכוונת בעצמה, אבל נגררת אליו. ואנחנו נאמר לא ירוחק שתהיה התנועה נפש הגמר, ולא דרישת המקום, ומה המשקר לזה.

Likewise, Shem-ṭob Falaquera quotes in the name of Avempace a view which corresponds exactly to that advanced here by Crescas. Furthermore, he claims that Aristotle himself has three different views with respect to the motion of the sphers, one of which is identical with that of Crescas. *Moreh ha-Moreh* II, 4, pp. 80–82:

"Avempace states that 'Aristotle's view is that the celestial sphere is moved *per se*.' And it is thus stated in *De Caelo et Mundo* that motion is natural to the celestial sphere and is one of its properties just as upward motion is natural to fire and downward motion to earth . . . .

We find that Aristotle has three statements in explanation of the motion of the celestial sphere: First, that the celestial sphere is moved by nature . . . . Second, that it is moved by a soul . . . . Third, that it is moved by an infinite force which acts as a motive agent after the manner of an object of desire, as has been explained above. In view of this, there are some people who find these statements contradictory to each other. But Aristotle himself has cleared the matter up in the *Metaphysics* where he says 'And the proximate cause of the motion of the spheres is not nature nor an Intelligence but rather a soul. The remote principle of its motion, however, is an Intelligence.' "

וכתב בן אלצאיג, כי דעת אריסטו שהגלגל מתנועע מעצמו ע"כ. וכן כתב בספר השמים והעולם, כי התנועה טבעית לגלגל ומסוגלת לו, כמו תנועת האש למעלה ותנועת הארץ למטה...

ומצאנו שאמר אריסטו בסבת תנועת הגלגל שלש לשונות: האחת כי הגלגל מתנועע בטבע.... והשנית שהוא מתנועע בנפש.... והשלישית שהוא מתנועע בכח שאין לו תכלה, ויניע כמו שיניע החשוק, כמו שנזכר למעלה. ועל כן יש אומרים שיש בדבריו סתירה, ובאר זה בספרו באלקיות, ושם כתב והמניע הקרוב לגלגלים אינו טבע ולא שכל אלא נפש, וההתחלה הרחוקה שכל.

It will have been noticed that Crescas uses here three terms in describing the motion of the spheres: (a) voluntary, רצונית, (b) appetent, תשוקיית, both of these attributed by him to Aristotle, and (c) natural, טבעית, thus corresponding to the three views which Shem-ṭob Falaquera has found in Aristotle. My insertion of "or" between "voluntary" and "appetent" in the text is based upon that consideration.

Among the Jewish philosophers Saadia also seems to have been of the opinion that the motion of the spheres was natural. Cf. *Emunot we-Deot* I, 3, הדעת השמיני, and VI, 3. See commentary *Shebil ha-Emunah, ad loc.*

This view is also shared by Judah ha-Levi (*Cuzari* IV, 1, cf. *Moscato's* commentary *Ḳol Jehudah, ad loc*) and Isaac ibn Latif (*Sha'ar ha-Shamayim* quoted in Isaac 'Arama's *'Akedah*, Sha'ar II, and by Moscato, *op. cit.*).

Isaac 'Arama (*op. cit.*), who lived after Crescas, argues in favor of this view, claiming, however, to have found no support for it among Jewish philosophers except in Isaac ibn Latif. For this he has been called to account by Moscato (*op. cit.*). But Moscato himself fails to make any mention of Saadia and Crescas.

12. Hebrew בשחרות אשר בנשם. This phrase was undoubtedly meant to be a quotation from the proposition. In the proposition, however, following Isaac ben Nathan's translation of Altabrizi, Crescas has בשחרות שהוא בנשם. This variation is probably due to the influence of a lingering reminiscence of Ibn Tibbon's translation, which reads: בשחרות אשר בזה הגשם.

13. The point of Crescas' criticism is as follows: From Maimonides' illustration of accidental motion it would seem that accidental motion is possible only in the case of accidental qualities, as, e. g., color, whereas there can be accidental motion in something which is not an accidental quality, namely, the extreme point of a line.

Crescas does not explain why the motion of the extreme point of a line along with the line should be called 'accidental' motion rather than motion 'according to part,' which are treated by Maimonides as two distinct classes in this proposition. It would seem that Maimonides would have put the motion of the extreme

point of a line under motion according to part rather than under
accidental motion. He could cite Aristotle as his authority.
Physics VI, 10, 240b, 8–13: "These things being demonstrated,
we say that the impartible cannot be moved, except according
to accident; as, for instance, the body being moved, or the magnitude in which the impartible is inherent: just as if that which is
in a ship should be moved by the motion of the ship, or a part by
motion of the whole. But I call that impartible, which is indivisible according to quantity."

Cf. *Intermediate Physics* VI, 12: "I say that that which is indivisible cannot have essential motion, as is the case of a mathematical point in the opinion of the geometricians. If something
indivisible is moved at all, it is only accidentally so; after the
manner of parts which are moved along with the motion of the
whole and of man who is moved by the motion of the ship."

ואומר שמה שאי אפשר שיחלק אי אפשר שיתנועע בעצם, כמו שידמו זה המהנדסים
בנקודה. אבל אם היה זה הנה הוא במקרה במדרגת החלקים אשר יתנועעו בתנועת
הכל והאדם המתנועע בתנועת הספינה.

Crescas is constantly insisting upon the use of "accidental
motion" in the sense of "motion according to part." See above
n. 6, and Proposition VII, Part I, n. 18.

14. Hebrew המפורסם היתי לדעת. I take המפורסם as qualifying
לדעת, despite their disagreement in gender. The surrogate "the
Greek" is similarly applied to Aristotle by Crescas' teacher Nis-
sim ben Reuben: כמו שחשבו המתפתים והנמשכים אחר היוני, (quoted
by Isaac Abravanel in *Mif'alot Elohim* I, 3, p. 6b).

15. Cf. Prop. I, pp. 161, 410.

16. Cf. Prop. I, Part II, n. 21, p. 411.

17. Cf. Prop. I, Part II, n. 22.

18. This illustration is an unhappy one. Aristotle himself admitted that air has some gravity. The question was merely
whether fire has any gravity or is absolutely light. Cf. Prop. I,
Part II, n. 23.

19. Cf. Prop. I, Part II, n. 23.

20. Hebrew ‏ודי בזה ההערה בזה הפרק‎. This is the only chapter which ends with such a remark. Crescas has evidently meant by this remark to refer to his inclusion of the criticism of this proposition in the chapter dealing with its proof instead of putting it in a separate chapter, as he has done in other propositions. My translation of this remark runs accordingly.

PROPOSITION VII

Part I

1. The first part of the proposition reads alike in Crescas, in Ibn Tibbon's translation of the *Moreh* and in Isaac ben Nathan's translation of Altabrizi. The last part reads in Ibn Tibbon: ‏וכל מה שלא יתחלק לא יתנועע ולוה אי אפשר שיהיה גשם כלל‎ and in Isaac ben Nathan: ‏וכל מה שלא יחלק לא יתנועע ולוה לא יהיה גשם כלל‎. Crescas's reading agrees with neither. But within the text of Altabrizi's commentary there is another version of this part of the proposition: ‏ואולם הטענה הד', והוא שכל מה שלא יתחלק לא יתנועע ולא יהיה גשם בהכרה‎. Evidently Crescas has combined these two versions of the latter part of the proposition.

2. Altabrizi divides this proposition into four parts, which are designated in Isaac ben Nathan's translation by ‏טענות‎ and in the anonymous translation by ‏בקשות‎, i. e., *theses, questions, problems* (see Prop .I, Part II, p. 457, n. 81). But they are referred to later, in the course of discussion, by the term ‏הקדמה‎, which has been adopted here by Crescas. Altabrizi: "Know that this proposition contains four theses." Isaac ben Nathan's translation: ‏דע שאת‎ ‏ההקדמה מקפת על טענות ארבעה‎. Anonymous translation: ‏דע כי זאת‎ ‏ההקדמה כוללת ארבע בקשות‎.

3. So also in Altabrizi: "Now for the fourth thesis, namely, 'anything that is indivisible cannot have motion and cannot be a body.' After having shown in the second proposition that 'everything divisible is movable,' and as it is known that every body is divisible either potentially or actually, it follows by the method of the conversion of the obverse that 'anything that is indivisible cannot have motion and cannot be a body.'"

ואולם הטענה הד', והוא שכל מה שלא יתחלק לא יתנועע ולא יהיה נשם בהכרח,
הנה לפי שהוא קיים בהקדמה השנית שכל מתנועע מתחלק, וידוע שכל נשם מתחלק
אם בכח אם בפעל, הנה יחוייב בדרך הפך הסותר שאשר לא יתחלק כלל לא
יתנועע ולא יהיה נשם.

Similarly in Narboni: "This is known by the conversion of the
obverse." וזה נודע מהפוך הסותר.

The expression הפוך הסותר reflects Aristotle's ἡ κατὰ τὴν
ἀντίφασιν ἀκολούθησις ἀνάπαλιν γινομένη. (Topics II, 8, 113b,
25–26). This kind of inference is called ἀντιστροφή σὺν
ἀντιθέσει by Alexander and conversio per oppositionem or conversio per contrapositionem by Boethius (cf. Sir William Hamilton,
Logic (1866), Vol. I, p. 264). Thus הפוך represents ἀνάπαλιν
γινομένη, ἀντιστροφή, and סותר represents ἀντίφασις, ἀντίθεσις.

In the anonymous translation the expression used is התהפכות
הסותר. But in both translations once the term הפוך occurs
without סותר. Isaac ben Nathan: ואולם הטענה הרביעית הנה היא הפך.
ההקדמה הקודמת. Anonymous: ואמנם הבקשה הרביעית... יתחייב בדרך
ההפוך.

4. A body, σῶμα, is that which has three dimensions and is a
magnitude, ποσόν. (Cf. De Caelo I, 1, 268a, 7 ff.; Metaphysics
V, 13, 1020a, 7.). A magnitude is a continuous quantity (ibid.),
and a continuous quantity is "divisible into things always
divisible," διαιρετὸν εἰς ἀεὶ διαιρετά, מתחלק אל מה שיתחלק תמיד
(cf. Physics VI, 1, 231b, 16, and De Coelo I, 1, 268a, 6). We thus
have the proposition: every body is divisible. By converting the
obverse of that proposition, we get the fifth proposition mentioned here by Crescas, namely, anything that is indivisible cannot be a body. This proof is a development of a suggestion made
by Altabrizi. Cf. quotation above n. 3.

5. So far Crescas has been following Altabrizi. In his subsequent
proofs of the first and second propositions, however, Crescas no
longer follows him. These proofs are rather based upon Averroes'
works: Long Commentary on Physics VI, iii, 1 (Latin, p. 265 ff),
Intermediate Physics VI, 7, Epitome of the Physics VI (p. 30 ff),
where the entire discussion of Crescas is to be found. The views of
Alexander, Themistus, and Avempace are also to be found there.

The expression נתחבטו בה המפרשים, used here by Crescas seems to reflect the Long Commentary which reads in Latin: "Et ideo *expositores ambiguunt* in responsione in isto loco."

6. Cf. *Physics* VI, 4, 234b, 10 ff., and *Intermediate Physics*, VI, 7.

7. Crescas' statement here seems to be based upon the Long Commentary on *Physics* VI, iii, 1 (Latin, p. 265vb): "Sed si hoc modo fuerit intellectus iste locus, excipiuntur tunc transmutationes, quae fiunt non in tempore, et ista transmutabilia sunt diuisibilia, et corporalia, et sic demonstratio erit particularis, et deberet esse universalis."

In *Intermediate Physics* VI, 7, this objection is quoted in the name of Theophrastus: "Against this proof an objection has already been raised by Theophrastus. He maintains that the argument employed in it is applicable only to a certain kind of changeable things, namely, things whose change takes place in time; but with reference to things whose change takes place in no-time, it cannot be truthfully said that some parts of them are in the *terminus a quo* and others in the *terminus ad quem*."

זה המופת כבר ספק עליו תפרסיות ואמר שאמנם יתאמת על קצת המשתנים, והם המשתנים בזמן, ואמנם הדברים אשר ישתנו בזולת זמן, הנה לא יצדק עליהם שיאמר שקצתם במה שממנו וקצתם במה שאליו.

The foregoing passage in the *Intermediate Physics*, as will have been noticed, does not contain Crescas' concluding remark that "the demonstration will thus be of particular application." It occurs, however, in another passage in the same chapter in the *Intermediate Physics*:

"Inasmuch as it is evident that Aristotle does not mean by his statement 'from one thing to another' from one contrary to another, for in that case the demonstration would then be particular and not universal, i. e., applying only to certain changes, such as are in time, but not to all changes, it follows that what he means by that phrase is from one state of rest to another."

וכאשר היה מבואר שלא ירצה אריסטו באמרו מדבר אל דבר מהפך אל הפך, לפי שהבאור אז יהיה חלקי לא כולל, ר"ל לקצת השנויים יהיו בזמן לא בכלם, הוא מבואר שאמנם רצה באמרו מדבר אל דבר מדבר נח אל דבר נח.

As for the meaning of "particular" and "universal" demonstration, see Prop. I, Part II, p. 462, n. 96.

8. Again based upon the Long Commentary (*ibid.*): "Et ideo expositores ambigunt in responsione in isto loco, et dicunt quod Alexander exponit quod omnis transmutatio est in tempore, sed quondam latet sensum." Cf. *Intermediate Physics* VI, 7: "But Alexander, in his answer to this question, is reported to have maintained that everything that is changeable is changeable in time and that if anything is said to be changeable in an instant it is only because the time in which the change takes place escapes the notice of people."

ואמנם אלכסנדר הוא השיב במה שזכרו כי חשב שכל משתנה הוא משתנה בזמן, ולזה אשר יאמר בו שהוא משתנה בעתה אמנם הוא להעלם הזמן אשר ישתנה בו מבני אדם.

9. Crescas is simply re-echoing Averroes' summary dismissal of Alexander's view: "It does not behoove us to enter into such subtle discusions with Aristotle as to be led to say that the ends of the processes of change take place in time as did Alexander. Heavens! unless Alexander did not want us to include the ends of changes in the proposition that every change is in time, considering them to be not changes but rather the limits of changes . . . . This is probably what Alexander has meant, for that man is of too great eminence and distinction to be ignorant of such an important point in Aristotle's doctrine and to try to answer for him by an impossible statement, namely, that the ends of motion take place in time."

ואינו ראוי גם כן שנעמיק להתעצם עם אריסטו עד שנאמר שהגעת אותם השנויים הם בזמן, כמו שעשה אלכסנדר, האלהים, אם לא שירצה אלכסנדר באמרו שכל שנוי בזמן שאותם אינם שנויים ואמנם הם תכלית שנויים... ואולי אלכסנדר הוא מה שרצה, כי האיש ההוא גדול המעלה והשעור מאשר יעלם ממנו זה הענין המעלה מדברי החכם עד שיתנצל בדבר הוא בטל, והוא שיהיו תכליות השנוים מניעות בזמן.

10. Crescas' paraphrase of Themistius's view does not correspond with what we have of it in the *Intermediate Physics*. It is not impossible that Crescas has derived his knowledge of Themistus from some supercommentary on Averroes.

Intermediate Physics VI, 7: "Themistius has discussed this view of Alexander and has arrived at the conclusion that there are things changeable which are changed in no-time. His answer to the difficulty in question is that Aristotle did not intend that his proof be applied to this kind of change, i. e., change in no-time.

He saw no need for mentioning this exception because it is
self-evident that such changes are indivisible, for when we say
that certain things are changed suddenly we mean that they meet
with a sudden change in all their parts."

ואמנם תמסטיוס דבר בזה ואמר וקבל שקצת המשתנים בזולת זמן. והשיב בזה
הספק שההחכם לא יכוין בזה המופת לזה המין מן השנויים, ר"ל אשר ישתנו בזולת
זמן. ואמנם עזב החכם זכרם, לפי שהדבר מבואר בם שהם בלתי מתחלקים, אחר
שהיה הענין אמרנו בם שהם משתנים פתאום, ר"ל שהם יפנשו ההשתנות פתאום
בכל חלקיהם.

Cf. Themistius, *In Aristotelis Physica Paraphrasis* (ed. Schenkl),
p. 197.

11. Hebrew כחול הצורה בחומר. The word חול is used in philosophic Hebrew as a technical term in describing the act of the
entrance of any kind of form into any kind of matter, corresponding to the Arabic حل (cf. *Cuzari* II, 14: כאשר השכל צופה למי שנשלמו
טבעיו ונשתתה נפשו ומדרותיו שיחול (אן יחל) בו. It reflects the Greek
ἔπειμι as in *Enneads* II, iv, 8: ἔπεισι τοίνυν τὸ εἶδος αὐτῇ.

That the change of form is timeless is also confirmed by the
following passage in *Moreh* II, 12: "Every combination of the
elements is subject to increase and decrease, and this comes-to-
be gradually. It is different with forms; they do not come-to-be
gradually, and have therefore no motion; they come-to-be or
pass-away without time."

כל מזג מקבל התוספת וההסרון, והוא יתחדש ראשון ראשון, והצורות אינם כן,
שהם לא יתחדשו ראשון ראשון, ולזה אין תנועה בהם, ואמנם יתחדשו או יפסדו
בלא זמן.

Cf. Averroes' *Epitome of the Physics* V, p. 21b: "But the last
actuality in them, namely, form, arrives without time."

אבל השלמות האחרון בהם, והוא הצורה, מגיע בזולת זמן.

12. *Intermediate Physics* VI, 7: "Avempace has solved this difficulty by contending that the Philosopher did not mean by the
term 'divisible' the divisibility of magnitudes at the end of their
motion but rather the divisibility of something changeable during
the interval between two contraries existing in it, i. e., between
the *terminus a quo* and the *terminus ad quem*. For Avempace
believes that the latter kind of divisibility is peculiar to that

which is changeable in time whereas the divisibility at the extremities of motion applies to both kinds of changeable objects, namely, those which change in time and those which change without time."

ואמנם אבובכר בין אלצאיג השיב מזה הספק בשהחכם לא רצה בהחלק החלק הגדלים בתכליות, ואמנם רצה החלק המשתנה בשני העניינים המקבילים אשר ימצאו בו בין מה שממנו ומה שאליו. וזה שהוא חשב שזה החלק הוא מיוחד במשתנה בזמן. ואמנם החלוק בתכליות הוא כולל לשני המינים יחד ממיני המשתנים, ר"ל המשתנים בזמן ובזולת זמן.

13. *Intermediate Physics* VI, 7: "This being so, it is clear that this proposition includes all the kinds of change that occur within the qualities and forms that are generated, whether they be change from one contrary to another, as, e. g., the motion from whiteness to blackness, or from non-being to being, as, e. g., the change of generation and corruption. But would that I knew whether the timeless changes are changes of independent existence or only ends of changes and whether they are from one state of rest to another. It is evident that they are ends of changes, seeing that they are timeless, and that they are not from one state of rest to another."

וכאשר היה זה כן, הוא מבואר שזה המאמר יכלול כל מיני השנויים הנמצאים בעצמותם המחודשים, הן שהיו מהפך אל הפך, כמו התנועה מהלובן אל השחרות, או מהעדר אל מציאות, כמו השנוי בהויה והפסד. ומי יתן ואדע אם השנוים אשר יהיו בזולת זמן אם הם שנוים נמצאים בעצמם או תכלית שנוים]ואם[הם ממנוחה אל מנוחה. והוא מבואר שהם תכליות שנוים, אחר שהיו מזולת זמן, ולא היו ממנוחה אל מנוחה.

14. According to Aristotle, if a thing is becoming to be in time A, the process of becoming is actually completed in the extremity of A. Cf. *Physics* VIII, 8, 263b, 28–264a, 3: "For if D was becoming to be white in the time A . . . . . it was generated, and it is the last point of the time in which it was becoming to be."

15. Crescas' proof for the third proposition differs from that given by Altabrizi.

16. Cf. definition of place above Prop. I, Part I (p. 153).

17. Quality and quantity are accidents residing in a body. Consequently qualitative and quantitative changes imply the

existence of a body. In substantive change, too, the subject that undergoes the change from being into non-being must contain matter which is the persistent substratum of the change (cf. *Metaphysics* VIII, 1, 1042b, 1–3, and above Prop. IV, p. 512, n. 8).

18. This comment of Crescas is based upon the following passages in Altabrizi:

"As for the second thesis, namely, 'everything movable is divisible,' that, too, may be doubted. For when a body is moved, its motion necessarily causes the motion of its surface and of the extremity of the surface, i. e., the line, and of the extremity of the latter, i. e., the point. So that the point is moved along with the motion of the body even though it is indivisible."

ואולם הטענה השנית, והוא שכל מתנועע מתחלק, בו גם כן ספק. חה שהגשם כאשר תתנועע הוא יתנועע בתנועתו השטח וקצהו, והוא הקו, וקצהו, והוא הנקודה, בהכרח. ואצל תנועת הגשם התנועע הנקודה גם כן, עם שהיה בלתי מתחלקת.

"As for the explanation of the second thesis, know that by 'movable' is meant here that which is movable essentially to the exclusion of that which is movable accidentally. By this the objection from the motion of the point falls to the ground, for the point is moved only accidentally but never essentially."

ואולם הטענה השנית, דע שהרצון במתנועע המתנועע בעצמות, לא המתנועע במקרה. ונפלה ממנה הסתירה בנקודה, כי הנקודה אמנם במקרה התנועע לא בעצמות.

Strictly speaking, the motion of a point is according to Aristotle accidental only in the sense of "according to part." See Prop. VI, p. 539, n. 13.

PART II

19. The assumptions underlying this statement are as follows: All knowledge originates in sense-perception. The sense data, however, before they become pure objects of knowledge, must pass through the faculty of imagination, whence they emerge as imaginative forms. It is these latter upon which the Active Intellect operates, transforming them into intellectual conceptions. Hence the statement here that the mind derives its knowledge from sense perception and imagination. Cf. *De Anima* III,

3, 427b, 14–16: "Imagination, too, is different from sensation and discursive thought. At the same time, it is true that imagination is impossible without sensation, and conceptual thought, in turn, is impossible without imagination."

Milḥamot Adonai I, 9: "Because the Active Intellect makes of the forms of the imagination actual objects of the intellect after they have been only potential objects of the intellect."

מפני שהשכל הפועל הוא משים הצורות הדמיוניות מושכלות בפועל אחר שהיו
מושכלות בכח.

Crescas, however, has taken his entire comment from Altabrizi: "As for the first thesis, namely, 'everything changeable is divisible', it contains a difficulty . . . . . The rational soul, as will be shown later, is an indivisible substance, and still it is subject to all kinds of changes, as, e. g., it is without knowledge and then becomes possessed of knowledge, and similarly universal forms are generated in it as a result of its preoccupation with imaginary and perceptual forms. And so also there is a change with respect to the qualities of the soul, such as appetite, desire, joy, fear, anger, and their like. Thus the essence of the soul is susceptible to all these changes and still is indivisible. How then can it be asserted that 'everything changeable is divisible.' "

אולם הטענה הראשונה, והוא אמרו כל משתנה מתחלק, הנה בו ספק... הנפש
המדברת כאשר יראה אחרי כן עצם בלתי מתחלק ויהיו לה שנויים, כמו שתהיה
סכלה ותשוב יודעת, ויתחדשו בה ציורים כוללים נקנים מהשמוש במדומות והמוחשות.
וכן האיכות הנפשיות, כמו התשוקה והחשק והשמחה והפחד והכעס חולתם, ואם
כן עצם הנפש מקבל לאלה השנויים, עם שהוא בלתי מתחלק, ואיך יצדק שיאמר
כל משתנה מתחלק.

20. Hebrew אשר יהיו בזולת זמן. This phrase does not occur in Altabrizi. Crescas has added it himself for a very significant reason. In *Physics* VII, 3, 247a, 16–b, 1, Aristotle states that while the emotions of pleasure and pain are qualitative changes, the habits of the intellective part of the soul undergo no change. To the explanations advanced by Aristotle as to why the acquisition of knowledge is not a qualittive change, Simplicius adds another one. It is due, he says, to the fact that qualitative change must always take place in time, whereas the act of the mind's acquiring knowledge is without time. (Cf. *Simplicius in Physica*,

ed. Diels, p. 1075, l. 23—p. 1076, l. 15. Cf. Taylor's translation of the *Physics*, p. 416, n. 5).

A statement like that of Simplicius is also found in Averroes' *Intermediate Physics* VII, 4: "It seems also that the action of the intellect in attaining knowledge is not a motion, inasmuch as it does not take place in time."

וגם כן יראה שהגעת השכל אינו תנועה מאשר הגעתו לא תהיה בזמן.

Similar statements to the same effect occur in the writings of Jewish philosophers:

Likkute Sefer Mekor Hayyim III, 30: ופעל השכל השגת כל הצורות המושכלות בלא זמן ובלא מקום, of which the following is the Latin in *Fons Vitae* III, 48 (p. 187): "Actio autem intelligentiae est apprehensio omnium formarum intelligibilium in non-tempore et in non-loco"

Cuzari V, 12: "Although the activity of the intellect in framing syllogisms by means of careful consideration appears to require a certain time, the deduction of the conclusion is not dependent on time, reason itself being above time."

והשכל ואף על פי שנראה מעשהו בזמן בהרכבת ההקשות בעיין ובמחשבה,
הנה הבנתו לתולדת אינו נראית בזמן, אך עצם השכל מרומם מהזמן.

Thus according to Aristotle, the acquisition of knowledge is not, properly speaking, a qualitative change, inasmuch as it does not take place in time. But as for that matter, Crescas seems to argue, it may still be called timeless change, for the proposition, according to the interpretation adopted by Crescas, includes both change in time and change in no-time.

But see quotation from *De Anima* below in n. 22, where the act of thinking is called motion by Aristotle himself.

21. While Crescas uses here the expression "motions of the soul," Altabrizi in the corresponding passage (quoted above n. 1) uses the expression "qualities of the soul." In Aristotle himself the emotions of fear, anger, and their like are described both as "qualities" ποιότητες (*Categories*, 8, 9b, 36) and as "motions" κινήσεις (*De Anima* I, 4, 408b, 4). Cf. next note.

22. That the emotions of pleasure and pain are changes, and hence in time, is asserted by Aristotle in *Physics* VII, 3, 247a, 16–17: "Pleasure and pain are changes in the quality of the sensitive part [of the soul]. Cf. also *De Anima* I, 4, 408b, 2–4: "The soul is said to feel pain and joy, confidence and fear, and again to be angry, to perceive and to think; and all these states are said to be motions." Cf. also *Topics* IV, 1, 121a, 30 ff., where Aristotle discusses the question whether motion is the genus of pleasure. But a direct statement on this point is found in *Likkute Sefer Mekor Ḥayyim* III, 30: ופעל הנפש החיונית שהיא תרגיש בצורות הגשמים העבים בזמן, of which the following is the Latin in *Fons Vitae* III, 48 (p. 187): "Actio animae animalis est sentire formas grossorum corporum in tempore."

The main point of Crescas' argument is this: The soul suffers change both in its rational and sensitive faculties. In the former it is change without time and in the latter it is change in time. And yet the soul itself is indivisible. It will be remembered that Crescas has interpreted the proposition to include both change in time and without time. That the soul is indivisible was generally accepted on the authority of Aristotle. Cf. *De Anima* I, 5, 411a, 26-b, 30.

A refutation of Crescas' criticism is found in Shem-ṭob Ibn Shem-ṭob's supercommentary on the *Intermediate Physics* VI, 7: "By the same reasoning may be answered the objection raised by Rabbi Ibn Ḥasdai in his book, where he argues against Aristotle, contending that the intellect is something that undergoes a change in passing from ignorance to knowledge, and still it is indivisible. But we may answer him in the same way by saying that the intellect can only be said to have been changed, for its change takes place suddenly, inasmuch as there is no intermediate between ignorance and knowledge, but it cannot be said that the intellect is undergoing a change."

ובזה בעצמו יושב ספק ספפק הרב ן' חסדאי בספרו שספק על אריסטו באמרו שהשכל דבר ישתנה מהסכלות אל הדעת והוא אינו מתחלק. אבל נשיבהו בזה גם כן ונאמר שהשכל יאמר בו שהוא כבר השתנה, לפי שהשתנה פתאום, כי אין בין הידיעה והסכלות בדבר אמצעי, אבל לא יאמר בו שהוא משתנה.

23. Altabrizi: "The answer to the first objection is that we mean here by 'changeable' that which is changeable with reference to the qualities of the body, as, e. g., heating, cooling, which are called alteration, whereas the objection raised was from the example of the qualities of the soul."

התשובה מהספק הראשון, שאנחנו נרצה במשתנה הנה המשתנה באיכות הגשמיות,
כמו החמום וההקרה, והיא השתנות, והסתירה נפלה באיכות הגשמיות.

24. That is to say, if the Proposition, whether taken according to
the interpretation of Avempace or according to that of Averroes,
means, as is maintained by Altabrizi, that only corporeal objects
that are changeable or movable must be divisible, it is entirely
superfluous, for it is generally known that corporeal objects are
divisible.

This objection has been anticipated by Altabrizi himself, and
he answers it: "Shouldst thou say that, when the term 'changeable' is taken as referring only to corporeal qualities, then the
object so changeable is self-evidently a body, and hence necessarily divisible, and there was therefore no need for a special
proposition, my answer is as follows: By 'divisible' is not meant
here that which is potentially divisible, in which case the proposition would be self-evident, but rather that which is actually
divisible. The meaning of the proposition is accordingly as follows: That which is changeable with a corporeal change is actually
divisible. The proposition so interpreted is not self-evident.
Quite the contrary, it needs to be demonstrated, for the elements,
which are simple bodies, are one in reality, just as they appear to
the senses, and still they are not actually divisible but only
potentially."

ואם אמרת, כאשר חזקתה השני באיכות הגשמיות הוא המשתנה הוא הגשם והוא
המתחלק בהכרח, ואין זה צורך אל שומה מההקדמות המופרדות, אמרתי אין
הנרצה במתחלק, המתחלק בכח, עד תהיה ההקדמה הכרחית, אבל הרצון בה
המתחלק בפועל. ויהיה שעור זאת ההקדמה כן: המשתנה בהשתנות הגשמיות
מתחלק בפועל. ואין זאת ההקדמה הכרחית, אבל היא צריכה אל הראיה, שהגשם
הפשוט אחד באמת, כמו שהוא אצל החוש, ואינו מתחלק בפועל אבל בכח לבד.

25. In *Moreh* II, 1, First Speculation, Maimonides proves from
this proposition that since God is immovable he must likewise be
unchangeable and indivisible. Now if, according to Altabrizi's
interpretation, the term changeable in this Proposition refers only
to physical qualities, Maimonides could not prove thereby that
the First Cause of motion is free of any kind of change, even of
such change as does not refer to physical qualities.

26. Cf. *Or Adonai* II, vi, 1

PROPOSITION VIII

Part I

1. The Hebrew text of the proposition is taken from Isaac ben Nathan's translation of Altabrizi, except that Altabrizi has אותה התנועה המקרית (Ibn Tibbon: התנועה ההיא המקרית) in place of Crescas, התנועה המקרית. I have translated it here in accordance with the original Arabic reading which is faithfully reproduced both in Ibn Tibbon and in Altabrizi. The significance of "*that* accidental motion" will appear later in the discussion as to what kind of "accidental" motion is meant here in this proposition.

2. Cf. *Physics* VIII, 5, 256b, 9–10: οὐ γὰρ ἀναγκαῖον τὸ συμβεβηκός, ἀλλ᾽ ἐνδεχόμενον μὴ εἶναι. Cf. below n. 4.

3. That is to say, since accidental motion has only possible existence, i. e., it may and may not exist, both these possibilities, existence and non-existence, must be realizable, for, according to Aristotle, "it cannot be true to say that this thing is possible. and yet will not be" (*Metaphysics* IX, 4, 1047b, 4–5). Cf. also *Metaphysics* IX, 8, 1050b, 11–12. "That, then, which is possible to be may either be or not be; the same thing, then, is possible both to be and not to be."

4. On this proposition Crescas had before him several different interpretations all turning about the meaning of the term "accidental." First, Altabrizi, who takes the term "accidental" in the sense of "violent" motion. Second, Hillel of Verona and Isaac ben Nathan, the translator of Altabrizi, who take the term "accidental" in its ordinary sense of the motion of an accident inherent in a subject. Third, Narboni, whose view will be quoted by Crescas later.

The source of these differences of interpretation, it seems to me, is the ambiguity of the term בעצמותו, "in its own essence," used by Maimonides in the proposition. We have seen above (Prop. VI, n. 3) that in Aristotle there is a difference between καθ᾽ αὐτό and ὑφ᾽ αὐτοῦ, the former meaning to be itself essentially translated as a whole from one place to another, contrasted with the motion of color in a body or of a part with the whole, the latter meaning to have the cause of its motion in itself, contrasted with having the cause of motion external to itself. In Hebrew no less than in English it is difficult to translate accurately the difference between the two Greek prepositions, κατά and ὑπό, though, as I have pointed out, in the *Intermediate Physics* one is translated by בעצמו and the other by מפאת עצמו or מצדו. Now, in this proposition it is not clear what Maimonides' בעצמתו represents, whether the καθ' αὐτό or the ὑφ' αὐτοῦ. Altabrizi seems to take it to represent the latter, and therefore takes its opposite "accidental" in the sense of having the cause of motion external to itself, i. e., violent motion. Hillel of Verona and Isaac ben Nathan, on the other hand, seem to take it in the sense of the former, and therefore take "accidental" in the sense of the motion of accidental qualities. As for Narboni's interpretation, we shall take it up later.

Altabrizi: "You already know, from what has been said before, the meaning of 'accidental motion' and 'essential motion' and their subdivisions, and in the light of this the intention of the author in this proposition will not be hidden from thee."

כבר ידעתי במה שקדם ענין התנועה המקרית והתנועה העצמתיית ומיניה, ולא
יעלם עליך רצונו מזאת ההקדמה.

Upon this Narboni comments: "The learned Mohammed ben Zechariah (see Steinschneider, *Uebersetzungen*, p. 361, n. 764) Altabrizi, the Persian, the commentator of the Propositions of the *Guide*, in his explanation of this proposition takes the term 'accidental' in the sense of 'violent,' for 'violent motion' is one of the subdivisions of accidental motion, as he has explained in the sixth proposition . . . . But the translator of Altabrizi's commentary Rabbi Nathan ben Isaac [read: Isaac ben Nathan, see Steinschneider, *Uebersetzungen*, p. 362, n. 769] of Xativa, in answer to the difficulty raised by Altabrizi said that while it is true that violent motion is called accidental, the Master does not use here the term accidental in the sense of violent but rather in the sense in which blackness is accidental to a body."

והחכם מחמד בן זכריה אלתבריזי הפרסי מפרש ההקדמות המורה פירש זאת
ההקדמה על שלקח המקרה מקום ההכרח, למה שהיה ההכרח אחד ממיני מה
שבמקרה, כמו שבאר בהקדמה הששית. ...והמעתיק הפירוש ההוא, החכם ר'
נתן ב"ר יצחק ור' יצחק בר נתן] משיאטיבא, כאשר ראה זאת הקושיא כתב עליו

ותירץ, כי אם הוא אמת שהתנועה ההכרחית כבר תקרא מקרית הנה לא רצה הרב
במקרית שהיא ההכרחית, אבל רצה אשר במקרה כשחרות לנשם.

(Isaac ben Nathan's answer referred to by Narboni is not found
in the printed edition of Altabrizi).

Hillel of Verona in his commentary, *ad loc.:* "This proposition
hardly needs a proof, for an accident is that which disappears and
does not continue to exist in the same state. An accident is
defined as that the existence and the passing-away of which are
conceivable without having to conceive the passing-away of its
subject, as, e. g., the color in a garment." אין צריך ביאור, כי המקרה
יסור ולא יעמוד על ענין אחד, וגדר המקרה הוא דבר שידומה הוייתו והעדרו מבלי
שידומה העדר נשאו, המשל בזה הצבע עם הבגד.

If we assume with Altabrizi that the term 'accidental' is to be
taken in the sense of 'violent motion' then the source of the
proposition is the following passage in *De Caelo I*, 2, 269b, 6–9:
"If, on the other hand, the movement of the rotating bodies about
the centre is *contrary to nature*, it would be remarkable and indeed
quite inconceivable that this movement alone should be continuous and eternal, being nevertheless contrary to nature." In the
Arabic versions of the *De Caelo*, the Greek 'contrary to nature,
παρὰ φύσιν, must have been replaced by 'accidental'. Thus in
Averroes' *Intermediate De Caelo* I, iv (Latin, p. 274va, H) the
passage quoted is paraphrased as follows: "For *accidental* motion
cannot be perpetual and infinite, and to assume this is beyond the
bounds of all reasoning, for we observe that all things perish and
disappear." כי התנועה המקרית אי אפשר שתמצא תמידית אין תכלית לה,
ושהצעת זה יוצאה מכל הקש, כי אנחנו רואים הדברים המקריים כלים אובדים.

In the *Moreh ha-Moreh* (p. 67) this passage of the *De Caelo* is
used as the explanation and hence the source of the proposition,
and this view is followed by Munk (*Guide* II, p. 9, n. 3).

Crescas, however, seems to place the source of the proposition
in *Physics* VIII, 5, 256b, 3–13, for his proof of the proposition is
based upon that passage, and in this he is following Narboni,
whose proof is likewise based upon that passage.

Aristotle's own argument in proof of this proposition may be
outlined as follows: Starting with the major premise that motion
is eternal and that there is a first mover, Aristotle tries to prove
that the first mover cannot itself be moved. If the first mover,

he argues, is assumed to be moved, the question is whether it is moved accidentally (κατὰ συμβεβηκός) or essentially (καθ' αὐτό). If you say it is moved accidentally, then it may be possible that at some time or other it will not be moved, "for accident is not necessary and it may not exist" (*Physics* VIII, 5, 256b, 9–10). But if the first mover may at some time cease to be moved, it may also cease to move, since it is now assumed that it is of such a nature that it must be moved while it moves. But that motion should come to an end is impossible, according to our major premise.

Averroes' Long Commentary on *Physics* VIII, ii, 3, p. 375vb, K: "Cum posuerimus quod iste motor non movet, nisi moveatur, et posuerimus ipsum moveri per accidens, possibili est ut aliqua hora veniat, in qua non movebitur, quod enim est per accidens, non est semper neque necessarium. Et cum fuerit possibile ut non moveatur, erit possibile ut non moveat, cum sit ita, quod suum moveri est necessarium in suo movere."

The text in the *Intermediate Physics* VIII, iv, 4, 2, upon which Crescas' proof is directly based, reads as follows: "That not every mover must necessarily be moved becomes evident by the following argument. For if every mover were moved, it would have to be moved either essentially or accidentally, as in the case of the sailor who causes the ship to move and is himself moved accidentally by the motion of the ship. But if every mover were moved accidentally, and its being so moved were a condition in the existence of the mover as a mover, then, inasmuch as that which is accidental may not continue to exist, for that which is accidental does not continue eternally, it will follow that the first mover may not continue to exist as a mover, and if the first mover may cease to exist, motion may cease to exist. But this is a logical absurdity, for it has been shown that motion cannot cease to exist. And any premise that gives rise to an impossibility is itself impossible, and of such a nature would be the statement that every mover must be moved accidentally."

ואמנם שלא יחוייב שיהיה כל מניע מתנועע, זה יראה ממה שיאמר אותו. וזה שאם היה כל מניע מתנועע, הנה אם שיהיה זה בעצם ואם במקרה, כמו המלח אשר יניע הספינה והוא מתנועע ממנה במקרה. ואם היה זה במקרה, והיה תנאי במציאות המניע מניע, והיה מה שבמקרה כבר אפשר שלא ימצא, אחר שאינו מתמיד,

הנה כבר אפשר שלא תמצא המניע המניע הראשון מניע. וכאשר אפשר שלא ימצא, הנה
כבר אפשר שלא תמצא תנועה. זה שקר, לפי שכבר התבאר שהתנועה אי אפשר
שתעדר. ומה שיחוייב ממנו הבטול, הוא בטל, והוא שכל מניע מתנועע במקרה.

PART II

5. The term כדור, literally, "sphere" or "globe" and גלגל,
literally, "circle" 'or "orb" represent the Arabic كرة and فلك.
respectively, but on the whole they are indiscriminately used by
Maimonides with reference to all the different varieties of the
celestial spheres (see Friedlander, *Guide of the Perplexed* I, 72, p.
291, n. 1, and II, 4, p. 32, n. 1). Here Crescas and Altabrizi (see
below n.' 6) use כדור with reference to "fire," and by implication
with reference to all the other sublunar elements, and גלגל with
reference to the celestial spheres. In *Cuzari* V, 2 (end), however,
the author speaks of פלך אלנאר, גלגל האש, "fire sphere", פלך
אלמא, גלגל המים, "water
sphere," but כר"ה אלארץ, כדור הארץ, "terrestial globe." Similarly
in *Cuzari* II, 6, אלפלך אלאעלי, הגלגל העליון, "uppermost sphere"
but כר"ה אלארץ, כדור הארץ, "terrestial globe."

6. This criticism as well as the illustration is taken from Altabrizi:
"As for the truth of this proposition, I know of no proof for it.
Quite the contrary, it is possible for one body to be set in motion
accidentally by another body, and if the other body is moved
essentially for ever and the two bodies are linked together as
cause and effect, the accidental motion of the body moving
accidentally will also continue for ever. An illustration for this is
the globe of fire which is moved by the motion of the celestial
sphere, and inasmuch as the motion of the sphere continues for
ever the accidental motion of the globe of fire continues for ever."

ואולם קיום אמתתה הנה לא בא מופת אצלי באמתתה, כי מהאיפשר שיהיה גשם
מתנועע תנועה מקרית מגשם אחר, יהיה הוא המתנועע בעצמות תמיד התנועה,
ויהיו השני גשמים מתחייבים במציאות והתמיד התנועה המקרית לגשם המתנועע
במקרה, כמו כדור האש, כי הוא מתנועע בתנועת הגלגל, ובעבור שהיתה התנועה
לגלגל תמיד, היתה התנועה המקרית לכדור האש תמיד.

Strictly speaking the illustration used by Altabrizi is a species of
'violent' motion rather than of 'accidental.' But we have seen

above (n. 4) that Altabrizi takes the term 'accidental' in the proposition in the sense of 'violent.'

7. By the parts of the sphere he means the spheres that are within the spheres. Cf. *Mishneh Torah, Yesode ha-Torah*, iii, 2: "Every one of the eight spheres containing stars is divided into several spheres." כל גלגל וגלגל משמונה הגלגלים שבהן הכוכבים נחלק לגלגלים הרבה. *Moreh* II, 4: "Though in some of these spheres there are several orbs." ואף על פי שבקצת הכדורים ההם גלגלים רבים. Crescas undoubtedly alludes by this to the illustration used by Gersonides in the second passage quoted in the next note.

8. These two illustrations, one from the superficies of the celestial sphere and the other from its parts, are not found in Altabrizi. They are based respectively upon the following two passages of Gersonides.

A. Supercommentary on the *Intermediate Physics* VIII, iv, 4: "Says Levi, Would that I knew, when something accidental is the consequence of something essential, why should not the accidental continue for ever as a result of the continuity of the essential? To illustrate: If we assume that there exists a certain body that is moved eternally, such as has been shown before, but that its surfaces are moved accidentally, shall we then say that those surfaces may on that account come to rest, which will mean that the body itself will of necessity have to come to rest? In general, it is not impossible that something accidental should continue forever in consequence of the continuity of something essential."

אמר לוי, מי יתן ואדע, כאשר היה מה שבמקרה נמשך למה שבעצם, למה לא יהיה מתמיד בהתמדת מה שבעצם, והמשל אם נניח שיש הנה נשם מתנועע תמיד, כמו שהתבאר, האם מפני ששטחיו מתנועעים במקרה נאמר שתהיה אפשרית בו המנוחה, וינוח הנשם ההוא בהכרח? ובכלל הנה אינו נמנע במה שבמקרה שיהיה מתמיד בעבור שבעצמות.

B. Supercommentary on *Intermediate De Caelo* I, 4:

" 'For accidental motion cannot be continuous and infinite' . . An objection may be raised against this proposition by showing that accidental motion can continue for ever, as, e. g., the diurnal revolution of the sun which is caused by something external, for of itself it has only the annual motion. That it should be so is quite explicable, for this accidental motion of the sun is caused by an eternal and natural circular motion, namely, the motion of the diurnal sphere. This, to be sure, is not an objection against the principle which Aristotle has meant to establish by this proposition, for, after all, this accidental motion is consequent to a natural, circular motion, but it is an objection against Aristotle's wording of the proposition. Some philosophers have been led to say that it is not inconceivable that something may be possible with reference to itself and necessary with reference to its cause, according to which view there may be continuity in that which is moved accidentally. Averroes, however, rejects this view. But this is not the place to discuss this matter."

כי התנועה המקרית אי אפשר שתמצא לו תמידית אין תכלית לה... וכבר אפשר
שיסופק על זה ויאמר, שהתנועה המקרית כבר תהיה תמידית, כאלו תאמר תנועת
השמש היומית, שהוא לו מצד זולתו, כי התנועה אשר לו מצד עצמו בשנה. ויהיה
זה כן, לפי שהסבה בזאת התנועה המקרית תנועה סבובית נצחית טבעית, והוא
תנועת הגלגל היומי. אלא שאין זה ספק על מה שהוליד אריסטו הנה, כי על כל
פנים זאת התנועה המקרית תמשך לתנועה טבעית סבובית, אבל הוא ספק על
ההקדמה אשר חייבה אריסטו. ויאמרו קצת הפילוסופים שאינו נמנע שיהיה כבר
אפשר בבחינת עצמו מחוייב בבחינת סבתו, ועל זה הצד יהיה התמידית במה
שבמקרה. ואבן רשד ימאן זה, ואין הנה מקום החקירה.

An argument similar to that contained in the second quotation is also raised by Simplicius on *Physics* VIII, 6, 259b, 28–31 (ed. Diels, p. 1261, 11. 14–19, and Taylor's translation of the *Physics*, p. 479, n. 1): "Aristotle having said, that in things which are immovable, indeed, but which move themselves according to accident, it is impossible to move with a continued motion, it becomes doubtful how the celestial orbs, since they are self-motive animals, and have a mover essentially immovable, and not moving itself according to accident, but accidentally moved by another; for the planets are moved by the inerratic sphere with the motion of that sphere,—it becomes doubtful, how they are at the same time moved with a continued motion."

There is also a similarity between the answer mentioned by Gersonides in the name of some philosophers (probably Avicenna; see below n. 15) and that offered by Simplicius, as will be shown below in n. 11.

9. I take this comment to refer only to the last two cases of participative motion borrowed from Gersonides and not to the first case of violent motion borrowed from Altabrizi (see above n. 6). These last two cases, strictly speaking, are motion 'according to part' and not 'accidental motion.' But Crescas justifies himself here for calling them accidental motion by alluding to Maimonides' statement in Prop. VI that motion according to part "is a species of motion according to accident." See Prop. VI, n. 1. The direct reference of במשלו, *in his illustration*, is to the statement וכל מחובר יתנועע בכללו יאמר שחלקו כבר התנועע "and similarly, when something composed of several parts is moved as a whole, every part of it is likewise said to be moved" in Prop. VI.

10. By "others" Crescas undoubtedly refers to Narboni, whom he mentions later in the course of his discussion, and to Gersonides, from whom, as I have suggested, he must have taken his last two illustrations (see above n. 8). It may also allude to the answer attempted by Altabrizi's translator quoted above in n. 4.

11. What Narboni wants to say is this: The term 'accidental' in the proposition does not refer to violent motion, nor to motion according to part, nor to the motion of accidental qualities. It refers only to one particular kind of motion, namely, the motion produced accidentally in a mover as a result of its being itself the cause of motion in something else. It is quite clear from this that Narboni did not take this proposition to reflect Aristotle's statement in *De Caelo* I, 2, 269b, 6–9 but rather the statement in *Physics* VIII, 5, 256b, 3–13 (see above n. 4).

Narboni's text reads as follows: "What the divine Rabbi Moses meant by this proposition is as I shall state. The expression 'everything that is moved accidentally,' concerning which he says in this proposition that it 'must of necessity come to rest,' is meant by him to refer to everything that is moved accidentally, by any kind of accidental motion, in so far only as it is moved accidentally. If, for instance, we assume a certain mover to be moved accidentally but that accidental motion therein is the result of the very motion of which it is the cause, then that mover must of necessity come to rest, be it a force distributed throughout the body and divisible or an indivisible force, as, e. g., the human soul in man and the Intelligence, according to the Master's view

(cf. *Moreh* II, 1; below Prop. XI, n. 5, p. 605; above p. 267).
When this proposition is thus interpreted, namely, that, everything that is moved accidentally is, to be taken in a restricted
sense, i. e., in so far as it is moved by the motion of the body of
which it is itself the cause, it becomes self-evident that it must of
necessity come to rest, unless there be outside of it another
immaterial mover, as is the case of the soul of the sphere, which
continues to be moved perpetually by the perpetual motion of the
sphere, even though it is moved accidentally, the reason for this
being that the soul of the sphere acquires its perpetuity of motion
from the eternal immaterial mover."

ואשר כוון האלהי רבינו משה הוא כפי שאומר: המתנועע במקרה אשר אמר
בו בזאת ההקדמה שינוח בהכרח, איזה מין שיהיה ממה שבמקרה, במה הוא מתנועע
במקרה, עד שאם היה זה המתנועע במקרה מניע וסבה לתנועתו על שיתנועע במקרה
בזאת התנועה שהיא סבתה, ינוח בהכרח, יהיה כח מתפשט בו [ומתחלק או כח]
בלתי מתחלק, כנפש האדם באדם והשכל כפי דעת הרב. וכאשר יובן שזאת
ההקדמה לו זה הענין, ר"ל שכל מתנועע במקרה מקושר, ר"ל במה הוא מתנועע
במקרה בתנועת הגשם שהיא סבתה, הוא מבואר בעצמו שהוא ינוח בהכרח, אך
אם לא יצטרף לשם מניע אחר זולתו, יהיה נבדל. חה כי נפש הגלגל הוא על זה
התאר, והיא מתנועעת תמיד בהתמדת תנועת הגלגל, ואם היא מתנועעת במקרה,
כי תקנה הנצחיות מהמניע הנצחי הנבדל.

Narboni's answer, as will have been observed, is practically
based upon a distinction between a mover that is moved accidentally by itself and one that is moved accidentally by an external
cause. This corresponds exactly to the answer offered by Simplicius to the same question (quoted above in n. 8): "He solves
this doubt, therefore, by saying that it is not the same thing for
any being to be moved accidentally by itself and to be moved by
another" (ed. Diels, p. 1261, 11. 19–21). And this is exactly the
same distinction implied in the answer mentioned by Gersonides
in the name of some philosophers (see above n. 8). As we shall
see, it is adopted also by Crescas here (see below n. 15).

It should also be noticed that Narboni's interpretation of the
term "accidental" corresponds exactly to the use made of the
term in the passage from Averroes quoted above in n. 4. where it
is illustrated by the motion caused accidentally in the sailor as a
result of his setting the ship in motion.

12. Hebrew בלתי מחוייב. ‏והנה כשנשתדל בזה נמצאהו בלתי מחוייב‎. Literally the Hebrew ‏השתדל‎ is the equivalent of the Arabic ‏جهد‎, *exert one's self, make efforts* (see Steinschneider, *Uebersetzungen*, pp. 279, 339, n. 252). But it is not impossible that here it reflects the Arabic ‏استدلال‎, *have a thing shown to one's self, ask for an argument.* In the *Maḳasid al-Falasifah* II, p. 82, however, ‏يستدلّون‎ is translated by ‏ירחקו‎, *shrink from, keep away from,* or ‏ירחיקו‎, *repudiate, reject.* See Prop. X, n. 9.

13. Hebrew ‏הקשר מציאות או הקשר עירוב‎. These two expressions which describe two different views as to the relation of the rational soul to body may be traced to Aristotle. The expression ‏הקשר מציאות‎ reflects the view that the soul "is not body (σῶμα), but something belonging to body (σώματος δέ τι) and therefore existing (ὑπάρχει) in the body" (*De Anima* II, 2, 414a, 19–22). Thus the term ‏מציאות‎ in this expression represents the Greek ὑπάρχειν, *inesse, inexistence, inbeing.* The term ‏עירוב‎ represents the Greek κρᾶσις, μεῖξις (*De Anima* I, 4, 407b, 31; 408a, 14). These two views with regard to the relation of soul to body are mentioned by Bruno and are designated by him by the same terms as in Hebrew: "Questa forma non la intendete accidentale, ne simile alla accidentale, ne come *mixta* alla materia, ne come inherente á quella: ma *inexistente,* associata, assistente" (*De la Causa, Principio, et Uno,* II, ed. Lagarde, p. 240, 1. 40—p. 241, 1. 2).

14. The criticism against Aristotle's proposition raised here by Crescas, including his rejection of Narboni's answer, is reproduced by Pico Della Mirandola in *Examen Doctrinae Vanitatis Gentium* VI, 2: "Falsum quoque et illud esse Hebraeus Hasdai contendit, quickquid ex accidenti movetur, quandoque necessario quiescere. Nam ex Aristoteleo dogmate sphaera ignis ex accidenti mota, videlicet ad orbis superioris motum, non qiescet coelo agitato: quod noluit Aristoteles posse quiescere, superficies quoque coeli extima, et partes ipsius semper agitatae, non ex se, sed ex accidenti ad motum corporis in quo sunt moventur. Nec responsio Moysis Narbonensis quicquam suffragatur, ut illud ex accidenti quatenus, ex accidenti vim exemplorum imminuat. Animae enim dum motu corporum moventur, ut coniunctae sunt moventur, et aeterno motu coeli anima ex eius sententia movet." (Cf. Joël, *Don Chasdai Creskas' religionsphilosophische Leheren*. p. 83).

15. I take this conclusion to be Crescas' own attempt to remove the objection raised against the proposition, by pointing out that the proposition is not meant to include the kind of accidental motion which proceeds by necessity from something that moves essentially. In a similar way Gersonides solves the difficulty in the two passages quoted above in n. 8. In the second of those passages he justifies the exclusion of this kind of accidental motion from this proposition on the ground that such accidental motion, brought about by necessity by something that moves essentially, is to be considered as a "necessary" rather than a "possible" motion, according to the Aristotelian view as interpreted by Averroes. It is only Avicenna, he says, who would call such an accidental motion possible. We have already seen that the proof of this proposition, namely, that every accidental motion must be transient, rests upon the principle that everything accidental is possible (see above notes 2, 3, 4). Consequently, if an accidental motion cannot be called possible, such, for instance, as the accidental motion necessitated by some essential motion according to Averroes, it will have to be excluded from this proposition.

As to the controversy between Avicenna and Averroes on the meaning of the term possibility, see notes on Prop. XIX.

PROPOSITION IX

Part I

1. The Hebrew text of this proposition is taken from Ibn Tibbon's translation of the *Moreh*.

2. This comment of Crescas is based upon the following passage of Narboni: "Motion may be produced by either one of two causes, one of them acting as a final cause and the other acting as an efficient cause. By the mover in this proposition is meant that which acts as a proximate and efficient cause, for a mover which acts as a final cause, not being proximate, is not moved, as, e. g., fire, for when air is moved upward in quest of its natural locality and ascends as high as fire, it is acted upon by the latter as a final cause. But that which produces motion as an efficient cause, whether by pushing or by drawing, produces that motion only by contact . . . . . and hence must necessarily be moved."

התנועה לה שתי סבות, אחת מהם אשר על דרך התכלית, והשנית על דרך הפועל. והנרצה הנה במניע הסבה הקרובה אשר על דרך הפועל, כי המניע אשר על דרך התכלית לא יתנועע בקרוב, כאש, בעלות האויר, בדרשו מקומו הטבעי ויעלה לאש על דרך התכלית. אבל הפועל לתנועה, אם דוחה ואם מושך, אמנם יניע בשימשש. . . . . ויתנועע עמו בהכרח.

Narboni's comment, as will have been observed, contains two points. First, that only movers which act by *contact* are themselves moved in producing motion. Second, that movers that act by contact produce motion either by impelling or by drawing. Both these points are traceable to Aristotle.

The first point is based upon *Physics* III, 2, 202a, 3–7, (which seems to be the direct source of Maimonides' proposition and not the lengthy discussion in *Physics* VIII, 5, referred to by Shem-ṭob and Munk): "But as we have said, everything which moves is moved, being movable in capacity, and of which the immobility is rest: since the immobility of that to which motion is present is rest. For to energize with respect to that which is movable, so far as it is movable, is to move. But it effects this by contact: so that at the same time also it suffers."

The distinction between a cause which acts by contact and one which does not act by contact is elaborately developed by Maimonides in *Moreh* II, 12 (see below n. 5).

The second point is based upon *Physics* VII, 2, 243a, 16–17, and the corresponding passage in *Intermediate Physics* VII, 3, where Aristotle enumerates four ways by which an external agent can produce motion in an object: (1) drawing, ἕλξις, משיכה. (2) pushing, ὦσις, דחיה. (3) carrying, ὄχησις, משא. (4) rolling, δίνησις, סבוב.

3. Hebrew אבן המגניטס, حجر المغناطيس ἡ Μαγνησία λίθος. Hebrew translations of magnet are: 1. אבן השואבת (*Moreh* II 12. cf. Sanhedrin 107b). 2. אבן המושכת (*Epitome of the Physics* VII, p. 37a). 3. אבן הזוחלת (Anonymous translation of Altabrizi, Prop. IX). Cf. I Kings 1, 9. But in Hebrew זחל is intransitive, meaning

creep, crawl. Its use by the anonymous translator of Altabrizi in a transitive sense, as synonymous with שאב and משך, is probably due to the influence of the Arabic زَحَل, *take* or *draw from a place*. The connection between the two words has already been pointed out by Ibn Janaḥ in his *Sefer ha-Shorashim*.

4. Cf. *Intermediate Physics* VII, 3: "A certain difficulty has been raised in the case of motion by drawing, for there are things which appear to move by drawing without being themselves moved, as in the case of the motion caused by the Magnesian stone which attracts iron."

וכבר יושג במשיכה ספק מה, וזה שבכאן דברים יראה מענינם שהם ימשכו מבלתי שיתנועע, הנמשך באבן המגניטס, שתמשוך הברזל.

5. These two explanations are quoted by Averroes (*Intermediate Physics* VII, 3) in the name of Alexander:

"Alexander in his commentary on this passage answers this objection in two ways: First, that it is doubtful concerning these things whether their motion is brought about by drawing or not by drawing, for one may argue that the iron is moved of itself toward the stone by reason of a certain disposition which accrues to it from the stone, but that the stone does not draw the iron. Second, if we admit that it is done by drawing, this drawing may be explained by the fact that certain particles are emitted from the object which draws and come in contact with the object that is drawn and then draw it toward the former object."

ואלכסנדר ישיב בזה המקום על זה הספק בשתי תשובות: האחת, שאלו הדברים מסופק מענינם האם תנועותיהם משיכה אם אינה משיכה, כי לאומר שיאמר שהברזל מתנועע בעצמו אל האבן במוז אשר יקרה מהאבן, לא שהאבן תמשך הברזל. והתשובה השנית, שאם קבלנו שהוא משיכה, הנה אמנם יהיה זה בשיותכו מהמשך נשמים ימששו הנמשך וימשכוהו אל המשך הראשון.

The second of these explanations represents the general view of the Atomists (see Zeller, *Pre-Socratic Philosophy*, Vol. II, p. 230, n. 1), which is fully described by Lucretius, *De Rerum Natura* VI, 11. 998–1041. It is also followed by Maimonides, *Moreh* II, 12: "In the natural sciences it has been shown that a body in acting upon another body must either directly be in contact with it, or indirectly through the medium of other bodies . . . . . The magnet attracts iron from a distance through a certain force communicated to the air which is in contact with the iron."

וכבר התבאר בחכמת הטבע, כי כל נוף שיעשה מעשה אחד בנוף לא יעשה בו רק כשיפעשהו או יפגוש מה שיפגשהו... עד שהאבן השואבת אמנם תמשוך הברזל מרחוק בכח שתתפזר ממנו באויר הפוגש הברזל. Efodi significantly explains Maimonides "force" to mean a certain "quality emanating from the magnet," איכות מה שיוצא מהאבן השואבת i. e., the "parti-cles" of Alexander's second explanation.

Pico Della Mirandola's discussion of the magnet in *Examen Doctrinae Vanitatis Gentium* VI, 18, is evidently based directly upon Averroes, and is not taken from Crescas, though the latter is mentioned immediately before that discussion in some other connection.

PART II

6. Hebrew מזג משכנת המגניטס. There is a subtle suggestion of a contrast in the choice of words here, for מזג and שכנית are two contrasting terms, denoting two different kinds of composition, one consisting of a harmonious blending of ingredients and the other of simply a juxtaposition of ingredients. (Cf. הרכבת מזג and הרכבת שכנית in Samuel ibn Tibbon's *Perush me-ha-Millot Zorot*). Now, if the iron is to acquire a new characteristic or tendency it must be the result of a new harmonious blending of its ingredients or qualities. Hence Crescas argues: How can the iron acquire a new characteristic out of its mere juxtaposition to the magnet?

7. Hebrew אשר לכל אחד כח טבעי שעור גדול. My translation of this passage is conjectural and it has necessitated the insertion prior to it of a statement which is not found in the text. The passage, however, lends itself also to the following three translations:

(1) "which is apparent to everybody that it must be a natural force of considerable strength."

(2) "which would require on the part of either one of them (i. e., the iron and the magnet) a natural force of considerable strength."

(3) "which would require on the part of every piece of iron a natural force of considerable strength."

8. Hebrew למה שהוא נלוי מעניינם היותם קשי ההפעלות מאד. All the MSS. and editions agree upon having a plural pronominal suffix in both עניינם and היותם. A change to the singular, would make these pronouns refer to the act of acquiring a new disposi-

tion on the part of the iron. What the plural pronominal suffixes refer to is hard to determine. My translation is conjectural and is dependent upon my other conjectural translation of the preceding passage. The plural may also refer to the iron and the magnet or to every piece of iron, if either one of the last two translations of the preceding passages suggested in n. 7 is correct.

It is not impossible that both this passage and the preceding passage are misplaced. Another instance of a misplaced passage we have already met in Prop. I, Part I, n. 104 (p. 374). Cf. also Prop. I, Part II, n. 120 (p. 469). The order of the text here may be rearranged to read as follows:

הנה השני פנים אשר זכרו ממה שיראה ממשיכת אבן המגניטס הברזל, אשר לכל אחד כח טבעי שעור גדול, מבוארי הנפילה בעצמם, למה שהוא גלוי מעניינם היותם קשי ההפעלות מאד. כי שיקנה הברזל מזג משכונת המגניטס הוא רחוק קרוב לנמנע.

"The two methods mentioned by them in explanation of the phenomenon of the power of the Magnesian stone to attract iron which, according to either one of the suggested methods, is a natural force of considerable strength, are self-evidently ground-less, inasmuch as it is clear from their nature that both these methods are very difficult of performance. That the iron should acquire from the magnet, through its proximity to the latter, a new disposition, is a far-fetched assumption and well-nigh impossible."

9. Hebrew מי יתן ואשער. See Prop. I, Part II, p. 417, n. 30.

10. In opposition to the two explanations advanced by Alexander, Crescas argues that the attraction of iron by a magnet is not due to a new property which the iron acquires from the magnet nor to corporeal particles emanating from the magnet but rather to a certain natural disposition or tendency in the iron itself. This natural tendency, תנועה טבעית, he describes as being either due to האותות, suitableness, i. e., the fact that the magnet is the proper place to which the iron belongs and consequently tends towards it, just as the natural elements according to Aristotle move in different directions because they have different proper localities, or to a סגולה, a certain peculiar property within the nature of the iron itself, just as the natural elements, according to Crescas' own view (see Prop. I, Part II, p. 456, n. 76), move in different directions because of a peculiar property in their own nature.

Crescas' explanation of the motion of iron toward a magnet and its analogy to the natural motion of the elements can be traced to the following passage in Gersonides' supercommentary on the *Epitome of the Physics* VII: "The motion produced by the magnet may be considered as an action produced by a final cause, in the same manner as the elements are moved toward their proper places by reason of agreeableness and likeness."

זה שתנועת אבן המושכת היא על צד התכלית, כמו שיתנועעו הגשמים אל מקומם על צד הערבות והדמיון.

The passage in the *Epitome of the Physics* VII, p. 37a, upon which the foregoing quotation from Gersonides is a comment reads as follows: "For the magnet and its like produce motion as a final cause in the same manner as the water circumference causes earth to move toward it."

האבן המושכת והדומים לה יניעו על צד התכלית, כמו שיניע הקף המים לארץ.

It must have been to this passage of Averroes that Gersonides' father, Gershon ben Solomon, referred in his following explanation of magnetic attraction. *Sha'ar ha-Shamayim* II, 3: "Of the amber stone, i. e., the magnet, which attracts iron, some say that it is of the nature of iron, but [what we call iron is] of an imperfect nature and hence it desires to unite itself with iron that is perfect [i. e., the magnet]. This is the view of Averroes."

אבן אלענברי היא אלמגניטס, והיא המושכת לברזל, ויש שאומרים כי היא מטבע ברזל שלא נשלם שלמות טבעו, ולזה הוא חושק להדבק לברזל השלם. וכן דעת בן רשד החכם.

Literally the passage reads that the magnet is an imperfect kind of iron and hence is attracted by iron. But that obviously is not what the author meant to say.

We thus have three explanations of magnetic attractions, the two recorded by Averroes in the name of Alexander and Crescas' explanation, which, we have seen, can be traced to Averroes. I believe there is still another explanation discernible in certain passages of Jewish philosophic writings. This explanation, like that of Crescas, attributes magnetic attraction to a certain unknown power or peculiar property. But unlike Crescas' explanation, it places that power or peculiar property not in the iron but in the magnet.

Sha'ar ha-Shamayim III, 1: "In this all philosophers agree, namely, that plants have a vegetative soul, except Galen, who claims that what they have is not a soul but only a power like that which exists in a magnet."

ובזה הסכימו כל החכמים, כי יש לצמחים נפש צומחת, חוץ מגאלינוס, שאומר שאין להם נפש אלא כח אחד כמו אבן המושכת.

Joseph Zabara's *Sefer Sha'ashu'im* IX, 11 (ed. Davidson, p. 104):

"And he said: 'Knowest thou whence comes the juice of the food into the liver, seeing that the intestines have no aperture through which it could exit nor is there an aperture in the liver through which it could enter?'

I said: 'By that peculiar power which in the land of Arabia is called *ḥaṣṣiyat*, but which no man is able to understand, for it is not a physical force. It is analogous to the action of the load-stone which attracts iron not by a physical force nor by means of anything, but by that peculiar power'."

ויאמר, התדע מאין יבא מיץ המאכל אל הכבד, ואין במעים נקב שיצא ממנו ולא בכבד להכנס בו?

אמרתי, בכח הנפלא אשר בארץ ערב קוראים אותו כאציה, ואין כל אדם יכל לדעתו, כי איננו טבע, כמו כח האבן השואבת אשר תמשוך הברזל בלי טבע ובלי דבר אבל בכח הנפלא.

The expression כח נפלא in this passage is intended to be a translation of خاصّية, which, in addition to meaning *peculiarity*, *property*, i. e., סגולה, also means *particular efficacy*, *power*, *energy*. I have therefore rendered כח נפלא by "peculiar power" instead of "wonderful power."

The same explanation is also suggested in the following passage in Altabrizi, Prop. IX:

"Know that when one body moves another body, it moves it either because it is a body or because it is a [peculiar kind of] body, that is to say, it moves the other body either because of its very corporeality or because of a certain peculiar property it possesses. If the second explanation is accepted, then the real cause of that motion is the peculiar property it possesses and it is not the body *qua* body, and consequently the body under such circumstances must not necessarily be moved itself while causing motion in something else. As an illustration we may take the magnetic stone which causes motion in iron not by its corporeality but by a certain peculiar property it possesses, on which account it is not moved itself while causing the iron to be moved."

דע שכל גשם יניע גשם אחר, אם שיניעוהו לפי שהוא גשם, או לפי שהוא גשם
(וזה) [זה], לפי שהנעתו לו אם לנפש נשמותו או ליחוד בו. ואם היה השני, הנה
עלת אותה ההנעה באמת אמנם הוא אותו היחוד, לא הגשם מאשר הוא גשם, ולכן
לא יחוייב מהנעתו זולתו בשיתנועע הוא גם כן בעצמו, כמו אבן אלמגניטס, כאשר
הניע הברזל, כי הוא אמנם יניעהו ליחוד בו, לא לנשמותו, ויניעוהו מבלתי שיתנועע
הוא בעצמו.

The term יחוד in this passage I again take to be a translation of خاصّة as the כח נפלא in Zabara's passage.

This last type of explanation seems to reflect the view attributed by Plato to Thales who is said to have affirmed the load-stone to possess a soul, because it attracts iron." (De Anima I, 2, 405a, 19–21.) Plato himself explains magnetic attraction by a power (δύναμις) which not only the stone itself possesses but it imparts to others (Ion, 533D). Thus the "power" of the Sha'ar ha-Shamayim, the "peculiar power" of the Sefer Sha'ashu'im and the "peculiar property" of Altabrizi are all heirs of the 'soul' of Thales and the "power" of Plato.

11. Hebrew אשר לא נשער אלא שאמתהו החוש. The printed editions and some MSS. read here אשר לא נשער אלא עד שאמתהו החוש which would mean: "the nature of which we shall not know until it will have been verified by sense perception." This would lead one to credit Crescas with a vision of a future experimental science. But the real meaning of the passage becomes clear by a comparision with the following passage in 'Ikkarim IV, 35: "Just as the existence of the Magnesian stone attracting iron is indisputably true, even though it cannot be demonstrated by reason, but since it is warranted by experience." כמו שמציאות אבן המגניטס תמשוך הברזל הוא אמת גמור, אף על פי שלא יורהו ההקש, הואיל ויעיד עליו הנסיון. I have therefore adopted here the reading which omits עד and translated the passage accordingly.

PROPOSITION X

PART I

1. The Hebrew text of the proposition down to this point follows Isaac ben Nathan's translation of Altabrizi.

2. This part of the text follows Ibn Tibbon's translation of the *Moreh*, except that Ibn Tibbon uses אן, as does also Isaac ben Nathan, in place of Crescas' second אם.

3. In the passage following Crescas reproduces Aristotle's argument for the deduction of matter and form, as given in *Physics* I, and *Metaphysics* XII, 2–4. Crescas deals again with the same argument later in Propositions XXII and XXV.

4. Aristotle himself has grouped together all the views of his predecessors with regard to the composition of corporeal substance into two classes; (a) the pluralists, among whom are included the Atomists, and (b) the monists, who are identified with the Ionian school. Cf. *De Gen. et Corr.* I, 1; *Physics* I, 2–4.

In Arabic philosophy this classification has been preserved. Thus Algazali enumerates three views with regard to the composition of body, the Atomistic, the Ionian and the Aristotelian. *Kawwanot ha-Pilosofim* II (*Maḳaṣid al-Falasifah* II, pp. 85–86): "Concerning the difference of opinion with regard to the composition of body . . . There are three different views. Some say that body is composed of parts which are not divisible either in thought or in actuality. These parts are called atoms and of these body is composed. Others say that body is not composed at all, but its being is one in reality and definition and without any number in its essence. Still others say that body is composed of matter and form."

בחלוף אשר בהרכבת הגשם... וכבר התחלפו על שלשה סברות. הנה מהם
יאמר שהוא מורכב מחלקים לא יתחלקו במחשבה ולא בפועל, ויקראו אותם החלקים
עצמיים פרדיים, והגשם מחובר מאותם העצמים. ומהם יאמרו שהוא בלתי מורכב
כלל, אבל הוא נמצא אחד באמתות והגדר, אין בעצמותו מספר. ומהם יאמרו
שהוא מורכב מחומר וצורה (MS. Adler 978)

There is one characteristic which is common to both the one element of the Ionians and the atoms of the Atomists. Both the element of the former and the atoms of the latter are essentially simple in their essence. Whatever changes may occur in the one element or whatever differences may be discovered between one atom and another are due only to some unessential quality. Maimonides thus lays down as one of the tenets of Arabic atomism the proposition that "there exists nothing but substance and accident, and the physical forms of things belong also to the class of accidents" (*Moreh* I, 73, Prop. VIII). שאין נמצא אלא עצם ומקרה, ושהצורות הטבעיות גם כן מקרים. Similarly Algazali says of the same school (*Maḳaṣid al-Falasifah* II, p. 82) that according to their opinion "form is an accident related to the existence of the 'abode'." כי הצורה אצל המדברים מקרה נמשך למציאות המשכן.

Crescas' characterization here of the pre-Aristotelian theories as to the composition of body may therefore apply to both the Atomistic and the Ionian schools. It will be noted, however, that the first part of Crescas' characterization resembles in its wording Algazali's description of the Ionian view whereas the second part resembles the proposition quoted from Maimonides.

5. Aristotle's refutations of the views of his predecessors are found in *Physics* I, 2–4, and in *De Gen. et Corr.* I, 2. These arguments are all reproduced in the corresponding places in Averroes' commentaries, with which Crescas was acquainted. The arguments against atomism are also reproduced by Algazali in *Maḳaṣid al-Falasifah* II, p. 86 ff. and by Altabrizi in Prop. XXII. Furthermore, we shall see that Crescas' subsequent reproduction of Aristotle's argument for the distinction of matter and form is based upon Abraham ibn Daud's *Emunah Ramah*. Hence the significance of Crescas' reference here to the commentators of Aristotle.

6. Hebrew הכו על קדקד. This expression occurs in *Moreh* I, 74, The Seventh Argument: "Abu Naṣr Alfarabi has already knocked on the head of this proposition." וכבר הכה אבונצר אלפראבי על קדקד זאת ההקדמה. Maimonides himself, in a letter to Samuel ibn Tibbon, explains this expression as the Arabic دمغ which literally means "to strike someone on the head or brain so as to cause him to die" but is used idiomatically as the Talmudic מחו לה אמוחא (Megillah 19b) which literally also means "they struck it on the head or

brain" but idiomatically is used in the sense of refuting and rejecting somebody's opinion. See Munk, *Guide* I, 74, p. 438, n. 1.

7. The following is a brief summary of Averroes' presentation of the arguments advanced by Aristotle in *Physics* I, 7, in deducing the existence of matter and form and establishing their relation to each other. The logical order of these arguments may be restated as follows:

A. From the phenomena of change and becoming it is evident that the principles (ἀρχαί, התחלות) must be more than one, and that they must be contraries (ἐναντία, הפכים), namely, non-being and being.

B. These contraries alone cannot be the sole principles of becoming, for nothing can come out of nothing. We must therefore assume the existence of a substratum (ὑποκείμενον, נושא, מונח) to which both non-being and being equally belong. That substratum is matter.

C. Of these three principles, substratum, non-being and being, only the first and the third are true principles. The second, non-being, is merely privation and is called principle only in an accidental sense.

Intermediate Physics I, iii, 1–3 (Latin, p. 438va): "First, wherein he reproduces the well-known arguments proving that the principles must be contraries and that they must be more than one.

Second, wherein he reproduces the well-known arguments proving that the contraries alone are not sufficient as principles and that it is impossible but to admit a *tertium quid* which constitutes the subject.

Third, wherein he shows that the principles in truth are only two, matter and form, and that privation which is the contrary of form is not matter but only an accident of matter, and if privation be a principle it is so only accidentally."

הראשון, יזכור בו המאמרים המפורסמים אשר יחייבו שההתחלות הפכים ושיחוייב
שתהיינה יותר מאחד.

השני, יזכור המאמרים המפורסמים אשר יחייבו שההפכים לא יספיקו להיות
התחלות, ושאי אפשר מבלתי הכנס טבע שלישי, והוא הנושא.

השלישי, יבאר בו שההתחלות באמת אמנם הם שתים בלבד, ההיולי והצורה,
ושההעדר המקביל לצורה אינו החומר, אבל הוא דבר קרה לו, ושאם היה ההעדר
התחלה הנה הוא במקרה.

Cf. *Moreh* I, 17: "You are aware that the principles of generable
and corruptible things are three, namely, matter, form, and the
particular privation which is always joined to the matter, for,
were matter unaccompanied by privation, it would be incapable
of receiving form. It is from this point of view that privation is
included among the principles."

ואתה יודע כי התחלות הנמצאות ההווה הנפסדות שלשה, החומר והצורה וההעדר
המיוחד אשר הוא מחובר לחומר לעולם, ולולא התחברות ההעדר לחומר לא הגיעה
אליו הצורה, ובזה הצד היה ההעדר מן ההתחלות.

Cf. *Metaphysics* XII, 2, 1069b, 32–34: "The causes and principles, then, are three, two being the pair of contraries of which
one is definition and form and the other is privation, and the
third being the primordial matter."

This Aristotelian method of deducing the existence of matter
and form from the transmutation of the elements is already found
in Abraham ibn Daud's *Emunah Ramah* I, 2. From an analogy
of many expressions it may be inferred that Crescas' discussion
here is taken from the *Emunah Ramah*.

The corresponding passage in the *Emunah Ramah* reads as
follows: "We thus know by observation that these elements are
changed into one another . . But . . . . it is inconceivable that
the form, after passing away, should become the recipient . . . .
Hence we infer that they have a common underlying matter,
which matter we call first matter."

ונדע מזה בחוש שאלה היסודות ישתנו קצתם אל קצת. . . אמנם. . . לא יתכן
שיהיה הצורה הנעדרת היא המקבלת. . . ולכן נדע שיש להם חמר משותף, הוא אשר
נקראהו החומר הראשון.

The assertion made by both Crescas and Abraham ibn Daud
that that which no longer is cannot be the recipient of that which
is coming to be reflects Aristotle's principle that "from nothing
nothing is produced" (*Physics* I, 4, 187a, 28–29). Cf. also *ibid*.
187a, 32–34: "For it is necessary that whatever is generated
should be generated either from beings or from non-beings, and
it is impossible that things should be generated from non-beings."

The immediate source of this method of deducing the existence
of matter and form from the reciprocal transformation of the

elements would seem to be the discussion in *De Gen. et Corr.* II, 1–4.

8. That is to say, matter must be substance inasmuch as it is a substratum.

The definition of substance implied in this statement is based upon the identification of substance with substratum, which is the first of the four meanings of the term substance enumerated by Aristotle in *Metaphysics* V, 8. In Aristotle this definition of substance reads as follows: "All these are called substance because they are not predicated of a subject" (*ibid.* 1017b, 13–14). In Algazali's *Maḳaṣid al-Falasifah* II, p. 82, the reading of this definition is as follows: "Substance is an appellative for that which does not exist in a subject." עצם הוא מליצה מכל נמצא לא בנושא. Thus, argues Crescas, matter must be substance in the sense of substratum.

The corresponding passage in *Emunah Ramah* I, 2, p. 11, reads as follows: "We shall now prove that matter is substance. For why should it not be substance? seeing that it never passes away." אחר כן נאמר אמנם באור היות היולי עצם. הנה איך לא תהיה עצם? והיא לא נעדר לעולם. The same statement occurs also in II, iv, 3, p. 64.

Cf. *Metaphysics* VII, 3, 1029a, 10–12: "And further, on this view, matter becomes substance. For if this is not substance, it is beyond our power to say what else it is. When all else is taken away, evidently nothing but matter remains."

Cf. also *Metaphysics* VIII, I, 1042a, 32–34: "But clearly matter also is substance, for in all the opposite changes that occur there is something which underlies the changes."

9. That is to say, form also is substance. The reason given here by Crescas for the substantiality of form reflects again mediaeval as well as Aristotelian discussions on the subject. Though form cannot be called substance in the sense of substratum, still, it is argued, it must be called substance by reason of its being the cause of the existence of a thing and also of its being that which limits the character of a thing and constitutes its essence. *Kawwanot ha-Pilosofim* II (*Maḳaṣid al-Falasifah* II, p. 82): "The upshot of this discussion is that the philosophers apply the term form in a general sense to that which is an 'abode' and also to

574   CRESCAS' CRITIQUE OF ARISTOTLE   [259

that which resides in an 'abode.' On this last point the Mutakallimun disagree, for in their opinion form is an accident related
to the existence of the 'abode.' But the philosophers repudiate
this view and say, how can form not be substance when it is that
through which substance itself persists and in which it has its
nature and essence?''

והגיע מזה שהם שלחו שם העצם על מה שהוא משכן ועל מה שהוא שוכן גם כן.
וחלקו בזה המדברים, כי הצורה אצל המדברים מקרה נמשך למציאות המשכן.
ואלה ירחקו [يَسْتَدِلُّون, ירחיקו זה: 1500 MS. Adler] ויאמרו, ואיך לא תהיה
הצורה עצם, ובה תעמוד עצמות העצם ותעמיד אמיתתו ומהותו (978 MS. Adler).

This new meaning of substance corresponds to the other three
senses in which the term substance is used according to Aristotle,
to wit, (1) as the internal cause of the being of things, (2) as the
limits which define the individuality of bodies, and (3) as the
essence of things. Form is substance, according to Aristotle, in
all these three senses: "And of this nature is the shape or form of
each thing" (*Metaphysics* V, 8, 1017b, 25–26). It will be noted
that the three terms used by Crescas here in proving that form
is substance correspond exactly to these three senses in which
the term substance is applied by Aristotle to form, to wit, (1)
בו יאמר שהדבר הוה, through form a thing is said to have its being,
(2) ומוגבל, it is limited through form, (3) ובו נתעצם, it has its
essence in form.

That form is substance but not in the sense of substratum but
rather in the other senses of the term substance is also the implication of the following passage in *Sefer ha-Yesodot* I, p. 12:
"Should any one be tempted to think that the first form is an
accident and not a substance, we shall prove the falsity of his
opinion from the analogy of man. Man is composed of soul and
body. His body is analogous to matter and is related as a subject
to his form. His soul is his form and the cause of the preservation
of his species. And still the soul is not an accident."

ואולי החושב יחשוב שהצורה הראשונה הוא מקרה ולא עצם, אם כן נודיעהו
הפסד מחשבתו מהאדם, כי האדם מורכב מנפש וגוף, וגופו יסודו וחמרו הנושא
צורתו, ונפש צורתו וקיום בעל מינו, והנפש איננה מקרה.

The corresponding passage in *Emunah Ramah* I, 2, p. 11, reads
as follows: "As for the proof that form is substance, why should
it not be substance?, seeing that it is form which transforms

something that does not exist in actuality into something that
does exist in actuality." ואמנם באור היות הצורה עצם, הנה איך לא
תהיה עצם? והיא חשית הבלתי נמצא בפועל נמצא בפועל. The same statement occurs also in II, iv, 3, p. 64.

Aristotle's definition of substance is discussed by Hillel of
Verona, in Prop. XXV, as follows: "It is well-known that substance has no true definition, for a definition is composed of a
genus and a specific difference, whereas substance, being a
summum genus, is only part of a definition, and the parts of a
definition are prior to the definition. Substance, however, has six
properties which constitute its description, so as to differentiate it
from accident. To begin with, it exists by itself and not with
reference to something else, it is not in a subject, it is the cause
of the existence of all other beings and is prior to them in nature.
As for the other properties, there is no need of repeating them
here."

ידוע כי העצם אין לו גדר אמתי, בעבור שהגדר מורכב מסוג ומהבדל, והעצם
הוא סוג הסוגים, אם כן הוא חלק מהגדר, וחלקי הגדר הם קודמין לגדר. אמנם יש
לו שש סגולות הם אליו כמו חוק למען הבדילו מן המקרה. אחת מהם היא שהוא
נמצא סתם בעצמו ולא בערך אל דבר, ואינו בנושא, ושהוא סבת כל שאר ההויות
וקודם להם בטבע. ושאר הסגולות הם בלתי צריכות להכתב בכאן.

Crescas has thus enumerated two substances, matter and form.

According to Aristotle, the following are substances: matter,
form, and the concrete thing composed of matter and form. Cf.
Metaphysics VII, 3, 1029a, 1–3; VII, 10, 1035a, 2; VIII, 1,
1042a, 26 ff.; XII, 3, 1070a, 9 ff.; XII, 4, 1070b, 13–14.

In Arabic philosophy, with the introduction of the Separate
Intelligences, of Neo-Platonic origin, these, too, were added to
the substances. Thus Algazali enumerates the following four
substances: matter, form, the concrete thing composed of matter and form, and the Separate Intelligences. Cf. *Kawwanot
ha-Pilosofim* II, (*Maḳaṣid al-Falasifah* II p. 82). וחלוק העצם
ארבעה מינים: ההיולי, והצורה, והגשם, והשכל הנבדל העומד בעצמו.

Abraham ibn Daud has further subdivided them into six corporeal substances and six incorporeal substances. *Emunah Ramah*
II, iv, 3 (pp. 64–65): "At first they discovered by perception six
kinds of bodies: a celestial body, an elementary body, a mineral
body, a vegetable body, an irrational animal body, an animal

body endowed with reason. Then by reasoning they inferred the existence of three incorporeal substances, namely, the common matter underlying the four elements . . . . form . . . . soul . . . . the active intellect . . . . Intelligences . . . . First Mover . . . . Thus the incorporeal substances are six in kind and the corporeal substances are six in kind."

וראו תחלה לעין ששה מיני גשמים: גשם שמימי, וגשם יסודיי, וגשם מחצביי, וגשם צמחיי, וגשם חיוני בלתי מדבר, וגשם חיוני לאותו החי שכל. ואחר כן עלו על ידיעת שלשה עצמים בלתי גשמים, והם החומר המשותף ליסודות הארבעה... והצורה... נפש... השכל הפועל... שכלים... מניע ראשון... אם כן היו העצמים בלתי גשמים ששה מינים, והעצמים הגשמים ששה מינים.

10. Cf. *Metaphysics* VIII, 1, 1042a, 27–28: "And by matter I mean that which, not being a 'this' actually, is potentially a 'this'."

11. According to Aristotle there are three kinds of changes, that which is from a non-subject to a subject, that which is from a subject to a non-subject, and that which is from a subject to a subject. In Averroes' Intermediate Commentary, the terms *existence* and *non-existence* are used synonymously with the terms *subject* and *non-subject* (see Prop. IV, n. 8, p. 514). The first kind of change is generation; the second kind is corruption; the third kind is simply change or motion. Cf. *Physics* V, 1, 225a, 7–14, 17–18; 225b, 2.

12. Hebrew צורה טבעית. As for the meaning of this term, see below n. 16.

Crescas has thus explained the second part of the proposition, namely, that the natural form is the cause of the existence of body.

13. Hebrew צורה נשמית. As for the meaning of this term, see below n. 16.

The corresponding passage in *Emunah Ramah* I, 2, p. 11, reads as follows: "As for the accidents, they apply only to that which happens to the body after it has become something definite."

אך המקרים, אמנם יאמרו על מה שישיג הגשם אחר היותו מעוין.

14. See definition of substance above notes 8, 9.

15. By this comment Crescas is trying to explain the particular sense in which Maimonides uses the term 'force,' כח, قوّة, in this proposition. The term כח usually means 'potentiality' as opposed to 'actuality.' Here, however, according to Crescas' explanation, Maimonides uses it in the sense of 'inaliety,' 'in-an-other-ness,' 'existing in something else,' as opposed to 'perseity,' 'in-itself-ness,' 'existing in itself' (cf. Munk, *Guide* II, p. 11, n. 4). In the same sense is the term used by Maimonides in Propositions XI, XII, XVI.

According to this explanation Maimonides considers both accident and form as "forces" existing in something else. In this he follows the conventional method generally employed in stating the difference between matter, form, and accidents. Thus Algazali divides *being*, מציאות, وجود; into that which requires something in which to abide and that which does not require anything for its abode.

The former class is called "accident" in a general sense, and includes both form and accident proper. The latter class includes matter. Since form, however, is the cause of the actual existence of matter, unlike accident, it is called substance, even though it abides in matter. Matter is therefore called with respect to accident נושא, موضوع, *subject*, whereas with respect to form it is called משכן, محل, *abode*. (Cf. *Maḳaṣid al-Falasifah* II, pp. 80–82; Shahrastani, pp. 364–365).

Altabrizi (Prop. X) calls both accident and form by the general term ענין or תאר and he designates both the subject, נושא, of the accident and the matter, חומר, of the form by the term בעל הענין or מתואר. Thus Maimonides' כח here is the equivalent of Altabrizi's ענין. Unlike Altabrizi, however, Maimonides uses the term ענין, معنى, with reference to both matter and form (cf. Propositions XXI, XXII). Hence Altabrizi's ענין=حال (cf. p. 517).

16. Preliminary to the explanation of this passage we shall try to define the terms which are used here by Crescas and incidentally to give some of their equivalents.

(a) חומר is used here in the sense of חומר ראשון, היולי, πρώτη ὕλη, *first matter*, which in *Emunah Ramah* 1, 2, is also designated by היולי המושכל, ὕλη νοητή, *intelligible matter*. As for the meaning of ὕλη νοητή in Aristotle, see Ross's commentary on the *Metaphysics* (VII, 10, 1036a, 9–10), Vol. II, p. 199.

(b) צורה גשמית, *corporeal form*. So it is also designated by Simplicius, Avicenna and Shahrastani (see below n. 18, pp. 582, 583). Crescas calls it later in his criticism of this proposition and in Prop. XI צורת הגשמות and צורת הגשמיות *form of corporeity*, the *forma corporeitatis* of Thomas Aquinas. It is also called צורת הגשם, *form of the body*, and צורה ראשונה, *first form* (see *Sefer ha-Yesodot* I, p. 11, and *Emunah Ramah* I, 2). Plotinus and the Iḥwan al-Safa call it simply "quantity" (see references below in n. 18, pp. 582, 580). As for the history of this kind of form, see below n. 18.

(c) גשם *body*. The term is used here in the specific sense of the compound of the *first matter* and the *first form*. In the Iḥwan al-Safa (see below n. 18, p. 580) and *Emunah Ramah* I, 2, it is more precisely called גשם משולח, *absolute body*.

(d) צורה טבעית, *forma naturalis*, by which is meant here the forms of the four simple elements which have as their matter the גשם or גשם משולח of (c). This form is also known by the following names. צורה מיוחדת, *proper form* (Crescas above, p. 262, l. 2); צורת היסודות, *forma elementorum* (*Emunah Ramah* I, 2); צורה יסודית, *forma elementalis* (Abravanel quoted below in n. 18 p. 590); צורה מינית, *forma specifica* (Altabrizi, Prop. X); צורה עצמית *forma essentialis* (Altabrizi, Prop. X; Abravanel quoted below in n. 18 p. 590).

(e) מקרה, *accident*. It is also called צורה מקרית, *forma accidentalis* (*Emunah Ramah* I, 2).

Now it will be noticed that in the proof adduced by Crescas for the existence of matter and form the terms used are חומר and צורה, i. e., *first matter* and *first form*, whereas in Maimonides' proposition the terms used are גשם and צורה טבעית, i. e., *body* and *natural form*. It is Crescas' purpose here to show that everything he has said about the relation between *first matter* and *first form* may be also applied to the relation between *body* and *natural form*.

The main point of Crescas' observation then is that the term matter is always to be taken as relative to the term form and that there is an analogy between the relation of the *first matter* to the *first form* and the relation of any subsequent matter to a

respective subsequent form. The source of Crescas' observation
may be found in the following passages.

Emunah Ramah I, 2, p. 10: "That which all the elements have
in common serves them as matter, even though *first matter* is only
that which is matter of *absolute body*, but absolute body, which is
somewhat like hyle to the elements, is not hyle in the true sense
of the term, for it has form, namely, conjunction. From these
elements are generated the composite things, and of these, too,
some may be considered as matter in relation to others."

והענין אשר הם מסכימים בו הוא להם כחומר, עם היות שהחומר הראשון אמנם
הוא חומר הגשם המשולח, אבל הגשם המשולח, אשר הוא כדמות היולי ליסודות,
אינו על דרך האמת היולי, לפי שבו צורה, והוא ההתדבקות. ואחר כן נתחדשו
המרכבים, וקצתם גם כן יחשב שהם חומר לקצת.

Liḳḳute Sefer Meḳor Ḥayyim II, 1: "Thus the relation of
corporeality to the matter, which is its subject, is analogous to
the relation of the universal form, i. e., figures and colors, to the
corporeality which is the subject of these figures and colors."

ויהיה הקש הגשמות ליסוד הנושא אותה הוא הקש הצורה הכללית, כלומר התבניות
והגוונים, אל הגשמות הנושא להן.   Cf. *Fons Vitae* II, 1, p. 21, ll. 15–18.

PART II

17. Cf. below n. 24.

18. Hebrew דבקות השלשה רחקים. The term דבקות, اتصال, in this
connection is translated into Latin by the usual "continuatio"
(*Epitome of the Metaphysics* II, Arabic, p. 76, l. 17; Latin, p. 373va
l. 17; cf. below Prop. XIII, Part I, n. 6, and Prop. X, Part II,
n. 23). But "cohesion" or "cohesiveness," i. e., that which makes
for mass, would seem to be a more exact translation, especially
when the term is used in connection with the views of Avicenna
and Algazali which will be explained in the course of this note. By
the term "cohesion" is meant here the characterization of matter
as having "mass" or "bulk," עובי, and "rigidity" or "resistance,"
מקשיות. This is the definition of "cohesion" as given in a passage
in *Emunah Ramah* I, 2, which will be quoted later in this note. It
will also be gathered from our subsequent discussion that this
"cohesion" or "mass" was conceived by Avicenna and Algazali as

something which by itself is not tridimensional but which is capable of becoming tridimensional.

With this preliminary remark about the meaning of the term "cohesion" we shall now trace the origin and history of the idea of "corporeal form" which is introduced here by Crescas.

The corporeal form of which Crescas is speaking here is the first form in the successive stages of matter and form. In the Encyclopedia of the Iḥwan al-Safa it is also called "quantity," الكمية. The compound of this corporeal form with first matter is "absolute body," جسم مطلق, or "second matter." It is this second matter that is the proximate matter underlying the four elements. Cf. *Emunah Ramah* I, 2; Dieterici, *Die Lehre von der Weltseele bei den Arabern*, p. 25, *Einleitung und Makrokosmos*, pp. 176–177, *Die Naturanschauung und Naturphilosophie der Araber*, pp. 2–3. *Die Abhandlungen der Ichwân Es-Safâ*, p. 25. Cf. above n. 16.

According to Isaac Abravanel there is no mention of the corporeal form in Aristotle, though he says, it is made much of by his commentators. He further indicates that the reason for the introduction of the corporeal form was the general belief that Aristotle's first matter could not itself be corporeal, that is, it could not be an extended body, and hence extension or corporeality had to be postulated as a form of first matter.

She'elot Saul X, p. 18a-b: "There is no statement in Aristotle with regard to the corporeal form . . . . But the commentators upon his works have advanced many views concerning it. One thing upon which they all agree is that the corporeity of a thing is not the first matter, for if corporeity were identical with matter, then matter would be something actual, and as a result all the forms that settle upon it would be accidents, for of such nature is substance: when it is actual it becomes a subject in which all things exist as accidents. Second, corporeity is a term applied to form and not to matter. Third, corporeal substance is a genus under which are included species. But it has been shown in the *Metaphysics* that matter is not a genus. Hence corporeity is not identical with matter. Fourth, Aristotle argues that matter is indivisible not only actually but even potentially, because matter, he contends, has no dimensions and is without

parts at all, and therefore it is not actually divisible except by
means of the forms which settle upon it. Since, then, matter is not
capable of division *per se*, matter cannot be identical with corporeity, but the latter is joined to it rather as a form, by means
of which it becomes capable of division. And just as they are all
agreed that corporeity is not identical with matter so they are
also all agreed that corporeity is not one of the essential forms
which are generated in a compound object, for just as the first
matter is not divisible *per se* so also the essential forms are not
divislble *per se*. Divisibility is due to corporeity which is [a form]
placed between the first matter and the essential forms. Thus
according to the view of all of them, the corporeal form is the
first form that settles upon the first matter."

הנה לא נמצא לאריסטו מאמר בצורה הגשמית. ... אבל מפרשי ספריו הרבו
בעניינה הדעות, וממה שהסכימו בה כלם הוא שהגשמות בדבר אינו החומר הראשון,
שאם היה הגשמות הוא עצם ההיולי היה בכאן היולי בפועל, ויהיו כל הצורות
החלות עליו מקרים, שכן הוא טבע כל עצם שבהיותו בפועל ינשאו עליו כל המקרים
כלם, גם שהגשמות הוא שם לצורה לא לחומר. ועוד שהגשם הוא סוג ויכנסו תחתיו
מינים, וכבר התבאר במה שאחר הטבע שאין ההיולי סוג, אם כן אין הגשמות ההיולי.
ועוד שאריסטו יבאר שההיולי אינו בלתי מתחלק בפועל, כי גם בכח לא יתחלק,
לפי שאין לו מרחקים ולא חלקים כלל, ולכן לא יתחלק בפועל, (ולא) ואלא]
באמצעות הצורות שיחולו בו, וכיון שאין החלוק להיולי מצד עצמו, אם כן אין
ההיולי עצם הגשמות, אבל יתחבר אליו (הצורה) [כצורה], באמצעותה יקבל הוא
החלוקה. וכמו שכלם הסכימו שאין הגשמות עצם ההיולי, כן נמנו וגמרו שאין
הגשמות אחת מהצורות העצמיות המתחדשות במורכב, לפי שכמו שהחומר הראשון
אינו מתחלק מפאת עצמו, כן הצורות העצמיות אינם מתחלקות מפאת עצמן, אבל
יהיה החלוק בגשמות שהוא ממוצע בין ההיולי הראשון והצורות העצמיות. הנה
אם כן לדעתם כלם הצורה הגשמית היא הראשונה שתחול בהיולי הראשון.

The reasons leading to the introduction of corporeal form may
also be gathered, I believe, from Simplicius' commentary on the
Physics (ed. Diels, pp. 227–233; cf. Taylor's translation of the
Physics, notes on p. 71 ff.). Simplicius finds a contradiction in
Aristotle's conception of matter. On the one hand, he finds that
Aristotle's proof for the existence of matter from the transmutation of the four elements would lead to the belief that matter is
corporeal and extended. "For Aristotle and Plato, first introducing matter from the mutation of things which are changed, were
of the opinion that the qualities of the elements are the hot and
the cold, the moist and the dry; but these, having a common sub-

ject body, are changed about it, so that the first matter will be body" (Diels, p. 227, ll. 26–30). But, on the other hand, he finds many statements in Aristotle which explicitly affirm that first matter is not body and has no magnitude. He furthermore shows by many arguments that matter cannot be body, the last of which arguments reads: "Body also is defined by three intervals; but matter is perfectly indefinite" (Diels, p. 230, l. 14).

As a way out of this difficulty he suggests that the matter immediately underlying the four elements is not identical with the first matter of Aristotle, that the former is extended but the latter is inextended and that between these two matters there is a corporeal form which endows the first matter with extension. "May we not, therefore, admit that body is twofold, one kind, as subsisting according to form and reason, and as defined by three intervals; but another as characterized by intensions and remissions, and an indefiniteness of an incorporeal, impartible, and intelligible nature; this not being formally defined by three intervals, but entirely remitted and dissipated, and on all sides flowing from being into non-being. Such an interval as this, we must, perhaps, admit matter to be, and not corporeal form (σωματικὸν εἶδος), which now measures and bounds the infinite and indefinite nature of such an interval as this, and which stops it in its flight from being" (Diels, p. 230, ll. 21–29).

In a similar manner Plotinus mentions two views with regard to matter, one of which attributes to it magnitude and hence considers it as a body, and another which does not consider it as a body (*Enneads* II, iv, 1). He then proves that matter cannot have magnitude (*Enneads* II, iv, 8). Finally he concludes that magnitude is imparted to matter by quantity which is a form ὅτι εἶδος ἡ ποσότης (*Enneads* II, iv, 9). It will be noted that what Simplicius calls "corporeal form" is called by Plotinus "quantity," the same term, as we have seen, that is used by the Iḥwan al-Safa.

Thus the corporeal form was introduced. But what is the nature of that form? It is on this point that the views of Avicenna, Algazali and Averroes differ.

Avicenna—Matter itself, though incorporeal, has a predisposition to receive corporeal dimensions. This predisposition, and

not the dimensions, is the corporeal form. The dimensions themselves are added to matter as accidents. That this represents
Avicenna's view, says Narboni, may be gathered from the former's
Al-Shafa and *Al-Najah*. Cf. Horten's translation of the *Al-Shafa*
under the title of *Die Metaphysik Avicennas*, p. 101, "Das eigent-
liche Wesen der Körperlichkeit, die aufnahmfähig ist für die Art
and Weise der drei Dimensionen...." Cf. also *Al-Najah*, p. 55.
Sharastani likewise says of Avicenna's definition of corporeal form
(الصورة الجسمية) that it is a predisposition (طبيعة) not identical with the cohesion (ed. Cureton, p. 366).

Narboni's statement in full reads as follows:

"Avicenna, however, believes that the corporeal form is not
identical with cohesion nor is it something to whose nature
cohesion is essentially necessary. But it is something different
from either of these, though it is joined to matter and is never
separable from it. He reasons thus: The corporeal form must be
either something to which cohesion is essentially joined in
such a manner that it cannot exist without necessarily having
the differentia of cohesion, or something identical with cohesion.
If it is identical with cohesion, then body will have to remain
coherent even after it has become divided. It follows, therefore,
that there is undoubtedly something that has a potentiality for
both cohesion and division, namely, matter. Hence cohesion itself
qua cohesion is not the recipient of division. Rather is it that
which is a recipient of cohesion that is also the recipient of
division, namely, matter, inasmuch as the recipient must remain
with that which is received. Nor can that recipient be something
to whose nature cohesion is essentially necessary, inasmuch as
that cohesion may pass away. Nor is it, as has been said, identical
with cohesion.

Hence it seems that there is a substance unidentical with the
corporeal form, and it is that substance to which both division
and cohesion happen as accidents. That substance must be
conjoined with the corporeal form; it cannot exist without it nor
can it change it for another form. Hence the corporeal form is
not identical with cohesion nor is it something to whose nature
cohesion is essentially necessary, inasmuch as the underlying
matter can become divided and thus have the cohesion disappear. It is that matter that is the recipient of unity through

the corporeal form, and it becomes a unified body by virtue of
the corporeal form which causes it to exist, or that unity comes
to it necessarily from the corporeal cohesion of which it is the
recipient. The corporeal form has no existence but in matter,
which matter is a substance, being the first abode in which other
things exist and itself does not exist in anything else. This is the
view of Avicenna in *Al-Najah* and *Al-Shafa*."

ואמנם אבן סינא חשב שאינה הדבקות, ולא טבע יחייב לו הדבקות בעצמותו,
אבל מה שזולת זה, והוא מחובר אל היולי ולא יפרד ממנו לעולם, כי הוא אמר
שהצורה הגשמית הנה לפי זה אם שתהיה עצם הדבקות טבע דבוק בה, עד לא
תמצא היא אם לא שהבדל הדבקות חוייב לה, ואם שתהיה עצם הדבקות. ואם
היתה עצם הדבקות, הנה כבר ימצא הגשם מתדבק אחר יפרד, ויהיה הנה בלי ספק
הוא בכח כל שניהם, והוא ההיולי, הנה אין עצמות הדבקות במה הוא דבקות מקבל
לפירוד, לפי שמקבל הדבקות הוא מקבל הפרוד, והוא ההיולי, כי המקבל הוא
שישאר עם המקובל, ולא הוא ג"כ, ר"ל המקבל, טבע יחייב לו הדבקות לעצמותו,
אחר שהנה כבר יסתלק הדבקות, וגם כן אינה עצם הדבקות.

הנה נראה שהנה עצם בלתי הצורה הגשמית, הוא אשר יקרה לו הפרוד והדבקות
יחד, והוא מחובר לצורה הגשמית לא יעמד בלתה ולא ימירה, ולכן אין הצורה
הגשמית עצם הדבקות, ולא טבע יחייב לו הדבקות לעצמותו, אחר שהוא כבר
יפרד ויסתלק הדבקות, והוא אשר יקבל ההתאחדות בצורה הגשמית, וישוב גשם
אחד למה שיעמידהו, או יחוייב לו מהדבקות הגשמי אשר יקבלהו. ואין קיום לצורה
הגשמית אלא בחומר, והחומר עצם לפי שהוא המשכן הראשון, ולא יחול בדבר
כלל. זהו דעת אבן סינא באלנגאה ובאלשפא.

A restatement of Avicenna's view is given also by Abravanel,
who informs us that among those who adopted Avicenna's view
should be included Abu Bekr ibn Tufail. *She'elot Saul*, p. 18b:
"Another group believes that the corporeal form is not identical
with the three dimensions, either the determinate or the indeterminate dimensions, for both of these kinds of dimensions are of
the same nature, both being accidents and unessential. Nor is
the corporeal form identical with cohesion. It is rather an
essential form which settles upon matter before the dimensions
settle upon it. It is the dimensions that are transformed, increased
and diminished and not the first form, for the latter is eternal,
and is not one of the forms of the elements or of the substances
composed of the elements. Of this view was Avicenna. Also Abu
Bekr ibn Tufail was of this view, except that he added that the
corporeal form is subject to generation and corruption."

וכת שנית תחשוב שאין הצורה הגשמית המרחקים השלשה, לא המוגבלים ולא
הבלתי מוגבלים, שענין כלם כלם אחד הוא, והם כלם, מקרים ולא עצם, ואינה גם כן
הדבקות, אבל היא צורה עצמית תחול בהיולי קודם שיחולו בו המרחקים, ושהם
יומרו ויתוספו ויחסרו לא הצורה הראשונה ההיא, כי היא נצחית, ושאינה מצורות
היסודות ולא מהמורכבים מהם. ומזה הדעת הוא היה בן סינא, וגם אבובכיר בן
אלטופיל מזה הדעת היה, אלא שהוסיף בעניינה שהצורה הגשמית היתה הווה ונפסדת.

According to Narboni on *Moreh* I, 69, Avicenna's view implies
that the dimensions are superimposed upon matter from without.
ואין שם שלוחים באים מחוץ כמו שחשב בן סינא.

Algazali—Matter indeed has no corporeality. Its corporeal
form, however, is not a mere predisposition. It is identical with
cohesion itself. The dimensions are, he agrees with Avicenna,
mere accidents.

Narboni: "According to Algazali the corporeal form is identical
with the cohesion itself." והצורה הגשמית לפי דעת אבוחאמד הוא הדבקות
בעצמו.

Abravanel: "But as to what is the corporeal form, I have found
among the commentators a variety of views. One group believes
that the corporeal form is identical with cohesion and that the
dimensions are only accidents. Of this group was Joseph ibn
'Aknin, and it was followed also by Algazali. Hence the latter defined body as that in which it is possible to posit three dimensions
intersecting each other at right angles."
האמנם מה היא הצורה הגשמית, הנה ראיתי למפרשים דעות חלוקות: כי הנה
כת אחת מהם חשבו כי הצורה הגשמית היא הדבקות, ושהמרחקים הם מקרים,
ושמזה היה אבו אל חנאז' יוסף יחייא הישראלי המערבי, ונמשך אחריו אבוחמד,
ומפני זה גדר הגשם שהוא שאפשר שינוחו בו שלשה שלוחים נחתכים על זויות
נצבות.

Altabrizi, too, seems to have adopted Algazali's view. Cf. his
commentary on Prop. XXII: "That recipient is matter and the
corporeal cohesion is form." ואותו המקבל הוא ההיולי והדבקות הגשמי
הוא הצורה.

Averroes—He disagrees with both Avicenna and Algazali. The
corporeal form to him is neither a predisposition for the cohesion
of the three dimensions nor the cohesion itself. It is rather
identical with the *dimensions*, not indeed the definite changeable
dimensions which constitute the quantity of an object, but absolute dimensionality as such, indeterminate and unlimited.

His argument in full is given by Narboni as follows:

"Thou seest that the reason on account of which they refrained from assuming that the dimensions themselves are the corporeal form is that the corporeal form is imperishable, being the cause of the existence of prime matter which is ungenerated and inde-structible, whereas the dimensions are subject to transformation and destruction. But the learned Averroes caught them up on this point, arguing that the determinate dimensions only are transformable, that is to say, their particular limits are altered, but not the indeterminate dimensions themselves. That some-thing non-dimensional should become dimensional is in truth the work of the corporeal form, which is the first form to settle upon the first matter and endow it with existence. It is this that the corporeal form is. It is not cohesion itself nor something to whose nature cohesion is essentially necessary, nor anything else, as was thought by Avicenna."

ואתה רואה כי הסבה אשר בעבורה ברחו מהניח שהמרחקים עצמם יהיו הצורה
הגשמית הוא שהצורה הגשמית לא תבטל, כי היא מעמידה החומר הראשון, אשר
הוא בלתי הוה ונפסד, והמרחקים יומרו ויופסדו. והחכם אבן רשד תפסם בזה
באמר, כי המרחקים המוגבלים הם אשר יומרו, ר"ל שהגבלתם תבטל, לא עצמות
המרחקים הבלתי מוגבלים, כי יתהוה מרחק מלא מרחק הם באמת הצורה הגשמית,
אשר תחול ראשונה בחומר הראשון המעמיד אותו, ואין הצורה הגשמית דבר זולתו,
לא עצם הדבקות ולא טבע שהדבקות יחוייב לו בעצמותו, ולא זולת זה, כאשר
חשב אבן סינא.

(Cf. the restatement of the views of Avicenna, Algazali and Averroes as given by Duhem, Le Système du Monde IV, p. 541 ff.)

Averroes' view of corporeal form seems to have been also held by Alfarabi. See his Mahut ha-Nefesh (Edelman's Ḥemdah Genuzah, p. 47a): "For corporeal form is defined as length and breadth and depth." כי הצורה הגשמית גדרה אורך ורוחב ועומק.

The original statement of Averroes' view is to be found in his Sermo de Substantia Orbis (מאמר בעצם הגלגל) where he also polem-izes against Avicenna. In a commentary on that treatise Narboni remarks that from Averroes' polemic against Avicenna it might be inferred that Algazali's identification of corporeal form with the cohesion is due to a misunderstanding on his part of Avicenna's position. He also adds that the Jewish philosophers Joseph ibn Yoḥai (i. e., Joseph ben Judah ibn 'Aḳnin, 1160–1226, disciple of

Maimonides, whose full name in Arabic is Abu al-Hajjaj Yusef
ibn Yaḥya ibn Shamʻun al-Sabti al-Maghrabi) had made the same
mistake: "This makes it evident that Avicenna assumes that the
corporeal form is other than the dimensions, and also that it is not
identical with cohesion, as was thought by Algazali and Joseph
ben Yoḥai."

הנה מבואר מזה שאבן סיני מניח שהצורה הגשמית היא זולת המרחקים ואינה
הדבקות, כמו שחשבו אבוחמד ויוסף בן יוחיי.

A similar reference to Joseph ibn ʻAḳnin, cited by his full
Arabic name, is made, as we have seen, by Abravanel in the
passage quoted above.

The original statement of Ibn ʻAḳnin reads as follows (ed.
Moritz Löwy, pp. 11–12; ed. J. L. Magnes, p. 8): "We say that
body is an appellative for the cohesion wherein may be posited
three dimensions intersecting each other at right angles. One of
these dimensions is called length, the other breadth and the third
depth, i. e., height. This is what is meant by corporeity, which
is the first [form] to be found in matter, while the latter is as yet
undistinguished by any other form, and this corporeity is not
identical with the dimension, for the latter is an accident of the
category of quantity, which may change and increase and
diminish in connection with any given matter . . . . Thus the
form is not the dimension itself but the cohesion wherein the
dimension may be posited."

ונאמר שהגשם מליצה מהדבקות אשר אפשר שיונחו בו שלשה שלוחים כריתותם
על זויות נצבות. ואחד השלוחים יקרא אורך, והאחר רוחב, והשלישי עומק, ר"ל
גובה. וזה הוא ענין הגשמות הנמצא בהיולי ראשונה בלתי בחינת צורה אחרת.
ואינו נפש השלוח, כי השלוח מקרה ממאמר הכמה, יומר ויוסיף ויחסר בחמר האחד...
הנה הצורה אינו השלוח אבל הדבקות אשר יונח בו השלוח.

It would seem that Algazali's view with regard to the identification of corporeal form with the cohesion itself was also adopted
by Abraham ibn Daud. *Emunah Ramah* I, 2, p. 10: "Then
God endowed matter with the form of body, i. e., the form
of an absolute body, which is not air, nor water, nor fire, nor
earth, but is only cohesion, by which we mean that thereby the
substance has a certain massiveness in which it is possible to
posit three dimensions intersecting each other at right angles."

אחר כן הקנה האל יתברך לחומר צורת נשם תחלה, רצוני צורת נשם בשלוח,
איננו אויר, ולא מים, ולא אש, ולא ארץ, אך היא ההתדבקות לבד, רצוננו לומר
שיהיה בה לעצם עובי, אפשר בו שיונחו שלשה התפשטויות נכרתים על זויות נצבות.
Cf. also *ibid.* p. 11: "You should also know that substance is
divided into corporeal and incorporeal. It is corporeal substance
which we are considering now. It is a substance which has
a certain mass and rigidity in which it is possible to posit
three dimensions intersecting each other at right angles. And
this is what we meant by saying that its form is the cohesion
and its matter is that which forms the substratum of the cohesion."
ועוד תדע שהעצם יחלק אל נשמי ובלתי נשמי, והעצם הנשמי הוא אשר נעיין
בו עתה, והוא עצם שיש לו מן העובי והמקשיות מה שבהם אפשר שיונחו בו שלשה
התפשטויות נכרתים על זויות נצבות. והוא אשר אמרנו שצורתו היא ההתדבקות
וחומרו הוא נושא ההתדבקות. It may, however, be argued that the
term התדבקות used in the *Emunah Ramah*, unlike the term דבקות,
does not mean "cohesion" but rather a "predisposition for cohesion," and Abraham ibn Daud would thus accurately reproduce the view of Avicenna.

(Cf. Plutarch, *De Placitis Philosophorum* I, 12: "A body is that
being which hath these three dimensions, breadth, depth, and
length;—or a bulk which makes a sensible resistance." Hence the
term עובי in the *Emunah Ramah* reflects the Greek ὄγκος, *bulk*,
mass, and the term מקשיות reflects ἀντιτυπία, *the resistance of a
hard body*.)

Joseph ibn Zaddiḳ, on the other hand, would seem to have
anticipated Averroes' conception of the corporeal form, namely,
that it is identical with the three dimensions. *'Olam Ḳaṭan* I,
iii, p. 13: "For the matter which is the substratum of these
four natural forms of the elements is something spatial, being
itself invested with the form of corporeity, which is identical with
length and breadth and depth." כי היסוד הנושא לארבעה הטבעים
האלו הוא עצם מחזיק מקום, בלבשו צורת הגשמות, שהוא הארך הרחב והעמק.
But, as we have shown before, Averroes' view had been held
by Alfarabi long before Joseph ibn Zaddiḳ.

It will be noticed that Crescas has reproduced here only one
definition of corporeal form and describes it as the view shared in
common by Avicenna, Algazali and their followers. He has phrased
his definition, however, is a vague and noncommital manner.
If he had simply said שהצורה הגשמית אצלם אינה זולת הדבקות, "for

they believe that the corporeal form is nothing but the cohesion," he would have been committing himself to Algazali's
view. If he had said שהצורה הגשמית אצלם אינה זולת השלשה רחקים,
"for they believe that the corporeal form is nothing but the three
dimensions," he would have been committing himself to Averroes'
view. By combining these two statements it is not clear which
of these two views he meant to espouse. Nor is there anything in
his statement to include or to exclude the view of Avicenna. It
is not impossible that Crescas has purposely used this vague or
rather cmposite language in order to leave the question open, as
if to say, the corporeal form is the cohesion of the three dimensions in whichever sense you prefer to take it. A similar
vagueness marks also his statement in Prop. XI, where he says
that the corporeal form is "the cohesion of the dimensions."
למה שצורת הגשמיות, שהיא דבקות הרחקים.

A few more data bearing upon the history of this problem are
contained in that correspondence between Saul ha-Kohen Ashkenazi and Isaac Abravanel.

Saul Ashkenazi's letter (pp. 9b–10b) contains a restatement of
Averroes' view from the latter's *Treatise on the Possibility of
Conjunction with the Active Intellect* (אגרת אפשרות הדבקות) and
Narboni's commentary on that work. The writer further gives an
account of the conflicting opinions held by Elijah Delmedigo,
Elijah Ḥabillo, Shem-ṭob, and Abraham Bibago.

In his answer (p. 18 ff.), Abravanel informs his correspondent
that the original sources of the discussion are Algazali's *Kawwanot*
and Averroes' *Epitome of the Metaphysics*. (See *Epitome of the
Metaphysics* II end. Arabic, p. 76, § 73 ff. Latin, p. 373rb ff.
Quirós Rodriques, p. 119 ff. Horten, p. 89 ff. Van den Bergh, p.
63 ff.) By the former reference he undoubtedly means Narboni's
commentary rather than the *Kawwanot* itself. He also ventures
to give his own view on the subject as well as that of his son Judah
Abravanel (Leo Hebraeus). The latter's view will be reproduced
below in n. 26. Isaac Abravanel's view is stated by him in the
following passage (pp. 19b–20a):

"I now turn my attention to another view which appears to me
to be the most plausible with reference to this problem, namely,
that the corporeal form in any body is identical with its sub-

stantial form [forma substantialis] . . . . And let not
this diversity of terms trouble you, viz., that the same form
should be called elemental form [forma elementalis] and also
corporeal form [forma corporeitatis] . . . . For the
truth of this view there are ten arguments."

חשבתי דרכי ואשיבה רגלי אל דעת אחרת, אותו ראיתי צדיק לפני בדרוש
הזה, והוא שהצורה הגשמית בכל נשם היא הצורה העצמית אשר לו... ולא יקשה
אצלך שני השמות שתקרא הצורה ההיא צורה יסודית ותקרא גם כן צורה נשמית...
וכבר יורה על אמתת הדעת הזה דברים עשרה.

There seems to have been a great deal of confusion among
Jewish students of philosophy in the Middle Ages as to the meaning of corporeal form. Narboni in his Commentary on the Kawwanot has the following justification for his lengthy discussion:
"We have dwelt at such length upon this subject, owing to the
abstruseness of the problem itself, the diversity of opinions about
it among the philosophers, the insufficient understanding on the
part of the philosophizers of our own time as to the proper distinction between these opinions, and, in addition to all this, the obscurity and confusion which characterize the discussions of those
commentators who attempted to explain it. It is for these reasons
that we have gone into all this trouble here to direct you to the
proper understanding of this problem."

והארכנו בבאור זה לעומק העניין, והתחלפות הפלוסופים בו, וקוצר הבנת
המתפלספים בזמננו זה להבדיל הדעות, עם שהעניין בספרים המבארים בבלבול
ומבוכה. ולכן הישרנוך בו הנה.

19. Hebrew שלשה רחקים מתחתכים על זויות נצבות. This corresponds
exactly to the definition of body as given by Algazali in Kawwanot
ha-Pilosofim II (Maḳaṣid al-Falasifah II, p. 83): فالجسم هو كل
جوهر يمكن ان يفرض فيه ثلاثة امتدادات متقاطعة على زواتا قائمة
which is translated into Hebrew as follows: (a) MS. Adler 1500:
הגשם הוא כל עצם אפשר שיונחו בו שלשה רחקים, ר"ל המשכים, נחתכים על זויות
נצבות. (b) MS. Adler 978: הגשם הוא כל עצם אפשר שיונחו בו שלשה
שלוחים נכרתים על זויות נצבות. See quotation from Abravanel above
in n. 18, p. 585. Cf. Emunah Ramah I, 2, p. 11: העצם הגשמי הוא
אשר נעיין בו עתה, והוא עצם שיש לו מן העובי והמקשיות מה שבהם אפשר שיונחו
בו שלשה התפשטיות נכרתים על זויות נצבות. Joseph ibn 'Aḳnin (ed. M.
Löwy, p. 11, ed. J. L. Magnes, p. 8): ונאמר שהגשם מליצה מהדבקות
אשר אפשר שיונחו בו שלשה שלוחים כריתותם על זויות נצבות.

The terms רחק, מרחק, המשך, שלוח, התפשטות, ‌ِ., امتداد, are all translations of διάστημα or διάστασις, *distance, interval, extension, dimension.* Cf. Prop. XV, Part I, n. 9 (p. 639).

20. Cf. below Prop. XI.

21. Hebrew השכל יחור, literally, *reason decrees.* Cf. the expression ἡ ἔννοια λέγει in *Enneads* III, vii, 4.

The expression, however, may also have an additional meaning, namely, that the distinction between matter and form is conceptual and not sensible. Algazali says in this connection as follows. *Kawwanot ha-Pilosofim* II (*Maķaṣid al-Falasifah* II, p. 90): "Matter and form cannot be distinguished from each other by perception but they can be distinguished from each other by reason."

ואי אפשר שיוכר אחד מן האחר ברמז החוש ואבל ברמז השכל יוכר אחד·מהם
מן האחר.

That prime matter is recognizable only by thought is stated by Aristotle in *De. Gen. et Corr.* II, 1. 329a, 24–26: "Our own doctrine is that although there is a matter of the perceptible bodies (a matter of which the so-called 'elements' come-to-be), it has no separate existence, but is always bound up with a contrariety."

22. In comparing the arguments for the deduction of matter and form reproduced here by Crescas with the argument reproduced by him above in his proof of the proposition, it will be noticed that while the two arguments are alike in logical form they proceed from different premises and employ different terms. The first argument takes as its premise the phenomenon of the transmutation of the elements and reasons from the antithesis of generation and corruption (הוה ונפסד), whereas this argument takes as its premise the definition of corporeal form and reasons from the antithesis of continuity and division (דבקות וחלוק). That the second argument is not merely Crescas' own verbal modification of the first argument may be shown by the fact that it has a long history behind it, appearing in Avicenna and running through the entire literature based upon Avicenna's writings.

Avicenna's own statement of the argument is to be found in his *Al-Najah, Metaphysics*, p. 55. It is reproduced in the name of Avicenna by Shahrastani (ed. Cureton. p. 366).

It occurs in Algazali's *Kawwanot ha-Pilosofim* II (*Maḳaṣid al-Falasifah* II, p. 90): "For the corporeal form is undoubtedly an appelative for cohesion, and the cohesive body is undoubtedly capable of being a recipient of division. Now, that which is capable of being such a recipient must inevitably be either the cohesion itself or something else. That it should be the cohesion itself is absurd, for the recipient must remain with that which is received, inasmuch as non-being cannot be said to be the antecedent of being, but cohesion cannot be the recipient of division. Hence there must be something else which is the recipient of both division and cohesion, and that recipient is called matter in the conventional (or technical) sense, and the cohesion, which is received, is called form."

כי הצורה הגשמית מליצה מן הדבקות בלי ספק. זה כי הגשם המתדבק מקבל
לפרוד בלי ספק, והמקבל לא ימנע אם שיהיה עין הדבקות או זולתו. ואם היה
עין הדבקות, הנה הוא שקר, כי המקבל הוא אשר ישאר עם המקובל, אחר שלא
יאמר העדר קודם הנמצא, והדבקות לא יקבל הפרוד. הנה אי אפשר מבלתי
ענין אחר הוא המקבל לפרוד והדבקות יחד. זה המקבל יקרא היולי בהסכמה
(والاصطلاح), והדבקות המקובל יקרא צורה.

It is used by Joseph ibn 'Aḳnin (ed. M. Löwy, pp. 12–13; ed. J. L. Magnes, p. 9): "For body is an appellative for cohesion, and cohesion is incapable of becoming the recipient of division for the recipient must remain at the receipt of that which is received, whereas cohesion does not remain at the receipt of division, but, quite the contrary, it passes away at its arrival. It cannot therefore be its recipient. Hence the recipient must be something different from either cohesion or division; it must be something to which both division and cohesion occur in succession."

שהגשם מליצה מהדבקות, והדבקות אינו מקבל הפרוד, והמקבל הוא אשר ישאר
עם הקבלה, והדבקות לא ישאר עם קבלת הפרוד, אבל יעדר ממנו, והוא בלתי
מקבלו. הנה המקבל דבר בלתי הדבקות, ועליו ישוב הפרוד והדבקות בבא זו
אחר זו.

It is similarly reproduced by Altabrizi, Prop. XXII: "Let us now prove that body is composed of matter and form. We say: Having established that a body is infinitely divisible but that its parts are actually finite, it must follow from the combination of these two propositions that if we have a body which appears to our senses as one in reality and that body becomes divided, then

the recipient of the division cannot be cohesion itself, for cohesion is the opposite of division, and a thing is incapable of
being the recipient of its opposite, the reason for this being
that the recipient must continue to exist together with that which
is received, and a thing cannot continue to exist when something
which is its opposite comes into being. Hence the recipient of the
division of a body which is one and coherent in itself must be
the recipient of both cohesion and division.   That recipient
is matter; the corporeal cohesion is form; the union of both of
them is body. Body is thus the compound of matter and form."

לבאר היות הגשם מורכב מן ההיולי והצורה. ונאמר, למה שקוים החלקים
האפשריים בגשם בב"ת, וקוים שהחלקים בפעל ב"ת, חויב ממחובר שתי אלה
ההקדמות שיהיה לנו גשם היה אחד באמתות כמו שהוא אצל החוש, וכאשר בא
עליו הפירוד, הנה המקבל לפירוד אי אפשר שיהיה הוא הדבקות, כי הדבקות
הפך הפירוד, ולא יהיה בדבר לעולם קבלת הפכו, כי המקבל לדבר יהיה נמצא
בעת מציאותו המקובל, והדבר לא ישאר בעת חדוש הפכו, ואם כן המקבל לפירוד
בגשם אשר הוא מתדבק בעצמותו דבר, ובעל הדבקות הוא המקבל לדבקות והפירוד
יחד, ואותו המקבל הוא ההיולי, והדבקות הגשמי הוא הצורה, ומקובץ שניהם הוא
הגשם. א"כ הוא מורכב מההיולי והצורה.

From all these quotations and references it may be gathered
that this argument is not a mere paraphrase by Crescas of the
first argument, and that while it is not altogether a new argument
it is a new version of Aristotle's argument for the deduction of
matter and form.

The question may now be raised, why was Aristotle's argument
given this new form?

The answer seems to me to be as follows: This new version was
purposely devised in order to prove not merely the distinction
of matter and form in general but the distinction between first
matter and corporeal form in particular. Aristotle's argument
from the transmutation of the elements, as we have seen above
(n. 18), established only the existence of the proximate matter of
the four elements as distinguished from the four natural forms of
the elements. This proximate matter, as we have also seen, was
generally taken to be dimensional and not identical with Aristotle's non-dimensional first matter. Now, Avicenna and his
followers were especially interested in proving the existence of
the first non-dimensional matter as distinguished from the first

or corporeal form. They therefore devised this new argument, or rather revised the old Aristotlelian argument, in order to make it answer the new requirement.

23. Speaking now of Averroes, Crescas again lapses into the vocabulary of the Aristotelian argument for the existence of matter and form.

24. That is to say, the celestial spheres are not composed of first matter and corporeal form. They have no first matter. They are pure corporeal form, or the cohesion of the triple dimensions. Of course, the spheres have each a specific form with reference to which their corporeal form may be considered as matter. But they have no indeterminate, unextended and purely potential matter.

Averroes' view may be found in *Intermediate Physics* VIII, vi: "After it has been shown that the celestial substance has no opposite and no substratum, it follows that it is simple and is not composed of matter and form. It is like matter in actuality in its relation to the separable forms. It is more like matter than form, though it has a resemblance to both of them. It resembles matter in so far as it is perceptible and is something definite and has a potentiality with reference to place and is a body. It resembles form in so far as it is actual and its essence is not potential."

אחר שנתבאר מענין זה אין לו הפך ולא מונח, הוא אם כן פשוט, בלתי מורכב מחומר וצורה, והוא כחומר בפועל לצורות הנפרדות, והוא יותר דומה בחומר ממה שידמה לצורה, ואף על פי שיש בו דמיון משניהם, כי הוא ידמה לחומר, מפני שהוא מוחש ונרמז אליו ושיש בו כח באנה ושהוא גשם, וידמה לצורה מצד שהוא בפועל ושאין עצמותו בכח.

Averroes has also written a special treatise *Sermo de Substantia Orbis* (מאמר בעצם הגלגל) in which he endeavors to prove the simplicity of the translunar substance.

A statement of Avicenna's view is to be found in his commentary on *De Caelo*: "Book IV. Wherein it is shown that the matter of the heavens and their forms are not subject to generation and destruction. It is already known that every body, including the body of the celestial spheres, has a matter and form of which it is composed and that every one of the four elements which are called simple [bodies] has that composition,"

השער הרביעי, יבחן בו שחומר השמים וצורתם לא יקבל הויה והפסד. כבר
נודע שכל נשם יש לו חומר וצורה, מורכב משניהם, אפילו הגלגלים, וכל אחד מהם
מהארבעה יסודות שנקראו פשוטים יש להם זאת ההרכבה.

This view is reproduced in all the philosophical treatises based
upon Avicenna's works. Algazali restates it in his *Happalat ha-
Pilosofim* IV, to which Averroes makes the following answer in
his *Happalat ha-Happalah* IV (*Tahafut al-Tahafut* IV, p. 70, l.
30—p. 71, l. 13; *Destructio Destructionum* IV, p. 70va-b; Horten,
p. 188):

"His statement that every body is composed of matter and
form does not agree with the view of the philosophers with regard
to the celestial body, unless the term matter is to be understood
in an equivocal sense. What he says represents only the view
of Avicenna . . . . The celestial bodies are, as said Themistius,
forms, or they have matter only in an equivocal sense. But I say
that they are either matter *per se* or matter having life *per se* and
not through an attribute of life."

אמנם אמרו שכל נשם מורכב מחומר וצורה אין זה דעת הפילוסופים בשם
השמיימי, אם לא שיהיה שם היולי בשתוף השם. ואמנם הוא דבר אמרו ן' סיני
לבד... ואם שיהיו כמו שאמר תמסטיוס צורות, ואם שיהיו להם חמרים בשתוף, ואני
אומר זה: ואם שיהיו החמרים עצמם או יהיו חמרים חיים בעצמם לא חיים בחיות.

It is this passage from the *Happalat ha-Happalah* that is
quoted in the *Moreh ha-Moreh* II, Prop. XXII, p. 71, in the
name of an "aforementioned philosopher," החכם המזכר, whom he
never names, but by which expression he means Averroes.

The last sentence of the quotation in the *Moreh ha-Moreh*
differs somewhat from our quotation above. It reads: ואני
אומר, או שיהיו הם החמרים עצמים ויהיו חמרים חיים בעצמם לא חיים (נצחיים)
(ובחיות] "or, as I say, they are matter itself and matter having
life *per se* and not through an attribute of life." The reading
in the *Moreh ha-Moreh* agrees with the Arabic text before
us. The reading in our quotation, however, is followed by the
Latin translation: "Ego vero dico, sive sint eaedem materiae,
sive materiae viventes ex se, non autem viventes vita." The
difference must have arisen in two different readings of the
Arabic. The Arabic text of the *Moreh ha-Moreh* read المواد
نفسها وتكون. Our Hebrew translation had before it the reading
المواد نفسها او تكون.

The *Moreh ha-Moreh* quotes also a passage from the *Metaphysics* with Averroes' comment thereon which has a bearing upon this discussion. "Aristotle says in the *Metaphysics* that all things have matter, but that some matter is not generable nor is it changeable except for the change from one place to another. These are his very words. In another place he says: It follows that there is no matter except in things that are generable and corruptible and are changeable into one another. Upon this the aforementioned philosopher says: Hence it follows that the celestial spheres consist of simple matter and are not composed of matter and form, for the spheres have only change of place, whereas it is change of substance that makes it necessary for a thing to be composed of matter and form."

ואמר אריסטו בספר מה שאחר הטבע, וכל הדברים יש להם חמר, אלא שאינו
הוה ולא משתנה אלא מאנה לאנה. זה לשונו. ואמר במקום אחר, ומהחיוב שלא
יהיה חמר כל, אלא לכל הדברים שיש להם הויה והפסד וישתנו קצתם לקצתם.
ואמר החכם הנזכר, ויתחייב שיהיו הגלגלים חמרים פשוטים, זולתי מורכבים מחומר
וצורה, מפני שלא ימצא להם השנוי אלא באנה, והשנוי בעצם אשר יחייב הוית
הדבר מחומר וצורה, ע"כ.

The passage in question seems to be *Metaphysics* XII, 2, 1069b, 24–26: "Now all things that change have matter, but different matter; and of eternal beings those which are not changeable but are movable in space have matter—not matter for generation however, but for motion from one place to another."

Averroes maintains that all the commentators upon Arisotle, Alexander, Themistius and Alfarabi, are agreed as to the simplicity of the celestial substance and that Avicenna's view was a misunderstanding of the Peripatetics.

Intermediate De Caelo I, x, 2, 8 (Latin, pp. 294vb–295ra): "On this account, i. e., by virtue of its being simple, the celestial body has no substratum and no contrary. Hence Aristotle maintains that it is ungenerated and incorruptible, seeing that it has no subject and no contrary. It is thus stated by him at the end of the first book of *De Caelo*. It is no surprise that this was overlooked by Avicenna, but what surprises us is that it should have been overlooked by Alexander, despite his admission that the celestial body is simple and not composed of matter and form, as is evident from a passage in his commentary on Book Lambda.

I believe that there is no difference of opinion among the commentators on this point, for it is very clear from Themistius' commentary on *De Caelo et Mundo* that the celestial body has no substratum. A similar view was expressed by Alfarabi in the name of Aristotle, i. e., that such was his own view."

ומזה הצד, ר"ל מצד היותו פשוט, היה הגשם הזה אין נושא לו ולא הפך. ולזה
יטען אריסטו לזה הגשם שהוא בלתי הווה ולא נפסד, מפני שאין נושא לו ולא הפך.
וכן הוא דבריו בסוף זה המאמר. ואין לתמוה מהתעלם זה העניין מאבן צינ"י, כי
אם התעלמו מאלכסנדר, והוא עם זה מודה שהגשם הרקיעי פשוט בלתי מורכב
מחומר וצורה. חה נגלה ממאמר בפירוש מאמר אל לאם ואני חושב שאין חלוף
בין המפרשים בזה, כי הוא מבואר מאד ממאמר תמסטיוס בפירושו לשמים והעולם,
ר"ל שהגשם הרקיעי אין נושא לו. וכמו כן נגלה דעתו בזה אבונצר בשם אריסטו,
ר"ל שזאת היא דעתו.

Averroes' reference to Themistius is to be found in Themistii *De Caelo*, ed. Landauer, Hebrew text, p. 9, ll. 26–27: ואין לו דבר
מונח, חה שהוא התבאר במקום אחר שהוא אין חומר לו. Latin text, p. 14, ll. 13–14: "nec ullum subiectum habet, (alibi enim declaratum est materia id carere)."

Happalat ha-Happalah III (*Tahafut al-Tahafut* III, p. 63, l. 16; *Destructio Destructionum* III, p. 64ra, A; Horten, p. 177): "The view that the celestial body is composed of form and matter like the other bodies has been erroneously attributed by Avicenna to the Peripatetics."

המאמר בשהגשם השמימי מורכב מצורה וחומר כשאר הגשמים הוא טעה בו בן
סיני על המאשאים.

Isaac Abravanel suggests that Avicenna's view was derived from Plato's theory of creation. · *Mif'alot Elohim* II, 3, p. 12b: "For Plato says that the heavens were generated of that eternal matter which had been in a state of disorderly motion for an infinite time until it was invested with order at the time of creation. Consequently, by their own nature the heavens are corruptible just as they have been generated, and it is only God who implanted in them eternity, as it is written in the *Timaeus*. It is from this view that Avicenna has inferred that the celestial sphere is composed of matter and form and is corruptible and possible by its own nature but necessary and eternal by virtue only of its cause."

כי אפלטון אמר שהשמים נתהוו מאותו חמר קדום שהיה מתנועע תנועה בלתי
מסודר זמן בלתי בעל תכלית, ובעת הבריאה קבלה הסדר, ושהיו השמים כפי טבעם

נפסדים כמו שהיו הווים, אלא שהאל יתברך נתן בהם הנצחיות, וכמו שכתב בספרו
טימיאוס. ומכאן לקח אבן סיני שהיה הגרם השמימי מורכב מחמר וצורה היה
נפסד ואפשרי מעצמו אבל היה מחוייב ונצחי מפאת סבתו.

The following passages in the works of Jewish philosophers
indicate the influence of Avicenna's view:

Ḥobot ha-Lebabot I, 6: "Composition and combination are visible in the entire universe and in all the parts thereof, in its roots
and its branches, in its simple elements and its composite beings,
in its above and its below."

וההרכבה והחבור נראים בכל העולם ובכל חלקיו, בשרשיו ובענפיו, בפשוטו
ובמורכבו, בעליונו ובתחתונו.

Emunah Ramah I, 2: "Inasmuch as conjunction and that which
is joined are also to be found in the celestial bodies, it follows that
they have matter and form."

ואחר שההתדבקות והמתדבק הם בגשמי שמים גם כן, הנה יש בהם חומר וצורה.

Moreh Nebukim I, 58: "Thou who readest this book knowest
that this heaven . . . . though we know that it must consist
of matter and form, is not of the same matter as ours."

ואתה האיש המעיין במאמרי זה יודע כי זה הרקיע. . . עם היותנו יודעים שהוא
בעל חומר וצורה בהכרח, אלא שאינו זה החומר אשר בנו.

For further Hebrew sources bearing upon problem, see *Tagmule ha-Nefesh* I, 3, pp. 4b–5a; Shem-ṭob on *Moreh* II, Introduction, Prop. XXII; *Neveh Shalom* VII, i, 3.

25. See explanation of this expression above Prop. I, Part II,
n. 30.

26. In Averroes' view, as may have been gathered, there is the
following distinction between the sublunar and translunar substances. The sublunar substances are composed of (1) the first
matter, (2) the corporeal form, and (3) the natural or specific
form. The celestial substance, he maintains, is without first matter. It is composed of (1) corporeal form and (2) the specific form
which each of the spheres possesses, the former being related to
the latter as matter to form, but even without the latter, the
former is not pure potentiality but has actual existence.

Hence Crescas' argument, which may be restated as follows:
It is true, as Aristotle maintains, that there must be three principles: (1) non-being, (2) being, and (3) a substratum (see above

n. 7). But why should these principles be identified with (1) the privation of any form, (2) the first form, and (3) a first matter which has no actual existence by itself. It is that purely potential first matter that Crescas is trying to eliminate. Why should not the substratum or first matter be the so-called corporeal form, i. e., tridimensionality, the same as Aristotle is reported by Averroes to have held in the case of the celestial spheres, and the first form be the natural or specific form of the elements, and privation be the privation of that natural form? As a result of this, the first matter, being identical with tridimensionality, will not be pure potentiality but will have actual existence, like the so-called matter of the celestial spheres in Averroes' theory.

The main point of Crescas' argument, then, is to show that first matter has actual existence. He is thus reviving the theory held by Ibn Gabirol, who likewise maintained the actual existence of what he called universal matter (cf. *Likkuṭe Meḳor Ḥayyim*, I, 6; *Fons Vitae* I, 10, p. 13, l. 15), though Ibn Gabirol's universal matter is not identical with corporeal form (cf. *Likkuṭe Meḳor Ḥayyim* II, 2; *Fons Vitae* II, 1, p. 24, ll. 15–22.'

We may get a better appreciation of the drift of Crescas' argument if we only recall that in his argument for the deduction of matter and form in his commentary on this proposition, Crescas followed Abraham ibn Daud's *Emunah Ramah* (cf. above notes 5, 7, 8, 9, 13, 16).

Now, Abraham ibn Daud, after deducing the existence of matter and form and defining the nature of the former, quotes Ibn Gabirol's theory of universal matter and criticizes it. His main objection against the universal matter as conceived by Ibn Gabirol is its independent actual existence. What Crescas does here, therefore, after reproducing Abraham ibn Daud's proofs for the existence of matter and form, is to defend Ibn Gabirol's universal matter against Ibn Daud's criticism. He does this by introducing the analogy of Averroes' conception of the celestial substance. That this is the intention of Crescas' argument is still further evidenced by the fact that his subsequent description of his proposed theory of first matter corresponds almost verbally with the description of Ibn Gabirol's universal matter as found in the *Emunah Ramah*. Cf. below notes 27, 30.

The view which Crescas advocates here, that first matter should be identical with corporeal form, has later found its exponent in Leo Hebraeus, as reported by his father Isaac Abravanel in *She'elot Saul* X, p. 20b:

"And know that my son Don Judah Abravanel has not been in this country for these two years, for he has been in Naples together with the Great Captain and the King of Spain who had been visiting there. Now that both the king and the Great Captain had returned to Spain my son has come here to my house. But on the way he fell ill with a high fever, and has arrived home very ill and weak. Still, disregarding his weakness, in order to comply with your request, I discussed with him this problem—he being beyond any doubt the most accomplished philosopher in Italy at the present time. Out of the fulness of his knowledge he told me that the view of Averroes is open to more doubts and refutations than all the other views. His own view is that the first matter is corporeity itself. He advanced arguments to prove it and cited as evidence passages from Aristotle in the fifth book of the *Metaphysics*. Inasmuch as I could not bring myself to accept his opinion, I mentioned here only my own view, and 'Every way of a man is right in his own eyes, but the Lord pondereth the hearts' [Prov. 21, 2]."

ואתה תדע שבני דון יהודה אבראבניל לא היה בארץ הזאת שנתיים ימים, כי
היה בנאפ'ולי"ש עם הקאפי'טאניו גר'נדו ועם מלך ספרד שבא שמה, ועתה שהלכו
שניהם המלך והשר צבא לארצו ספרד, בא בני פה אל ביתי, וקראוהו בדרך קדחת
חדית, ובא חולה וחלוש מאד, ועם כל חלשתו, למלאות רצונך דברתי עמו בדרוש
הזה, כי הוא בלי ספק מבחר הפילוסופים שבאיטאלייה בדור הזה, ויורני ויאמר
לי שהיה דעת בן רשד יותר רב הספיקות והבטולים מכל שאר הדעות. ודעתו
הוא שהחומר הראשון הוא הגשמות, ועשה על זה טענות, ומביא ראיות מדברי אריסטו
בחמישי ממה שאחר הטבע. ומאשר לא לבי לבי הלך בעצתו, לא זכרתי פה כי אם
דעתי. וכל דרך איש ישר בעיניו ותוכן לבות ה'.

27. So likewise the universal matter of Ibn Gabirol has actual and independent existence.

Emunah Ramah I, 2, p. 11: "And when Ibn Gabirol wanted to describe it, he said in the first book of the *Fons Vitae*, that if all things were to have a universal matter, it would have to possess properties as follows: that it has existence, that it exists in itself, that it is one in essence, that it underlies all the changes, and that it gives to everything its essence and name."

וכאשר רצה אבן גבירול לרשום אותו, אמר במאמר הראשון ממקור החיים, אם יהיה לדברים כולם יסוד כולל, חוייב לו מהסגולות שיהיה נמצא, עומד בעצמו, אחד בעצמות, נושא החלופים, נותן אל הכל עצמותו ושמו.

Cf. *Likkuṭe Meḳor Ḥayyim* I, 6: ואם היה לדברים כולם יסוד כללי יתחייב לו מהסגולות שיהיה נמצא, אחד העצם, עומד בנפשו, נושא לחלוף, נותן לכל עצמו ושמו. *Fons Vitae* I, 10, p. 13, ll. 14–17: "Si una est materia universalis omnium rerum, haec proprietates adhaerent ei: scilicet quod sit, per se existens, unius essentiae, sustinens diversitatem, dans omnibus essentiam suam et nomen."

28. Cf. *Job*, 16, 19. But compare also expression והמעיד האל Arabic ואלה שהיד, and האלהים יודע ועד in Maimonides' אגרת השמד and האלהים יודע ונביאיו ובחירים in *Cuzari* III, 49, all quoted in Steinschneider's *Uebersetzungen*, p. 56, n. 75.

29. Having thus refuted the accepted theory of matter, Crescas now takes up Maimonides' proposition. Maimonides, as Crescas has pointed out previously in his commentary, uses the term body, i. e., the compound of first matter and corporeal form, in the sense of matter in its relation to the specific or natural form of the elements. Again, Maimonides asserts that this compound of first matter and corporeal form has no independent, actual existence without the specific form. Against this Crescas argues that it is not so, for the corporeal form, as he has shown from the analogy of the celestial substance, may have actual and spatial existence without the specific form.

30. Hebrew אבל הצורה הגשמית הוא הנושא בפועל והמעמדת הצורה המיוחדת. So is also the universal matter of Ibn Gabirol. Cf. above n. 27.

31. Crescas is now trying to forestall a possible objection. The contention that the corporeal form should have actual existence, independent of the specific form, would seem to lead to the conclusion that the specific form would be a mere accident. For the specific form, unlike all other substances, has no independent existence. It cannot exist without matter. It is called substance only for the reason that it is the cause of the actual existence of matter. In fact, a certain school of philosophers, the Mutakalli-

mūn, consider form as a mere accident (see above n. 9). And so, if we say that the corporeal form could have actual existence without the specific form, the latter would have to be an accident.

32. That is to say, each of the four elements has a proper natural locality where it is at rest, when within it, and towards which it is moved, when outside of it. Cf. above Prop. I, Part I, p. 157.

PROPOSITION XI

1. As for the meaning of this term in Maimonides, see Prop. X, Part I, n. 15, (p. 577.)

2. The Hebrew text of the proposition follows Ibn Tibbon's translation of the *Moreh* except for the substitution of the term גשם for Ibn Tibbon's גוף. The term גשם is used in Isaac ben Nathan's translation of Altabrizi.

3. This entire comment is based upon the following passage of Altabrizi: "Know that things which are dependent upon a body fall into four classes. First, those which are divisible by the division of the body.... as color in a body.... Second, those which, though existing in a body, are not divisible by the division of the body.... as, e. g., the surface, the line and the point.... As for point, it is indivisible in an absolute and unrestricted sense. As for line and surface, their indivisibility with the division of the body applies only to some of their dimensions, thus in surface, it applies only to height but not to the other two dimensions, and in line, it applies only to width and height but not to length.... Third, things which constitute the existence of body and are divisible with the division of body, as, e. g., matter and the corporeal form, for both constitute the existence of body and they are divisible by the division of that body. For when a body happens to become divided and disjoined, the recipient of the disjunction is not the corporeal continuity itself, (i. e., the corporeal form), for continuity is the opposite of discontinuity and a thing cannot be the recipient of its opposite. Since the corporeal form is not the true recipient of the disjunction, matter must therefore be its recipient. Hence it follows that when the

body happens to become divided matter must likewise become divided. As for the [corporeal] form, it cannot be the recipient of an actual division, for the reason we have already mentioned, but it can become the recipient of a conceptual kind of division.... Fourth, that which constitutes the essence of the body and is not divisible by the division of the latter, as, e. g., the intellect."

דע שהדברים שלהם התלות בגשם על שני חלקים. אחד מהם יחוייב מחלוק
הגשם חלוקו... במראה בגשם... והשני מה שיעמוד בגשם ולא יחוייב מחלוק
הגשם חלוקו... בשטח והקו והנקודה.... אולם הקו והשטח הנה אצל חלוק הגשם
מבלתי צד חלוקם, אולם בשטח הנה בגבה בלתי שני הנשארים, ואולם הקו הנה
ברוחב ונבה בלתי הארך.... והחלק השלישי מה שיעמיד, ויחוייב מחלוק הגשם
חלוקו, חה כמו ההיולי והצורה הגשמית, כי שניהם מעמידים לגשם ויחוייב מחלוק
הגשם חלוקם. חה שכאשר קרה לגשם חלוק והפרדה, הנה המקבל להפרדה אינו
הדבקות הגשמי, כי הדבקות הפך ההפרדות, ולא יהיה בדבר כח קבלת הפכו
לנמרי. ואחר שאין המקבל להפרדה באמת הצורה הגשמית, הנה הגשם המקבילו
הוא ההיולי. הנה כבר יחוייב מהגעת החלוק על הגשם הגעתו על ההיולי. ואולם
הצורה, אי אפשר שתקבל החלוקה הפרודית, למה שזכרנוהו, אבל היא תקבל
החלוקה המחשבית... והחלק הרביעי, מה שיעמיד הגשם ולא יחוייב מן הגעת
החלוק על הגשם הגעתו על אותו המעמיד, כמו השכל.

It will have been noticed that while Crescas mentions two illustrations of accidents which participate in the division of body, color and magnitude, מראה ושעור, Altabrizi mentions only one, color, מראה. But in addition to color Altabrizi also discusses the case of the geometric figure of a body. It is not exactly divisible with the division of the body, he argues in effect, for to be divisible in the case of geometric figure would mean that the same geometric figure would be divided into many similar geometric figures, but "it does not necessarily follow that, by the division of a square body into parts, every one of the parts would likewise be a square differing only in size from the first square," He then concludes: "While the geometric figure of a body, on the division of the body, is not necessarily divided into parts which are similar to the whole, the geometric figure may still be said, in a general sense, to be divided with the division of the body, even though it is divided into parts which are dissimilar with the whole."

וכמו שיקרה לגשם בסבת קצות התמונה שתקרה לו בסבת השטח ברבוע, כי לא
יחוייב שיחלק בחלוק הגשם חלוק יהיה כל אחד מחלקיו רבוע בחלוף החלק הראשון,
... ואין היזק שיאמר התמונה, ואם לא תחלק בחלוק הגשם אל חלקים דומים לכלם,
הנה היא תחלק בחלוקו בכלל, ואם היה אל חלקים מתחלפים לכלם.

Crescas may have thus added שׁעוּר, *magnitude*, *size*, as a substitution for Altabrizi's "geometric figure" and as an improvement thereon.

4. The following preliminary remarks will be helpful to the understanding of the text:

The term נפשׁ ordinarily has the generic meaning of soul, including all the faculties, the vegetative, the animal, and the rational. The term שׂכל usually refers to the rational faculty of the soul, and also to the Separate Intelligences, identified with the angels of the Scriptures, which are considerd as the cause of the motion of the spheres. In this proposition, the terms נפשׁ and שׂכל are both used. It would at first thought seem that by the former term is meant the vegetative and the animal faculties of the human soul and by the latter the rational faculty. This interpretation, however, could not be construed with the text, for the vegetative and animal faculties are generally admitted to be divisible with the body (cf. Shem-ṭob's commentary on *Moreh*, *ad loc.*). Altabrizi, therefore, suggests that the terms נפשׁ and שׂכל are used here by Maimonides as a hendiadys, the term שׂכל thus limiting the term נפשׁ in order to make it unmistakably clear that the latter term refers here to the rational faculty.

"Notice how the author of this work has joined here the term soul with the term intellect. Soul is not the cause of the essence of body *qua* body nor is it the cause of its existence. It is rather a first entelechy of bodies, and it brings about their perfection by endowing them with life and what is implied by life, such as sensation, motion and their like. Soul thus constitutes the cause of the perfection of bodies and not that of their essence and existence. The division of the body does not involve the division of the separable souls, such as the rational souls, which are neither bodies nor anything belonging to body. As for the bodily souls, such as the animal and vegetable souls, they are necessarily divided by the division of the body. It is in this sense, i. e., by taking 'soul' here in the sense of separable soul, which is the cause of the perfection of body in its life, essence and existence, that the author's use of the term soul as an illustration of the case of indivisibility can be justified."

ודע שבעל הספר חבר זכר הנפש בשכל בזה המקום, ואין הנפש עלה למהות
הגשם מאשר הוא גשם ולא למציאותו, אבל הוא שלמות ראשון לגשמים ומשלימם
בהשפעת החיות והנמשכים אליה מהחוש והתנועה וזולתם הנה היא מעמדת הגשמים
בשלמותם בלתי מהותם ומציאותם. ולא יחוייב מחלוק הגשם חלוקם, ר"ל הנפשות
המופשטות אשר אינם גשם ולא גשמיות, כנפשות המדברות. ואולם הנפשות הגשמיות,
כנפשות החיוניות והצומחות, הנה יחוייב מחלוק הגשם חלוקם. ועל זה האופן,
והוא שירצה בנפש הנפש המופשטת בהשלמת הגשם בחיותו ומהותו ומציאותו, יתאמת
המשלו בנפש בזה החלק.

This interpretation, it seems to me, may be re-enforced by a
passage in *Moreh* II, 1, Speculation I, Fourth Case, where Maimonides himself explains the terms נפש ושכל of this proposition
by the phrase נפש האדם באדם "the human soul in man." Now,
the "human soul" is only another expression for the "rational
soul", הנפש המדברת.

Crescas follows Altabrizi's explanation, namely, that the purpose of the proposition is to state that the human soul, and more
particularly the hylic intellect of man, though existing in the
material body, is still indivisible. He adds, however, that this is
Maimonides' own peculiar theory whereas, according to what he
considered to be the genuine view of Aristotle, the rational soul
cannot be said to exist in body at all.

5. The entire passage, in which Crescas discusses here the distinction between Maimonides and Aristotle, is a paraphrase of Narboni's commentary on the *Moreh* (*ad loc.*). It would seem that the
passage was added by Crescas as an afterthought, after having
first stated that he would discuss it later.

The underlying assumption of the entire discussion is that
there is an analogy between the relation of the soul to the body
and that of the Intelligences to the spheres. Another allusion to
the interdependence of these two problems is made by Crescas
in Prop. VIII, Part II.

The differences between Maimonides and Aristotle, or rather
Averroes, as to these problems may be summarized as follows:

A. Maimonides:

(1) The spheres, like all material objects, are composed of
matter and form (see Prop, X, Part II, n. 24, p. 594), and, like
all animate rational beings, possess souls, נפשות, which are the efficient cause of their motion, and Intelligences, שכלים, which are the

final cause of their motion (see *Moreh* II, 4). Both the souls and the Intelligences, though not distributed through the body of the spheres as physical forces, are still said to exist in the sphere. Maimonides describes them as "an undistributed force within the sphere, כח בו בלתי מתפשט (*Moreh* II, 1, First Proof). In *Moreh* I, 72, he similarly says ויהיה דמיון הכח הדברי כשכלי הגלגלים אשר בגופות, which Shem-ṭob paraphrases as follows: "The rational faculty of man is analogous to the Intelligences of the spheres, which exist in bodies." ויהיה ענין הכח הדברי כשכלי הגלגלים אשר הם בגופות. Inasmuch as the Intelligences are assumed by Maimonides to exist in bodies, he also maintains that they must be moved accidentally while setting the spheres in motion.

(2) Since the Intelligences, in Maimonides' opinion, are subject to accidental motion, he could not identify God with the first of these Intelligences, to whom the expression "first mover" was originally applied (see above pp. 461–2). To the proof of this point he devotes much of the first chapter of the second part of the *Moreh*. His final conclusion is that God is beyond the "first mover", being its cause, and, unlike it, is absolutely outside of, or "separate" from, the sphere, thus not being subject even to accidental motion. God is therefore not to be called the First Mover, המניע הראשון, but rather the First Cause, הסבה הראשונה. Cf. *Moreh* II, 4 end: "It is impossible that the Intelligence which moves the uppermost sphere should be identified with Him of necessary existence." ולא יתכן שיהיה השכל המניע הגלגל העליון הוא המחוייב המציאות. Again, *ibid.* II, 1: "And that is God, praised be His name, that is to say, the first cause which sets the sphere in motion." וזהו האלוה יתעלה שמו, רוצה לומר, הסבה הראשונה המניעה לגלגל.

Corresponding to this theory is Maimonides' view on the relation of the human soul, both the hylic and the acquired intellect, to the human body.

(3) Maimonides' view as to the nature of the hylic intellect is a matter of doubt, for he has never stated it explicitly. According to Narboni's interpretation, Maimonides is following Alexander Aphrodisiensis, believing the hylic intellect to be a mere disposition, but going even further than Alexander, declaring it to be commingled with the body. Cf. Narboni on *Moreh* I, 68:

"Rabbi Moses follows in the footsteps of Alexander on this question, except that he believes that this predisposition within us is commingled, for he has stated that the rational faculty is corporeal.'' ורבנו משה הלך בעקבות אלכסנדר בזה הדעת, רק שהאמין שההכנה הנמצאת בנו מעורבת, כי הוא אמר שכח המדבר כח גופני. Whether this is an accurate representation of Maimonides' view may be questioned. Shem-ṭob is uncertain about it. Cf. his commentary on *Moreh* I, 68: "For all the philosophers are of the opinion that the human intellect is not force in a body with the exception of Maimonides who says in two places that the intellect is a force in a body, though he himself says in another place that the intellect is only a predisposition as is maintained by Alexander.'' שכלם הסכימו שהשכל האנושי אינו כח בגוף, זולת הרב שאמר בשתי מקומות שהשכל הוא כח בגוף, עם שאמר במקום אחר כי השכל אינו אלא הכנה לבד כדעת אלכסנדר. Cf. also Shem-ṭob on *Moreh* I, 1. Abraham Shalom scornfully repudiates Narboni's suggestion that Maimonides considered the hylic intellect to be commingled with the body. Cf. *Neveh Shalom* VIII, 3, p. 125b. Maimonides is, however, explicit as to what he considered to be the relation of the hylic intellect to the human body. It exists in the body, indivisible to be sure, but related to it as the Intelligences are to the spheres. Cf. *Moreh* I, 72, quoted above under (1).

(4) The acquired intellect, however, in no sense exists in the body. It stands related to the body as God to the world. Cf. *Moreh* I, 72, quoted above under (1).

B. As against all these points Aristotle, or rather his interpreter Averroes, maintains as follows:

(1) The spheres are simple substances and are not composed of matter and form. Nor do they possess souls in addition to Intelligences. They have only Intelligences as the sole cause of their motion. These Intelligences do not exist *in* the spheres, but rather *with* the spheres, being related to them by a nexus of inexistence, and are therefore *separate* forms. The Intelligences are, however, called "souls" in a loose sense, by virtue of their being the cause of the motion of the spheres, for the soul is the cause of motion in animals (cf. *De Anima* III, 9, 432a, 15–17). This is the significance of Crescas' (i. e., Narboni's) remark here: "Still that Intelligence, though separate, being the principle of the sphere's motion, is in a sense the latter's soul." ולהיותו מניעו הוא נפשו.

Furthermore, the Intelligences can in no sense be said to exist within the body of the sphere. They are related to the sphere by a "nexus of inexistence" rather than a "nexus of admixture" (as for the meaning of these expressions see Prop. VIII, Part II, n. 13, p. 560). As a result of this view, the Intelligences are not said to be moved accidentally by the motion of the spheres.

(2) Since the Intelligences have no accidental motion, God is identified with Aristotle's First Mover.

(3) and (4) The hylic intellect as well as the acquired intellect is related to the human body as the Intelligences are to the spheres. Neither of them is said to exist within the body in any sense whatsoever. All of these are related to their respective bodies as God, according to Maimonides, is related to the world.

With these preliminary remarks the meaning of the text becomes clear. In the translation I have supplied within brackets all the phrases that are necessary for the understanding of the text.

The original text of Narboni reads as follows:

"Rabbi Moses is of the opinion that the human soul and intellect are forces in the body but not divisible [with the body], inasmuch as they are not distributed through it. But there is this to be urged against him. First, they are not forces in a body, for if the intellect were a force in a body, it would not have power over matter, and consequently the latter would be able to transform the object of the intellect into something of a material nature. Second, every force that is in any way related to body, must be either mixed with the body or not mixed with it. If it is mixed with the body, then it will also have to be divisible [with the body] and distributed [through it]. If it is not mixed with the body, then its connection with it must of necessity be that of inexistence rather than that of admixture, and consequently it is not to be called a force *in* a body but rather a force *with* a body. Nor is it to be moved, for the Intelligence of the sphere is exactly in such a manner related to the sphere, being connected with it after the manner of a separate form, that is to say, by a nexus of inexistence rather than by that of admixture, and because of that it is assumed to be incapable of being moved even accidentally. And of the same description is also the acquired intellect according to Maimonides himself, for he compares the relation of the acquired intellect to man to the relation of the separate Intelligence to the universe as a whole.

You must know that Maimonides was led to this difficult position by his view that the sphere is composed of matter and that it possesses an Intelligence in addition to the separate Intelligence. As a result of this he further believes that it is only the separate Intelligence that is not in a body and hence not moved either essentially or accidentally. As for the Intelligence [of the spheres], it is a force in a body, though not distributed through the body, analogous in every respect to the case of the intellect of man. And since the Intelligence [of the sphere] is a force in a body, he maintains that it is moved accidentally, again as in the case of the human soul. As for the natural forms which are distributed [through the body] and as for the other distributed accidents, they are all not only moved accidentally but are also divisible with the division of the body. It is for this reason that Maimonides uses one argument to prove that the Intelligence of the sphere is not the mover [*par excellence*], for, being moved accidentally, it must come to rest, and he uses another argument to prove that a distributed force cannot be the mover [*par excellence*], for, being divisible with the division of the body, it must be finite and thus its activity must be finite, as you may find it in the first chapter of the second part.

Aristotle's way of viewing these problems is entirely different. He believes that the sphere is simple, inasmuch as everything composite is corruptible. The matter of the sphere is thus a simple substance existing by itself in actuality and having no potentiality except with reference to motion. He further believes that the separate Intelligence is separate only in the sense that it is not a force in a body and is not distributed through a body and is not divisible with the division of a body, inasmuch as it is not commingled or entangled with body. But still it is connected with the body by a nexus of inexistence though not by one of admixture, for it is a form of body, by reason of its being the cause of the perfection of body and the cause of its motion, and being the cause of its motion, it is its soul. Consequently the sphere may be said to contain one part which is moved by itself, but, inasmuch as that part is separate from the sphere, the sphere is not said to be moved according to part, but is rather said to be moved by itself in the true sense of the expression. He proves that the Intelligence must be 'separate' on the ground of its special activity, i. e., motion, which is assumed to be infinite, for were it not separate it would be a force in a body, distributed through the body and divisible with its division, and would thus be finite and its activity would be finite.

This is the way of Aristotle. And because of the importance of this problem I have tried to set you aright as to the Philosopher's view in addition to my trying to set you aright as to Maimonides' view, for by this, i. e., by a knowledge of the distinction between different views, the words of the author will become understandable according to their true meaning. It was his preoccupation with the doctrines of Avicenna as set forth in the *Al-Najah* and other works that led the Master to adopt such fantastic views and to consider them as the way of Aristotle. 'But this is not the way, neither is this the city' [2 Kings, 6. 19].''

רבינו משה סובר שנפש האדם והשכל הם כח בגוף בלתי מתחלק, כי אינם
משותפים וצריך לתקן: מתפשטים[בו. והפלא ממנו, ראשונה, כי אינם כח בגוף,
כי אם היה השכל כח בגוף, לא היה גובר על ההיולי, והיה משנה את המושכלות
אל טבע החמר. שנית, שכל כח מתיחס לגוף, הנה הוא מעורב או בלתי מעורב.
אם מעורב, הנה הוא מתחלק ומתפשט. ואם בלתי מעורב, הנה הוא נקשר בו הקשר
מציאות לא הקשר ערוב, ואם כן אינו כח בגוף כי אם עם הגוף, ואינו מתנועע,
כי השכל זה ענינו, שהוא נקשר עם הגלגל הקשר צורה נפרדת, ר״ל הקשר מציאות
לא ערוב, ואיננו מתנועע במקרה. וככה השכל הנקנה, לפי דעת רבינו משה,
אשר חבר שיחסו לאדם יחס השכל הנבדל אל איש העולם.

ואשר צריך שתדעהו שכל זה הביאו אליו למה שחשב הרב כי הגלגל מורכב
מחומר ושכל זולת השכל הנבדל, וחשב כי השכל הנפרד הוא אשר איננו בגוף
כלל, ולכן לא יתנועע לא בעצם ולא במקרה, כי השכל הוא כח בגוף, רק בלתי
מתפשט, כענין בשכל האדם. ולפי שהוא כח בגוף, יתנועע במקרה, כענין בנפש
האדם. והצורות הטבעיות המתפשטות ושאר המקרים המתפשטים יתנועעו במקרה
ויתחלקו בהתחלקו. ובעבור זה יתאחד מופת על שֹשכל הגלגל אינו המניע, כי
יתנועע במקרה וינוח, וייחד מופת על שהכח המתפשט אינו המניע כי יהיה בעל
תכלית, אחר שיתחלק בהתחלקו, ויהיה פעלו בעל תכלית, כמו שתראה בפרק
הראשון.

ודרך אריסטו אינו זה, אבל יאמר שהגלגל פשוט, כי כל מורכב הוא נפסד,
וכי חומר הגלגל הוא עצם פשוט נמצא בפועל בעצמו ואיננו בכח רק אל התנועה,
וכי השכל הנבדל הוא נבדל במה שאינו כח בגוף ולא מתפשט בו ולא מתחלק

בהתחלקו, כי לא יעורב בו ולא יסתבך, אבל הוא נקשר בה הקשר מציאות לא
הקשר עירוב, כי הוא צורתו, על שהוא משלימו ומקנה לו התנועה, והוא נפשו במה
הוא מניע לו, עד שיהיה הגלגל מחובר מחלק מתנועע מעצמו, ולפי שהוא נבדל,
לא יתנועע מפני חלק ממנו, ולכן היה מתנועע בעצמו באמת. וביאר על שהוא
נבדל מצד פעלו המיוחד שהוא בלתי תכלית והוא התנועה. ואם לא יהיה נבדל
יהיה כח בגוף ומתפשט בו מתחלק בהתחלקו, ויהיה בעל תכלית ופעלו בעל תכלית.
זהו דרך אריסטו, וליוקר הדרוש העמדתיך על דברי הפילוסוף את העמדה
על דעת הרב, גם כי בזה יובן דעת המחבר על אמתתו, ר"ל הידיעה בהפרש
הסברות. והעיון בדברי אבן סיני ובאלני חולתו הביא הרב אל אלו הדמיונות
וחשבם דרך אריסטו, ולא זאת הדרך ולא זאת העיר.

6. The passage as it stands is impossible, even though the reading
occurs in all the MSS. and printed editions, for it ascribes to
Maimonides the view that the Intelligences are divisible. Maimonides, however, never held such a view. Quite the contrary, he
has definitely stated that the Intelligences, though existing in
the spheres as a force, are indivisible. כח בו בלתי מתחלק. I have
therefore emended the reading by introducing, on the basis of
the underlying passage of Narboni, an additional statement. Cf.
Flensberg's commentary Oẓar Ḥayyim on Or Adonai, ad loc.

To understand the full meaning of this passage, it is necessary
to take it in connection with Maimonides' reasoning in his first
proof for the existence of God (Moreh II, 1). Maimonides tries
to show that the first cause of motion must inevitably be one of
the following four things: (1) A corporeal being outside the
sphere. (2) An incorporeal being outside the sphere. (3) A
force distributed throughout the sphere and divisible with the
division of the sphere. (4) An indivisible force. He then eliminates all but the second alternative. His arguments against the
third and fourth alternative, to which the passage here has reference, reads as follows: "The third case, viz., that the moving
object be a force distributed throughout the body, is likewise
impossible. For the sphere is corporeal, and must therefore be
finite (Prop. I); also the force it maintains must be finite (Prop.
XII), since each part of the sphere contains part of the force
(Prop. XI): the latter can consequently not produce an infinite
motion, such as we assumed according to Proposition XXVI,
which we admitted for the present. The fourth case is likewise
impossible, viz., that the sphere is set in motion by an indivisible

CRESCAS' CRITIQUE OF ARISTOTLE [267-273]

force residing in the sphere in the same manner as the rational
faculty resides in the body of man. For this force, though indivisible, could not be the cause of infinite motion by itself alone;
because if that were the case the prime motor would have an accidental motion (Prop. VI). But things that move accidentally
must come to rest (Prop. VIII), and then the thing comes also to
rest which is set in motion.''

PROPOSITION XII

PART I

1. The Hebrew text of the proposition is taken from Isaac ben
Nathan's translation of Altabrizi.

2. Cf. *Physics* VIII, 10, 266a, 24 ff., and *Intermediate Physics*
VIII, vi, 2, of which the entire chapter here is a paraphrase.
   This proposition is also given by Abraham ibn Daud in *Emunah
Ramah* I, 4, p. 17.

3. Hebrew הניעתו, so also in *Intermediate Physics, loc. cit.* In the
Vienna edition it has become corrupted into תנועתו, *its motion*.

PART II

4. See above Prop. I, Part II.

5. See above Prop. I, Part II, n. 13 (p. 403).

6. Hebrew הידוע אצל הטבע. See above Prop. I, Part II, n. 14
(p. 409).

7. This distinction between the two senses in which the expression infinite force may be used is repeated by Crescas in his
criticism of Maimonides' first proof of the existence of God (*Or
Adonai* I, ii, 15) and also in his discussion of the omnipotence of
God (*ibid*. II, iii, 2). The distinction is evidently borrowed
from Averroes, who advances it in his *Ma'amar be-'Eẓem ha-
Galgal* III (*Sermo de Substantia Orbis*, Cap. 3, p. 9va, G): "We
say briefly, that the term infinite may be applied in two senses.

First, in the sense of a force of infinite action and passion in time but finite in itself, that is, in velocity and intensity. Second, in the sense of a force of infinite action and passion in itself.''

ונאמר בקצור, שמאמרנו בלתי מכולה יאמר בשני עניינים: אחד/מהם כח בלתי מכולה הפועל וההפעלות בזמן, ואם מכולה בנפשו, ר"ל במהירות והחזק, והשני כח בלתי מכולה הפעל וההפעלות בנפשו.

It occurs also in the *Intermediate De Caelo* I, x, 2, 8 (Latin, p. 293vb, K): "In answer to this difficulty we say that a body may be said to have a finite force in two senses. First, that its motion is finite in intensity and speed. Second, that its motion is finite in time.''

ונאמר אנחנו בהתר זה הספק, שהגשם יאמר שיש בו כחות ב"ת על שני עניינים: האחד מהם מציאות התכלית לתנועתו בחזק ובקלות, והענין השני מציאות התכלית לה בזמן.

It is similarly adopted by Altabrizi in the following passage: "As for the second way in which a force may be said to be finite or infinite, namely, with reference to the motion it produces, it may mean three things, in intensity, in number, and in time.''

ואולם השני, והוא שינשא עליו התכלית או לא תכלית בבחינת הכחות וצריך לתקן: ההנעות] עליו, הנה זה מג' פנים: החזק והמספר והזמן. But whereas Altabrizi tries to prove the impossibility of the existence of an infinite force in a finite body in any of these three senses, Crescas argues for the possibility of the existence within a finite body of a force finite in intensity but infinite in time.

This distinction between these two senses of the expression "infinite force" is also made use of by Bruno ("infinitá estensiva", "infinitá intensiva'') in *De l'Infinito Universo et Mondi* II, ed. Lagarde, p. 318.

8. That is to say, the argument merely proves the impossibility of a mover which is infinite in intensity, but not of one which is infinite in the duration of its motivity.

9. That is to say, since circular motion is not by propulsion alone nor by traction alone and does not take place between two opposites, its velocity is uniform and unmitigated and can therefore be eternal. See below Prop. XIV, Part I.

10. Thus also Averroes, after drawing the distinction quoted above (n. 7) between infinite intensity and infinite duration con-

cludes that an infinite force of the former kind is impossible at all
whereas that of the latter kind is found to exist in the celestial
spheres. *Ma'amar be-'Eẓem ha-Galgal* III, (*Sermo de Substantia
Orbis*, Cap 3, p. 9va, G): "As for a force of infinite action and passion in itself, it does not exist in any body at all, be it celestial or
generable and corruptible.... But as for the existence of a
force of infinite action and passion in time, it must necessarily be
assumed to exist in the celestial spheres."

ואמנם הכח הבלתי מכולה בפועל וההפעלות עצמו, הנה לא ימצא בגשם כלל,
בין שתהיה שמימי או הוה נפסד... ואולם מציאות הכחות הבלתי מכולות בפועל
וההפעלות בזמן, הנה הוא הכרחי לגרמים השמימיים.

11. *De Caelo* I, 3, 270b, 1–4.

Intermediate De Caelo I, v–vi, (Latin, 272ra, G; p. 274vb; p.
275rb): "Summa V. To show that this celestial body is neither
heavy nor light. Summa VI. To show that it is neither generated nor corruptible, that it is susceptible to neither growth nor
diminution, nor change, nor passion, and that, in general, it is
susceptible to none of the qualities that are related to change and
passion, such as health, disease, youth, senility,"

הכלל הה', לבאר שזה הגשם איננו כבד וקל. הכלל הו' לבאר שהוא בלתי הוה
ובלתי נפסד, ולא יקבל הגדול והחסרון, ולא השנוי ולא ההתפעלות, ובכלל לא
יקבל מן האיכיות מה שהיה נמשך לשנוי וההתפעל, כמו הבריאות והחולי והבחרות
והזקנות.

12. That is to say, if to the fact that the spheres are not subject
to destruction we also add the fact that their circular motion is
natural to them and is not caused by any psychic principle, we
could still more forcibly argue that their eternal motion need
not be explained by the postulate of an internal motive force. Cf.
above Prop. VI, n. 11 (p. 535).

PROPOSITION XIII

PART I

1. The Hebrew text of the proposition is taken from Isaac ben
Nathan's translation of Altabrizi.

2. The discussion here is based upon *Physics* V, 4, 227b, 3–228a, 6, and VII, **1**, 242a, 33–242b, 8. Motion, says Aristotle, may be called *one* in three different senses:

(1) One in genus (γένει, בסוג), thus all kinds of locomotion may be called generically one, inasmuch as they all belong to the category of place. Qualitative change and spatial change are generically two.

(2) One in species (εἴδει, במין), thus all objects that are becoming white may be said to be moved with a motion that is specifically one, inasmuch as white is a species under the genus quality. The motions of whitening and blackening are specifically two.

(3) One in number (ἀριθμῷ, במספר), thus the walking of a certain man at a certain time may be called a motion that is numerically one. The walking of two men at the same time or of the same man at different times is not numerically one.

Intermediate Physics V, iv, 1–2: "Chapter I. We say that motion is described as one in three senses. It is one in genus, in species, or in number. Motion is one in genus when it takes place in one of the three categories, as e. g., in place or in quality. Such a motion in one category is called one in genus because the *terminus ad quem* in one category is one in genus. Motion is called one in species when it takes place in one species within any one of the given categories, and the reason for this is again to be found in the fact that the *terminus ad quem* of objects moved within one species is one in species, that is to say, those objects are divisible only with reference to individuals, as, e. g., objects which are moved from blackness to whiteness, for the whiteness, which is the completion of that motion, is one in species but many in individual.... Chapter II. For motion to be one in number three conditions are necessary. First, the object which is moved must be one in number, as, e. g., a certain man or a certain stone. Second, the motion by which it is moved must be one in number, as, e. g., the motion of a certain quality or in a certain place. Third, the time in which the motion takes place is also one in number,"

הפרק הראשון. ונאמר שהתנועה האחת תאמר על ג' טעמים, זה שהיא תהיה
אחת אם בסוג ואם במין ואם במספר. והתנועה האחת בסוג היא אשר תהיה בסוג
אחד מהמאמרות השלשה, כמו התנועה באנה ובאיך. ואמנם היתה התנועה אשר
במאמר אחד אחת בסוג, מצד מה שאליו התנועה במאמר אחד אחת בסוג. והתנועה

האחת ב מ י ן, היא אשר במין אחד ממיני המאמרות, והסבה בזה, שמה שאליו התנועה
בדבר האחד במין אחת במין, ר״ל שהם בלתי מתחלקים רק אל האישים, כמו
הדברים אשר יתנועעו מן השחרות אל הלובן, כי הלובן אשר הוא שלמותם האחרון
הוא אחד במין רבים באיש ... הפרק הב׳. ואמנם התנועה האחת במספר הנה היא
צריך אל ג׳ תנאים. אחד מהם, שיהיה המתנועע אחד במספר, כמדרגת האדם
והאבן. והתנאי השני, שיהיה הדבר אשר בו התנועה אחד במספר, והוא אם איכות
ואם מקום. והשלישי, ושיהיה הזמן אשר בו התנועה אחד במספר.

3. Cf. *Physics* VIII, 7, 261a, 31 ff., the purpose of which passage
is explained in the Latin translation of Averroes' Long commentary (p. 401rb, D) as follows: "Intendit in hoc sermone
declarare, quod motus successivi, qui inveniuntur in eodem moto,
qui sunt idem genere, et diversi specie, non sunt continui,"

4. Crescas fails to carry out his line of reasoning, and does not
state why the second alternative, namely, that change is timeless,
is impossible (but see below n. 5). Altabrizi, however, reasons
it out as follows:

"For change is either instantaneous or gradual. In the case
of instantaneous change, it is quite obvious that it cannot be
continuous and durable, for if only one single instantaneous
change is assumed, it undoubtedly can have no continuity and
duration, and if several instantaneous changes are assumed, one
following after the other, it is likewise impossible for them to
form a continuum, for these changes are now assumed to be each
taking place in an instant, and if the succession of such instantaneous changes could form a continuum, it would follow that the
succession of instants would likewise form a continum. But
this is absurd."

שהשנוי אם שיהיה פתאום או לא פתאום, והשנוי שיהיה פתאום אי אפשר שיתמיד
מתדבק, שאם לקח שנוי אחד, אין ספק בהעדר דבקותו והתמדתו, ואם לקח שנוים
רבים, כל אחד מהם אחר האחר, אי אפשר גם כן שיהיה מתדבק, כי כל אחד
מהם יתחדש בעתה, ולו התדבקו נמשכים חוייב שימשכו העתות, והוא שקר.

5. Hebrew ואם לא, היה הזמן מחובר מעתות, literally, "and if not, time
would be composed of instants." The passage may also be
rendered "and if change were timeless, time would be composed
of instants." Thus rendered, it would carry out the reasoning
against the second alternative. See above n. 4.

6. In the preceding passage Crescas interpreted the term מתדבק in the proposition to mean continuous in the sense of an unbroken connection of parts as opposed to discrete, διωρισμένον, and was therefore forced to maintain that the proposition could not apply to change in one species. Now, however, Crescas suggests that the term מתדבק may mean continuous in the sense of eternity and endlessness, in which case the proposition would also apply to change in one species, for no rectilinear motion, even if in one species, can be eternal.

Crescas' latter interpretation seems to be the right one. For the source of Maimonides' proposition is *Physics* VIII, 7–8, where Aristotle discusses the problem whether there is any continuous (συνεχής, 260a, 22) motion. In the course of his discussion he makes it clear that by συνεχής he means infinitely continuous.

This latter interpretation of Crescas may be further supported by the fact that the corresponding Greek term συνεχής likewise has the meaning of eternity. Thus in the following passage Aristotle uses the adverb συνεχῶς in the sense of endless and eternal continuity whereas the adjective συνεχής is used in the sense of *continuous* as opposed to *successive*. *Physics* VIII, 7, 260b, 19–21: ὥστ' ἐπεὶ κίνησιν μὲν ἀναγκαῖον εἶναι συνεχῶς, εἴη δ' ἂν συνεχῶς ἢ ἡ συνεχής ἢ ἡ ἐφεξῆς... In the Latin translation of Averroes' Long Commentary (p. 397ra, B) συνεχῶς of this passage is correctly translated by *aeternus* and συνεχής by *continuus*. "Quia igitur est necessarium ut motus sit aeternus, et non aeternus, nisi, aut quia est continuus, aut quia est successivus..."

A similar interpretation of the term "continuous" in this proposition is given also by Hillel of Verona (p. 36a): "The term 'continuous' here is to be understood in the sense of 'everlasting'." פירוש מדובק הוא בכאן כאמרו מתמיד.

7. From here to the end of the chapter, Crescas, commentary is a paraphrase of *Intermediate Physics* VIII, v, 1–4, corresponding to *Physics* VIII, 7–9.

8. The argument following is taken from Averroes' interpretation of Aristotle's argument contained in *Physics* VIII, 7, 261a, 31–261b, 22.

Intermediate Physics VIII, v, 2: "The question as to which kind of locomotion is eternal will be answered by us after we shall have first shown that none of the genera of motion can be eternally continuous except locomotion. The argument is as follows: All the other three kinds of motion must be from one opposite to another, and two opposite motions between two opposite poles cannot form a continuous motion, for a continuous motion is one motion, and opposite motions cannot be one motion. To assume that opposite motions are one motion would mean that that which is becoming white is becoming white and black at the same time and that which is generated is being generated and corrupted at the same time. Since therefore opposite motions must be two motions, there must of necessity be some interval of time between them.

In view of this, if the change is of the kind that is called motion, then indeed the object undergoing the change must of necessity come to rest between the two opposite motions. But if the change is of the kind that is not called motion, as, e. g., change from non-being to being and from being to non-being, then while there is no actual object in existence of which it can be said to come to rest, inasmuch as in this kind of change there is no actual object which bridges the entire change from beginning to end as in the other changes which constitute true motion, still, even in this kind of change, i. e., the change from non-being to being, there must be some interval of time between the two opposite changes during which interval the object is not undergoing either one of the changes, for it is absurd to assume that the generation of an object is continuous with its corruption without there being any interval of time between them.

This being evident in the case of generation, namely, that it cannot be continuous with corruption, the same must also be true with respect to the other motions, for the nature of things undergoing change is the same in every case."

ואמנם איזו העתקה היא הנצחית, הנה אנו נבאר זה אחר שנבאר תחלה שאי אפשר
שתהיה תנועה אחת מדובקת נצחית בסוג מסוגי התנועה מלבד תנועת ההעתק. וזה
שמיני השנויים השלשה אמנם יהיו מהפך אל הפך, ושני התנועות ההפכיות אשר נפלו
בשני ההפכים אי אפשר שתהיה תנועה מדובקת, לפי שהתנועה המדובקת אחת,
ואי אפשר שתהיינה התנועות ההפכיות אחת. וזה שאלו היתה אחת היה הוא אשר
תלבן יתלבן וישתחר יחד, ואשר הוא יתהוה יתהווה ויפסד יחד.

וכאשר היו התנועות ההפכיות שתים, היה ביניהם זמן בהכרח. ואם היה השנוי
מדבר היה מסוג התנועה, היה הדבר המשתנה נח בהכרח בין שתי תנועות ההפכות,
ואם היה השנוי אינו מדבר הוא מסוג התנועה, כמו השנוי אשר יהיה מהעדר
אל מציאות וממציאות אל העדר, לא יהיה בכאן דבר מתואר במנוחה, אחר שהיה
אין בזה השנוי דבר נושא בפעל התחלת השנוי עד סופו כמו מה שעליו הענין בשאר
השנויים האחרים, ר"ל אשר הם תנועות באמת, אבל יהיה בהכרח בין שני אלו
השנויים ההפכיים בזה הסוג מן השנוי, ר"ל אשר יהיה מהעדר אל מציאות, זמן לא
יהיה בו הדבר משתנה באחד משני מיני השנויים ההפכיים, כי מהמנונה שיאמר
שהוית ההוה מדובק בהפסדו מבלתי שיהיה ביניהם זמן כלל.

וכאשר היה זה מבואר מענין ההויה, ר"ל שלא תדבק בהפסד, הנה כמו כן יחוייב
שיהיה הענין בשאר התנועות, וזה שטבע המשתנים טבע אחד.

9. Cf. *Physics* V, 5, 229a, 25–27. "And every motion is denominated rather from that into which it is changed, than from
that from which it is changed. Thus that is called becoming well
which tends to health, but a becoming ill which tends to disease."

10. Corresponds to Aristotle's argument contained in *Physics*
VIII, 7, 261b, 22–24: "Again, in generation and corruption, it
may be seem to be perfectly absurd, if it is necessary that what
is generated should immediately be corrupted, and not remain
at rest for any time."

Intermediate Physics VIII, v, 2: "That is to say, between
non-being and being there must be a certain time during which
the object suffers neither of the two contrary changes, for it is
an absurdity to affirm that the generation and corruption of a
generable object form one continuous change, without there
being any interval of time between them."

ר"ל אשר יהיה מהעדר אל מציאות זמן לא יהיה בו הדבר משתנה באחד משני
מיני השנויים ההפכיים, כי מהמנונה שיאמר שהוית ההוה מדובק בהפסדו מבלתי
שיהיה ביניהם זמן.

11. Corresponds to the next class of Aristotle's arguments in
Physics VIII, 8, 261b, 27–263a, 3, intended to prove that locomotion in a right line cannot be infinitely continuous.

12. Cf. *Physics* VIII, 8, 261b, 28–29: "For every thing which is
locally moved, is either moved in a circle, or in a right line, or
that which is mixed of both of these," Also *ibid.* VIII, 9, 265a,
14–15 and *De Caelo* I, 2, 268b, 17–18.

Intermediate Physics VIII, v, 3: "For every motion in place
must be either rectilinear or circular or composed of both of
these. And as it will be shown that the first of these two simple
motions, namely, the rectilinear, cannot go on continually, it
will become clear that that which is composed of both of these
motions cannot go on continually, for that which cannot be
continual when simple cannot be so also when combined with
something else."

זה שכל תנועת העתק אם שתהיה ישרה ואם סבובית ואם מורכבת משניהם,
וכאשר יתבאר באחת משתי אלו התנועות הפשוטות שאי אפשר שתהיה מדובקת,
שהיא הישרה, גלוי שהמורכבת משניהם בלתי מדובקת, לפי שאשר בפשוטה מהעדרי
ההתדבקות ימנע שימצא במורכב.

13. Corresponds to Aristotle's argument "that a thing which is
locally moved in a finite right line, cannot be moved continually",
contained in *Physics* VIII, 8, 261b, 31–262a, 17. Aristotle
characterizes these arguments as being supported by sense perception (ἐπὶ τῆς αἰσθήσεως, *ibid.* 262a, 18).

Intermediate Physics VIII, v, 3: "That rectilinear motion
cannot be continual, that is to say, that one and the same object
that is locally moved, step after step, over a certain distance,
could not continue to be so moved without ever having to come
to a stop, can be demonstrated in several ways."

ואמנם שהתנועה הישרה אי אפשר שתהיה מדובקת כאשר היה הנעתק האחד
בעצמו יתנועע על הגודל האחד בעצמו, פעם אחר פעם, מבלי שיחדל מן התנועה,
זה יראה מפנים.

14. Corresponds to Aristotle's argument from reason (ἐπὶ τοῦ
λόγου) contained in *Physics* VIII, 8, 262a, 19–262b, 28.

The text here is an abridgment of the following passage in
Intermediate Physics v, VIII, 3:

"In every finite continuum there are three things, a beginning,
an end and a middle. The middle is one in subject but two in
definition (במאמר, λόγῳ), that is to say, it is the end of one of the
two parts into which it divides the continuum and the beginning
of the other, for the middle exists in a continuum in a twofold
respect: first, potentially, and, second, actually. It is evident that
when anything is moved with a finite continuous motion over a
finite magnitude, in so far as it is moved and continues its

motion uninterruptedly, it does not register an actual point in the
middle of the continuum. It is only when the moving object stops
and thereby divides the continuous magnitude over which it
moves into two halves that it registers an actual point on the
latter, which is at once both a beginning and an end, i. e., the end
of the prior part of the motion and of the prior part of the distance, and the beginning of the posterior part of the motion and
of the posterior part of the distance... To illustrate: Let A move
over the continuum BC with a continuous motion. I say that A
will not register an actual point, say point D, on BC unless A
stops somewhere between B and C. B_____D_____C If A does
not stop at D, there can be no actual point in the interval between
B and C, unless we assume that a line is composed of points. . . .

Inasmuch as it clear that when the moving object does stop,
it does register an actual point, I maintain that the contrary must
be equally true, namely, that when the moving object registers
an actual point, it must be inferred that it has come to a stop.
Assuming, for instance, that A in its motion over magnitude
B_____D_____C has registered an actual point D so that
it marks the end of motion BD and the beginning of motion
DC, I maintain that A must have come to a stop at D. For its
being at D is not the same as its being beyond D, and these two
points at which the moving object successively is, i. e., the actual
point D and a point beyond D, mark the end of two contrary
motions, [one toward D, and the other away from D]. Inasmuch
as the moving object must have performed two opposite motions,
when at first it moved toward D and then it moved away from
D, these two opposite tendencies could not have existed in it in
actuality except in two different instants, for only by way of
potentiality could they have existed in it in one instant. And
since these two tendencies imply two instants, there must necessarily have been some interval of time between them. . . .

As it has thus been established that when a moving object
registers an actual point it must have come to a stop, and as it is
further evident that a moving object, when it returns over the
same distance, registers on its return an actual point which is
the end of the prior motion and the beginning of the posterior
contrary motion, for were it not so, the two contrary motions
would be one, it follows that these two motions, redoubled over

the same distance, are not continuous, inasmuch as there must
have been some rest between them, and every rest is in time.
This is one of the proofs by which is established that the motion
of that which returns is not continuous, inasmuch as an interval
of rest must interrupt the two motions,"

זה שלמה שהיה שימצא בכל מתדבק בעל תכלית שלשה דברים, התחלה ותכלית
ואמצע, והאמצע אחד בנושא שנים במאמר, זה שהוא תכלית לאחד משני חלקי
המתדבק והתחלה לשני. זה האמצע נמצא במתדבק על שני צדדים, אחד מהם
בכח והשני בפעל. זה מבואר מענין המתנועע תנועה ב"ת מדובקת על גודל ב"ת
שלא יחדש המתנועע עליו נקודה הוא האמצע בפועל מצד שהוא מתנועע, זה מה
שהתמיד מתנועע על הדבקו, אלא כאשר עמד וחלק הגודל בשני חציים, כי באותו
העת יחדש נקודה על גודל בפועל הזה התחלה ותכלית, אם תכלית לנקודה
הראשונה והמהלך הראשון ואם התחלה לתנועה השנית והמהלך השני... משל זה
שהמתנועע א' יתנועע על גודל ב"ג תנועה מדובקת, הנה אומר שלא תחדש מתנועע
א' על גודל ב"ג נקודה בפועל בין ב' וג' כמו נקודה ה' אלא אם יעמוד מתנועע
א' בין ב' וג'. _____ ה _____ ג. ואמנם בשלא יעמוד בה', אין שם נקודה ב
בפועל, אם לא יהיה הקו מרכב מנקודות...

וכאשר נלוי מענין המתנועע כי כאשר יעמד יחדש נקודה בפועל, אומר שהפוך
זה גם כן מחוייב, והוא שהמתנועע, כשיחדש נקודה בפועל על המהלך, כבר נח.
וניח מתנועע א' כבר חדש בעת תנועתו על גודל ב"ג נקודה ה' בפועל, עד שתהיה
התכלית לתנועת ב"ה והתחלה לתנועת ה"ג, הנה אומר כי א' כבר נח בה' בהכרח.
זה שמציאותו בה' בלתי מציאותו נבדל מה', והם מציאויות הפכים למתנועע אשר
הוא א', ר"ל היותו בה' בפועל ונבדל מה'. וכאשר היה המתנועע היותו בה' והיותו
נבדל מה' שתי תנועות מתחלפות, אי אפשר שתהיינה נמצאות לו שתי אלו התכונות
בפעל רק בשתי עתות מתחלפות, לא בעת אחת, אלא אם היו שתי אלו התכונות
למתנועע נמצאות בכח, וכאשר היה בשתי עתות, לכל שתי עתות הנה ביניהם זמן...

וכאשר התישב שהמתנועע כשיחדש נקודה בפועל הנה הוא כבר נח, והיה מבואר
מהמתנועע החוזר על המהלך האחד בעצמו שהוא יחדש בחזירתו נקודה בפועל
היא תכלית התנועה הראשונה והתחלה לשנית ההפכית לה, ואם לא, היו שתי התנועות
ההפכיות אחת, הוא מבואר ששתי אלו התנועות, ר"ל הנכפלות על גודל אחת,
אינן מדובקות, אחר שהיתה מפסקת ביניהם מנוחה וכל מנוחה הוא בזמן. הנה זה
אחד מהבאורים יראה מהם שהמתנועע החוזר אין תנועתו מדובקת, אחר שיפסיק
בין זמני התנועות זמן מנוחה.

15. Cf. *Physics* VIII, 8, 261a, 28–31: "The like also takes place
in a circle...Hence if neither of these motions is continuous,
neither can that be continuous which is composed from both
of them."

16. Hebrew חלזוני, حلزونى, ἑλικοειδής, i.e., spiral-shaped, the name given to a line composed of straight and circular lines. See T. L. Heath, *The Thirteen Books of Euclid's Elements*, Vol. I, pp. 159–160, on the classification of lines. The term ἕλικος occurs also in *Physics* V, 4, 228b, 24, as a description of motion in a spiral line.

17. Corresponds to Aristotle's conclusion contained in *Physics* VIII, 8, 265a, 7–9: "But the arguments now employed universally show of all motion that it is not possible to be continually moved with any motion except that which is circular."

18. That is to say, every given point in circular motion is at once the *terminus a quo* and the *terminus ad quem* of the motion. Cf. *Physics* VIII, 8, 264b, 18–19: "For motion in a circle is from the same to the same, but the motion through a right line is from the same to another."

Intermediate Physics VIII, v, 4: "For that which is moved circularly is moved from and toward the very same thing, so that the *terminus a quo* and the *terminus ad quem* are the same, for in circular motion there are no opposite limits."

שהמתנועע בסבוב אמנם יתנועע מהדבר מה שממנו ומה שאליו אחד בעצמו אל אותו דבר בעצמו, חה שאין שם שני קצות מקבילות.

19. Cf. *Physics* VIII, 8, 265a, 10–12: "Thus much, therefore, has been said to prove that there is neither any infinite mutation, nor any infinite motion, except that which is in a circle."

Intermediate Physics VIII, v, 4: "That circular motion can be continual and perpetual and that it is prior in nature to rectilinear motion, we shall prove as follows."

ואמנם שהתנועה הסבובית אפשר שתהיה מדובקת ותמידית ושהיא קודמת בטבע אל ההעתקה הישרה אנחנו נאמר בזה.

PART II

20. This is a refutation of the first argument, viz., that between two specifically different changes, like whitening and blackening, there must be an instant of actual rest. Crescas' line of reasoning may be restated as follows: There is no instant of rest between the opposite changes of whitening and blackening. The time in which both these opposite motions take place is one and continuous, the instant in which the change from whitening to blackening takes place being the end of the past and the beginning of the future time. But while that instant, in so far as it pertains to the time of the change, is common to both the past and the future, still in so far as it pertains to the object undergoing the change from whitening to blackening it belongs only to the *terminus ad quem*, namely, blackening. Thus the object would not be whitening and blackening at the same time. For let ABC be the time, and D the object undergoing the change. Let D be whitening in A and blackening in C. B will then be the *now*, which has no extension, and will be at once the end of past time A and the beginning of the future time C. Still it must not necessarily follow that in B both whitening and blackening would take place at once, for in this respect B belongs to the posterior change, marking only the beginning of the blackening process.

The force of Crescas' argument is primarily due to the fact that Aristotle himself makes the same distinction in the case of a single continuous motion. Take for instance the motion from black to white. It is a single motion and is admitted by Aristotle to be continuous. Now, let ABC be the time and D the object undergoing the change. Again, let D be black in A and white in C. Now, since B, the *now*, is common to both past time A and future time C, would not the object in the instant B be both black and white at once? But Aristotle solves the difficulty in the manner we have just described, namely, that with reference to the object in change the instant B belongs to the posterior only. To quote Aristotle's own words: "It is also evident that unless the point of time by which prior and posterior are divided, is always attributed to the posterior, the thing itself being considered, the same thing will be at the same time being and non-being, and when it will be in generation, or becoming to be, will not be in generation. The point, therefore, is common to both the prior and the posterior, and is one and the same in number, but is not the same in definition; for it is the beginning of the one and the end of the other. But so far as pertains to the thing it is always of the posterior passive quality.'' (*Physics* VIII, 8, 263b, 9–15).

Intermediate Physics VIII, v, 3: "If we assume that the instant, which is the end of the existence of a thing and the beginning of its non-existence, is at once a part of the actual existence of the thing and of its actual non-existence, . . . then a thing will be existent and non-existent in one and the same instant. Take, for example, the case of Socrates who was alive during a certain past time and dead during a certain future time. If we assume that he was alive at the end of the past time and dead at the beginning of the future time, then, inasmuch as the end of the past time and the beginning of the future time is one in subject and is indivisible, . . . it will follow that Socrates will have been at once alive and dead in one and the same instant. Hence it must be inferred that an instant has nothing actual about it but that it is only a dividing point between opposite kinds of existence, just as it is only a dividing point between the past and the future, but when viewed with respect to the past it is more properly to be regarded as the end of the past rather than as the beginning of the future, and when viewed with respect to the future it is more properly to be regarded as the beginning of the future."

ואלו הנחנו שהעתה חלק ממציאות הנמצא בפועל, מצד מה שהוא תכלית
למציאות, והיה בו גם כן חלק מהעדרו בפועל, מצד שהוא גם כן התחלה להעדרו...
היה הדבר נמצא ונעדר יחד בעתה אחד. ומשל זה, שסוקראט למה שהיה נמצא חי
בכל זמן העובר ומת בהתחלת הזמן העתיד, והיה הדבר אשר הוא תכלית הזמן
העובר והתחלת העתיד אחד בנושא ובלתי מתחלק ... הוא מבואר שיתחייב מזה
שיהיה סוקראט חי ומת בעתה אחד יחד. ולזה מה שיחויב שלא יהיה בעתה דבר
בפועל אבל הוא מבדיל בין המציאות ההפכים כמו שהוא מבדיל בין העובר והעתיד,
אלא שכאשר הוקש בזמן העובר היה יותר ראוי שיהיה תכלית לעובר מאשר יהיה
התחלה לעתיד, וכאשר הוקש בזמן המתחדש היה יותר ראוי שיהיה התחלה לעתיד.

And so Crescas seems to argue that since Aristotle draws that distinction in a single motion, why not apply it also to opposite motions and prove thereby their continuity?

Crescas' argument against the proposition is reproduced by Pico Della Mirandola in *Examen Doctrinae Vanitatis Gentium* VI, 2: "Non recipitur et illud, solum motum orbicularem esse continuum, atque rationes Aristotelis quibus id probare sategerat fabulas appellat Hasdai, et nigrum cun movetur ad albedinem, licet non quiescat in ea, sed denigretur, non tamen sequitur propterea ut dealbetur simul et denigreatur, sed ratione diversa, hoc est, quatenus dealbatur potest id asseri, et quatenus denigratur hoc etiam potest affirmari: nec absurdum est ullum,''

21. Cf. above Prop. VII, p. 243, n. 8.

22. This is the refutation of the second argument, viz., that between two opposite rectilinear motions, like upward and downward, there must be an instant or rest. A similar refutation of the argument, containing a similar illustration of two objects, one rising and the other falling, may be found in Joannes Versor's *Quaestiones Physicarum, Liber* VIII, *Quaestio* XI.

"Question XI. Whether that which returns in its motion must come to rest at the point of its returning.

It would seem that it is not so. For if a small pebble is thrown upward, while a stone of the size of a millstone is coming downward in the opposite direction, the pebble will have to return downward without having first come to rest at all, for, were it not so, the millstone will have to come to rest too, but that is impossible.

Second, if we assume that the pebble which was thrown upward had come to rest prior to its beginning to come down, it will follow that a heavy object will remain at rest in a place above without anything supporting it, but that is impossible,"

השאלה הי"א, אם כל מה שהוא חחר בתנועתו הוא נח בנקודה בחזרה, ויראה
שאיננו כן. חה כי אם נשליך אבן קטן למעלה, ויהיה יורד למטה אבן
גדולה כריחיים, הנה תשוב האבן הקטנה למטה מבלי כל מנוחה כלל, ואם לא
כן, יחוייב שינוחו הריחיים, חה נמנע, א"כ, וכו'.

שנית, אם היתה האבן הנשלכת למעלה הונח קודם שתרד הנה, הנה יתחייב
שיהיה הדבר הכבד נח במקום במעלה מבלי יפסיק כלל, אשר זה נמנע וכו'.

This argument of Crescas is also reproduced by Pico Della Mirandola: "Illud quoque falsum inter duos contrarios motus necessario quietem intercedere, alioqui sequeretur ut pondus ingens, ut mons altissimus, super re levissima ascendere procumbens, sisteret motum et quietis interponeret morulam, et ipso in aere conquiesceret," (*Examen Doctrinae Vanitatis Gentium* VI, 2).

A similar argument by Descartes, *Oeuvres*, ed. Cousin, IX, pp. 71, 77, is referred to by Julius Guttmann in his "Chasdai Creskas als Kritiker der aristotelischen Physik," *Festschrift zum siebzigsten Geburtstage Jakob Guttmanns*, p. 43, n. 1.

23. The argument contained in this passage may be interpreted as follows:

In Prop. XIV, Maimonides states that generation and corruption are always preceded by a change in quality. As we shall see later (Prop. XIV, n. 1 p. 628) by the terms generation and corruption Maimonides means relative generation and corruption, i.e., the substantial change undergone by an actually existent object in passing from one form to another. That concomitant qualitative change, which must always precede a relative substantial change, must not necessarily be in opposite directions. It may as well be in one direction. Thus when water changes from cold to hot, with reference to *coldness-and-heat*, it is one continuous qualitative change in one direction, but with reference to *cold-water* and *hot-water*, it is a relative substantial change, the corruption of *cold-water* and the generation of *hot-water* (cf. Prop. IV, n. 8, p. 513). Now, Crescas seems to argue, if you say that between the corruption of *cold-water* and the generation of *hot-water*, or, as he suggests to call it, the end of one generation and the beginning of another generation, there must be an actual instant of rest, you will also have to assume the existence of an actual instant of rest in the concomitant continuous qualitative changes from *coldness* to *heat*. But this is absurd. Hence, Crescas would expect us to conclude, that there is no actual instant of rest between generation and corruption.

PROPOSITION XIV

PART I

1. The Hebrew text of the proposition follows Isaac ben Nathan's translation of Altabrizi.

The proposition is based upon the following passage in *Physics* VIII, 7, 260a, 26–260b, 5: "But since there are three motions, one according to magnitude, another according to passive quality, and another according to place, which we call lation, it is necessary that lation should be the first; since it is impossible there should be increase unless alteration had a prior subsistence... If also a thing is changed in quality, it is necessary there should be that which produces the change in quality... It is evident, therefore, that the thing which moves does not subsist similarly but at one time is nearer and at another time more remote from that which is changed in quality. But this cannot subsist without lation.''

It will have been noticed, however, that, unlike Maimonides, Aristotle makes no mention of the priority of locomotion and qualitative change to generation and corruption. He only speaks of the priority of locomotion to qualitative and quantitative change.

The discrepancy between Maimonides and Aristotle has been pointed out by Shem-ṭob in his commentary on the *Moreh*. Munk, in an attempt to justify Maimonides, takes the term "alteration", השתנות, in this proposition not in its usual sense of qualitative change (see Prop. IV, n. 3, p. 500) but in the sense of substantial change or generation (cf, *Guide* II, p. 14, n. 2). From Crescas' discussion of this proposition, however, where he uses the expression "motions of quality", תנועות האיך (p. 282) for Maimonides' "alteration", השתנות, it is clear that he understood the latter term in its usual sense. In this sense it is also taken by Narboni and Hillel of Verona.

It seems, therefore, that the term "alteration" is to be taken in its usual sense. Still it is possible to remove the discrepancy between Maimonides and Aristotle by taking the expression "generation and corruption" in the proposition to refer to relative generation and corruption, i. e., to the generation and corruption which marks the substantial change from one subject to another (see Prop. IV, n. 8, p. 513) This kind of generation and corruption is always concomitant with the other three changes and is preceded by alteration (see Prop. IV, n. 14, p. 519). In Crescas himself we have a definite statement, apropos of something else, that by "generation and corruption" in this proposition is meant "relative generation", הויה נמשכת (p. 582, l. 8). In the same sense the expression seems to have been understood by Narboni and Hillel of Verona.

2. Hebrew בחפוש. The same term is used by Narboni: זה מבואר בחפוש. Averroes uses in this connection the term בחקירה (see quotation below in n. 3). The characterization of the proof as "inductive" is based upon the following statement in *Physics* VIII, 7, 261a, 27–28: "That lation, therefore, is first of motion, is from these things evident (φανερὸν ἐκ τούτων)''.

3. Cf. *Physics* VIII, 7, 260b, 16–19: "For that which is first, as in other things, may be predicated multifariously: for that is said to be prior, without which other things will not be, but which can itself exist without others (i.e., what he calls later priority *in nature*, φύσει, cf. below n. 4); that also is said to be prior, which is first in time (χρόνῳ), and that which is first in essence (κατ' οὐσίαν)." He then proceeds to show that locomotion is prior to all the other motions in all the senses enumerated.

Intermediate Physics VIII, v, 4: "That it must be the first of all the kinds of translation and that it must be prior to them in nature and in time may be shown in several ways." ואמנם שיחוייב שתהיה ראשונה למיני העתק ושהיא קודמת עליהן בטבע ובזמן יראה גם כן מפנים. Again: "For when the other motions exist, this one must exist, whereas when this motion exists the other motions must not necessarily exist. This is the definition of prior in nature, as has been explained in its proper place. But that it must exist when other motions exist, can be demonstrated by *induction*." לפי שכאשר נמצא שאר התנועות חוייב שתמצא היא, וכאשר נמצאת היא לא יחוייב שתמצאנה שאר התנועות, חהו גדר הקודם בטבע כפי מה שנגדר במקומו. ואמנם שתמצא בהמצא שאר התנועות מבואר עניינה בחקירה.

Crescas seems to intimate here that in the proposition the term קודמת, Arabic אקדם, refers to "priority in time" whereas the term ראשונה מהם, Arabic אולאהא, as explained by Maimonides himself, means "priority in nature."

4. Cf. *Physics* VIII, 9, 265a, 16–23: "And the motion in a circle is prior to that which is in a right line because it is simple and more perfect... The perfect is prior by nature (φύσει), by reason (λόγῳ, i.e., κατ' οὐσίαν, cf. above n. 3), and by time (χρόνῳ) to the imperfect."

5. Cf. *Physics* VIII, 9, 265a, 27–32: "But it happens reasonably, that the motion in a circle is one and continued, and not that which is in a right line: for of the motion which is in a right line, the beginning, middle, and end are bounded, and it contains all these in itself; so that there is *whence* that which is moved began, and *where* it will end; for everything rests in boundaries, either from *whence* or *whither* it is moved; but these in circular motion are indefinite."

6. Hebrew ולא ישתנה שני, literally, "and no change occurs to it."
But I take it to refer to the uniformity of the velocity of the
circular motion of the spheres rather than to the unchangeability
and incorruptibility of their substance (see Prop. XII, Part II,
n. 11, p. 614), thus reflecting the statements contained in the
following passages:

Physics VIII, 9, 265b, 11–14: "Further still, the motion alone
in a circle can be equable (ὁμαλῆ); for things which are moved
in a right line, . . . by how much farther they are distant from that
which is at rest, are moved by so much the swifter.''

Intermediate Physics VIII, v, 4: "Furthermore, circular motion
can be equable. . . for the rectilinear natural motions undergo
variation with reference to swiftness and slowness.

ועוד כי התנועה הסבובית אפשר בה שתהיה שוה... מה שהתנועות הישרות
הטבעיות יכנס בהם החלוף במהירות ואיחור.

Altabrizi: "Circular motion is always of the same order, and
no variation occurs to it as it does to rectilinear motion, for the
latter, when natural, becomes stronger in the end, and, when
violent, becomes stronger in the middle and weaker at the end,
thus proving that rectlinear motion suffers variation.''

ותהיה תמיד על סדר אחד, ולא ידבק לה החלוף כמו שידבק לתנועה הישרה,
כי היא, אם היתה בטבע, הנה היא תתחזק באחרית, ואם היתה בהכרח, הנה היא
תתחזק באמצע ותחלש בסוף, ותהיה התנועה הישרה מתחלפת.

7. That is to say, the celestial sphere.

8. Hebrew אבל ענינו דומה אל הפעל הגמור. The term פעל may be
taken here either as a noun, meaning *actuality*, or as a participle,
meaning *agent*.

In the former sense, which I have adopted in the translation of
the text, it occurs in the *Moreh ha-Moreh*: "Locomotion may be
like perfect actuality in which there is no admixture of potentiality. An instance of such locomotion is to be found in the case
of the spheres.'' ודומה לפעל הגמור שלא יתערב בו כח, כמו שהוא הענין
בגלגלים. Similarly also Altabrizi: "This kind of motion, i. e.,
the circular, is the most important of all the motions for another reason, for it occurs to its subject in a manner implying a
perfection in its essence.'' זאת התנועה, ר"ל הסבובית, יותר נכבדת משאר
התנועות גם כן, כי היא אמנם תקרה לנושאה אחר שלמות עצמו. All these state-

ments about the actuality and perfection of circular motion reflect the following statement in *Physics* VIII, 9, 265a, 16–17: "And the motion in a circle is prior to that which is in a right line, for it is simple and more perfect."

If the term פעל is taken here in the other sense, the passage should be translated as follows: "but that in everything it is like the Perfect Agent [from which it proceeds]." It would thus reflect the following statement of Altabrizi: "But as for circular motion, it does not undergo any change at all, proceeding, as it does, from the action of a single force." ואולם הסבוביית הנה לא
.תתחלף כלל, כאשר סודרה מכח אחד

PART II

9. Hebrew הויה נמשכת. The term נמשך occurs as a translation of two Greek words: (1) ἀκόλουθος, *consequent upon* or *incident to* (see Prop. IV, n. 2, p. 497). (2) ἐφεξῆς, *successive* (see Prop. I. Part I, n. 113, p. 376). The two meanings of this word are so much alike that it is hard to tell in which sense it is used in any particular place. It is of greater importance always to discover what the term means to emphasize.

Here the emphasis is upon the fact that the generation is *consequent upon* something or *successive to* something in the sense of its being *preceded by* something as opposed to generation out of nothing.

In the following passage of *Or Adonai* I, ii, 20, the emphasis is upon *the succession of one thing after the disappearance of another*. "It is possible that the spheres are generated and destroyed in succession." חה שכבר אפשר שיהיה הוה ונפסד בהמשכות.

In Altabrizi (Prop. VI) it is used in the sense of a *necessary consequence* of a cause as opposed to an act of volition and choice. "But if the cause of that motion is something within the body, the latter is said to be moved of itself. But this is subdivided into two parts. If the motion proceeds from the cause by design and choice, it is called voluntary motion; if without design and choice, it is called sequential motion."
ואולם אם היתה סבת אותה התנועה דבר בנפש אותו הגשם, הנה יאמר לו שהוא
מתנועע בעצמות. והוא אם שתהיה מסודרת ממנו בכונה ובחירה, והיא התנועה
הרצונית, או מבלתי כונה ובחירה, והיא התנועה הנמשכת.

10. Cf. *Or Adonai* III, i.

11. The point of Crescas' comment is this: If we assume the world as a whole to be eternal, there being no first generation, it is true that with reference to each generated being within the ungenerated world, arising as they all do from one another (הויה נמשכת), locomotion must be the first of all motions. But if we assume the world to be generated, having been created in time, then the act of generation will have to be the first motion.

This comment of Crescas is based upon a passage of Aristotle, in which, after having stated that locomotion is the first of all motions, he proceeds to show that that statement does not hold true unless the world is assumed to be ungenerated. Cf. *Physics* VIII, 7, 260b, 30-261a, 10: "In each of these things which have generation, however, it is necessary that lation should be the last motion. For after a thing is generated, it is first necessary that there should be change in quality and increase; but lation is the motion of things which are now perfect. But it is necessary that something else should be prior, which is moved according to lation, and which is also the cause of generation to generated natures, not being generation itself; as that which generates is prior to that which is generated. But generation may seem to be the first of motions, because it is necessary that a thing should first be generated. This indeed takes place in each of the things which are generated; but it is necessary that something else should be moved prior to things which are generated, itself subsisting without being generated; and it is necessary that there should be something else prior to this. But since it is impossible that generation should be first (for if it were the case, everything that is moved should be corruptible), it is evident that no one of the successive motions can be prior."

12. For the common underlying shapeless matter first receives its four distinct specific forms, namely, the forms of the four elements, in consequence of which it is moved in space either upward or downward. See *De Caelo* IV, 3, 310b, 33–34: "A token of which is this, that locomotion belongs to things that are entire and complete, and is last in generation of motions." Cf. quotation from the *Physics* above in n. 11.

Gersonides' commentary on Interm. *De Caelo* I, vi: "We say . . .
that the first matter receives first the first qualities, i.e., heat, cold,
moisture, dryness, and these are related to it as form, and it is
for this reason that these qualities are called the forms of the
elements, as will be shown in *De Generatione et Corruptione*."
ונאמר... שהחומר הראשון יקבל ראשונה האיכיות הראשונות, והם החום והקור
והלחות והיובש, והם יהיו ממנו במדרגת הצורה. ולזה היו אלה האיכיות צורות
היסודות כמו שיתבאר בספר ההויה.

13. Hebrew הכמה בשלוח. By this is obviously meant the 'corporeal form' which is called by Plotinus and the Iḥwan al–Safa
simply 'quantity' (cf. Prop. X, notes 16, 18). The expression is
the exact equivalent of ποσὸν καθόλου *quantum-in-general* (*De
Generatione et Corruptione* I, 5, 322a, 16).

PROPOSITION XV

Part I

1. The Hebrew text of the proposition follows Ibn Tibbon's
translation of the *Moreh* except for the expression נמשך לתנועה in
which it follows Isaac ben Nathan's translation of Altabrizi. Ibn
Tibbon has נמשך אחר התנועה.

2. Crescas' analysis of the proposition is based upon Altabrizi
and Averroes, though it does not follow them throughout (see
below n. 5). Altabrizi says here: "Know that this proposition
contains three problems, ודע שהקדמה הזאת מקפת על שלש חקירות.
Averroes gives the following outline of Aristotle's discussion of
time. *Intermediate Physics* IV, iii; "The purpose of this summa
is to discuss the essence of time and the instant; the kind of
existence that time has; and if time belongs to those things which
exist in a subject, what its subject is, and in what way does it
exist in that subject."
זה הכלל כונתו במהות הזמן והעתה; ואיזה מציאות מציאותו; ואם היה ממה
שיאמר בנושא, מה הנושא לו, ואיך מציאותו בנושא.

It will have been noticed that in place of Crescas' הקדמות,
Altabrizi uses חקירות (Anonymous translation בקשות and also חפוש).
See Prop. VII, Part I, n. 2 (p. 540).

3. Altabrizi: "First, to prove what time is," אחד מהם, בביאור מהות הזמן.

4. Altabrizi: "Second, to prove that time and motion are joined together in such a manner that they can in no way be separated from each other." והשנית, בביאור היות הזמן עם התנועה דבקים לא יפרד אחד מהם מן האחר כלל.

5. This is not found in Altabrizi. Crescas, however, has made a special topic of it in order to use it later as his main point of attack on Aristotle's definition of time. His own definition, as will be shown subsequently (below n. 23), divorces the idea of time from motion.

6. Altabrizi: "Third, to prove that that which is immovable does not come under time." והשלישית, בביאור שאשר לא יתנועע לא יפול תחת הזמן.

7. Before giving his own definition of time, Aristotle says: "In the first place, then, it will be well to doubt concerning it, through exoteric reasons, whether it ranks among things or among non-entities; and in the next place to consider what its nature is" (*Physics* IV, 10, 217b, 31–32). Proving first that time has existence, Aristotle then summarizes the views of the ancients with regard to time: "For some say that it is the motion of the universe; but others that it is the sphere itself..... But the sphere of the universe seemed to those who made that assertion to be time, because all things are in time and in the sphere of the universe" (*ibid*. 218a, 33–218b, 7).

Intermediate Physics IV, iii, 1 and 3: "Wherein we shall mention the doubts raised by the dialecticians as to the existence of time..... The views held by the ancients with regard to time are two... First, the view of him who believes that time is the motion of the universe, i.e., the rotation of the whole heaven. Second, the view of him who believes that we are all in time and that all things are in the sphere."

בשמכיר הספקות אשר היו מספקים בם הנצוחיים במציאות הזמן... והדעות אשר היו לקדמונים בזמן שתי דעות.. אחד, דעת מי שראה שהזמן הוא תנועת הכל, ר"ל סבוב כל השמים, והשני, דעת מי שיראה שכלנו בזמן והדברים כלם בזמן.

Simplicious in his comment on this passage says that the first view mentioned by Aristotle is that which "Eudemus, Theophrastus, Alexander, conceived to be the opinion of Plato.''Simplicius himself, however, denies that Plato identified time with motion, and argues that Plato, like Aristotle, held time to be only the measure of motion. As to the second view mentioned by Aristotle, he says that it is that of "the Pythagoreans, who perhaps derived it from the assertion of Archytas who said that the universal time is the interval of the nature of the universe.'' (Cf. Simplicius in *Physica*, ed. Diels, p. 700, ll. 16–22, and Taylor's translation of the *Physics*, p, 242, n. 4).

These two ancient views mentioned by Aristotle, supplemented by Aristotle's own view, form the basis of Plotinus' threefold classification of the various theories of time. *Enneads* III, vii, 6: "For time may be said to be either (a) motion, or (b) that which is moved, or (c) something pertaining to motion.'' He then continues: "Of those, however, who say that time is motion, some indeed assert that it is every motion; but others, that it is the motion of the universe. But those who say it is that which is moved, assert it to be the sphere of the universe. But those who say that it is something pertaining to motion consider it either as extension of motion, or as its measure, or as some consequence of motion in general or of regulated motion.''

The classification of the various views on time given by the Iḫwan al-Safa (cf. Dieterici, *Die Naturanschauung und Naturphilosophie der Araber*, pp. 14–16; Arabic text, *Die Abhandlungen der Ichwân Es-Safâ*, p. 35) is evidently based upon the discussions of Aristotle and Plotinus. They enumerate four views. First, the popular view that time is the passage of years, months, days, and hours. Second, the view which we have already met with in Aristotle and Plotinus, that time is the number of the motion of the celestial sphere. Third, a view which we shall discuss subsequently and show that it can be traced to Plotinus' own view (see below n. 23). Fourth, the view discussed by Aristotle (see above n. 7) that time does not belong to the realm of existing things.

In Altabrizi three views are mentioned in addition to that of Aristotle: "We say that the ancients differed as to the essence of time according to four views. First, that time exists in itself, is neither a body nor anything belonging to body, but is something which has necessary existence in virtue of itself. Second, that it is the body that encompasses all the bodies of the universe, namely, the celestial equator. Third, that it is the motion of the celestial equator."

ונאמר חלקו הראשונים במהות הזמן על ארבע דעות: אחת מהן שהוא נמצא עומד בעצמו, בלתי גשם ולא גשמי, והוא מחוייב המציאות לעצמותו. והשנית, שהוא גשם המקיף בכל גשמי העולם והוא גלגל משוה היום. והשלישי, שהוא תנועת משוה היום.

(גלגל משוה היום, دائِرة مَعدل النَّهار, ἰσημερινὸς κύκλος, *equidiurnal circle, equator*).

Here, again, the second and third views are those reported by Aristotle and Plotinus, whereas the first view we shall show to reflect Plotinus' own conception of time (see below n. 23).

8. Hebrew ההפסד מבוארי להיותם. Reflects the following statement in *Intermediate Physics* IV, iii, 3: "Whence has been demonstrated the untenability of what the ancients have said concerning the essence of time." הנה התבאר מזה הפסד מה שאמרוהו הקדמונים בעצם הזמן.

9. Hebrew שהוא מספר הקודם והמתאחר בתנועה. This is rather an imperfect reproduction of Aristotle's definition of time in *Physics* IV, 11, 219b, 1–2: "For time is this, the number of motion according to prior and posterior." τοῦτο γάρ ἐστιν ὁ χρόνος, ἀριθμὸς κινήσεως κατὰ τὸ πρότερον καὶ ὕστερον. Crescas' version of the definition, however, is found in the following places:

Averroes' *Epitome of the Physics* IV, p. 18a: הנה הזמן הוא בהכרח ספור הקודם והמתאחר הנמצא בתנועה.

Narboni on *Moreh* I, 73, Prop. III: כי הזמן הוא ספור הקודם והמתאחר מהתנועה.

An accurate translation of Aristotle's definition is given by Maimonides himself in his letter to Samuel ibn Tibbon. *Kobez Teshubot ha-Rambam we-Iggerotaw* II, p. 27b: "Time is the measure of motion according to prior and posterior *in motion*." והזמן הוא שעור התנועה בקודם ומתאחר בתנועה.

A somewhat freer, but still accurate, rendering of this definition occurs in *Moreh* I, 52: "For time is an accident joined to motion, when the latter is viewed with reference to priority and posteriority and is numbered accordingly." כי הזמן מקרה דבק לתנועה כשיביטו בה ענין הקדימה והאיחור ותהיה נספרת.

It will have been noticed that in Maimonides' two renderings of Aristotle's definition one uses the term "measure" while the other uses the term "number." This point will be discussed below in n. 24.

It will also have been noticed that in the first of these renderings, which was evidently meant to be an accurate translation of Aristotle, the expression "according to prior and posterior" is qualified by the phrase ''in motion.'' Similar qualifying phrases occur in the following translation of the definition.

Intermediate Physics IV, iii, 1: "It is evident that the definition of time agreed upon is that it is the number of motion according to prior and posterior *in its parts*.'' הוא מבואר שגדר הזמן המוסכם עליו הוא שהוא מספר התנועה בקודם ומתאחר בחלקיה.

Altabrizi, Prop. XV: "Fourth, that time is the measure of motion according to the priority and posteriority *that are not conjoined*.'' והרביעית, שהוא שעור התנועה מצד הקדימה והאיחור אשר לא יתחברו.

Narboni's commentary on *Kawwanot ha-Pilosofim* III, iv: "Aristotle has defined time as the number of motion according to the prior and posterior *in motion*.'' אריסטו גדר הזמן בשהוא מספר התנועות מפני הקודם והמתאחר בתנועה. Again: הנה כשאמר אריסטו שהזמן. הוא מספר התנועה מפני הקודם והמתאחר בה.

The reason for these additional qualifying phrases may be stated as follows:

Aristotle's definition in its original wording, namely, that time is the number of motion according to prior and posterior, was felt to be somewhat ambiguous, for place, too, has the distinction of prior and posterior. In fact, Aristotle himself points out this analogy (*Physics* IV, 11, 219a, 14–19). But there is the following difference between the prior and posterior of place and those of time. In the former case, they are co-subsistent; in the latter case they are successive. It was in order, therefore, to make it unmistakably clear that the phrase prior and posterior used in the definition of time is the successive kind that the phrase 'in motion', or some similar phrase, was added as a qualification of 'prior and posterior.'

Cf. Narboni's commentary on the *Kawwanot ha-Pilosofim* II, iv: "Motion, as has been shown, is said to be measured in a twofold respect. First, with reference to the distance traversed. Second, with reference to time. Consequently, when we use the expression 'the number of motion with reference to prior and posterior,' the 'prior and posterior' may also refer to the parts of the distance, for those parts likewise are the measure of the motion which is performed over them, but these prior and posterior are in position and are generally known not to be in time, inasmuch as they do not measure motion with reference to the nature of succession that exists in it or with reference to the character of possibility that it possesses. It is therefore necessary to include in the definition the phrase 'in motion' [after 'prior and posterior'], for that phrase constitutes the final differentia by which time is distinguished from the other measure of motion which is not time."

ולפי שלתנועה גם כן, כמו שהתבאר, שעור משני פנים: אחד מהם מצד הדרך,
והשני מצד הזמן, והיה אמרנו מספר התנועה בקודם ובמתאחר כבר יאמר על
חלקי הדרך, כי הם ישערו לתנועה אשר עליהם, ויהיו קודם ומתאחר בהנחה,
וכבר נודע שאינם בזמן, אחר שלא ישערוה בהמשך מציאותה ואפשרות כל תנועה,
הוכרח לזה להוסיף בגדר מלת בתנועה, כי הוא ההבדל האחרון יבדילהו מהמשער
השני אשר לתנועה אשר אינו הזמן.

Similar explanations are given by Averroes, *Epitome of the Physics* IV, p. 17b, and Altabrizi, Prop. XV.

The additional qualifying phrase, however, is often omitted as, e.g., in the following translations of Aristotle's definition:

Abraham bar Ḥiyya, *Megillat ha-Megalleh*, p. 10: אין הזמן אלא
מין החלוף בנקדם ומאוחר.

Gersonides, *Milḥamot Adonai* VI, i, 21, p. 386: מפני מה שהתבאר
מגדר הזמן שהוא מספר התנועה בקודם ומתאחר.

All the above-quoted passages are direct versions of Aristotle's formal definition of time. But in both Hebrew and Arabic philosophic texts we find another definition of time, which, while assuming with Aristotle that time is not independent of motion or of objects which are in motion, is phrased differently from Aristotle's definition.

We find such a definition in Saadia, who says that "time is nothing but the extension of the duration of bodies" (*Emunot we-Deot* II, 11), وكان الزمان انّما هو مدّة بقاء الاجسام, והזמן אינו כי (Arabic text, p. 102) or that "The essence of אם מדת קיום הגשמים

time is the duration of these existent things" (*ibid* I, 4). وانّما

אבל אמתתו השארות הנמצאות האלה; حقيقته بقاء هذه الموجودات (Arabic text, p. 71). Cf. Guttmann, *Religionsphilosophie d. Saadia*, p. 80.

Similarly Abraham bar Ḥiyya defines time as ושאיננו כי אם אמירה [נוסח אחר: אמידת] מעמדת הנמצאות. (*Hegyon ha-Nefesh* I, p. 2a). In this last quotation, if we accept the reading אמירה and take it as the equivalent of the Arabic عبارة, usually translated by רמז, מליצה (see below quotation from Altabrizi), the definition would mean that time "is nothing but a term signifying the duration of existent things," thus corresponding to Saadia's second definition. But if we emend the dubious אמירה or אמידת to read מדה, then it would correspond to Saadia's first definition.

A similar definition is also found in Algazali: "Time is a term signifying the duration of motion, that is to say, the extension of motion." اذ الزمان عبارة عن مدّة الحركة اى عن امتداد الحركة
(*Makaṣid al-Falasifah* III, p, 192). כי הזמן מליצה מעת התנועה, ר״ל מהמשך התנועה (MS. Cambridge University Library, Mm. 6.30). כי הזמן רמז למדת התנועה, ר״ל התפשטות התנועה (MS. *ibid.*, Mm. 8. 24).

In the same passage, however, Algazali reproduces Aristotle's definition that "time is a term signifying the measure of the motion of the spheres according to its division into prior and posterior." הנה הזמן מליצה משעור תנועת הגלנלים אשר חלוקו אל קודם ומתאחר.

The common element in all these definitions is the use of the term extension (Saadia: מדה מד״ס, Algazali: امتداد, המשך, התפשטות) and "duration" (Saadia: بقا, קיום, השארות, Abraham bar Ḥiyya: עמדה), and this extension or duration is said to be either of "bodies" (Saadia) or of "existent beings" (Saadia, Abraham bar Ḥiyya) or of "motion" (Algazali), all of which mean the same thing. That it is not a mere coincidence that they all happen to use this definition but that there must be some common literary source to account for it, is not unreasonable to assume. That source, I believe, is to be found in a definition which is attributed to various Greek philosophers.

According to Plutarch, time is defined by Plato as "the extension (διάστημα) of the motion of the world." (*De Placitis Philosophorum* I, 21).

Simplicius reports that Zeno defined time as the extension (διάστημα) of motion, and that Chrysippus defined it as the extension of the motion of the world (Zeller, *Stoics, Epicureans, and Sceptics*, p. 186, n. 6).

Similarly Plotinus reports that those who say that "time is something pertaining to motion consider it either as the extension (διάστημα) of motion, or as its measure." (*Enneads* III, vii, 6).

All these definitions make use of the term διάστημα which undoubtedly underlies the Arabic ة‍ـمد, امتداد and بقا, and their Hebrew equivalents, used by Saadia, Abraham bar Ḥiyya and Algazali. All these definitions are essentially the same as Aristotle's, in so far as they make time dependent upon motion or upon the existence of things which have motion. It can, therefore, be readily seen how easy it was to have Aristotle's definition merged with this new definition.

10. Hebrew כמו הדברים שלא יצטרכו אל נושא, which is an indirect way of saying "substances." See definition of substance in Prop. X, Part I, notes 8, 9 (p. 573).

11. Crescas is restating here the successive steps which lead up to Aristotle's definition of time.

In the first place, he proves that it must exist in some other subject. His proof is taken from the following passage of Aristotle: "That time, therefore, in short, is not, or that it scarcely and obscurely is, may be suspected from the following considerations. One part of it was, and is not; another part is future, and is not yet; but from these parts infinite time and that which is always assumed is composed. That, however, which is composed from things that are not, does not appear to be ever capable of participating of essence" (*Physics* IV, 10, 217b, 32–218a, 3).

Intermediate Physics IV, iii, 1: "One of the reasons that leads one to doubt the existence of time is as follows. Time is divided into past and future. Either of these parts is non-existent, for the past is already completed and gone, the future is not yet come. But that whose parts are non-existent, is itself non-existent. Hence time does not exist."

והדברים אשר יספקו במציאות הזמן, אחד מהם, שהזמן יתחלקו חלקיו אל עובר ועתיד, וכל אחד משני אלו בלתי נמצא. זה שהעבר כבר נפסק ונשלם, והעתיד לא בא עדיין. וכל מה שהיו חלקיו בלתי נמצאים הנה הוא בלתי נמצא. הנה אם כן הזמן בלתי נמצא.

This Aristotelian reasoning underlies the following passage in
Abraham bar Ḥiyya's *Megillat ha-Megalleh*, p. 6: "Time has no
more stability and permanency than the turn of the wheel. The
part of time that has past, i. e., that which has gone before, as
yesternight, yesterday, the day before yesterday and so forth, is
already past and gone and is nothing and nil. The part of time
that is yet to come, as the next day, tomorrow, in the future and
so forth, exists only in potentiality and has not yet come into
existence. The part of time that now is has no continuance of
existence but flows and rolls on and on like water flowing down
the slope."

והזמן אין לו עמידה ולא קיימא כאשר אין להקפת הגלגל עמידה. אבל העובר
מן הזמן והוא הנקדם, כמו אמש אתמול שלשום וכל אשר לפניהם, כבר חלף ועבר
והוא אין ואפס, ואשר הוא עתיד לבוא מן הזמן, כגון מחרת ומחר ולהיום ולהבא,
וכל אשר אחריהם, הם בכח ולא יצאו לידי מעשה, ואשר הוא ממנו בעת הזאת
איננו עומד אבל הוא ניגר ומתגולל והולך כמים המוגרים במורד.

The simile of flowing water is also mentioned by Hillel of
Verona in Prop. IV: "The parts of time are three, or rather two,
namely, past and future. . . . The future continues for ever
infinitely like the rushing of the water of an overflowing river.
This comparison between water and time is found in the works of
the philosophers."

וחלקי הזמן הם שלשה, או שנים לפי האמת, עבר ועתיד . . . זה העתיד ימצא
בה תמיד לאין סוף כמו מרוצת מימי הנהר השוטפים, כי זה המשל מתמשל לזמן
בספרי הפילוסופים.

12. Having shown that time cannot be an independent substance,
again like Aristotle, Crescas endeavors now to show that time
cannot be identical with motion. Aristotle as well as Averroes
produce two arguments to disprove this identification (cf. *Physics*
IV, 10, 218b, 9–18). Of these two arguments Crescas reproduces,
in modified form, the second argument, which is found in *Physics*
IV, 10, 218b, 13–18: "Besides, every change is swifter and
slower; but time is not: for the slow and the swift are defined by
time; since that is swift which is much moved in a short time;
and that is slow which is but a little moved in a long time. But
time is not defined by time, neither because it is a certain quantity, nor because it is a certain quality. It is evident, therefore,
that time is not motion."

Intermediate Physics IV, iii, 1: "The second argument is that every change is swift or slow, but in time there is no swiftness or slowness. Now, the swiftness and slowness of motion are defined by time, for we say the swift is that which traverses a certain distance in a short time, and the slow is that which traverses the same distance in a longer time. Consequently, if time were identical with motion, the term motion would be included in the definition of swift and slow motion, ... but while we say that a certain motion takes place in a long time or in a short time, we do not say that motion takes place in motion."

והמופת השני, שכל שנוי יהיה מהיר ומאוחר ולא ימצא לזמן מהירות ואיחור:
הנה המהירות והאיחור בתנועה אמנם יונבלו בזמן כאשר נאמר שהמהיר הוא מה
שיחתוך המהלך האחד בזמן קצר, והמאוחר אשר חתכו בזמן יותר ארוך, ואילו
היה הזמן הוא התנועה, היתה התנועה לקוחה בגדר התנועה המהירה והמאוחרת...
כי אנו נאמר זאת התנועה בזמן ארוך וקצר ולא נאמר כי התנועה בתנועה.

13. Having already shown that time cannot be a substance nor identical with motion, Crescas now endeavors to prove that time must in some way or other belong to motion or, more specifically, that it is an accident of motion. Here, too, Crescas closely follows Aristotle's method of procedure, for Aristotle, too, after having shown that time is not identical with motion proceeds to prove that time nevertheless cannot be perceived without motion (cf. *Physics* IV, 11, 218b, 21ff.) and concludes with the statement that "Since, therefore, it is not motion, it is necessary that it should be something belonging to motion" (*Physics* IV, 11, 219a, 9–10).

Intermediate Physics IV, iii, 1: "Having been made evident that time is not identical with motion and that it is also not without motion, it becomes clear that it must be one of the properties of motion. We must therefore investigate what that property is, for when we know what that is, we shall know what time is."

ואחר שנגלה שהזמן אינו תנועה, ולא ימצא ריק מתנועה, הוא גלוי שהוא משיג
ממשיני התנועה. ונעיין מה זה המשיג, כי כאשר ידענו מה הוא, ידענו עצם הזמן.

The proof given here by Crescas, however, differs from the one found in Aristotle and Averroes. Aristotle proves that time must belong to motion by showing first that magnitude, motion, and time are all interrelated, and then by further showing that the distinction of prior and posterior, which primarily subsist in place, or magnitude, must also be found in motion and time.

Physics IV, 11, 219a, 14–19: "But prior and posterior primarily subsist in place; and here indeed in the position of the parts. Since, however, there are prior and posterior in magnitude, it is also necessary that these should be in motion, analogous to the prior and posterior which are there. Moreover, there are also prior and posterior in time, because one of these is always consequent to the other."

Intermediate Physics IV, iii, 1: "Inasmuch as prior and posterior are something belonging to magnitude and distance, they must also belong to motion, that is to say, prior and posterior are to exist in motion, for it is self-evident that the prior and posterior of motion are not identical with motion but are rather a pair of its properties, just as the prior and posterior in magnitude are not identical with magnitude but are a pair of its properties."

למה שהיה הקודם והמתאחר אחד ממה שישיג השעור והרחק, חוייב בהכרח שישיגו התנועה, ר"ל שימצאו בה הקודם והמתאחר, כי הוא מבואר בעצמו שהקודם והמתאחר בתנועה אינם התנועה, ואמנם הוא משיג ממשיגיה, כמו הקודם והמתאחר בשעור אינו השעור אבל משיג ממשיגיו.

Crescas, as will have been noticed, has slightly departed from his sources. He tries to show the connection between time and motion by "swiftness and slowness" rather than by "priority and posteriority." The change is immaterial. That it was, however, done intentionally is clear from Crescas' subsequent reference to it. Cf. below n. 16.

The reason for Crescas' departure from his original sources may be conjectured as follows: By proving that time belongs to motion on the ground of its being the measure of the swiftness and slowness of motion, he could immediately conclude his main point "that time must also be an accident adjoined to motion," inasmuch as swiftness and slowness are accidents of motion. Had he followed the original argument of Aristotle and Averroes, he would have had to go through several processes of reasoning before reaching that conclusion. First he would have had to identify time with the prior and the posterior of motion. Then he would have had to show that the prior and the posterior are not identical with motion. Finally he would have had to prove from the analogy of space that the prior and the posterior must be the accidents of motion.

14. See quotation above in n. 12.

15. Cf. *Intermediate Physics* IV, iii, 1: "For motion, as has been said, is related to magnitude, and time is related to motion... Consequently time is the measure of motion."

שהתנועה, כמו שנאמר, תמשך לשעור, והזמן תמשך לתנועה... ולזה הענין היה
הזמן אמנם תשער לתנועה.

16. That is to say, whether you prove that time must be an accident of motion by showing first that it is the prior and the posterior of motion and then that the prior and the posterior are accidents of motion, as did Aristotle and Averroes, or by showing more directly that swiftness and slowness which are accidents of motion are in fact measured by time, as did Crescas himself— in either case, time is shown to be the measure of motion. It is thus Crescas' own allusion to his departure from Aristotle and Averroes in reproducing their discussion above. See above n. 13.

17. *Physics* IV, 12, 221a, 9–11: "To have subsistence in time is one of two things: one of which is then to be when time is; and the other, just as we say, that certain things are in number." The first of these meanings of being in time is rejected by Aristotle, who finally concludes: "But since that which is in time is as in number, a certain time may be assumed greater than everything which is in time. Hence it is necessary that all things which are in time should be comprehended by time, just as other things which are comprehended in anything; as, for instance, that which is in place by place" (*ibid.*, 221a, 26–30).

Intermediate Physics IV, iii, 3: "For their relation to time must inevitably be conceived in either one of two ways. It may mean that they are when time is. Or, it may mean that time comprehends them and is equal to the duration of their existence and it measures them, just as we say, that a certain thing is in number,... which means two things: First, that it is a part of number or one of its properties or differentiae. Second, that it is enumerated by a certain number.... Similarly in time there are these two relations. The relation of the instant to time is like the relation

of the unit to number, which is a part of it. The relation of the
prior and the posterior to time is like the relation of the even and
the odd to number, for by the prior and posterior and by the
even and odd time and number are respectively divided in a
primary sense and in them they have their primary differentiae.
But the relation of all other things to time is like the relation of
that which is numbered to number, or of that which is comprehended to that which comprehends it, or of that which is in place
to place. Consequently, just as in the case of any number it is
possible to conceive a number greater than it, so also in the case
of anything which exists in an equal time, it is possible to conceive a time transcending it on both ends.''

חה שיחסם אל הזמן לא ימנע מאחד משני ענינים: אם שנרצה בזה שהם נמצאים
עם מציאות הזמן, ואם שנרצה בזה שהזמן מקיף בם, ושוה למציאותם ומשער אותם,
כמו שנאמר שהדבר במספר.... על שני פנים: אחד מהם, כאשר יהיה חלק מהמספר
או משיג ממשיגיו והבדל מהבדליו, והשני כאשר היה ספור מה.... ובזמן יש שני
אלה היחסים, יחס העתה אליו הוא יחס האחד אל המספר, שהוא חלק ממנו; יחס
הקודם והמתאחר אליו, הוא יחס הזוגות והנפרדות אל המספר, יען כי בהם יחלקו
ראשונה, והם ההבדלים הראשונים. אבל יחס שאר הדברים אל הזמן, הוא כיחס
הספור אל המספר, הנכלל אל הכולל, או מה שבמקום אל המקום, ואם כן כאשר
יהיה כל מספר כבר אפשר שימצא יותר ממנו, הוא מבואר שכל מה שהיה בזמן
שוה, הנה כבר אפשר זמן יעדיף עליו משני קצותיו.

18. *Physics* IV, 12, 221b, 3–4: "So that it is evident that eternal
beings, so far as they are eternal, are not in time.''

Intermediate Physics IV, iii, 5: "As for the eternal, everlasting
beings, they are not in time, inasmuch as time does not transcend
them nor comprehend them.'' הנה הדברים הנצחיים המתמידיים, הנה
אינם בזמן, חה שהזמן לא יעדיף עליהם ולא יכללם אותם.

19. *Intermediate Physics* IV, iii, 5: "And if those things are said
to be in time, it is because time measures them, and it does
measure them in so far only as they are moved or in so far as they
are at rest, when their rest implies a corresponding motion.
But this applies only to such beings as are capable of motion.''
וכאשר היו אלו הענינים אמנם יאמר בם שהם בזמן מצד שהזמן ישערם, והוא
אמנם ישערם מצד שהם מתנועעים או נחים וידומה בהם התנועה, והם הדברים
שמדרכם שיתנועעו.

20. Cf. Simplicius in *Physics* (ed. Diels, p. 741, 11. 19–26, and Taylor's translation of the *Physics*, p. 266, n. 4): "What then shall we say of perpetual motion? for a circular motion will be demonstrated by Aristotle to be perpetual. Is this, therefore, in time or not? for if it is not in time, time is not the number of every motion. But if it is in time, how is that in time which time does not transcend? To this we reply, that because there is always another and another motion, and never the same according to number, on this account, it is possible to assume a time greater than that which is assumed.''

Cf. *Moreh ha-Moreh* II, Prop. XV: "The eternal motion, i.e., the motion of the sphere, is not in time as a whole. It is, however, said to be in time with reference to its parts. Hence the sphere does not exist in time at all. It is in time only in so far as it is in motion. But then, too, while any given part of its motion is in time, the whole of its motion is not in time.''

והתנועה המדובקה, תנועת הגלגל, אינה בכללה בזמן, אבל יאמר שהוא בזמן
בחלקיה. ועל כן בגוף העגול אין מציאותו בזמן כל עיקר, אבל הוא בזמן מצד
שהוא מתנועע, אלא שחלק תנועתו בזמן אבל תנועתו בכללה אין בזמן.

21. Cf. above n. 18.

Intermediate Physics IV, iii, 5: "It is thus clear that that which is said to have neither motion nor rest is not in time. Consequently, those beings which continue to exist forever and those non-entities which can never come into existence are not in time."

ומבואר שמה שיאמר שהוא בלתי מתנועע ולא נח אינו בזמן. ולזה היו הדברים
המתמידי המציאות והנעדרים הנמנעים המציאות אינם בזמן.

PART II

22. Throughout this chapter Crescas speaks of time being measured by motion or rest when we should expect him to say that time is the measure of motion or rest. A justification for this may be found in the following passage in *Physics* IV, 12, 220b, 14–16: "We not only, however, measure motion by time, but time by motion, because they are bounded by each other."

Aristotle himself admits that time is not only the measure of motion but also of rest. But he qualifies this statement by explaining the term rest to mean only the privation of motion in the case of such beings as are capable of being moved but not the absolute negation of motion as in the case of beings which are incapable of being moved.

Physics IV, 12, 221b, 7–19: "But since time is the measure of motion, it is also the measure of rest according to accident: for all rest is in time: for it does not follow that as that which is in motion must necessarily be moved so also that which is in time, since time is not motion but the number of motion. But in the number of motion there may also be that which is at rest, for not every thing movable is at rest, but that is at rest which is deprived of motion when it is naturally adapted to be moved, as we have before observed."

Intermediate Physics IV, iii, 5: "Furthermore, it is evident that time measures the things which exist in it whether they be moved or at rest, for inasmuch as it is the measure of motion it must also be the measure of rest, for opposites are measured by the same criterion just as they are perceived by the same faculty, as, e.g., light and darkness are perceived by the sense of sight and sound and silence by the sense of hearing. Still, inasmuch as time is the measure of motion and not of rest, it measures motion primarily and essentially and it measures rest secondarily, by the computation of the measure of a corresponding motion... When we describe a thing which is at rest as being in time it is not necessary that it should also be in motion, i.e., being actually moved, for time is not motion but the number of motion, and as a rule it does not necessarily follow that a thing [i.e., the object at rest] which exists in something [i.e., in time] which is an accident to something else [i.e., motion] should also exist in that something else [i.e., in motion]."

ועוד שהוא גלוי שהזמן ישער הדברים הנמצאים בו מצד מה שהם מתנועעים או נחים, חה שלמה שהיה משער התנועה היה מחוייב גם כן שישער המנוחה, כי בדבר אחד ישוערו ההפכים, כמו שישגו המקבילות בכח אחד. משל זה, האור והחושך אשר יושג בחוש הראות, והקול והשתיקה בחוש השמע, אלא שלמה שהיה הזמן הוא מספר התנועה לא מספר המנוחה, היה שערו לתנועה ראשונה ובעצמית, וישערו למנוחה שנית בשער התנועה השוה לה. ... ולא יחייב תארנו שהנח אמנם ינוח בזמן שיהיה הנח בתנועה, ר"ל מתנועע, כי הזמן אינו תנועה, אמנם הוא מספר התנועה, ואין כל מה שימצא בדבר יקרה לדבר יהיה מחוייב מציאותו בזה הדבר.

As against this statement of Aristotle, the following series of counter statements are made by Crescas in this chapter: (a) First, arguing from Aristotle's own point of view, he says that even if the time of rest is measured by our imagining a coresponding motion, time does not require the actual existence of motion. (b) Then, arguing against Aristotle's point of view, he maintains that the time of rest can be measured independently and without our having to imagine a corresponding motion. (c) He also states that rest can be measured as great and small (גדול וקטן, but once, loosely, רב ומעט, *much and few*; see Prop. I, Part II, n. 33), without our having to imagine a corresponding motion. (d) Again, seemingly following Aristotle, he speaks of rest as a privation (העדר) of motion. (e) Finally, throughout this chapter he maintains that time has existence and that rest is measurable without our having to imagine (בציורנו) a corresponding motion, and still, in his refutation of the third premise, he admits that by defining time in terms of rest we indirectly form a conception (נשכיל) of motion.

It seems to me that all these statements of Crescas can be combined to form a connected argument as follows:

What Crescas is trying to establish in opposition to Aristotle is the principle that for an object to be in time it is not only unnecessary for it to be actually in motion but it is also unnecessary for it to be capable of motion. In Crescas' terminology both an object that is immovable because it is incapable of motion and an object that does not happen to be moved, though capable of motion, are described as being at rest. In both cases, then, rest may be considered in a general way as a privation of motion. But there is the following difference between these two kinds of rest. The former kind of rest is an absolute privation, implying not only the absence of motion but also the impossibility of it, the latter kind is relative privation, implying only the absence of motion but not its impossibility. (On this distinction between the two kinds of privation, see *Moreh* I, 58). When Crescas, therefore, describes rest of the former kind as a privation of motion, he means absolute privation.

Furthermore, both these kinds of rest, according to Crescas, are measurable, or, to use his own words, they can be described as long and short. But here, again, there is the following difference. In the case of the rest of an object capable of motion, the time during which the object is at rest is measured by our imagining a corresponding motion in the same object. In the case of the rest of an immovable object, the time of the rest is measured without our having to imagine a corresponding motion in the same object. But how is it measured? The answer to this question may be found in a comparison of Crescas' statement here as to the measurability of rest, which is the privation of motion, with his statement elsewhere as to the measurability of the vacuum, which is the privation of body, for in both cases he uses the same expressions. A vacuum is also said by Crescas to be independently, and without our imagining of its being itself occupied by a body, described as great and small, provided it is conceived as being enclosed within another body (see Prop. I, Part II, p. 189). Thus while we need not imagine the vacuum itself to be occupied by a body in order to measure it, we must conceive of the existence of another body to enclose it. So also here in the case of the rest of an immovable body, while we can measure it without having to imagine the same body to be in motion, still we must conceive of the existence of motion as a concept in order to determine thereby the length and the shortness of the rest of the immovable body. Hence, says Crescas, while it is not necessary for us to imagine that the body that is in time must itself be capable of motion, we must conceive of the existence of motion as a mere concept in order to provide a criterion of measurement for the rest of the immovable body. In our subsequent discussion of Crescas' definition of time (below n. 23) we shall see the significance of this distinction.

A refutation of this argument of Crescas is found in *Neveh Shalom* XII, i, 3, p. 204a: "From this argument of his one can see the scantiness of his knowledge of philosophy, for if time is measured by rest it is only in an accidental sense, in virtue of its being measured by motion primarily and essentially, but were we to have no perception of motion, we could never have an awareness of time, for time is an accident related to motion."

מאמרו זה יתבאר מעוט בקיאותו בחכמת הפילוסופיא, כי אם ישוער הזמן במנוחה הוא במקרה, למה שישוער בתנועה בראשונה בעצם, ולולי נרגיש זו התנועה לא נשער הזמן לעולם, להיות הזמן מקרה נמשך לתנועה.

An allusion to this passage of Crescas occurs in Isaac ben Shem-
ṭob's *second* supercommentary on the *Intemediate Physics* IV,
iii, 4:

"One may raise the following objection. Inasmuch as Aristotle states in the next chapter that time measures rest by the
computation of the measure of a corresponding motion, why then
did he not define time as the number of both motion and rest....

In answer to the twenty-fifth objection we repeat what we
have already said in answer to the preceding objection that true
time does not exist in rest.. This being so, it cannot be argued
that rest should be included in the definition of time, as has been
thought by *one of the philosophers* in his discussion of this subject."

ועוד יש למספק שיספק, שיאמר שאחר שאריסטוטולוס אומר בפרק הבא אחר
זה שהזמן משער למנוחה בציורו לתנועה השוה לה, למה זה ועל מה זה לא אמר
בגדרו אריסטו מספר התנועה והמנוחה...

ונאמר בהתרת הספק הכ"ה, שכבר באת בהתרת הספק שעבר שאינו נמצא
אמתת הזמן במנוחה... ואחר שזה כן אי אפשר שנאמר שיהיה ראוי שתלקח המנוחה
בגדר הזמן כמו שכבר חשב חכם אחד מן החוקרים בזה המקום.

The answer referred to by Isaac ben Shem-ṭob reads as follows:
"Time is possession, rest is a privation, and no possession can be
the measure of a privation." זמן הוא קנין ומנוחה הוא העדר, ואין קנין
משער העדר.

Crescas, however, as we have seen, does not use 'rest' in the
sense of privation of motion but rather in the sense of immovability.

Crescas' argument is also reproduced by Pico Della Mirandola
in *Examen Doctrinae Vanitatis Gentium*, VI, 3: "Neque autem
omnia recenseo, nam cunctas fere de naturalibus principiis
Aristotelis doctrinas evertere tentarunt multi, inter quos etiam
R. Hasdai Mosi Aegyptio minime assensus, qui propositiones
Peripateticas tanquam solido nixas fundamento receperat, inter
quas illam: tempus esse numerum motus. Quiete namque men-
surari tempus affirmat, etiam si nunquam motus inveniretur,
magnam siquidem quietem vocari saepe numero est advertere,
cum quicquam longo tempore conquiescit......quare falsum
affirmat esse ut tempus dicatur motui iunctum, quando et quieti
quae illi opponitur non minus aptetur."

It will have been noticed that in the quotation from the *Intermediate Physics* in this note there occurs the following statement: וישערו למנוחה שנית בשער התנועה השוה לה. The corresponding statement in the quotation from Isaac ben Shem-ṭob's supercommentary reads: שהזמן משער למנוחה בציורו לתנועה השוה לה. Thus while the ישערו of the former passage is retained in the משער of the latter, the term בשער is changed for בציורו.

The explanation seems to be as follows: The Hebrew שער is a translation of the Arabic قدّر, which has many meanings, two of them being (1) *to measure* and (2) *to suppose*. Now, in both passages quoted, the ישערו of the *Intermediate Physics* and the משער of Isaac ben Shem-ṭob are used in the sense of measuring. The בשער of the *Intermediate Physics*, however, stands for *supposing*. The same word is therefore correctly rendered in Isaac ben Shem-ṭob by בציורו. In my translations of these passages I have used in both cases the expression "by the computation of the measure" which combines the two meanings.

Crescas' use of the terms שעור and ציור may be illustrated by the following quotations from this chapter:

(1) ואם היה שנשער המנוחה בציורנו שעור המתנועע בה.

(2) וכל שכן שהמנוחה, בזולת ציורנו בתנועה, כבר תתחלף.

(3) מי יתן ואדע למה לא ישוער הזמן בה בזולת ציורנו התנועה.

(4) ואמנם שוער בתנועה ומנוחה למה שציורנו בשעור התדבקותם הוא הזמן.

(5) ולזה יתחייב שיהיה הזמן נתלה בציורנו שעור התדבקות אם בתנועה ואם במנוחה.

(6) שכבר ימצא זמן בזולת תנועה והוא המשוער במנוחה או בציור התנועה.

(7) למה שאין מהכרח הזמן מציאות התנועה בפועל, אלא ציור שעור התנועה או המנוחה.

In all these passages שעור seems to be used in the sense of *measuring* and ציור in the sense of *supposing*.

In the statement כשנשער הזמן במנוחה נשכיל התנועה, the term נשכיל seems to be used in the sense of נצייר.

23. Hebrew ולזה הגדר הנכון בזמן יראה שהוא שעור התדבקות התנועה או המנוחה שבין שתי עתות. Literally: "Time is the measure of the continuity of motion or of rest between two instants." As thus defined, Crescas' conception of time would seem to differ from that of Aristotle in the following three respects: (1) It is the meas-

ure and not the number of motion (but see below n. 24). (2)
Furthermore, it is the measure not only of motion but also of rest.
(3) Finally, it is not the measure of motion "according to prior
and posterior" but it is the measure of the continuity of motion
or of rest between two instants.

The external form of this definition would seem to be based
upon Gersonides' following discussion of the nature of the instant
and time.

The instant, says Gersonides, has two aspects. "First, it distinguishes the prior from the posterior. Second, it sets off a
certain definite portion of time or of motion, as, e.g., one day or
one hour, for a day is that which is set off by two instants which
limit it on both ends, and so is also an hour. But if an instant
served only as a division between the prior and the posterior in
time, then three days and three hours would mean one and the
same thing, for both are numerically the same, if by their number
is meant the number of instants which distinguish the prior from
the posterior, for in either case there are only two instants.
If there is a difference between three days and three hours, it is
only because there is a difference in the [number of the equal]
parts into which they may be divided, and the difference between
the number of the parts of these two intervals of time is due to
the difference in the respective distances between the instants
which limit them, for the distance between the two instants which
determine a day is greater than the distance between the two
instants which determine an hour. This being so, it is clear that
the instant has a twofold manner of existence. First, it is that by
which a certain number is generated, in which sense it distinguishes the prior from the posterior. Second, it is that by which a
certain continuous quantity is limited, in which sense it sets
off a certain portion of time" (*Milḥamot Adonai* VI, i, 21, p. 387).

ובכלל הנה אנחנו נראה מענין העתה שיש לו ש נ י צדדים מהמציאות, הצד
הא ח ד הוא חלוקת הקודם מהמתאחר והצד ה ש נ י הוא הגבלת החלק הרמוז
מהזמן או מהתנועה, כאלו תאמר יום אחד או שעה אחד, זה כי היום יהיה מוגבל
אליו מצד שתי העתות אשר יגבילוהו, וכן השעה. ואם לא היה ענין העתה אלא
חלוקת הקודם מהמתאחר בזמן, היה אמרנו שלשה ימים או שלשה שעות דבר אחד
בעינו, כי הספירה בכל אחד מאלו הזמנים היא אחת בעינה מצד העתות אשר
יחלקו הקודם מהמתאחר, כי הם שתי עתות בכל אחד מאלו הזמנים, ואולם היה
ההתחלפות אלו הזמנים מצד חלוף חלקי אלו הזמנים, וחלוף חלקי אלו הזמנים

הוא מצד מה שיתחלפו קצתם מקצת במרחק אשר בין העתות אשר ישימם מוגבלים,
וזה שהמרחק אשר בין שתי העתות אשר יגבילו היום הוא יותר גדול מהמרחק אשר
בין שתי העתות אשר יגבילו השעה. ובהיות העניין כן, הוא מבואר שהעתה ימצאו
לו שני צדדים מהמציאות, הא ח ד הוא אשר יחודש בו מספר, והוא חלוקת
הקודם מהמתאחר, וה א ח ד הוא הגבלת הכמות המתדבק, והוא הגבלת החלק
האחד מהזמן.

Finally, on the basis of this distinction and after a long discussion, Gersonides concludes that "time is the measure of motion
as a whole according to the instants which form the boundaries
of motion but not according to the instants which only distinguish
the prior from the posterior" (*ibid.*, p. 388).

הוא מבואר שהזמן הוא משער התנועה בכללה מצד העתות אשר הם תכליות
התנועה, לא מצד העתות שיחלקו בה הקודם מהמתאחר לבד.

Gersonides' distinction between the two functions of the instant
as well as his revised definition of time can be traced to Aristotle's
own discussion in *Physics* VI, 11, 219a, 22–30: "We likewise know
time when we give a boundary to motion, distinguishing prior
and posterior: and we then say there has been time when we
receive a sensible perception of prior and posterior in motion.
But we distinguish them only by apprehending them to be different from one another, and also by conceiving that there is
something between, different from these: for when we understand
that the extremes are different from the middle, and the soul
says that there are two instants, the prior and the posterior, then
we say that this is time: for that which is bounded by instants
appears to be time. And let this be admitted." What Gersonides seems to have done was merely to develop one part of
Aristotle's discussion as to the nature of time and the instant in
order to refute thereby the latter's contention elsewhere that
time must be eternal on the ground that an instant, by its nature
of being the common limit of the past and the future, can never
be conceived as a first instant or a last instant in time. Essentially Gersonides follows Aristotle in making time dependent upon
motion.

Still, while it must be admitted that Crescas' definition of time
is not altogether free from the influence of Gersonides, at least
in its phraseology, it must be assumed to contain some new element, for if Crescas merely meant to reproduce Gersonides' definition as against that of Aristotle, he has failed to establish his

main contention, namely, the absolute independence of time from motion. His addition of the phrase "or of rest" hardly achieves that purpose, and in fact it is a meaningless phrase, for, if time is the measure "of the continuity of motion", it must be dependent upon motion, and it cannot therefore be the measure "of the continuity of rest," unless we take rest in the sense of a privation of motion and not in the sense of immovability, which is the sense in which Crescas would like us to understand that term.

It seems to me, therefore, that Crescas' definition is not a mere paraphrase of the definition advanced by Gersonides, but is to be understood in an entirely new sense. The key to the understanding of it is to be found in the word התדבקות, which is to be taken here not in the general sense of *continuity* but in the specific sense of *duration*. Elsewhere we have seen how Crescas himself interprets the term מתדבק in Maimonides in the sense of eternal duration and we have shown how the corresponding Greek συνέχεια also has these two meanings "continuity" and "duration" (see Prop. XIII, Part I, n. 6, p. 617). By taking the term התדבקות in the sense of duration, the definition assumes an entirely new aspect, and it falls at once in the line of a philosophic tradition which runs through many mediaeval philosophers, such as Bonaventura, Duns Scotus, Occam, Suarez, and many modern philosophers, such as Descartes, Spinoza and Locke. We shall first discuss what may be considered as the origin of this new definition of time, then we shall show that this new definition was not unknown to Arabic and Jewish philosophers, and, finally, in the light of this new definition we shall try to interpret the definition of Crescas.

In Plotinus we have the clearest and probably also the first statement on the identification of time with duration. He starts out with a denial of all views that make time dependent upon physical motion, showing that it is not (a) that which is movable, nor is it (b) motion itself. (c) It is not the extension of motion, (d) it is not the measure or number of motion, and (e) it is not an accident or some consequence of motion (*Enneads* III, vii, 6–9).

Instead of making time dependent upon physical motion he connects it with the motion or the activity of the life of the universal soul. He says that time is produced by the extension (διάστασις, III, vii, 10) of the life of the soul, that it is the "length of the life" (μῆκος βίου, III, vii, 11), and that that length implies a continuity or duration of action (συνεχὲς τῆς ἐνεργείας, ibid.). This *extension* or *length* or *continuity* or *duration* of the life or action of the universal soul is, according to Plotinus, the essence of time. As such, however, it is unmeasured and undetermined, it is invisible and incomprehensible (III, vii, 11). In order to get a definite portion of time, it must be measured by the motion of the sphere. Still, while the motion of the sphere is the measure of definite time, it does not thereby become the cause of the existence of time. "Hence that which is measured by the revolution of the sphere, viz. that which is indicated, but not generated, by it, will be time" (III, vii, 11). Unlike Aristotle, therefore, Plotinus declares that time is not the measure of motion but, quite the contrary, motion is the measure of time (III, vii, 12). But see above n. 22 (p. 646).

What we get then in Plotinus is above all a distinction between indefinite time and definite time. Indefinite time is in its essence *the extension* or *continuity* or *duration* or *length* of the life and activity of the universal soul. Definite time, too, remains in its essence that *extension* or *continuity* or *duration* or *length* of the life and activity of the soul, but its definiteness is determined by the motion of the spheres.

This view of Plotinus is reproduced anonymously by the Iḥwan al-Safa. We have already mentioned the four views with regard to time enumerated by them in their Encyclopedia (see above n. 7). The third of these four views reads, "Or, it is said that time is a duration which becomes numerically determined by the motion of the celestial sphere." وقد قيل انه مدّة تعدّها حركات الفلك (Dieterici, Die *Naturanschauung und Naturphilosophie der Araber*, pp. 14–15; Arabic text: Die *Abhandlungen der Ichwân Es-Safâ*, p. 35). The correspondence of this definition with Plotinus' conception of time as we have outlined it above is so striking that it needs no further comment.

That Plotinus' definition of time was not unknown to other Arabic and Jewish philosophers can be equally established.

First, there is the following passage of Saadia in *Emunot ve-Deot* I, 4: "Perhaps somebody might argue from the case of time and say, before these bodies came into being, how could

time have existed without the existence of anything within it?
Such an argument, again, could not be raised except by one who
is ignorant of the definition of time and imagines that time is
external to the sphere and that it contains the world within it.''

ושמא יחשב גם כן בזמן ויאמר, קודם שהתחדשו הגשמים האלה, איך היה הזמן
ההוא ערום מהנמצאות כלם? חה עוד אין אומר אותו כי אם מי שהוא סכל בגדר
הזמן ויחשוב כי הוא דבר יוצא חוץ לגלגל ושהעולם כלו בו. The contention of the unnamed opponent cited in this passage is quite clear.
While bodies are to co-exist with time from eternity, time is
assumed to be by its nature independent of body. This is exactly
the view of Plotinus.

Second, the first of the four views of time reported by Altabrizi
reads: ''Time exists in itself, is neither a body nor anything belonging to a body, but is something which has necessary existence
in virtue of itself'' (see above n. 7).   Here, again, the assertion
that time is independent of body reflects the view of Plotinus.

Finally, Albo's discussion of time in 'Ikkarim II, 18.   There
are two kinds of time, according to Albo.   One ''is unmeasured
duration, which is conceived only in thought and has perpetual
existence, having existed prior to the creation of the world and
continuing to exist after its passing away.''   This kind of time
is called by him ''absolute time'' (זמן בשלוח), in which there is no
distinction or equality and inequality.   The other kind of time
is that which is ''numbered and measured by the motion of the
sphere and in which there is the distinction of prior and posterior,
of equal and unequal.''

המשך הבלתי משוער המדומה במחשבה, שהוא נמצא תמיד, קודם בריאת העולם
ואחר העדרו... ויהיה הזמן לפי זה שני מינים: ממנו נספר ומשוער בתנועת הגלגל,
ויפול בו הקודם והמתאחר והשוה והבלתי שוה, וממנו בלתי נספר ומשוער, והוא
המשך שיהיה קודם מציאות הגלגל, שלא יפול עליו השוה והבלתי שוה.

The similarity between Albo and Plotinus and the Iḥwan al-
Safa is again strikingly obvious.

If Plotinus' conception of time was not unknown to Albo, we
have good reason to believe that it was not unknown also to his
teacher Crescas.   In fact there are many points in Albo's discussion of time which sound like an echo of his master's teachings.   By taking, then, the term התדבקות in Crescas' definition
in the sense of ''duration,'' the equivalent of Albo's המשך, we
can reconstruct the meaning of the definition in all its fulness.

To begin with, Crescas takes time in the absolute as being pure duration. Such duration does not depend upon motion or upon material objects for its existence; it depends upon a thinking mind. Plotinus finds the source of its existence in the activity of the universal soul. Albo says that it exists in our thought. But inasmuch as indefinite time or duration existed, according to Albo, prior to the existence of the world and consequently prior to the existence of our thought, we may be justified in assuming that Albo conceived it to be the activity of God's thinking just as Plotinus conceived it to be the activity of the universal soul. And this view expressed by Albo may with good reason be also attributed to his teacher.

The essence of time, according to Crescas, will thus be pure duration. But pure duration, as was pointed put by Plotinus and Albo, is indefinite. It becomes definite only when it is measured by motion. Time, i.e., some definite portion of duration, could consequently be defined by Crescas as duration measured by motion. But evidently wishing to retain the conventional formula used in the definition of time ever since Aristotle and following the phraseology of Gersonides which, as we have seen, is derived from Aristotle, Crescas defines time as the measure of the duration of motion between two instants, which is practically the same as saying that time is duration measured by motion between two instants.

Furthermore, by conceiving time-in-general to be duration, and independent of motion, it follows that it is not necessary for a thing to be actually in motion or even to be capable of motion in order to be in time. All things are in time, in the indefinite sense of that term, in so far as there is always a thinking mind, the thinking activity of God. And all things are also in definite time, whether they are themselves movable, inasmuch as their duration can always be measured by a conceptual motion. Thus the Intelligences, even though assumed to be immovable, will be in time. Similarly time existed prior to the creation of the world, even though there was no motion then. Crescas therefore includes in his definition of time the phrase "and of rest," meaning by "rest" not merely the relative privation of motion but absolute immobility. Cf. above n. 22.

It seems, however, that there is the following difference between Albo and Crescas. According to Albo, pure duration is not true time. True time is only that which is measured by physical motion. Unmeasured duration is only what Maimonides describes as suppositive and imaginary time (שער זמן או דמות זמן, *Moreh* II, 13; '*Ikkarim* II, 18), and it has not that order and succession which are implied in the old rabbinic expression "the order of the divisions of time" (סדר זמנים, *ibid.*). According to Crescas, pure duration, even though not measurable by physical motion, can still be called true time, inasmuch as it can be measured by conceptual motion. To that extent, too, pure duration has order and succession. We thus find that while Crescas states, in opposition to Maimonides, that the order of time existed prior to the creation of the world, Albo maintains, evidently in opposition to Crescas, that the order of time did not appear until after the creation of the celestial spheres (see below n. 33).

In framing this definition of time Crescas has thus attained his main purpose, namely, the separation of time from motion. Even the definite time of objects which are in motion is essentially duration and independent of motion; it is only its definiteness that is determined by motion. With Plotinus he would say that time is not generated by motion; it is only measured by it. And thus immediately after laying down his own definition of time, he directly challenges Aristotle by stating: "Consequently it may be inferred that the existence of time is only in the soul" (see below n. 28). Being absolutely independent of motion, magnitude and space, time could have been conceived by a mind even had there been no external world in existence. We thus find Crescas, again in consequence of his definition of time, challenging Maimonides by maintaining that the statement of Rabbi Jehudah bar Rabbi Simon that the order of time has existed prior to creation should be taken in a literal sense (see below n. 33).

A literal translation of Crescas' definition of time is given by Pico Della Mirandola: "Definit autem ipsum ita (ut eius verbis agam) mensura continuitatis vel motus vel quietis quae inter duo momenta" (*Examen Doctrinae Vanitatis Gentium* VI, 3).

24. This criticism is unjustified. Aristotle himself states it quite clearly that the term *number*, used in the definition of time, is not be taken in the ordinary sense of a discrete quantity. *Physics*

IV, 11, 219b, 4–9: "Since, however, number is twofold, for we call
both that which is numbered and that which is numerable number, and also that by which we number; time is that by which is
numbered, and not that by which we number. But that by
which we number is different from that which is numbered.''

This passage is reproduced in Averroes' works as well as in the
works of Hebrew authors dealing with the subject of time.
Narboni, in his commentary on Algazali's *Kawwanot ha-Pilosofim*
III, iv, has the following long statement:

"Averroes has explained that the term number is used in two
senses, in the sense of absolute number, i.e., that wh'ch numbers
but is not numbered essentially, and in the sense of both that
which numbers and that which is numbered... Know also that
the term number applies likewise to that which measures, so that
everything that is divided is incidentally measured by those
parts into which it is divided, and this is especially true in cases
where the division is only conceptual. Thus the parts are the
number of the things into which the object, i.e., the aggregate,
is divided, and are therefore to be included under the second kind
of number, which is both that which numbers and that which is
numbered. Consequently, when Aristotle says that 'time is the
number of motion according to the prior and posterior in it,' he
means by 'number' the second kind of number, i.e., the material
number, which is both that which numbers and that which is
numbered, but he does not mean thereby number *per se*, for
absolute number belongs to discrete quantity whereas time belongs to continuous quantity. What he means by 'number,'
then, is that which is numbered, that is, the parts of the motion,
not indeed in so far as they are parts only, for in this respect they
may all be co-existent, but in so far as they are prior and posterior.''

ופרש בן רשד ואמר כי המספר יאמר על שני מינים: מספר מוחלט, ר"ל מונה
ולא מנוי בעצם, ומספר הוא מונה ומנוי... ודע גם כן כי המספר כבר יאמר על
המשער, ויקרה לכל דבר נחלק שישערדו הדבר אשר אליו חולק, וביחוד כאשר
תהיה החלוקה מפני הנפש, הנה אם כן החלקים הם מספר הדברים הנחלק, שהוא
הקבוץ, ויהיה זה נכנס תחת המין השני מהמספר, ר"ל שהוא מונה ומנוי. הנה כשאמר
אריסטו שהזמן הוא מספר התנועה מפני הקודם והמתאחר בה ירצה בו המין השני
מהמספר, ר"ל המספר החמרי, שהוא מונה ומנוי, ולא ירצה בו המספר עצמו, כי

המספר המוחלט מהכמה המתחלק, והזמן המתדבק. אבל אמנם רצה בו הספור,
שהם חלקי התנועה, ולא במה הם חלקים לבד, כי כבר ילקחו יחד, אבל במה
הם קודמים ומתאחרים.

Furthermore, Aristotle himself, having once explained his peculiar use of the term number, uses afterwards the term measure.
Physics IV, 12, 221b, 7: "Since, however, time is the measure
(μέτρον) of motion..."

We have also seen above (n. 9) how Maimonides, following
Aristotle, uses both terms in the definition of time. Similarly
Plotinus, in his reproduction of Aristotle's definition, uses the
term measure (see above n. 7). The same is also to be observed
in the works of Arabic philosophers.

The question as to the applicability of the term number to
time discussed by many Scholastics, as, e. g., Joannes Versor,
Quaestiones Physicarum, quaestio XIII (Hebrew title: *She'elot
Tibe'iyot* XIII): "Whether the definition given of time is a
proper definition, viz., that time is the number of motion according to prior and posterior. It seems that it is not a proper
definition, for time, belonging to continuous quantity, cannot
be number, seeing that number belongs to discrete quantity....
As for the first objection, I say that time is not absolute
number, but it is the number of motion in a sense in which it may
be taken as a genus, for in this way, in virtue of itself, number is
continuous. It is only in virtue of the act of numbering that
number is a discrete quantity."

השאלה הי"ג, אם גדר הזמן הוא גדר נאות לה, והוא אשר נאמר בו כי הזמן הוא
מספר התנועה כפי הקודם והמתאחר. ויראה שאינו גדר נאות, כי הזמן הוא מהכמה
המתדבק, אם כן אינו מספר, כי המספר הוא מהכמה המתחלק...
אל הטענות, אל הראשונה אומר, שהזמן הוא מספר בהחלט, אבל מספר התנועה
באופן שמספר התנועה יונח במדרגת הסוג, כי מצד עצמו הנה הוא מדובק, ואמנם
מצד פעולו הנה עניינו כן כעניין המתחלק.

25. Cf. *Physics* IV, 11, 220a, 24–26: "That time, therefore, is the
number of motion according to prior and posterior, and that it is
continuous, for it is of the continuous, is evident."

26. Cf. Prop. I, Part II, n. 35.

27. Hebrew סוג בלתי עצמי וראשון, "an unessential and unprimary
genus." This statement reflects Aristotle's theory that a de-

monstration as well as a definition must contain a *universal*
(καθόλου, Crescas' סוג, *genus*, here), which universal must be *essential* (καθ' αὐτό, עצמי) and *primary* (πρῶτον, ראשון).   Cf. *Anal.
Post.* I, 4.

Crescas' argument is reproduced by Pico Della Mirandola as
follows: "Ut genus sit ipsa mensura, viderique iure affirmat numerum genus esse primo non posse, cum sit dicretae quantitatis,
mensura continuae" (*Examen Doctrinae Vanitatis Gentium* VI, 3).

28. According to Aristotle, time is partly real and partly conceptual. In so far as it is consequent on motion, it is real, inasmuch
as the magnitude, which is the subject of the motion, is real.
But in so far as it is the number of motion, it is conceptual.

Physics IV, 14, 223a, 16–23: "It deserves also to be considered
how time subsists with reference to soul; and why time appears
to be in everything; in the earth, in the sea, and in the heavens.
Shall we say it is because time is a certain passive quality or
habit of motion, since it is the number of it?....It may, however,
be doubted whether if soul were not, time would be or not: for
when it is impossible for that which enumerates to be, it is also
impossible that there should be anything numerable."

Intermediate Physics IV, iii, 7: "In one respect time is in the
soul, but in another respect it is outside the soul.   In so far as it
is number, it is in the soul, for without that which enumerates
there can be no number, and without an instant there can be no
prior and posterior.   But motion itself is outside the soul...
Similarly, if you only think of time as a concept, it is in the soul,
but its matter is outside the soul."

הזמן הוא מצד בנפש ומצד חוץ לנפש. מצד היותו מספר הוא בנפש, כי באין
מונה אין מין, ובאין עתה אין קודם ומתאחר. אבל התנועה בעצמה היא חוץ
לנפש. . . וכמו כן, אם תצ״ירהו הנה הוא נמצא בנפש, ואמנם חמרו הוא חוץ לנפש.

Crescas, however, having defined time as something essentially
different from motion and independent of body, maintains that
time is purely conceptual.   See above n. 23.

Cf. Abraham bar Ḥiyya, *Megillat ha-Megalleh*, p. 6: "Hence it
has been said concerning time that it is dependent upon existent
things and is consequent to them and that all creatures exist in it
but itself does not exist except in thought and is perceived only by
the mind's eye."

ומכאן אמרו על הזמן שהוא תלוי בנמצאות ונמשך אליהן, וכל היצורים נמצאים
בו, והוא אינו נמצא אלא בתוך הדעת ונראה בעין הלב.

Cf. Isaac ibn Latif, *Rab Pe'alim*, 18 (ed. Samuel Schönblum):
"Five things have their existence in the mind and not outside the
mind, namely, point, centre, number, species [i. e., universals],
and time."

חמשה דברים מצואים בשכל לא מחוץ לשכל, והם הנקודה והמרכז והמנין
והמינים והזמן.

29. While substance must not necessarily be a body, for there
are also immaterial substances, such as soul and the Intelligences,
still it must exist in itself (see Prop. X, Part I, notes 8, 9, p. 573).
Consequently, time is not a substance, for it does not exist in
itself, being the measure of something else.

It will be recalled, however, that Altabrizi, in defining time as
independent of body, also describes it as existing in itself. He
furthermore describes it as having necessary existence in virtue
of itself (see above notes 7, 23). The expression "necessary
existence in virtue of itself" is usually applied only to God. How
then does Altabrizi happen to ascribe it to time? The explanation
seems to me to be as follows: Altabrizi has confused here the term
time with eternity. Such a confusion may be explained as due
to the theory that time is the image of eternity, which from
Plato and Plotinus (*Timaeus* 37 D, *Enneads* III, vii, Introduction)
has found its way into the pseudepigraphic Theology of Aristotle
(see Dieterici, *Die sogenannte Theologie des Aristoteles*, German,
p. 109, Arabic, p. 107). Now, according to Plotinus, eternity is
identical with God (*Enneads* III, vii, 4: καὶ ταὐτὸν τῷ θεῷ.

30. This passage is reproduced by Pico Della Mirandola as
follows: "Motum autem et quietem dimetitur animus: quare
cum tempus accidens appelletur, ad eum ipsum referri iubet,
alioqui falsum essent, illud esse accidens extrinsecus, quoniam et
quietem consequitur quae privatio est, non autem persistens et
stata natura" (*Examen Doctrinae Vanitatis Gentium* VI, 3).

31. Cf. *Physics* IV, 12, 221b, 3-4: "So that it is evident that
eternal beings, so far as they are eternal, are not in time." By
'eternal beings' the Intelligences are meant here. See above n.
18, 21.

Pico Della Mirandola reproduces this passage as follows: "Falsum item, quod non habet motum, id sub tempore non contineri, quandoquidem quae sunt a materia seiuncta motu carent et sub tempore solent reponi" (*Examen Doctrinae Vanitatis Gentium* VI, 3).

32. The criticism applies only to Maimonides but not to Aristotle. For the latter believes not only in the dependence of time upon motion but also in the eternity of the world as well as of the Intelligences and of time. He furthermore maintains that to be in time means to be transcended by time (see above n. 17). Consequently, unless the meaning of the expression 'being in time' is changed, the Intelligences cannot be in time even if time is made independent of motion. Maimonides, however, unlike Aristotle, believes in the creation of the world as well as of the Intelligences. If time, therefore, is made independent of motion, as is done by Crescas, and is supposed to have existed prior to the creation of the world, the Intelligences can be in time even according to Aristotle's understanding of the expression 'being in time.'

33. This is a reference to the following passage of Maimonides in *Moreh* II, 30: "We find some of our Sages are reported to have held that time existed before the creation.... Those who have made this assertion have been led to it by a saying of one of our Sages in reference to the expressions 'one day,' 'a second day'.... Rabbi Jehudah son of Rabbi Simon said. 'Hence we learn that the order of time has existed previously.' "

Maimonides, to whom time is generated by motion, dismisses the statement of Rabbi Jehudah son of Rabbi Simon as a mere homiletic utterance. But Crescas, believing as he does that the essence of time is duration, its measurability only depending upon motion and that, too, not necessarily upon actual motion, takes the statement of the rabbi literally.

The same statement of Rabbi Jehudah son of Rabbi Simon is also discussed by Albo. Taking the expression "order of time" to apply only to time that is measured by physical motion, he interprets the statement of the rabbi to mean that time existed not prior to the creation of the world but rather prior to the fourth day of creation. *'Iḳḳarim* II, 18: "Inasmuch as the literal meaning of the scriptural verses might lead one to believe that the order of day and night did not come into existence until the fourth day, on which day the luminaries were hung out, Rabbi Jehudah son of Rabbi Simon explains that, by reason of the fact that the celestial sphere has been in motion from the first day on which it was created, the order of day and night existed prior to the fourth day.''

אלא שלפי המובן מן הפסוקים הוא שלא היה סדר היום והלילה נמצא עד היום הרביעי, שנתלו בו המאורות, אמר כי מיום הראשון שנברא הגלגל היה מתנועע, והיה נמצא סדר היום והלילה קודם יום הרביעי.

34. *Moreh* II, 30:

ולזה אמר בראשית, והבי"ת כבי"ת כלי, ופירוש זה הפסוק האמתי כן, בהתחלה ברא השם העליונים והתחתונים.

This passage has been variously interpreted in the commentaries on the *Moreh*. Crescas' paraphrae of it here is rather vague. But from his subsequent argument it becomes clear that he has understood it to mean that God as cause created the heaven and the earth. My translation runs accordingly.

35. That is to say, a necessary cause, acting without knowledge and design.

36. Cf. *Moreh* II, 13–27.

37. Cf. *Or Adonai* III, i, 2.

PROPOSITION XVI

Part I

1. The Hebrew text of the proposition is taken from Ibn Tibbon's translation of the *Moreh*.

2. Crescas endeavors to show that the first part of Maimonides' proposition is a restatement of Aristotle's theory of universals. He thus takes the term "force," כח, in the proposition as referring to the universal or, as he calls it, "the quiddity of the species," מהות המין. Now, the universal, according to Aristotle, has no distinct reality but exists in particulars, or, as the expression goes, *in re*. In Maimonides' proposition it is, therefore, described as a

"force in a body," כח בגוף. The universal is further characterized
by Crescas as being "one in species but many in number," אחד
במין רבים במספר. The significance of this phrase becomes clear when
contrasted with the phrase "one in number," אחד במספר, which
is used as a characterization of the Platonic idea, for the Platonic
idea, unlike the Aristotelian universal, has distinct reality and
does not become diversified by the particulars, the particulars
being only imperfect images of the idea. A description of the
Platonic idea couched in language which is antithetical to that
used here by Crescas is found in Narboni's commentary on
Kawwanot ha-Pilosofim II, i: "Know that the Platonic theory of
ideas is based upon the assumption that the idea of Zaid and of
Omar is identical and *one in number*. The idea comprehends a
plurality of individuals in the same manner as the sun comprehends in its light a number of different things. But just as the
sun is the same everywhere, so the idea is the same in every individual comprehended by it. Consequently the idea of one man
is exactly the same as the idea of another man, i. e., it is *one in
number*."

ודע כי הצורות האפלטוניות הן הנחת צורה א ח ת ב מ ס פ ר היא בעינה לזיד
ועמר, והיא תכלול אישים רבים על צד מה שיכלול השמש מספר, כן בכל איש מן
האישים, הצורה ההיא הכוללת. וצורת האיש האחד היא צורת האיש האחר בעינה,
ר"ל אחד במספר.

Judged by its vocabulary, Crescas' statement is based upon
the following passage of Altabrizi: "The purpose of this proposition is quite evident. Its purpose is to show that whenever individuals belonging to the same specific quiddity are numbered, the
cause of their being numbered is to be found in the numerability
of their matter and the diversity of their receptacle."

המכוון מזאת ההקדמה מבואר, שכל מהות מיניית ימנו האישים אשר תחתיה, הנה
סבת אותו המין אמנם הוא מין בחמרים ושנוי המקבלים.

Cf. *Kawwanot ha-Pilosofim* II (*Maḳaṣid al-Falasifah* II, pp.
107, 109): "The first proposition is that the idea called universal
exists in minds and not in things . . . The second proposition is
that the universal cannot have a plurality of particulars unless
those particulars are distinguished from one another by some
differentia or accident."

המשפט הראשון, שהענין (المعنى) הנקרא כולל מציאותו בשכלים לא בענינים
(الاعيان)... המשפט השני, שהכולל אי אפשר שיהיו לו חלקים רבים כאשר לא
יוכר כל חלק מהאחר בהבדל או מקרה.

Cf. also *Teshubot She'elot*, pp. XLVIII–XLIX: "Plurality is
inconceivable in one species except through the plurality of the
matter. Consequently, that which is immaterial can have no
plurality except by a specific difference, that is to say, by a certain
peculiarity which distinguished one from the other. This peculiarity cannot be an accident, for it would be impossible for anything immaterial to have an accident which does not exist in its
species. Consequently, being immaterial, it can have no plurality
except [through some distinction] in species."

והרבוי לא יצויר במין אחד אלא ברבוי החמר. ומה שאינו חמר לא ירבה אלא
בחלוף המין, והוא ההתיחדות בהבדל יובדל בו האחר. ולא יהיה מקרה, אחר
שיהיה שקר שיחוייב לדבר מקרה לא ימצא במינו, וכאשר לא יהיה חמר, לא יהיה
רבוי אלא במין.

All these statements reflect the following passage in *Metaphysics*
XII, 8, 1074a, 33–34: "But all things that are many in number
have matter."

3. Here Crescas begins to explain the second part of the proposition. While universals are only "forces in a body," there are beings which exist apart from a body. These are the Intelligences.
The term נבדל, مفارق, *separate*, is the Greek χωριστός, i. e.,
χωριστὸς τοῦ σώματος נבדל לגשם, *separated from body*; hence
incorporeal.

4. Cf. Prop. XV, Part I, n. 21 (p. 646).

5. For according to definition place implies the existence of one
body within another. Cf. Prop. I, Part I, p. 153.

6. The implication of this statement is that accidents cannot exist
apart from their material subject. Cf. *Physics* I, 4, 188a, 6: "For
affections are not separable." *Metaphysics* XII, 1, 1069a, 24: "Further, none of the categories other than substance can exist apart."

7. The theory that the Intelligences proceed from one another
and hence are related among themselves as causes and effects

represents the view of Avicenna. Averroes is opposed to this view. According to him, all the Intelligences proceed directly from God and are not related to each other as cause and effect. There is, however, between them a difference of degree with regard to their perfection and importance, and it is that difference which constitutes their individuality and makes it possible for them to be numbered. Cf. Shem-ṭob on Prop. XVI.

PART II

8. This is an allusion to Crescas' own theory of immortality as contrasted with that of Avicenna and his followers. Cf. *Or Adonai* II, vi, 1; III, ii, 2.

9. This is the Avicennean theory of immortality which has been adopted also by some Jewish philosophers. Cf. *Or Adonai* III, ii, 2.

10. Hebrew חושיו וכחותיו. Literally: "its senses and faculties." By "faculties" is probably meant here the "internal senses," especially "imagination," as contrasted with "senses" by which is meant the "external senses." Cf. the expression המוחשות והמדומות, "percepts and images" in Prop. VII, Part II, p. 246.

11. This is another allusion to the difference between himself and the philosophers as to the immortality of the soul. According to the accepted opinion of the philosophers, immortality is consequent to the soul's acquisition of intellectual conceptions. According to Crescas' own view, it is consequent to the soul's love for God as its attachment to Him. Cf. *Or Adonai* III, ii, 2.

12. Hebrew איׁשי העצם. Literally, "individual substances." Cf. Prop. XXV, n. 5 (p. 699). But the expression carries also the connotation of corporeality. Cf. Kaufmann, *Attributenlehre*, p. 12, n. 17; p. 13, n. 24.

13. This is the view of Alexander, Themistius and Averroes. Cf. *Milḥamot Adonai* I, 8.

14. Cf. *Or Adonai* II, vi, 1.

15. That is to say, the expression עניינים נבדלים, "separate (or "immaterial") beings," in the proposition refers to שכלים in the sense of the Intelligences of the spheres and not in the sense of the acquired intellects of man. On the two meanings of the term שכל, and the analogy between the Intelligences and the Intellect, see Prop. III, Part I, n. 6 (p. 486) and Prop. XI, n. 5 (p. 605).

PROPOSITION XVII

1. The Hebrew text of the proposition is taken from Isaac ben Nathan's translation of Altabrizi.

2. These opening remarks of Crescas are based upon the following passage of Altabrizi: "Know that our discussion here will deal with two problems. First, to prove the statement that everything that is moved must have a mover different from itself. Second, to classify the various kinds of movers and to explain the expression 'that which is moved by itself'."

דע שזה הדבור מקיף על שתי חקירות. אחת מהם בבאור שכל מתנועע לו מניע
זולתו. והב' מה שבו חלוק המניע ופירוש המתנועע מצדו.

Crescas, as will have been noticed, reproduces only the first part of Altabrizi's statement, thus confining himself only to the explanation of the first part of the proposition. His failure to explain the latter part of the proposition is discussed below in n. 7.

3. *Physics* VIII, 4, 254b, 12–14:, "Of those things, however, which are moved essentially, some are moved by themselves, and others by something else; and some by nature, but others by violence and contrary to nature."

Intermediate Physics VIII, iv, 4, 1: "As for those things which are moved essentially, they require some consideration. Some of these things are moved by themselves but others by something else, and some are moved by nature but others by violence and contrary to nature."

ואמנם מה שבעצם הם אשר ראוי לעיין בהם. ואלו מהם מה שיתנועעו מפאת
עצמם, ומהם מה שיתנועע מחוץ, וגם כן קצתם מתנועעים בטבע וקצתם מתנועעים
בהכרח (וקצתם מתנועעים) בתנועה חוץ מהטבע.

4. *Physics* VIII, 4, 254b, 24–28: "And it is especially obvious that a thing which is moved, is moved by something, in things which are moved contrary to nature, in consequence of their being moved by something else being evident. But after things which are moved contrary to nature, among such as are moved according to nature, those are more manifest which are moved by themselves as animals."

Intermediate Physics VIII, iv, 4, 2: "In the case of things which are moved by violence or contrary to nature, it is self-evident that they are moved by a mover which is something different from the things moved. It is equally self-evident in the case of animals that they are moved by something, namely, a soul."

זה שהענין בדברים אשר יתנועעו בהכרח או חוץ מן הטבע שהם יתנועעו ממניע
הוא דבר אחר זולתם הוא ענין מבואר בעצמו. וכמו כן הענין מבואר בעצמו בבעלי
חיים שהם יתנועעו מדבר מה, והוא הנפש.

Cf. *Intermediate Physics* VII, 1: "With reference to those things which are moved by an external agent, it is evident that they are moved by a mover which is different from that which is moved . . . But even in the case of animals, it will also become apparent that there is a distinction between that which is moved and that which moves."

זה שהמתנועעים יתנועעו מדברים מחוץ, הענין בהם מבואר שהם יתנועעו ממניע
יתחלפו למתנועע... ואמנם החי הנה כבר יראה מעניינו הבדל המתנועע למניע.

5. *Physics* VIII, 4, 254b, 33–255a, 5: "But it may be especially doubted concerning the remaining member of the last mentioned division; for of things which are moved by another, some we have considered as being moved contrary to nature; but others remain to be opposed, because they are moved by nature. And these last are the things which may occasion a doubt by what they are moved; as, for instance, things light and heavy; for these are moved by violence to opposite places; but to their proper places naturally, the light indeed upward, and the heavy downward. But it is no longer apparent by what they are moved, as it is when they are moved contrary to nature."

Intermediate Physics VIII, iv, 2: "But a doubt arises concerning the simple elements, that is to say, the heavy and light elements, as, e. g., in the case of the motion of fire upward and of the motion of a stone downward. For when these bodies are

moved by violence, it is quite clear that they are moved by something different from themselves, that is to say, by an external
force. But a doubt arises when these bodies are moved with their
natural motion, for, when fire is moved upward and earth downward, it seems that they are moved by themselves and that the
mover in them is identical with that which is moved."

הנה אשר בו הספק אמנם הוא בגשמים הפשוטים, ר"ל הנשמים הכבדים והקלים,
כמו תנועת האש למעלה והאבן למטה, חה שאלו הנשמים, כאשר יתנועעו בהכרח,
העניו בם מבואר שהם יתנועעו מזולתם, שהוא המכריח. אמנם יקרה הספק בענינם
כאשר יתנועעו תנועתם הטבעית, כי כבר יחשב שהאש כאשר תתנועע למעלה והארץ
למטה שהם יתנועעו מעצמותם ושהמניע בם הוא המתנועע.

Cf. *Intermediate Physics* VII, 1: "But of all these instances a
doubt arises concerning those things which are moved in place
without any mover external to them, and especially concerning
the simple elements, such as earth and fire, for of these it may be
thought that they are moved by themselves and that the mover
in them is identical with that which is moved."

ואמנם אשר יפול בו הספק מהם הם הדברים אשר יתנועעו במקום מבלתי דבר
מחוץ ובפרט הגשמים הפשוטים, כמו הארץ והאש, כי אלו כבר אפשר שיחשב בם
שהם יתנועעו מעצמם ושהמניע בם הוא המתנועע בעצמו.

6. Aristotle himself advances several arguments to prove that
the four natural elements are not moved by themselves. In one
of the arguments he tries to show that the diversity of direction
in the natural motion of the elements could not be accounted for,
if the elements were assumed to be moved by themselves. The
argument is contained in the following passage in *Physics* VIII,
4, 255a, 8–11: "I say, for instance, if anything is the cause to
itself of walking, it will also be the cause to itself of not walking;
so that since it is in the power of fire to tend upward, it is evident
that it is also in its power to tend downward. It is also absurd to
suppose that they should be moved by themselves with only one
motion, if they themselves move themselves."

This Aristotelian argument is reproduced, either singly or together with other arguments, in the following works:

Altabrizi, Prop. XVII, who offers it as the *second* of *four* arguments, not all of which are taken from Aristotle. "The proof with
regard to the first problem is as follows. When a body is moved,
it must be moved either because it is a body in the absolute or

because it is a certain kind of body. The first alternative is refutable on several grounds. First, . . . Second, if the body is moved by virtue of its being a body, then it must necessarily be moved either in one direction or in more than one direction . . . But if the body *qua* body must not necessarily be moved in one direction, but could be moved in any direction at all, then there is no reason why the elements should each tend toward one direction rather than toward another.''

החקירה הראשונה, ראיתה היא שהגשם כאשר התנועע הנה אם שיתנועע לשהוא
גשם משולח או לשהוא גשם מה. והראשון בטל מפנים. אחד... והשני מהפנים,
שהגשם, אם היה מתנועע במה שהוא גשם, לא ימנע אם שיהיה מכוון לצד מורגש או
לא יהיה... ואולם אם לא יהיה המתנועע מכוון לצד מעיין אבל עבר שיתנועע אל
איזה צד הזדמן, הנה אין התנועעו אל קצת הצדדים ראשון מהגעת אל שאר הצדדים.

Emunah Ramah I, 3, p. 14: ''Then we observe that the elements are moved in different directions. Thus fire tends upward as does also air, whereas earth tends downward as does also water. Now, if the elements were moved in their respective directions by their corporeality, [i. e., ccrporeal form, see Prop. X, Part II, n. 18, p. 579] they would all be moved in one direction, and a direction which would be common to all of them, just as corporeality is common to all of them. Similarly, if they were all moved by their matter, they would likewise to moved in one direction, for matter is common to all of them, as has been shown in the preceding chapter. Since the elements could not be moved in different directions by corporeality or matter, it follows that the cause of the motion of body is not body. This is an important principle. Bear it in mind.''

אחר כן נמצא צדדיהם מתחלפים ויתנועעו, האש עולה והאויר גם כן, והארץ
יורדת והמים גם כן, ואם התנועעו אל צדדיהם בגשמיותם, יתנועעו כלם אל צד אחד,
יהיה משותף, כמו שהגשמיות משותף להם, ואם התנועעו גם כן בחמרהם, יתנועעו על
צד אחד, לפי שהחומר משותף להם, כמו שכבר התבאר בפרק קודם זה. ולא
יתנועעו על הצדדים המתחלפים בגשמיות או החומר, אם כן מניע הגשם אינו גשם, וזה
שורש גדול ושמור אותו.

Kawwanot ha-Pilosofim III (*Maḳaṣid al-Falasifah* III, p. 239): ''There is no doubt that a body is not moved by itself by virtue of its being a body, for were it so, it would be moved perpetually and every body would be moved in the same direction.''

ואין ספק שלא יתנועע מעצמותו להיותו גשם, כי לו היה כן, היה תמיד, והיה
לכל גשם על אופן אחד.

Crescas' restatement of this argument contains certain expression which point to Altabrizi and the *Emunah Ramah* as his immediate sources. See below n. 7.

7. This conclusion does not occur in Altabrizi. But it occurs in the following other sources.

Kawwanot ha-Pilosofim, loc. cit. "The body is moved by something added to it, that something being called nature."

אבל לענין נוסף עליו, יקרא אותו הענין טבע.

Emunah Ramah I, 3, p. 14: "Hence the four elements are moved in their different directions either by their different forms or by their different accidents. But to say that the accidents cause the elements to be moved in their different directions is absurd . . . It is, therefore, the forms of the elements that cause them to be moved in the directions that are natural to them, and it is these forms to which the term nature is primarily applied. And thus we say that nature is a certain principle of motion and rest to that in which it is inherent, essentially and not according to accident."

וכבר נשאר שיתנועעו הגשמים הארבעה על הצדדים המתחלפים אם בצורותיהם המתחלפות ואם במקרים המתחלפים. רק שהמאמר בשהמקרים הם מניעים היסודות על מקומותיהם המתחלפים בטל... וצורות היסודות הם המניעות אותם אל צדדיהם הטבעיים להם, והם אשר שולח להם שם הטבע ראשונה.   ונאמר שהטבע הוא התחלה מה לתנועת מה שהוא בו ומנוחתו, בעצם ולא במקרה.

Cf. *Physics* II, 1, 192b, 20–23: "Nature being as it were a certain principle and cause of motion and rest to that in which it is primarily inherent, essentially and not according to accident." Another rendering of Aristotle's definition of nature occurs in *Cuzari* I, 73: "Nature is the principle and the cause by which the thing in which it is inherent, rests and is moved, essentially and not according to accident."

כי הוא ההתחלה והסבה אשר בה ינוח וינוע הדבר אשר הוא בו, בעצם ולא במקרה.

Narboni in Prop. XXV has the following rendering:

ולכן גדר ארסטו בטבע שהוא התחלה מה וסבה לאשר יתנועע וינוח הדבר אשר הוא בו ראשונה ובעצמות לא בדרך המקרה.

Cf. also the rendering reproduced by Hillel of Verona quoted above in Prop. IV, n. 18.

The view expressed here by Crescas that the form of the simple elements is the cause of their natural motion reflects the opinion of Avicenna and Algazali, as given by the former in *Al-Najah*, p. 25, (cf. Carra de Vaux, *Avicenne*, pp. 184–185) and by the latter in the *Maḳaṣid al-Falasifah* III, p. 239. In connection with this, Shem-ṭob, in his commentary on the *Moreh* (II, Prop. XVII) has the following statement: "Some people thought that in fire, for instance, the body is that which is moved and the form is that which moves. This is the view of Avicenna and Algazali."

ואנשים חשבו כי גשם האש הוא מתנועע וצורת האש הוא המניע. וזהו דעת ב'ס
ואבוחמ'ד.

According to this view, therefore, the cause of the natural motion of the elements abides within the elements themselves. The form is the cause of the motion of the elements just as the soul is the cause of the motion of animals. The elements are therefore said to be moved by themselves (ὑφ' αὐτοῦ), in the same way as animal beings.

Averroes' view, based upon his own interpretation of Aristotle, is opposed to this. According to him, all the elements, to be sure, contain within themselves a certain principle of motion, but not one of causing motion but rather one of receiving motion. The cause of the motion he contends, does not abide within the elements themselves. It is rather external to them. The elements therefore, unlike animal beings, are not said to be moved by themselves, ὑφ' αὐτοῦ.

Averroes' view is based upon Physics VIII, 4, 254b, 12–24, which is analyzed by him in his *Intermediate Physics* VIII, iv, 4, 1, as follows: "As for those things which are moved essentially (מה שבעצם, καθ' αὐτό), they require further consideration. Some of these things are moved by themselves (מפאת עצמם, ὑφ' αὐτοῦ) but others by something from without, and some are moved by nature but others by violence and contrary to nature. Of those which are moved by nature, some are moved by themselves, as, e. g., an animal, for an animal is moved by itself, though its body may be moved by nature and contrary to nature, but some are moved not by themselves as, for instance, the light and heavy elements."

ואמנם מה שבעצם הם אשר ראוי לעיין בהם. ואלו מהם מה שיתנועעו מפאת
עצם ם, ומהם מה שיתנועעו מחוץ. וגם כן קצתם מתנועעים ב ט ב ע וקצתם
מתנועעים ב ה כ ר ח (וקצתם מתנועעים) בתנועה חוץ מה ט ב ע. ואשר ב ט ב ע,

מהם מה שיתנועעו מ פ א ת ע צ מ ם, כמו החי, כי החי יתנועע בטבע מפאת עצמו,
ואמנם גופו הנה אפשר שיתנועע בטבע וחוץ מהטבע, ואמנם [מהם] מה שיתנועע לא
מפאת עצמו, כמו הדברים הקלים והכבדים.

The rest of the chapter contains an argument to prove that while
the natural motion of the elements is caused by a mover the mover
is not within themselves. Averroes concludes the argument with
the following statement: "Hence it is clear that these simple elements are not moved in place by themselves but rather by something from without.

וכאשר היה זה כן, מבואר שאלו הגשמים אינם מתנועעים במקום מפאת עצמם אבל
מדבר מחוץ.

Crescas, as will have been noticed, has explained only the first
part of Maimonides' proposition, namely, everything that is
moved has a mover. In his explanation, as we have seen, he has
followed the Avicennean view by showing that the mover in the
case of the natural motion of the elements is the form of the
elements. He does not, however, discuss the second part of the
proposition where Maimonides undertakes to explain the meaning of the expression "that which is moved by itself" (Arabic
אלמתחרך מן תלקאיה. Altabrizi and Crescas: מתנועע מצדו. Ibn
Tibbon and Al-Ḥarizi: מתנועע מעצמו, ὑφ' αὐτοῦ. See Prop. VI.
n. 3, p. 531). From the context of the proposition it is not clear
whether Maimonides has meant to use the expression only with
reference to animals or also with reference to the natural elements.
Among his commentators there is a difference of opinion on this
point.

According to one interpretation offered by Altabrizi, with which
he is in agreement, the expression is applied by Maimonides also
to the natural elements. "Some of them take the expression 'that
which is moved by itself' to refer to that whose motion is not
produced violently by some cause outside itself but whose cause
is either in itself or is dependent upon itself. The proponents of
these views are the truest philosophers. Accordingly the expression includes the sphere, vegetables, animals, and the simple
elements when moved according to nature, but it excludes all the
motions that are violent and compulsory. And this is what the
author of this book has meant by the expression."

ומהם מי שפרש ה מ ת נ ו ע ע מ צ ד ו במתנועע אשר לא תהיה סבת תנועתו חוץ
ממנו במכריח, אבל יהיה אם בתוכו או נתלה בו, והם אמתיים מן החכמים. ולפי זה
יכנס בו הגלגל והצמח והחי והפשוטים המתנועעים בטבע, ויצאו ממנו התנועות
ההכרחיית האנוסות.   והוא אשר רצה בה בעל הספר.

The same interpretation is evidently adopted by Efodi, who
in his comment on the last part of the proposition mentions the
natural form, הצורה הטבעית.

Shem-ṭob, on the other hand, maintains that Maimonides' last
statement about "that which is moved by itself" refers only to
animal beings and does not include the elements. He furthermore
maintains that Maimonides has purposely left out any mention
about the natural elements in this proposition, because he did not
want to commit himself as to the question whether the cause of
their motion is within them or outside of them.   "The view of
Avicenna and Algazali is untenable, for the body of the element
is not that which is moved nor is the form that which moves. Nor
in this view espoused here by the Master, for he does not say that
the elements are moved by themselves, he only says that the
animal is moved by itself.   This shows the pre-eminence and
superiority of the Master in all the branches of philosophy."

חה דבר בטל, כי המתנועע אינו הגשם ולא המניע הוא הצורה, ולא אמרו הרב
גם כן, כי לא אמר שאלו היסודות מתנועעים מעצמותם, אבל אמר שהחי הוא המתנועע
מעצמו.   חה יורה על גודל מעלתו ויתרונו בחכמות.

Again: "It is for this reason that the Master did [not] say that
the elements are moved by themselves, nor did he say that their
mover is from without, but he rather left them unmentioned, for
all this is a matter of fine-spun speculation among philosophers,
and it was the Master's intention to state only well established
views."

ולכן [ולא] אמר הרב שהיסודות מתנועעים בעצמם, ולא דבר בהם כלל, ולא
אמר שהמניע להם הוא מחוץ, למה שכל זה הוא עיון דק פילוסופי, וכונת הרב להניח
דברים מבוארים.

PROPOSITION XVIII

1. The Hebrew text of the proposition is taken from Isaac ben
Nathan's translation of Altabrizi.

Crescas' interpolation of the words "the author concludes this
proposition by saying," וחתם ההקדמה הזאת באמרו, before Maimo-

nides' last words, "and note this," והבן זה, has its precedent in
Narboni ("and the author says at the end, 'And note this'",
(ואמר בסוף והבן זה) and in Hillel of Verona ("and so on to the end
of the proposition which the Master concludes by saying 'and
note this'", עד סופה שחתם בה הרב ז'ל והבן זה). In the case of Narboni and Hillel of Verona, however, the interpolation was necessary, because they quote only the first part of the proposition.
But Crescas, in quoting the entire proposition, had no reason for
introducing this interpolation. It was probably used by him in
imitation of Narboni and Hillel of Verona. Or, he may have introduced this statement in order to indicate that the expression "and
note this" is part of Maimonides' original proposition and not a
comment by himself. In the absence of quotation marks it was
necessary to use some such expression to indicate the beginning
and end of a direct quotation. The interpolation here is thus the
equivalent of the expressions זה לשונו and עד כאן which usually
introduce and close a direct quotation. See Prop. III, Part II,
p. 226, l. 10.

2. The entire discussion in this chapter is based upon Altabrizi.
Crescas has only rearranged the parts of Altabrizi's discussion
and introduced a few slight changes, as will be pointed out in the
succeeding notes.

3. The three cases enumerated here by Crescas are based upon
the following statement of Altabrizi: "We say that whenever anything passes from potentiality to actuality, the passage takes
place according to a threefold manner." ונאמר אמנם מה שבא מן הכח
או הפעל יהיה על שלש מדרגות.

4. Altabrizi: "First, when something non-existent becomes existent, as e. g., when the heat which is non-existent in the water
but is capable of becoming existent is brought into existence by
an agent, the transition involved in the process is called a transition from potentiality to actuality."
הראשונה מהם שיהיה אותו הדבר נעדר וישוב נמצא, כמו שהחמימות נעדרת במים
אבל היא מקבלת המציאות, וכאשר תמציאה הפועל שבה נמצאת בו, ויאמר שהיא
יציאה מן הכח אל הפועל.

5. Crescas' argument here differs from the corresponding argument employed by Altabrizi. The latter's argument reads as

follows: "We say that whenever anything passes from potentiality
to actuality, according to the manner described in the first two
cases, there must be something to bring about that passage from
potentiality to actuality, for whenever a thing comes into existence after non-existence it must undoubtedly be with reference
to its own nature only possible of existence, and thus both existence and non-existence must bear to it the same relation. It
therefore needs something to determine the preponderance of
existence over non-existence. That something which determines
the preponderance of the existence of a thing over its non-existence is undoubtedly that which causes the thing to pass from
potentiality to actuality."

ונאמר כל מה שיצא מן הכח אל הפועל על שני פנים הראשונים לו מוציא מציאותו
מן הכח אל הפועל, לפי שאותו הדבר איפשר בעצמותו בלי ספק, ויחס המציאות
וההעדר אליו על השווי, ויצטרך אל מכריע יכריע מציאותו על העדרו, ומכריע
מציאות הדבר על העדרו מוציאו מהכח אל הפועל בלי ספק.

6. Altabrizi: "Second, as when, e. g., something existing actually
as a substance has the possibility of acquiring a certain attribute,
be it a form or an accident, which does not as yet exist in it. Such
an actually existent substance is said to be potential with reference to that attribute, as long as it has only the possibility of
acquiring it. But once it has acquired it, it is said to have become
actual with reference to that attribute. An illustration thereof
is the case of water which is an actually existent substance and
has the possibility of acquiring the attribute of heat. Before its
acquisition of heat, the water is said to be hot in potentiality, but
after its acquisition of heat, it is said to have become hot in
actuality."

והשנית, שיהיה הדבר נמצא בפועל בעצמותו ואיפשר שיהיה לו תאר מה, אם צורה
ואם מקרה, אבל היא לא תהיה נמצאת, ויאמר לאותו הדבר הנמצא בפועל, כפי
אפשרות הגעת אותו אותו התאר לו, שהוא בכח כך, וכאשר נמצא לו אותו התואר, יאמר
שהוא שב בפועל, כמו המים, כי היו נמצאים בפועל בעצמותם, ואיפשר שיתוארו
בחמימות, וקודם מציאותה להם יאמר שהמים בכח, וכאשר נמצאה לו, יאמר שהוא שב
חם בפועל.

7. Crescas' reasoning here differs from that of Altabrizi. Crescas
uses here the argument which is later used by Altabrizi in connection with the "case of a potentiality to impart action." Cf. below
n. 9.

8. Altabrizi: "Third, as when, e. g., a being which exists in actuality and is perfect as to its essence and complete as to its attributes creates something new, not in itself but outside itself. Before its creation of that something new, the creator is said to be the potential agent of its creation, but after the act of creation, it is said to have become its actual agent."

והמדרגה השלישית שיהיה הדבר נמצא בפעל שלם העצמות, תמים התארים, ואפשר
שיחודש ממנו דבר אחר לא נמצא בו אבל נפרד ממנו, ולפני חדושו ממנו יאמר לאותו
הומצא שהוא פועל לדבר האחר בכח, וכאשר חודש ממנו יאמר לו שב פועלו בפועל.

9. Altabrizi: "That determinant agent which causes the transition (see above n. 4) must be either outside the thing which is in potentiality, as, e. g., fire in its relation to water, or within the thing itself, as, e. g., the natural power which causes the growth of fruits and brings about their ripening. In the second alternative, if that power has never ceased to act, then we must consider that in which it exists to have always been in actuality and never to have been in potentiality, but our assumption now is that at one time it was in potentiality but later passed to actuality. And if that power was once inactive and then passed from potentiality to actuality, there is no doubt that its former lack of activity must have been due to the presence of some obstacle or to the absence of some condition. It thus follows that it must have had something external to itself which removed that obstacle or created that condition, and it is that something external which has brought about the removal of the obstacle or the creation of the condition which will have to be considered as the agent which has caused that power to pass from its potential activity to its actual activity. Take, for instance, the natural power that causes the growth of fruits and brings about their ripening. If it happens to fail to bring about that ripening it is only because of the presence of some obstacle, such as cold which causes the fruit to remain hard and unripe, or to the absence of some condition, such as the absence of the required temperature. But whenever the obstacle is removed or the required condition is created, as, for instance, when the cold disappears through the warming of the air by the sun, then it is the sun which causes that natural power to pass from its potential activity to its actual activity."

זה המכריע המוציא כבר יצא לחוץ מעצמות אותו הדבר אשר הוא בכח, כאש
ביחס אל המים, וכבר יהיה בתוכו מקיף, כמו הכח הטבעי המבשל לפרות ההוות בו.
והחלק השני מהפנים, שלא יתחלף ממנו פעלו, הנה הוא בשנעמיד מה שהוא נמצא בו
בפעל תמיד, ולא יהיה משכנו בכח בעת מן העתים, ודברנו במה שיהיה בכח, אחר
כן יצא אל הפעל, ואם נתחלף ממנו פעלו בשלא(?)ואם(?) מה שהוא מן הכח אל
הפועל ואין] ספק שיהיה אותו החלוף אם להקש מונע או לחסרון התנאי, ויצטרך
אל ענין חוץ ממנו יסיר אותו המונע או יניע אותו התנאי, ומסיר המונע או מניע התנאי,
שהוא חוץ ממנו מוציא לזה (שהוא חוץ) [הכח] אשר הוא בדבר בפעלו מהכח אל
אל הפעל, ככח הטבעי המבשיל הפרות ההוות בו, כאשר לא יניע ממנו אותו הבשול,
אם להקש בו מונע, כקור, משים אותם פנים בלתי מבושלים, או להפקד תנאי, כחמום
האויר, וכל עת סר המונע ההוא או הניע זה התנאי, כשמש כאשר התפשטה בו בחמום
האויר, הנה הוא מוציא הכח הטבעי בפעלו מהכח אל הפעל.

10. By this distinction Crescas means to obviate a difficulty with
regard to the creation of the world. If the world was created,
then it has passed from potential existence to actual existence.
God, being the cause of the transition, must have likewise passed
from a potential agent to an actual agent. Cf. *Moreh* II, 14:
"If God produced the universe from nothing, then before the
creation of the universe He was a potential agent and upon its
creation He became an actual one. Thus God must have passed
from a state of potentiality into that of actuality."

מהם שאמר, אם השם יתברך חדש העולם אחר ההעדר, אם כן היה הבורא קודם
שיברא העולם פועל בכח וכאשר בראו שב פועל בפעל, הנה כבר יצא השם מן
הכח אל הפעל.

The answer suggested here by Crescas does not agree with that
given by Maimonides. Maimonides' answer is based upon the
distinction between a corporeal and an incorporeal agent, the
latter exemplified by the active Intellect and God. An incorporeal
agent, he argues, may act only at times and still not pass from
potentiality to actuality. Furthermore, quite the contrary to
the explanation suggested here by Crescas, Maimonides maintains that while the occasional inactivity of the Active Intellect
may be due "to the absence of substances sufficiently prepared
for its action," the period of God's inactivity prior to the creation
of the world is not to be explained in the same way (*Moreh* II, 18).

Crescas' distinction is based upon Altabrizi's discussion which
is as follows: The activity of a perfect agent may be operated
either upon a material object or upon an immaterial object. In

the former case, he says, the change from inactivity to activity on the part of the agent "does not imply a change in the agent itself, for his transition from inactivity to activity is not due to an imperfection in the agent itself, which indeed would imply a change in its being, but rather to an imperfection in those which receive its action."

אבל זה לא יחוייב שנוי בפעל, כי אותו החלוף לא יהיה לחסרון ענין בפועל,
עד ייוחסו אליו שנוי, אבל לחסרון במקבלים.

Crescas, however, rejects this answer later in his discussion of the problem of creation. Or Adonai III, i, 4 (p. 66b).

PROPOSITION XIX

1. The Hebrew text of the proposition is taken from Isaac ben Nathan's translation of Altabrizi.

This proposition as well as propositions XX and XXI is taken from Avicenna. The Avicennean origin of these propositions has been recognized by all the commentators of Maimonides. Cf. Efodi, Shem-ṭob, Asher Crescas and Munk, ad. loc.

The principle which Avicenna is trying to establish by these propositions is that the term possible means to be caused and the term necessary means to be causeless (see below n. 4). Nothing, therefore, of which the existence is due to a cause can be said to have necessary existence, even though its existence may continue unchanged eternally. God alone, according to Avicenna, has necessary existence. The celestial spheres have only possible existence by their own nature, their eternity and hence necessity of existence are due only to their cause. The transient sublunar beings, on the other hand, are possible in every respect.

As against this view, Averroes denies that in eternal beings there is such a distinction as being possible by their own nature and necessary by their cause. According to him, things are said to be necessary when they eternally remain in the same state, either eternally existent (מוכרח המציאות) or eternally non-existent (מוכרח ההעדר). Things which have only transient existence are said to be possible because of their not remaining unchanged in the same state, for before their coming into existence they have the possibility of either coming-to-be or not coming-to-be and after their coming into existence they have the possibility of either passing-away or not passing-away.

Averroes' conception of "necessary existence" seems to be based upon the following passage in *Metaphysics* VI, 2, 1026b, 27–29: "Since, among things which are, some are always in the same state and are of necessity, not necessity in the sense of compulsion but that which means the impossibility of being otherwise . . ."

The origin of Avicenna's distinction in eternal beings between possibility by their own nature and necessity by their cause is, according to Averroes, to be found in his attempt to solve the following difficulty. No finite body, according to Aristotle, can possess an infinite force (cf. Prop. XII). Since the spheres are finite bodies, their motive force must be finite and consequently their motion must be finite. But still the spheres, according to Aristotle's theory of eternal motion, have a motion which is infinite in duration. In order to remove this difficulty Avicenna was compelled to distinguish within the spheres between a possibility with reference to their own nature and a necessity with reference to their cause. This distinction, again according to Averroes' testimony, was first suggested by Alexander. Averroes himself, however, answers the difficulty by distinguishing between a force which is infinite in time and a force which is infinite in intensity and maintaining that while the spheres, owing to their finitude, cannot have an infinite force of the latter kind, they can have an infinity force of the former kind.

Intermediate De Caelo I, x, 2, 8 (Latin, p. 293va, G–293vb, K): "There is room here for the following great doubt. It has been shown that nothing eternal has the possibility of being corrupted nor can there be in it a potentiality for corruption. But it has also been shown in this treatise that a body which is finite in magnitude cannot but have a finite force. Now, since the celestial sphere is finite in magnitude, the force within it must necessarily be finite. The inference must therefore be that while the sphere by its own nature has the possibility of being corrupted it must be free of corruption on account of the infinite immaterial force, outside the sphere, which causes its motion. That this is so is maintained by Alexander in a treatise of his, and he is followed by Avicenna, who says that to have necessary existence may mean either of two things. First, to have necessary existence by one's own nature.

Second, to have only possible existence by one's own nature but necessary existence by reason of something else . . . This being the case, it follows that that which is eternal may have a potentiality for corruption . . . Our own answer to this difficulty, however, is that a body may be said to have a finite force in two senses. First, in the sense that its motion is finite in intensity and speed. Second, in the sense that its motion is finite in time."

וממה שיש לו מקום ספק גדול הוא, שכבר נתבאר הנה שלא ימצא דבר נצחי שיהיה אפשר שיפסד ושאין בו כח על זה, ונתבאר עם זה בזה המאמר שכל גשם כחו בעל תכלית מפני שהוא בעל תכלית בשעור. ואם הדבר כן, הנשם הרקיעי בעל תכלית השעור, ואם הוא בעל תכלית הכח, הנה הוא אפשרי ההפסד מעצמו, בלתי נפסד מצד הכח הבלתי בעל תכלית אשר הוא בבלתי חטר, רוצה לומר המניע לו. וזה שכבר גלה דעתו בקצת מאמרו, ונמשך עמו אבן סיני, ואמר שהמחוייב המציאות שני חלקים: חלק מחוייב המציאות בעצמו, וחלק אפשרי המציאות בעצמו מחוייב בזולתו... ואם הדבר כן, יש בנצחי כח ההפסד... ונאמר אנחנו בהתרת זה הספק, שהגשם יאמר שיש בו כחות בעלי תכלית על שני עניינים. האחד מהם, מציאות התכלית לתנועתו בחזק וקלות. והענין השני, מציאות התכלית לה בזמן.

This passage of Averroes is reproduced in the *Moreh ha-Moreh* II, Prop. XII.

Cf. also *Mif'alot Elohim* II, 3, p. 12b: "For Plato says that the heavens were generated from that eternal matter which had been in a state of disorderly motion for an infinite time but at the time of creation was invested with order. Consequently by their own nature the heavens are corruptible just as they were generated, and it is God who implanted in them eternity, as it is written in the *Timaeus*. It is from this view that Avicenna has inferred that the celestial sphere is composed of matter and form and is corruptible and possible by its own nature but necessary and eternal by virtue of its cause."

כי אפלטון אמר שהשמים נתהוו מאותו חמר קדום שהיה מתנועע תנועה בלתי מסודר זמן בלתי בעל תכלית ובעת הבריאה קבלה הסדר, ושהיו השמים כפי טבעם נפסדים כמו שהיו הוים, אלא שהאל יתברך נתן בהם הנצחיות, וכמו שכתב בספרו טימאוס. ומכאן לקח אבן סיני שהיה הגרם השמימי מורכב מחמר וצורה והיה נפסד ואפשרי מעצמו אבל היה מחוייב ונצחי מפאת סבתו.

2. The entire chapter is based upon Altabrizi with the exception of the last statement which is based upon Narboni. See below n. 4.

3. Hebrew זולתו העדר בהעדר העדרו יחוייב לא. I take העדר here in the sense of "being non-existent" rather than in the sense of "ceasing to exist." The Hebrew העדר (Arabic عدم) is a translation of the Greek στέρησις, which means (a) *privation*, and (b) *depriva-tion*. The former meaning is implied in the first *three* senses of the term discussed by Aristotle in *Metaphysics* V, 22, 1022b, 22–31. The latter meaning is implied in the *fourth* sense of the term. *Ibid.* 31–32: "The violent taking away of anything is called priva-tion." Cf. IX, 1, 1046a, 34–35: "And in certain cases if things which naturally have a quality lose it by violence, we say they suffer privation." Similarly the Hebrew and Arabic terms have these two meanings. Thus in Mamonides' proposition נעדרו (Arabic עדמו) is used in the sense of *deprivation*, i. e., ceasing to exist, whereas here Crescas uses it in the sense of *privation*, i. e., being non-existent.

4. This last statement is based upon the following passage of Narboni: "This proposition does not mean to imply that that which owes its existence to a cause must have the possibility of passing away, for [if it had that possibility it could not be eternal, inasmuch as] that which is possible cannot be eternal, but, as a matter of fact, many of the things which owe their existence to a cause are eternal. What the proposition really means to affirm is that when a thing owes its existence to a cause, then the exis-tence of that thing, be it eternal or otherwise, is due to something else."

לא שיחוייב שיהיה בו אפשרות על ההעדר, כי האפשר לא ישוב נצחי; והרבה מן
העלולים הם נצחיים. אבל הרצון בזה שהמציאות שלו, אם נצחי או איזה שיהיה,
הוא מצד זולתו.

What Narboni and Crescas are trying to say is this: Possible existence does not mean corruptible existence, for it has already been shown in the discussion of Prop. VIII, Part II, n. 15 (p. 561), that accidental motion, i. e., possible motion, may be eternal if its cause is eternal. Possible existence simply means conditioned existence, i. e., existence dependent upon a cause.

Altabrizi's conclusion reads here as follows: "Everything which has a cause is with reference to the existence of that cause neces-sary of existence, with reference to the non-existence of that cause impossible of existence, but with reference to its own essence, ir-

respective of the existence or non-existence of its cause, possible
of both existence and non-existence."

שכל אשר לו סבה הוא בבחינת מציאות סבתו מחוייב המציאות, ובבחינת העדר
סבתו נמנע המציאות, ובבחינת עצמותו, עם הפסק העיון ממציאות סבתו והעדרו,
איפשר המציאות וההעדר.

PROPOSITION XX

1. The Hebrew text of the proposition is taken from Isaac ben
Nathan's translation of Altabrizi.

2. Similarly Altabrizi: "For we have already explained in the
proposition preceding this, that everything which has a cause is
in respect to its own essence possible of either existence or non-
existence, whence it follows by the method of the conversion of
the obverse that that which in respect to its own essence is not
possible of either existence or non-existence has no cause at all
but its existence is necessary in respect to its own essence."

כי אנחנו בארנו בהקדמה אשר לפני זאת שכל אשר לו סבה הנה הוא בבחינת
עצמותו איפשר המציאות וההעדר, ויחוייב מזה בדרך הפך הסותר שאשר לא יהיה
אפשר המציאות וההעדר לעצמותו הנה לא יהיה לו סבה כלל, אבל יהיה מחוייב
המציאות לעצמותו.

Cf. Prop. XIX, n. 4.

As for the expression הפך הסותר, *the conversion of the obverse*, see
Prop. VII, Part I, n. 3 (p. 541).

3. The question is raised by Altabrizi: "One may raise the following question. You have already shown in the proposition preceding this that everything which has a cause is in respect to its own
essence only possible of existence, whence this proposition is
deducible by the method of the conversion of the obverse. There
was therefore no need of making of it a separate proposition."

ולאומר שיאמר, אתם בארתם בהקדמה אשר לפני זאת שכל אשר לו סבה הנה הוא
אפשר המציאות בבחינת עצמותו, ותהיה זאת ההקדמה מחוייבת ממנה בדרך הפך
הסותר, ואין צורך לשומה הקדמה נפרדת?

On a marginal note in the Vienna Manuscript, signed אב׳א.
there is a reference to Altabrizi. The note is reprinted in the
Vienna Edition. It reads as follows: "This question has been
raised by Altabrizi, but the author of the *Moreh* has been justified

after the manner explained by that worthy commentator."

.הפלא הזה הפליאו תבריזי, וניצל הרב המורה ממנו כדרך המפרש החשוב

Altabrizi's answer reads as follows: "The answer to this question is as follows. Inasmuch as this proposition was found to be
very helpful on account of its manifold applicability, the author
saw no harm in making of the problem treated in it a proposition
by itself, so that the principle it establishes may be directly known
to the reader and exist in his mind in actuality, without there
being any need of deriving it from another proposition."

והתשובה ממנה שזאת ההקדמה למה שהיתה רבת השמוש, לרוב נפילת הצורך
אליה, הנה השאלות אשר ידברו בם אין פשע לזכרם בפני עצמם, כדי שיהיו ידועים,
נמצאים בשכל בפעל, ולא נצטרך להוציאם מהקדמה אחרת.

PROPOSITION XXI

1. The Hebrew text of the proposition is taken from Isaac ben
Nathan's translation of Altabrizi.

2. Cf. Altabrizi: "The proof of the proposition is as follows. The
existence of every composite object requires the existence of its
component parts, and those parts are something different from
the whole. Hence every composite object requires for its existence
something different from itself. Now that which requires for its
existence something different from itself, will disappear with the
disappearance of that something different. Hence the composite
must be possible in respect to its own essence and cannot be anything that is necessary of existence in respect to its own essence.
The conclusion is that nothing composite can be necessary of
existence in respect to its own essence."

ביאורו שכל מורכב מציאותו צריך אל מציאות חלקיו, וחלקיו זולתו. הנה כל
מורכב מציאותו מצטרך אל זולתו, וכל צריך אל זולתו הנה יסור בסור אותו הזולת,
הנה הוא אפשר לעצמותו. ויוליד שכל מורכב הוא אפשר לעצמותו, ואין דבר
מאשר הוא מחוייב המציאות לעצמותו. יוליד אין דבר מורכב מחוייב המציאות
לעצמותו.

3. Cf. Prop. XIX.

PROPOSITION XXII

1. The Hebrew text of the proposition is taken from Isaac ben Nathan's translation of Altabrizi.

2. Hebrew נמצא בפעל נרמו אליו, reflects the Greek τόδε τι. Cf. *Metaphysics* VIII, 1, 1042a, 27–29: "And by matter I mean that which, not being a 'this' actually, is potentially a 'this', and . . . by form, which being a 'this' . . ."

3. Cf. Prop. X, n. 7 (p. 571). This, as will have been noticed, is the Aristotelian proof for the deduction of matter and form. Altabrizi in this place reproduces the Avicennean proof. Cf. Prop. X, Part II, n. 22 (p. 591).

4. Crescas is trying to forestall the question why Maimonides mentions only the three accidents of quantity, geometrical form and position out of the nine accidents enumerated by Aristotle in his list of categories. His answer is based upon the division of accidents into "separable" and "inseparable," or "external" and "inherent," and the assumption that Maimonides confines himself here only to the latter.

A similar division of accidents is found in *Kawwanot ha-Piloso-fim* II, 1 (*Maḳaṣid al-Falasifah* II, pp. 97–98): "Accidents are divided into two classes. First, those the conception of whose essence does not require the conception of something external . . . as, e. g., quantity and quality . . . Second, those which require attention to something external. Of the latter are the following seven: relation, place, time, position, possession, action, passion."

במקרים... יתחלקו אל שני חלקים... לא יצטרך בציור עצמותו אל ציור חוץ ממנו... כגון הכמה והאיך... הצריך אל ההבטה אל ענין חוץ ממנו הנה הם שבעה: הצרוף, האנה, מתי, ההנחה, הקנין, שיפעל, שיתפעל.

The term "quality" is used by Algazali to include among other qualities also that which Maimonides calls here "figure" (see below n. 5). His inclusion of "position" among the "external" accidents is explained below in n. 8. As for similar attempts by modern scholars to classify Aristotle's nine accidents, see Zeller, *Aristotle*, Vol. I, p. 280, n. 2.

Unlike Crescas, Narboni does not consider the selection of these three accidents by Maimonides as being of any particular significance. "As for the accidents which occur to body . . . they are quantity, figure, position and others of the remaining categories according to their order."

ואמנם המקרים המשיגים אותו... הם הכמות, והתכונה וההנחה חולתם משאר המאמרות על מדרנתם.

In Altabrizi, however, there is a suggestion of Crescas' interpretation. "As for body, it cannot be without these three accidents, namely, quantity, figure and position."

ואם כל נשם לא ימנע מאלה המקרים השלשה אשר הם הכמה והתמונה והמצב.

5. Cf. *Categories*, 8, 10a, 11–12: "The fourth kind of quality is figure (σχῆμα) and the form (μορφή), which is about everything." *Intermediate Categories* II, iv, 5: וסוג רביעי הוא התמונה והתאר הנמצאים באחד אחד מן הדברים. This kind of quality is designated by Aristotle as "quality according to form," κατὰ τὴν μορφήν... ποιόν, ibid, 10a, 16. Avicenna designates it as "qualities inherent in quantity" (cf. Horten, *Die Metaphysik des Avicennas*, p. 219). Maimonides describes it as "quality which occurs to quantity qua quantity," איכות המשנת הכמות באשר הוא כמות (*Moreh* I, 52. Cf. Munk, *Guide* I, 52, p. 196, n. 5).

The underlying Arabic word for תמונה, "figure," here is شكل. This Arabic word is translated here by Ibn Tibbon by the term תכונה. The latter term usually translates the Arabic هيأة, διάθεσις, disposition, in which sense it is used by Ibn Tibbon himself in *Moreh* I, 52 (see Munk, *Guide* I, p. 195, n. 2). How he has come to use it here in the sense of "figure" or "form" may perhaps be explained as follows. The Hebrew תכונה, as a result of its use as a literal translation of the Arabic هيأة in the sense of *disposition*, has acquired all the other meanings of the Arabic term. Now, the Arabic هيأة, in addition to *disposition*, means also "exterior," "appearance," "form," and is thus the equivalent of شكل. Hence, Ibn Tibbon translated here شكل by תכונה. Cf. H. A. Wolfson, "The Classification of Sciences in Mediaeval Jewish Philosophy," *Hebrew Union College Jubilee Volume* (1925), p. 302, note.

Hillel of Verona, having before him the reading תכונה of Ibn Tibbon's translation, takes it refer to "such things as weight and lightness, smoothness, roughness, rareness, density, and

their like, for all these are called corporeal affections." ,תכונה
פירושה כמו כובד וקלות, חלקות, שעירות, ספוגיות, מקשיות, ורומיהם, שכל אלה
נקראים תכונות גופיות. From his list of examples it is clear that
he did not know that תכונה here represents the Arabic شكل and
is therefore to be taken in the sense of "figure." As to the particular sense in which Hillel understood the term תכונה in this passage, it can be determined by the examples he includes under it.
The quality of weight and lightness is described by Aristotle as
an "affection," πάθος (*Metaphysics* V, 21, 1022b, 15–18). Now
the particular kind of quality known as πάθος is usually translated
into Hebrew by הפעלות, אנפעאל (cf. *Categories*, 8, 9a, 29, and
Moreh I, 52). Hence, תכונה is used by Hillel of Verona partly in
the sense of הפעלות. The other four examples he mentions are
specifically stated by Aristotle not to be varieties of "quality"
but rather of "position." *Categories*, 8, 10a, 14–20: "The rare and
the dense, the rough and smooth, may appear to signify a certain
quality, but probably these are foreign from the division of quality, as each appears rather to denote a certain position (θέσιν)
of parts." By "a certain position of parts" Aristotle undoubtedly
means here what he calls elsewhere" disposition," διάθεσις.
Metaphysics V, 19, 1022b, 1–3: " 'Disposition' means the arrangement of that which has parts, in respect either of place or of
potency or of kind; for there must be a certain position, as the
word 'disposition' shows." Hence, it would seem that the term
תכונה is used here by Hillel of Verona partly in its original sense
of "disposition."

However, as against the last quoted statement from Aristotle
there is a statement by Maimonides which describes smoothness
and roughness, rareness and density as qualities. *Moreh* II, 21:
"We say that the necessary result of the primary qualities are
roughness, smoothness, hardness, softness, rareness and density."
מן האיכיות הראשונות התחייבו בו החלקות והפכו ואלכשונה=השעירות] והקושי
והרכות והספוגיות והפכו ואלכתאפה=מקשיות]. Similarly Algazali describes
roughness and smoothness as qualities. *Kawwanot ha-Pilosofim*
II (*Makaṣid al-Falasifah* II, p. 98): המין הב' האיכות... כמראים והטעמים
והריחות והשעירות והחלקות וההרכות והקושי הלחות והיובש והחום והקור.

6. Altabrizi: "For figure is a term applied to that which is contained by any boundary or boundaries." כי התמונה מליצה מדבר יקיף

בו גבול או נבולים. Cf. Euclid, *Elements* I, Def. XIV, and above Prop. I, Part I, n. 148 (p. 388).

7. Hebrew והמצב, Arabic ואלוצע. Ibn Tibbon: הנחה. Al-Ḥarizi: התכונה המיוסדת. The term תכונה is evidently used by Al-Ḥarizi here in the sense of "place" (see Ibn Ezra on Job 23, 3 and Fürst's *Wörterbuch*), and hence תכונה מיוסדת, "fixed place" or "position."

8. This description of "position" is based upon Altabrizi: "As for position, it is a term signifying the condition of a body which arises as a result of the relation of its parts to each other and their relation to other bodies on the outside. It is well known that every body has its parts related to each other after a certain manner and is as a whole variously related toward other bodies with reference to proximity and remoteness."

ואולם המצב הנה הוא מליצה מהתכונה המגעת לגשם בסבת יחס חלקיו קצתם אל קצת ויחסם אל הגשמים אשר חוצה לו, וידוע שכל גשם לו יחס מיוחד בין חלקיו ויחם אל הגשמים מהקורבה והרוחק.

The second part of the description of "position," which Altabrizi illustrates by the examples of "proximity and remoteness" is used by Algazali as a description of "relation," and is illustrated by him by the examples of "on the right" and "on the left" *Kawwanot ha-Pilosofim* II, *Maḳaṣid al-Falasifah* II, p. 98): "As for relation, it is a condition which happens to a substance by reason of something else, as . . . to be on the right of something or on its left . . ." אולם ההצטרפות, הוא ענין לעצם תקרה בסבת היותו זולתו... והיותו על הימין ועל השמאל. Similarly in *Emunah Ramah* I, 1, p. 7, it is used as a description of a special kind of "relation" characterized as "relation in position." "When you say 'on the right of Simeon' or on the left of Levi', the statement expresses a *relation in position*." וכאשר תאמר לימין שמעון לשמאל לוי הוא צירוף במצב.

"Position" itself is described in *Emunah Ramah* I, 1, p. 6, as follows: "It is the relation of the parts of the body to the parts of the place . . . This is what is advanced by some as a description of position. But others think that position is the relation of the parts of the body to each other."

המצב, והוא יחס חלקי הגשם אל חלקי המקום...זה הרושם שרשמו קצתם למצב. ומהם מי שיראה שהמצב הוא יחס אל חלקי הגשם קצתם לקצת. Of these two descriptions given in the *Emunah Ramah* of "position," the

second corresponds to the first given by Altabrizi and reproduced
here by Crescas. It occurs also in Algazali's *Kawwanot ha-
Pilosofim* II (*Maḳaṣid al-Falasifah* II, p. 98): "As for position,
it is the relation of the parts of the body to each other." אולם
המצב הוא יחס חלקי הגשם קצתם לקצת. The first description of "position" in the *Emunah Ramah* evidently reflects the following passage in Metaphysics V, 19: "Disposition means the arrangement
of that which has parts, in respect either of place or of potency
or of kind; for there must be a certain 'position,' as the word
'disposition' shows."

The fact that Algazali uses the term "position" in the sense of
the external relation of one body to another and not in the sense
of the inner arrangement of its parts may explain why he includes
"position" among the accidents which Crescas characterizes here
as "separable." See above n. 4.

PROPOSITION XXIII

Part I

1. The Hebrew text of the proposition is taken from Isaac ben
Nathan's translation of Altabrizi.

2. Based upon Altabrizi: "Know that on this proposition there
are two questions. First, to say of a thing that it is 'in potentiality' means the same as to say that it is possible of existence
but does not yet exist, as we have explained above. When the
author, therefore, has said 'everything that is in potentiality,'
we already know that it contains a certain possibility. What need
was there for him to explain his first statement further by saying
'and in whose essence there is a certain possibility.' "

דע שעל זאת ההקדמה שתי שאלות. אחת מהם שאמרו כל מה שהוא בכח
ענינו כל מה שהוא אפשר המציאות ואינו נמצא, כמו שכרנוהו, הנה האיפשרות ידוע
תכר בו בכללותו, ולמה חזקו פעם אחרת במאמרו ובעצמותו איפשרות מה

This difficulty is not unanswerable. It is discussed by Maimonides himself in his letter to Ibn Tibbon (*Kobeẓ Teshubot ha Ram-
bam we-Iggerotaw* II, p. 27b), where a distinction is made between
"potentiality" and "possibility." "A thing is said to be in potentiality when it is capable of receiving a certain form which as yet

does not exist in it, and the form, in that case, is said to exist in
the thing in potentiality, as when, e. g., a piece of iron is said to
be a sword in potentiality and a date seed is said to be a palm tree
in potentiality. When a thing is thus said to be something else in
potentiality, then the thing itself is said to contain a possibility
of becoming something else, as, e. g., a piece of iron is said to have
the possibility of becoming a sword. To grasp the distinction between 'potentiality' and 'possibility' requires great subtlety and
is a matter of utmost difficulty even to trained philosophers. A
good account of the distinction is given by Avempace at the
beginning of his commentary on the Physics."

יאמר כי הדבר הוא בכח, בהיות שום תואר מן התארים נעדר עתה מן הדבר
ההוא, אך הוא מוכן ומועד להתישב בו [להמצא] התאר ההוא, ויאמר בתאר ההוא
שהוא בדבר ההוא בכח, כאמרנו בחתיכת ברזל שהוא סייף בכח, וכאמרנו בגרעינה
של תמרה שהוא דקל בכח. והדבר אשר הוא בכח שום ענין יש בעצם הדבר ההוא
אפשרות להתישב בו הענין ההוא, כמו שתאמר בחתיכת ברזל שהוא אפשרי להיות
ממנה סייף. ולדעת ההבדל אשר בין הכח והאפשרות הוא דבר דק וקשה מאד על
הפילוסופים הבקיאים. וכבר דבר בזה הענין אבן אלצאיג בתחלת פירושו לשמע
הטבעי דבר טוב מאד.

Maimonides' reference to the difficulty of grasping the meaning
of the distinction is reproduced by Hillel of Verona (Prop. XXIV,
p. 39b) as follows: כי הוא דבר עמוק מאד וחמור אפילו אצל בקיאי הפילוסופים.

The distinction made by Maimonides between "potentiality"
and "possibility" may be traced to Aristotle's discussion of the
term "potentiality," δύναμις, in Metaphysics IX. The meaning
of the term "potentiality" is explained by Aristotle in the following passage: "Actuality means the existence of the thing, not in
the way which we express by 'potentially;' we say that potentially, for instance, a statue of Hermes is in the block of wood and
the half-line is the whole, because it might be separated out, and
we call even the man who is not studying a man of science, if he
is capable of actually studying a particular problem" (Metaphysics
IX, 6, 1048a, 30–35). This explanation, it will be noticed, corresponds exactly to the explanation given by Maimonides. Later,
Aristotle further explains and restricts the meaning of potential
existence. In the first place, it is not everything that can be
called potentially something else, for it is only certain things that
are capable of becoming certain other things. "But we must dis-

tinguish when a thing exists potentially and when it does not; for it is not at any and every time. E. g., is *earth* potentially a man? No—but rather when it has already become *seed*, and perhaps not even then, as not everything can be healed by the medical art or by luck, but there is a certain kind of thing which is capable of it, and only this is potentially healthy" (*Metaphysics* IX, 7, 1048b, 37–1049a, 5). "If, then, a thing exists potentially, still it is not potentially any and everything; but different things come from different things" (*ibid.*, XII, 1069b, 28–29). In the second place, even those things which are capable of becoming something else are not potentially that something else unless there is nothing external to hinder the actualization of that potentiality (*ibid.*, IX, 7, 1049a, 5–18). It is quite evident, then, that the "possibility" which according to Maimonides a subject must possess in order to be said to have a "potentiality" for something else refers to those conditions laid down by Aristotle as governing the meaning of potential existence and making its realization possible.

The distinction between 'potentiality' and 'possibility' is fully discussed by Hillel of Verona on this proposition. The most important statement in his lengthy discussion is the following: "When we say that the form of a man is in the seed, that potentiality, inasmuch as it exists in a subject, i. e., the seed, must be preceded by a certain disposition called possibility on the part of the subject."

כי אמרנו יש בזרע צורת האנוש ב כ ח, זה ה כ ח, מהיותו נמצא בנושא, ר"ל בזרע, צריך שיקדם לו תכונה אחת שנקראת א פ ש ר ו ת ותדבק לו.

Hillel of Verona then proceeds to explain the meaning of "possibility." His explanation is nothing but an outline of Metaphysics IX, 7. The term "possibility," he says, has two meanings. First, it means that the subject that is said to be potentially something else must be by its nature fit to become that something else, as it is not everything that is fit by nature to become that something else. Second, there must be all the conditions favorable for the realization of the potentiality of the subject to become something else.

Etymologically both כח, *potentiality*, and אפשרות *possibility*, are translations of the Greek δύναμις, but they represent two different senses of the Greek word. "Potentiality" represents

δύναμις as the opposite of ἐνέργεια *actuality*, whereas "possibility" reflects δύναμις as the opposite of ἀδυναμία, *impossibility*
and ἀνάγκη, *necessity*. Arabic: קוה ,אמכאן.

3. Again based upon Altabrizi: "Second, the predicate of a proposition must be something different from its subject, inasmuch as
there is nothing to be gained by the repetition of the same terms.
It is furthermore evident that the predicate must be something
external to the subject, for were it not so, its predication of the
subject would be self-evident and the proposition would require
no demonstration. But we are dealing here with propositions
which do require demonstrations."

והשאלה השנית היא נשוא הגזירה ראוי שיהיה זולת נושאה, אחר שאין תועלת
בהשנות הדבר, מופשט ושיהיה חוץ ממנו, ואם לא, יהיה קיימו לנושא מוסכל, ולא
יהיה מורה תדרש אמתחה במופת, ודברינו במזירות המופתיות.

4. In this passage Crescas reproduces and criticizes Altabrizi's
interpretation of the proposition. In his interpretation, Altabrizi
distinguishes first between the terms "potentiality" and "possibility" in the proposition. "Potentiality," according to him,
refers to something which does not yet exist but may come into
existence (cf. above n. 2). "Possibility" refers to something which
already exists but whose existence is conditioned by the existence
of a cause, so that the continuance of its existence is only possible.
Then he takes the expression כבר אפשר בעת מה שלא ימצא בפעל,
"may at some time *not exist in actuality*," to mean "may at some
time *cease to exist*," שיעדר בעת מה. On the basis of this interpretation, Altabrizi paraphrases the proposition as follows: Everything
that exists only potentially and, when it acquires actual existence,
its continuance of existence is only possible, may at some time
cease to exist.

Crescas criticizes this interpretation on two grounds: *First*,
the expression "and in whose essence there is a *certain* possibility"
cannot refer to the possibility of continuing to exist. *Second*, the
expression "may at some time *not exist in actuality*" cannot mean
"may at some time *cease to exist*."

My interpretation of Crescas' *second* criticism is based upon
the assumption that like his *first* criticism it is aimed at Altabrizi.
The obvious meaning of the *second* criticism, however, would seem

to imply that the interpretation under criticism takes the expression כבר אפשר בעת מה שלא ימצא בפעל in the sense of כבר אפשר בעת מה שלא יצא בפעל, "may at some time *not pass into* actual existence." But it seems to me unlikely that, after having aimed at Altabrizi's interpretation in his *first* criticism, Crescas' should aim at some unsponsored interpretation in his *second* criticism.

5. Maimonides own interpretation of this phrase in the proposition does not agree with the interpretation given here by Crescas. Cf. above n. 2.

6. The distinction drawn here by Crescas is the same as the distinction drawn by him in Prop. XVIII between the potentiality to act and the potentiality to be acted upon, i. e., between a potential agent and a potential patient.

7. Hebrew כבר אפשר בעת מה שלא ימצא בפעל, רוצה לומר שיהיה נעדר. The statement is rather vague. Its meaning may be made clear by the following considerations:

(1) The term העדר, according to Maimonides, applies both to absolute non-existence and to the absence of properties. Cf. *Moreh* III, 10.

(2) Then, again, the term העדר, as we have seen, means both "not to exist" and "to cease to exist." Cf. Prop. XIX, n. 3 (p. 683).

(3) Finally, form is the cause of the actual existence of anything. Without form matter has no actuality; it is pure privation.

Now, Crescas takes the expression שלא ימצא בפעל in the proposition as affirming that everything which contains a possibility within itself, i. e., matter, may be conceived as being without any form, inasmuch as none of its forms exist in it permanently, and thus it may be without actual existence (שיהיה נעדר).

A different interpretation of the proposition is given by Maimonides himself in his letter to Ibn Tibbon. "It is thus evident that everything that is potentially something else must not be actually that something else at some time, for a given piece of iron cannot be called potentially a sword unless it is not a sword at some time. Otherwise, its being a sword would not be potential but it would rather be actual all the time."

וכבר נתבאר כי כל אשר בכח דבר אחר בהכרח יהיה התואר ההוא נעדר בעת
מן העתים, כי זאת החתיכה של ברזל לא יאמר בה שהוא סייף בכח אלא כשלא
תהיה סייף עת אחת מן העתים. אמנם אם לא תסור לעולם מלהיות סייף, אינה
סייף בכח, אבל תהיה סייף בפועל לעולם ועד.

8. Hebrew עצם בעצם ההעדר סבת הוא המשתנה החמר כי. The term עצם
here is used in the sense of "corporeal substance." Cf. Prop.
XVI, Part II, n. 12.

Crescas' reasoning here reflects a statement by Maimonides in
which by a subtle change in the use of terms he seems to suggest
that matter is the cause of both "destruction," הפסד, فساد,
and "privation" העדר, عدم. Moreh III, 8: "All generated and
corruptible bodies are subject to destruction only through their
matter . . . The true nature of matter is such that it never
ceases to be associated with privation. It is for this reason that
matter does not retain permanently any single form but is always
taking off one form and putting on another." Cf. Prop. XIX, n. 3.
כל הגשמים ההוים הנפסדים לא ישיגם ההפסד רק מצד החמר שלהן... וטבע
החמר ואמתתו שהוא לעולם לא ימלט מחברת ההעדר, ומפני זה לא תתקיים בו צורה
אבל יפשיט צורה וילבש אחרת תמיד.

9. The passage to which Crescas refers reads as follows:
וזה לא ילך אל לא תכלית, ואי אפשר מבלתי הגיע אל מוציא מכח אל פעל
יהיה נמצא לעולם על עין אחד ואין בו כח כלל, רוצה לומר שלא יהיה בו בעצמו דבר
בכח, שאם היה בו בעצמו אפשרות. היה נעדר, כמו שנזכר
בשלשה ועשרים.

What Crescas means to say here is that the passage, quoted from
Maimonides' fourth proof for the existence of God, in which reference is made to Prop. XXIII, can be interpreted in conformity
with his own interpretation of that proposition.

Accordingly, the expression אפשרות בעצמו בו היה שאם in the
passage will be understood by Crescas as emphasizing the existence of the possibility *within the essence of the cause itself*, and
the expression נעדר היה will be understood by him in the sense of
remaining unrealized. The translation of the passage will therefore read as follows: "We must at last arrive at a cause of the
transition of an object from the state of potentiality to that of
actuality which exists always in the same state and in which there
is no potentiality at all, that is to say, in whose own essence there

is nothing potential, for *if there were any possibility in its own essence*, it might *remain unrealized*, as has been stated in the twenty-third proposition."

There is, however, nothing in the original text of that passage to exclude the other interpretations of the proposition. In fact both Altabrizi and Hillel of Verona, whose interpretations of the proposition differ from that of Crescas, refer to the same passage as an illustration of the use made by Maimonides of the proposition.

PART II

10. That is to say, if prime matter is identified with corporeal form, then matter is never without actual existence.

PROPOSITION XXIV

1. The Hebrew text of the propositions reads alike in Ibn Tibbon's translation of the *Moreh* and in Isaac ben Nathan's translation of Altabrizi.

2. Cf. Prop. XXIII, n. 8.

3. Hebrew ואם לא, לא היה הוא דבר אחד. That is to say, if there were no underlying actually existent substratum, every qualitative change would be the generation of something new, and it would thus be a change in substance. Cf. Prop. IV, n. 8 (p. 512), and Prop. X, Part I, n. 11 (p. 576).

Throughout this chapter there is a confusion of אחד and אחר in all the printed editions and manuscripts. But in the proposition itself there can be no doubt that the proper reading is אחד, for it represents the Arabic מא. I have therefore retained the same reading throughout the chapter.

It is not impossible that Crescas has taken the expression דבר אחד in the proposition to mean "one thing" as well as "a certain thing." Hence, the force of his argument here.

Most of the manuscripts read here (?) אחר (?) דבר אחד (?) היה הוא לא, ואם לא, in which case the last word is to be read אחר, and the passage is to be translated "for, were it not so, it would become another thing altogether."

4. Hebrew זנאר, زنجار.

5. The distinction drawn here by Crescas between the two appli-
cations of the term "possible" occurs in the following sources.

Hillel of Verona on Prop. XXIII: "The term potential is
applied in two ways. First, it is applied to a substance in which
something exists potentially. This is called 'the subject of the
potentiality.' Second, it is applied to a thing which exists potentially in a certain substance. This is called 'the potential' in the
true sense of the term. An example of the first kind is when we
say the seed is potentially a human form. An example of the
second kind is when we say that a human form exists potentially
in the seed."

יאמר גם הוא על שני פנים. האחד הוא עצם שיש בו דבר פלוני בכח, וזה נקרא
בעל כח. השני הוא הדבר הפלוני שהוא בעצם פלוני בכח, וזה נקרא מה ה בכח
באמת. המשל לראשון אמרנו יש בזרע צורת האנוש בכח. המשל לשני, אמרנו צורת
האנוש היא בזרע בכח.

Narboni on Prop. XXIV: "From this you may gather that the
term 'possible' may be applied in general to two kinds of things.
First, to that which receives, which may be named the sustaining
subject, and an example of this is prime matter, which is potential
with reference to form, and likewise body, which is potential with
reference to accidents. Second, to that which is received, which
may be named the material subject, and an example of this
is form [with reference to prime matter] or the accidents [with
reference to body]. The former is called potential with reference
to something else and is potential in a limited and relative sense.
The latter is called potential by its own essence and in an absolute
sense."

ותראה לך מזה כי האפשרי יאמר בכלל על שני מינים, על המקבל, והוא הנושא
המעמיד, והוא החמר הראשון, אשר הוא בכח אל הצורה, וכמי כן הגשם, אשר הוא
בכח אל המקרים, ויאמר על המקובל, והוא הנושא החמרי, והוא הצורה או המקרים.
הראשון יקרא בכח לדבר אחר, והוא בכח בקצת ובקשור, והשני יקרא בכח מצד
עצמו ובשלוח.

Averroes, Happalat ha-Happalah I, Fourth proof (Tahafut
al-Tahafut I, p. 32, l. 10; Destructio Destructionum I, p. 35rb, E;
Horten, p. 106, l. 27): "The possible is said both of that which
receives and of that which is received, or both of the subject and

that which inheres in the subject." ‏האפשר יאמר על המקבל והמקובל,‏
‏או הנושא והנשוא.‏

The same distinction is also implied in Altabrizi's distinction between the *first* and the *second* kind of transition from potentiality to actuality. See Prop. XVIII, notes 4 and 6.

In MSS. ‏ג, א, ב, p, ו, מ,‏ the text reads here ‏נשוא הנעדר‏ "non-existent predicate" instead of ‏נושא הנעדר‏ "non-existent subject." The former reading agrees with the expression ‏המקבל והמקובל או‏ ‏הנושא והנשוא‏ quoted above in this note from Averroes. The latter reading agrees with Narboni's expression ‏ויאמר על המקובל והוא‏ ‏הנושא החמרי‏ quoted also above in this note.

6. That is to say, the statement made in the Proposition that possibility must always inhere in matter is true only of what Crescas calls the possibility of an "existent subject" but not of what he calls the possibility of a "non-existent subject." See preceding note.

PROPOSITION XXV

1. The Hebrew text of the Proposition is taken from Isaac ben Nathan's translation of Altabrizi.

2. That is to say, in the process of generation and corruption which we observe in nature, the generation of a thing cannot be from absolute nothing but must be from something. Cf. Prop. X, Part I, n. 7 (p. 572).

3. *Physics* I, 5, 188a, 31–34: "In the first place, therefore, it must be assumed, that in the universality of things, nothing is naturally adapted to act casually upon anything; or be casually acted upon by anything, nor is anything disposed to be generated from anything, unless some one considers these things as taking place according to accident."

4. *Physics* I, 6, 189a, 34–189b, 1: "Hence, if some one should think that what is before asserted is true, and should also admit the truth of what is now said, it is necessary, if he wishes to preserve both assertions, that he should introduce a certain third thing as a subject to contraries."

Cf. *Metaphysics* XII, 1, 1069b, 3–9: "Sensible substance is changeable. Now if change proceeds from opposites or from intermediate points, and not from all opposites . . . but from the contrary, there must be something underlying which changes into the contrary state, for the contraries do not change. Further, something persists, but the contrary does not persist, there is, then, some third thing besides the contraries, viz. the matter."

5. Hebrew איש העצם. Cf. Prop. XVI, Part II, n. 12 (p. 667). Hillel of Verona in his commentary on this proposition explains the expression as referring to "an individual substance," עצם איש, which Aristotle designates as "primary substance," עצם ראשון, as distinguished from "universal substance," עצם כללי, or the genera and species, הסוגים או המינים, which Aristotle designates as "secondary substance" עצם שני. "Of substances there are two kinds, a primary substance and a secondary substance. Averroes in his commentary gives three reasons why the individual substance is more fit to be described as 'primary' than the universal, i. e., the generic or specific . . . Thus I have made known to thee what the Master has meant by the expression 'individual substance,' namely, that it refers to what is called by Aristotle 'primary substance.' "

ומן העצמים שני מינים, עצם ראשון ועצם שני. ואבן רשד אמר בפירושו שיותר
ראוי שיקרא ר א ש ו ן העצם ה א י ש י מן ה כ ל ל י, כלומר ה ס ו ג י או ה מ י נ י,
משלשה טעמים... ראה הודעתיך מה שרצה הרב באמרו ע צ ם א י ש י, בעבור
שהוא ר א ש ו ן אצל ארסטו.

The reference in Hillel of Verona's passage is to *Metaphysics* VII, 13, 1038b, 9–10: "For primary substance is that kind of substance which is peculiar to an individual." Aristotle, however, does not apply the expression 'secondary substance' to universals. He only denies that universals are substances. The term 'secondary,' however, is applied to them by Averroes.

6. Hebrew ואם היה ההעדר הקודם מן ההתחלות. Taken literally, the passage would seem to mean "though privation is the first of the principles." But, while it is true that in the enumeration of the three principles, privation, matter and form, the term 'privation' is usually mentioned first, it would be entirely pointless for Crescas to designate it as "the first of the principles." I therefore take the words ההעדר הקודם to stand by themselves as

an expression meaning "prior privation," that is to say, "privation which precedes form." As such an expression it is the equivalent of what Maimonides calls ההעדר המיוחד, אלעדם אלמכצוץ, "particular privation," by which is meant privation with reference to a certain form (*Moreh* I, 17; cf. Munk, *Guide* I, 17, p. 69, n. 1), as contrasted with "general privation," העדר כולל, i. e., the privation of all forms, and "absolute privation," העדר מוחלט, i. e., privation in the sense of non-being (cf. Shem-ṭob on *Moreh*, *loc. cit.*). Crescas' substitution of ההעדר הקודם for Maimonides' ההעדר המיוחד is due to the influence of Narboni in whose commentary on the *Moreh*, *loc. cit.*, the expression ההעדר המיוחד is paraphrased by ההעדר המיוחד הקודם לצורה המתהוה, "the particular privation which precedes the generated form."

7. *Physics* I, 7, 190b, 23–27: "The subject, however, is one in number, but two in species . . . But privation and contrary are accidents."

8. *Metaphysics* XII, 3, 1069b, 35–1070a, 2: "Next we must observe that neither the matter nor the form comes to be . . . For everything that changes is something and is changed by something and into something. That by which it is changed is the immediate mover (πρώτου κινοῦντος); that which is changed, the matter; that into which it is changed, the form."

The expression אלמחרך אלקריב, המניע הקרוב thus reflects the Greek πρῶτον κινοῦν in the preceding quotation, which otherwise, however, is translated by מניע ראשון, *prime mover*.

By the "immediate mover" Maimonides means here the celestial sphere, which is the source of every motion in the sublunar world. Cf. *Moreh* I, 72, and Hillel of Verona on this proposition.

9. Hebrew החומר לא יניע עצמותו. This statement is quoted from Maimonides' proposition where it is attributed to Aristotle. Cf. *Metaphysics* I, 3, 984a, 21–25: "For at least the substratum itself does not make itself change, e. g., neither the wood nor the bronze causes the change of either of them, nor does the wood manufacture a bed and the bronze a statue, but something else is the cause of the change." *Ibid.* XII, 6, 1071b, 28–30: "For how will there be movement, if there is no actual cause? Wood will surely not move itself—the carpenter's art must act on it." Cf. Munk, *Guide* II, p. 22, n. 5.

BIBLIOGRAPHY
AND
INDEXES

BIBLIOGRAPHY

I. Manuscripts and Editions of the Or Adonai

The text of the *Or Adonai* included in this work rests on the *editio princeps* of Ferrara, 1555, collated with eleven manuscripts. The rejected readings of the Ferrara edition are recorded in the critical notes together with the variant readings of the manuscripts. The variants found in the Vienna edition, 1859, are partly based upon the Vienna manuscript, which I have consulted directly, but in the greater part are the result of errors. Of the latter I have taken no notice. When in a few instances the readings of the Vienna edition are recorded, it is on the assumption that they represent readings of the Vienna manuscript which I may have overlooked. The Johannisburg edition, 1861, is a reprint of the Ferrara edition with some conjectural emendations on the part of the publisher. Of these I have taken no notice, although one of the emendations is discussed in the explanatory notes (p. 379). The first part of the propositions (*Ma'amar I, Kelal I*) printed with the commentary *Oẓar Ḥayyim* by H. J. Flensberg, Wilna, 1905–07, is likewise based upon the earlier editions with conjectural emendations by the editor. Of these, too, I have taken no notice.

In the critical notes I have recorded only such readings as I could check up at the time the text was prepared for publication. At that time, however, I had before me only three manuscripts in photostatic reproduction (MSS. ב, א, ג), whereas of the other eight manuscripts I had only a collection of variant readings copied in note-books. Consequently, whenever I decided to depart from the Ferrara edition and to record its reading in the critical notes, I had no way of assuring myself of the agreement between the rejected reading of the Ferrara edition and that of any of the eight manuscripts except the absence of any record to the contrary in my note-books. In such instances, which are comparatively few in number, rather than quote the manuscripts on the evidence of the silence of my note-books or else quote them with some query-mark, I thought it more advisable to omit them altogether and to record the reading in the name of the Ferrara edition only.

Neither the *editio princeps* nor any of the manuscripts seems to represent what may be considered a copy of an original definitive text. In fact, it may be doubted whether such a definitive text ever came from the hands of the author. The variants which are to be observed in the Ferrara edition and the manuscripts would seem to represent largely not so much corruptions of copyists as alternative tentative readings contained in the copies of the work made by students of Crescas to whom the *Or Adonai* was first delivered in the form of lectures and who participated in its composition (cf. above pp. 23, 29). The author's death which followed soon after the completion of the work precluded the possibility of a final revision and of the issuance of an authoritative text. On the basis of a colophon in the Jews' College manuscript Hirschfeld concludes that it is "probable that the MS. is an autograph of the author." But this manuscript, adorned with some notes by a student of Crescas, is with a few material exceptions (see, for instance, above p. 140, l. 14, and p. 338, n. 23; p. 180, l. 18; p. 352, l. 15), an exact duplicate of the Parma manuscript, and if both of them are not copies of a single manuscript, it would seem from internal evidence that the former is a copy of the latter. As for the colophon, see above p. 17, n. 61.

Some suggestions as to the relationship of the manuscripts are available. The Parma and the Jews' College manuscripts, as already mentioned, are of the same origin. The Paris and Vatican manuscripts have many readings in common. Occasionally they are followed by the Adler manuscript. In the same way there is a resemblance between the Bloch and Bamberger manuscripts. The Sulzberger manuscript comes nearer the Ferrara edition than any of the others. In four of the manuscripts, Sulzberger, Jews' College, Paris and Parma, there is an omission of an entire section in *Ma'amar III*, *Kelal I*, *Perek 4*, beginning with 'ואמנם הטענה הב and ending with the word preceding 'והמופת הד (Vienna edition, p. 66b, l. 41—p. 67b, l. 29).

The texts, arranged in the order in which I have consulted them, and the symbols by which they are designated in the critical notes, are as follows:

פ—Ferrara edition, 1555.

ש—Jewish Theological Seminary, New York, MS. Sulzberger. This consists of 246 folios, of which folios 197–246 (beginning early in *Perek 3* of *Ma'amar III*, *Kelal III*, Vienna edition, p. 73b, l. 4) are in a different hand. The first part of this manuscript is badly damaged by the corrosion of the ink, and of folios 93–129 only the margins are left.

מ—Munich. See M. Steinschneider, *Die hebräischen Handschriften der K. Hof-und Staatsbibliothek in München*, München, 1875, No. 301 (containing *Ma'amar I–II*) and No. 303 (containing *Ma'amar III–IV*).

ל—Jews' College, London. See H. Hirschfeld, *Descriptive Catalogue of the Hebrew MSS. of the Montefiore Library*, London, 1904, No. 281.

ז—Paris, Bibliothèque Nationale. See H. Zotenberg, *Catalogues des Manuscrits Hébreux et Samaritains de la Bibliothèque Impériale*, Paris, 1866, No. 737.

ו—Vienna. See A. Krafft und S. Deutsch, *Die handschriftlichen hebräischen Werke der k. k. Hofbibliothek zu Wien*, Wien, 1847, No. 78; A.Z. Schwarz, *Die hebräischen Handschriften der Nationalbibliothek in Wien*, Wien, 1925, No. 150.1.

ר—Rome, Vatican. See St. Ev. Assemanus et Jos. Sim. Assemanus, *Bibliothecae . . . Vaticanae Codd. MSS. Catal.* Rome, 1756, No. 261.

ד—De-Rossi Collection in Biblioteca Palatina, Parma. See *MSS. Codices hebraici Biblioth. I. B. De-Rossi*, Parma, 1803, III, p. 81, Cod. 1156; H. J. Michael, *Or ha-Ḥayyim*, Frankfurt a. M., 1891, p. 422.

ק—Oxford. See Ad. Neubauer, *Catalogue of the Hebrew Manuscripts in the Bodleian Library*, Oxford, 1886, No. 1351. 4; H. J. Michael, *Oẓerot Ḥayyim*, Hamburg, 1848, p. 33, No. 386. 4. This MS. ends with *Ma'amar I, Kelal III, Pereḳ 6.* In Neubauer this MS. is erroneously said to end with III, 6.

ב—Akademie für die Wissenschaft des Judentums, Berlin. Formerly. owned by Prof. Philipp Bloch.

א—Jewish Theological Seminary, New York, MS. Adler 1800. See *Catalogue of Hebrew Manuscripts in the Collection of Elkan Nathan Adler*, Cambridge, 1921, p. 55.

ג—Jewish Theological Seminary, New York, MS. Bamberger. "Written in beautiful Spanish characters in Lisbon, 20th of Shebat (Jan. 15), 1457, about half a century after the author's death, by a member of the famous Ibn Yaḥya family, Solomon b. David, for a Solomon b. Yeḥiel" (Prof. Alexander Marx in the Register of the Jewish Theological Seminary for 1928–1929, p. 139).

The MS. which once existed in Turin but is no longer extant is described in the following catalogues: Josephus Pasinus, *Codices Manuscripti Bibliothecae Regii Taurinensis Athenaei*, Taurini, 1749, p. 54, Codex CXLVI, a. v. 31; B. Peyron, *Codices Hebraici Manu Exarati Regiae Bibliothecae quae in Taurinensi Athenaeo Asservatur*, Taurini, 1880, p. 99, Codex CVII. A. 25; H. J. Michael, *Or ha-Ḥayyim*, p. 422. Cf. letter by A. Berliner to H. J. Flensberg in *Or Adonai* with *Oẓar Ḥayyim*, Wilna, 1905–07, p. 184.

The colophon of the Turin MS. is reproduced by Pasinus as follows: והיתה השלמה למאמרים בחדש זיו שנת ק"ע לפרט אלף השׁי ליצירה. The same reading is given by Michael. Peyron has ההשׁלמה למחבר instead of השׁלמה למאמרים and at the end of the colophon adds סרגוסא אשר במלכות ארגון. See above p. 17, n. 61.

II. MANUSCRIPTS AND EDITIONS OF WORKS CITED

This list, arranged alphabetically, contains only those works which are not adequately described when cited. They are entered here either by title or by author according as they happen to be referred to. A complete list of works cited will be found in the Index of Passages. The titles of Hebrew books, which are given throughout this work in transliterated form, are reproduced in Hebrew characters at the end of this list.

Albalag, Isaac, Commentary on Algazali's *Kawwanot* (*De'ot*) *ha-Pilosofim*. MS. Paris, Bibliothèque Nationale, 940. 3.

Albertus Magnus, Hebrew translation of his *Philosophia Pauperum*. MS. Cambridge University Library, Mm. 6. 32 (6).

Al-Najah, by Avicenna, published together with the *Kitab al-Kanon*, Rome, 1593.

Altabrizi, Commentary on Maimonides' twenty-five propositions. Isaac ben Nathan's translation, Venice, 1574; MS. Vienna (Krafft and Deutsch 74, Schwarz 150. 2). Anonymous translation, MS. Paris, Biblithèque Nationale, 974. 2.

Anonymous:

(1) Supercommentary on Averroes' *Intermediate Physics*. MS. Jewish Theological Seminary, New York, Adler 1744. 1.

(2) Supercommentary on Averroes' *Intermediate Physics*. MS. Jewish Theological Seminary, New York, Adler 1744. 2.

(3) Commentary on Averroes' *Epitome of the Physics*. MS. Bodleian 1387. 1. Neubauer describes it as on the "Large" commentary in the body of his Catalogue (p. 495) but as on the "paraphrase," i. e., Epitome, in the Index (p. 924). The latter is correct.

Aristotle, *Opera*, ed. I. Bekker, Berlin, 1831–1870. English translations: *Physics* by Thomas Taylor, London, 1812; *De Caelo* by Thomas Taylor, London, 1812, by J. L. Stocks, Oxford, 1922; *De Generatione et Corruptione* by H. H. Joachim, Oxford, 1922; *De Anima* by W. A. Hammond, London, 1902, by R. D. Hicks, Cambridge, 1907; *Metaphysics* by W. D. Ross, Oxford, 1908.

Avicenna, Commentary on *De Caelo*. MS. Cambridge University Library, Add. MS. 1197.

Azriel, *Perush 'Eser Sefirot* (= *'Ezrat Adonai*), ed. N. A. Goldberg, Berlin, 1850.

Bittul 'Ikkere ha-Nozerim, by Ḥasdai Crescas, ed. E. Deinard, Kearny, 1904.

Bruno, Giordano, *De l'Infinito Universo et Mondi*, in *Opere Italiane*, ed. P. de Lagarde, Gottinga, 1888; *De la Causa, Principio, et Uno, ibid.; De*

Immenso et Innumerabilibus, in *Opera Latina Conscripta*, I, 1–2, ed. F. Fiorentino, Neapoli, 1879–1884.

Cuzari, by Judah ha-Levi. Arabic and Hebrew texts, ed. H. Hirschfeld (German title: *Das Buch Al-Chazari*), Leipzig, 1887; Hebrew with commentary *Kol Yehudah* by Moscato and commentary *Oẓar Neḥmad*, Wilna, 1904.

Efodi. See *Moreh Nebukim*.

Emunah Ramah, by Abraham Ibn Daud, ed. S. Weil, Frankfurt a. M., 1852.

Emunot we-De'ot, by Saadia. Hebrew, with commentary *Shebil ha-Emunah* by Israel ha-Levi Kitqver, Yosefov, 1885. Arabic, *Kitâb al-Amânât wa'l-I'tiqâdât*, ed. S. Landauer, Leyden, 1880.

Epitomes of Aristotle's works by Averroes:

(1) *Epitome of the Topics*. נצוח in *Kol Meleket Higgayon, Riva di Trento, 1559*.

(2) *Epitome of the Sophistic Elenchi*, הטעאה, *ibid.* and MS. Bodleian 1352. 3 (included in the codex described in Neubauer's Catalogue as ספר הסבוא, *Isagoge*). My quotation follows the reading of the MS.

(3) *Epitome of the Physics*. Hebrew, *Kiẓẓure Ibn Roshd 'al Shema' Tibe'i le-Aristoteles*, Riva di Trento, 1560.

(4) *Epitome of the Meteorologica*. Hebrew. MS. Paris, Bibliothèque Nationale, 918.

(5) *Epitome of the Metaphysics*. Arabic, ed. Carlos Quirós Rodríguez, Madrid, 1919; Latin translation from the Hebrew, *Epitomes in Libros Metaphysicae*, in *Aristotelis omnia quae extant opera . . .* Venetiis, apud Iuntas, Vol. 8 (pp. 356–396), 1574; German translation by Max Horten, *Die Metaphysik des Averroes*, Halle, 1912; Spanish translation by Carlos Quirós Rodríguez, *Averroes Compendio de Metafísica*, Madrid, 1919; German translation by S. van den Bergh, *Die Epitome der Metaphysik des Averroes*, Leiden, 1924.

Fons Vitae, by Solomon Ibn Gabirol. *Avencebrolis Fons Vitae*, ed. C. Baeumker, Münster, 1895.

Gershon ben Solomon, *Sha'ar ha-Shamayim*, Roedelheim, 1801.

Gersonides' Supercommentaries on Averroes'

(1) *Intermediate Physics*. MS. Bodleian 1389; MS. Paris, Bibliothèque Nationale, 964. 1.

(2) *Epitome of the Physics*. MS. Paris, Bibliothèque Nationale, 962. 1.

(3) *Intermediate De Caelo*. MS. Paris, Bibliothèque Nationale, 919. 4; MS. Parma 805.

(4) *Epitome of De Caelo*. MS. Paris, Bibliothèque Nationale, 962. 2.

Happalat ha-Happalah. Hebrew translation of Averroes' *Tahafut al-Tahafut.* MS. Bodleian, 1354. Arabic original, Cairo, 1903. Latin translation from the Hebrew, *Destructio Destructionum,* in Aristotelis *omnia quae extant opera* . . . Venetiis, apud Iuntas, Vol. IX, 1573. Partly translated and partly paraphrased into German by M. Horten, *Die Hauptlehren des Averroes,* Bonn, 1913. See also *Happalat ha-Pilosofim.*

Happalat ha-Pilosofim. Hebrew translation of Algazali's *Tahafut al-Falasifah.* MS. Paris, Bibliothèque Nationale, 910. 1. Arabic original, Cairo, 1903. (The new edition by M. Bouyges, Beyrouth, 1927, was not available at the time this work was sent to the press). The first four "Disputations" are translated into French by Carra de Vaux (*Les Destruction des Philosophes*) in *Muséon,* 1899, 1900. The entire work is incorporated in Averroes' *Tahafut al-Tahafut.* See also *Happalat ha-Happalah.*

Hegyon ha-Nefesh, by Abraham bar Ḥiyya, ed. E. Freimann, Leipzig, 1860.

Hillel of Verona, Commentary on Maimonides twenty-five propositions, published together with *Tagmule ha-Nefesh,* ed. S. J. Halberstam, Lyck, 1874.

Ḥobot ha-Lebabot, by Baḥya Ibn Pakuda. Hebrew, Wilna edition. Arabic: *Al-Hidāja 'Ilā Farā'id Al-qulūb,* ed. A. S. Yahuda, Leyden, 1912.

'Iḳḳarim, by Joseph Albo, Wilna edition, with commentary *Eẓ Shatul* (divided into *Shorashim, 'Anafim,* and *'Alim*) by Gedaliah Lippschitz.

Intermediate Commentaries on Aristotle's works by Averroes:

(1) On the *Categories.* Hebrew. MS. Columbia University.

(2) On the *Physics.* Hebrew. Kalonymus ben Kalonymus' translation, MS. Paris, Bibliothèque Nationale, 938 (also 943); Zeraḥiha Gracian's translation, MS. Bodleian 1386. Latin translation from the Hebrew of Books I–III, in Aristotelis *omnia quae extant opera* . . . Venetiis, apud Iuntas, Vol. IV (pp. 434–456), 1574.

(3) On *De Caelo.* Hebrew. MS. Paris, Bibliothèque Nationale, 947. 1. Latin translation from the Hebrew, in Aristotelis *omnia quae extant opera* . . . Venetiis, apud Iuntas, Vol. V (pp. 272–326), 1574. In this translation the commentary is described as "Paraphrasis" instead of "Expositio Media."

(4) On *De Generatione et Corruptione.* Hebrew. MS. Paris, Bibliothèque Nationale, 939. 2.

(5) On *Meteorologica.* Hebrew. MS. Paris, Bibliothèque Nationale, 947.

(6) On *De Anima.* Hebrew. MS. Paris, Bibliothèque Nationale, 950. 2.

(7) On the *Metaphysics.* Hebrew. MS. Paris, Bibliothèque Nationale, 954.

Isaac Ibn Latif, *Rab Pe'alim*, ed. Samuel Schönblum, Lemberg, 1885.

Isaac ben Shem-ṭob:

(1) *First* supercommentary on Averroes' *Intermediate Physics*. MS. Trinity College Library, Cambridge, Cod. R. 8. 19. 3.

(2) *Second* supercommentary on Averroes' *Intermediate Physics*. MS. Munich, 45; MS. Cambridge University Library, Mm. 6. 25.

(3) *Third* supercommentary on Averroes' *Intermediate Physics*. MS. Trinity College Library, Cambridge, Cod. R. 8. 19. 2.

Joseph ben Judah Ibn Aknin, *Ma'amar R. Joseph ben Judah Ibn Aknin*.

(1) Hebrew text and German translation by Moritz Löwy (*Drei Abhandlungen von Josef b. Jehuda*), Berlin, 1879.

(2) Hebrew text with English translation by J. L. Magnes (*A Treatise as to . . . by Joseph Ibn Aknin*), Berlin, 1904.

Joseph Caspi, *'Amude Kesef*, ed. R. Kircheim, Frankfurt a. M., 1848.

Joseph Zabara, *Sepher Shaashuim*, ed. I. Davidson, New York, 1914.

Judah Messer Leon, Commentary on Averroes' *Intermediate Categories*. MS. Jewish Theological Seminary, Adler 1486.

Kawwanot ha-Pilosofim. Hebrew translation of Algazali's *Maḳaṣid al-Falasifah*. MS. Paris, Bibliothèque Nationale, 901; MSS. Jewish Theological Seminary, Adler 131, 398, 978, 1500.

Kitāb Ma'ânî al-Nafs, ed. Goldziher, Berlin, 1907; Hebrew translation, *Torat ha-Nefesh*, by I. Broydé, Paris, 1896.

Kol Meleket Higgayon, Niẓẓuaḥ. Averroes' *Epitome of the Topics* in the Hebrew translation of his *Epitome of the Organon*, Riva di Trento, 1559.

Ḳobeẓ Teshubot ha-Rambam we-Iggerotaw, Leipzig, 1859.

Liḳḳutim min Sefer Meḳor Ḥayyim. Hebrew version of Ibn Gabirol's *Fons Vitae*, in S. Munk's *Mélanges de Philosophie Juive et Arabe*, Paris, 1859.

Long Commentaries on Aristotle's works by Averroes:

(1) On the *Physics*. Hebrew. MS. Bodleian 1388. Latin translation from the Hebrew in Aristotelis *omnia quae extant opera* . . . Venetiis, apud Iuntas, Vol. IV, 1574.

(2) On the *Metaphysics*. Latin translation from the Hebrew, *ibid.*, Vol. VIII, 1574.

Ma'amar Yiḳḳawu ha-Mayyim, by Samuel Ibn Tibbon, Presburg, 1837.

Maḳaṣid al-Falasifah, by Algazali, Cairo, without date.

Megillat ha-Megalleh, by Abraham bar Ḥiyya, ed. Poznanski and Guttmann, Berlin, 1924.

Mif'alot Elohim, by Isaac Abravanel, Venice, 1592.

Milḥamot Adonai, by Gersonides, Leipzig, 1866.

Millot ha-Higgayon, by Maimonides, ed. D. Slucki, Warsaw, 1865.

Mishneh Torah, by Maimonides, Berlin, 1880.

Mizan al-'Amal, by Algazali, Cairo, A. H. 1328.

Moreh ha-Moreh, by Shem-ṭob Falaquera, Presburg, 1837.

Moreh Nebukim, by Maimonides. Hebrew: Samuel Ibn Tibbon's translation, with the commentaries of Efodi, Shem-ṭob, Abravanel, and Asher Crescas, Lemberg, 1866; Judah al-Ḥarizi's translation, ed. L. Schlossberg, 3 vols., London, 1851, 1876, 1879. Arabic and French by S. Munk, *Le Guide des Égarés*, 3 vols., Paris, 1856, 1861, 1866. English by M. Friedländer, 3 vols., London, 1881, 1885. Whenever possible I incorporated the phraseology of Friedländer's translation in my English translations of the passages from the *Moreh Nebukim* quoted in this work.

Moscato, *Ḳol Yehudah*. See *Cuzari*.

Moses ha-Lavi, *Ma'amar Elohi*. MS. Bodleian 1324. 5.

Mozene Zedek. Hebrew translation of Algazali's *Mizan al-'Amal*, ed. J. Goldenthal, Leipzig and Paris, 1839.

Narboni, Moses:

(1) Commentary on Maimonides' *Moreh Nebukim*, ed. J. Goldenthal, Vienna, 1852.

(2) Commentary on Algazali's *Kawwanot ha-Pilosofim (Maḳaṣid al-Falasifah)*. MS. Paris, Bibliothèque Nationale, 901.

(3) Commentary on Averroes' *Intermediate Physics*. MS. Paris, Bibliothèque Nationale, 967. 1.

(4) Commentary on Averroes' *Ma'amar be-E'zem ha-Galgal (Sermo de Substantia Orbis)*. MS. Paris, Bibliothèque Nationale, 918. 10.

Neveh Shalom, by Abraham Shalom, Venice, 1575.

'Olam Ḳaṭan, by Joseph Ibn Zaddiḳ, ed. S. Horovitz (German title: *Der Mikrokosmos des Josef Ibn Ṣaddiḳ*), Breslau, 1903.

Or Adonai, Vienna, 1859.

Pico della Mirandola, Giovanni Francesco, *Examen Doctrinae Vanitatis Gentium*, in *Opera Omnia*, Vol. II, Basel, 1573.

Plutarch, *De Placitis Philosophorum*, in *Scripta Moralia*, II, Paris, 1841. English: *Plutarch's Morals* by W. W. Goodwin, III, Boston, 1870.

Reshit Ḥokmah, by Shem-ṭob Falaquera, ed. M. David, Berlin, 1902.

Ruaḥ Ḥen, attributed to various authors, among them Jacob Anatolio, ed. D. Slucki, Warsaw, 1865.

Sefer ha-Bahir, pseudonymous, Wilna, 1883.

Sefer ha-Gedarim, by Menahem Bonafos, ed. I. Satanow, Berlin, 1798.

Sefer ha-Madda'. See *Mishneh Torah.*

Sefer ha-Shorashim, by Ibn Janaḥ, ed. W. Bacher, Berlin, 1897.

Sefer ha-Yesodot, by Isaac Israeli, ed. S. Fried (German title: *Das Buch über die Elemente*), Drohobycz, 1900.

Sermo de Substantia Orbis, by Averroes. Latin translation from the Hebrew *Ma'amar be-'Eẓem ha-Galgal* in Aristotelis *omnia quae extant opera* ... Venetiis, apud Iuntas, Vol. IX (pp. 3–14), 1573.

Shahrastani, *Kitab al-Milal wal-Niḥal,* ed. W. Cureton (English title: *Book of Religious and Philosophical Sects*), London, 1846. German translation by Th. Haarbrücker, Asch-Scharatâni's *Religionspartheien und Philosophenschulen,* Halle, 1850–1851.

Shamayim Hadashim, by Isaac Abravanel, Rödelheim, 1828.

Shebil ha-Emunah. See *Emunot we-De'ot.*

She'elot Saul, containing philosophic correspondence between Saul ha-Kohen Ashkenazi of Candia and Isaac Abravanel, Venice, 1574.

Shem-tob Ibn Shem-tob:

(1) Commentary on the *Moreh Nebukim,* Lemberg, 1866.

(2) Commentary on Averroes' *Intermediate Physics.* MS. Paris, Bibliothèque Nationale, 967. 4.

Tagmule ha-Nefesh, by Hillel ben Samuel of Verona, ed. S. J. Halberstam, Lyck, 1874.

Tahafut al-Falasifah. See *Happalat ha-Pilosofim.*

Tahafut al-Tahafut. See *Happalat ha-Happalah.*

Teshubot She'elot = Ma'amar Abu Hamid Algazali bi-Teshubat She'elot Nishe'al Mehem, ed. H. Malter (*Die Abhandlung des Abû Hâmid al-Gazzâlt. Antworten auf Fragen, die an ihn gerichtet wurden*), Frankfurt a. M. 1896.

Versor, Joannes, *She'elot Tibe'iyot,* Hebrew translation by Elijah Habillo of the *Quaestiones Physicarum,* MS. Paris, Bibliothèque Nationale, 1000.

Zeraḥiah Gracian, Commentary on Maimonides' twenty-five propositions. MS. Paris, Bibliothèque Nationale, 985. See also *Intermediate Commentaries.*

LIST OF HEBREW TITLES

הפלת ההפלה	בן קהלת	אגרת אל תהי כאבוחיך
הפלת הפלוסופים	בתי הנפש	אור ה'
		אמונה רמה
חובות הלבבות	דעות הפלוסופים	אמונות ודעות
יסודי התורה	הגיון הנפש	בטול עקרי הנוצרים

III. Selected list of Books, Articles and Other Items about Crescas

(Arranged in chronological order)

Johann Christoph Wolf, *Bibliothecae Hebraeae*, III, p. 274, Hamburg, 1727.

G. B. De-Rossi, *Bibliotheca Judaica Antichristiana*, pp. 24–25, 39–41, Parma, 1800.

G. B. De-Rossi, *Dizionario Storico degli Autori Ebrei e delle Loro Opere*, I, p. 192, Parma, 1802.

M. P. Jung, *Alphabetische Liste aller gelehrten Juden und Judeninnen, Patriarchen, Propheten und beruemten Rabbinen*, p. 101, Leipzig, 1817.

M. Steinschneider, *Catalogus Librorum Hebraeorum in Bibliotheca Bodleiana*, p. 841, Berlin, 1852–1860.

H. Graetz, *Geschichte der Juden*, VIII, pp. 98–101, 410–415, *et passim*, Leipzig, 1864. Hebrew translation by S. P. Rabinowitz, *Toledot 'Am Yisrael*, VI, pp. 92–96, 405–408, *et passim*, Warsaw, 1898.

M. Joël, *Don Chasdai Creskas' religionsphilosophische Lehren*, Breslau, 1866. Hebrew translation by Z. Har-Shefer, *Torat ha-Pilosofiyah shel Rabbi (Don) Ḥasdai Crescas*, Tel Aviv, 1928.

M. Joël, *Zur Genesis der Lehren Spinoza's*, Breslau, 1871.

Kalman Schulman, *Toledot Ḥakme Yisrael*, IV, pp. 38–43, Wilna, 1878.

Philipp Bloch, *Die Willensfreiheit von Chasdai Kreskas*, München, 1879.

Frederick Pollock, *Spinoza*, pp. 95–97, London, 1880.

Moritz Eisler, *Vorlesungen über die jüdische Philosophie des Mittelalters*, III, pp. 123–186, Wien, 1884.

Samuel Joseph Fuenn, "Ḥasdai Crescas," in *Keneset Yisrael*, pp. 373–374, Warsaw, 1886.

Gustav Karpeles, *Geschichte der jüdischen Literatur*, pp. 811–814, Berlin, 1886. Hebrew translation, *Toledot Safrut Yisrael*, pp. 467–470, Warsaw, 1888.

Isaac Hirsch Weiss, *Dor Dor we-Doreshaw*, V, pp. 142–148, Wien, 1891; *idem*, Wilna, 1904, V, pp. 138–144.

Heimann Joseph Michael, *Or ha-Ḥayyim*, pp. 420–423, Frankfurt a. M., 1891.

Philipp Bloch, "Die jüdische Religionsphilosophie," in Winter und Wünsche's *Die judische Litteratur*, III, pp. 776–786, Trier, 1894.

S. Bernfeld, *Da'at Elohim*, pp. 464–476, Warsaw, 1899.

J. Hamburger, *Real-Encyclopadie des Judentums*, III, v, pp. 44–45, Leipzig, 1900; III, vi, pp. 102–106, Leipzig, 1901.

E. G. Hirsch, "Crescas," in the *Jewish Encyclopedia*, IV, pp. 350–353, New York, 1903.

Julius Wolfsohn, *Der Einfluss Gazâlî's auf Chisdai Crescas*, Frankfurt a. M., 1905.

David Neumark, *'Iḳḳarim*, in *Oẓar ha-Yahadut, Ḥoberet le-Dugma*, pp. 69–71, Warsaw, 1906.

Ḥayyim Jeremiah Flensberg, *Sefer Or Adonai . . . 'im Perush . . . Oẓar Ḥayyim . . .* Part I (*Ma'amar I, Kelal I*), Wilna, 1905–1907.

David Neumark, "Crescas and Spinoza," in the *Year Book of the Central Conference of American Rabbis*, XVIII (1908), pp. 277–318. Separately printed, 1909. The same in Hebrew "Crescas u-Spinoza," in *he-'Atid*, II, pp. 1–28, Berlin, 1909.

S. von Dunin-Borkowski, *Der junge De Spinoza*, pp. 210–214, *et passim*, Münster, 1910.

Leo Niemcewitsch, *Crescas contra Maimonides*, Lublin, 1912.

M. Seligsohn, "Crescas," in *Oẓar Yisrael*, IX, pp. 237–238, New York, 1913.

A. Gurland, "Kreskas," in *Yevreiskaya Entziklopediya*, IX, pp. 846–849, St. Petersburg, (1910–1914).

714 · CRESCAS' CRITIQUE OF ARISTOTLE

Julius Guttmann, "Chasdai Creskas als Kritiker der aristotelischen Physik," in *Festschrift zum siebzigsten Geburtstage Jakob Guttmanns*, pp. 28–54, Leipzig, 1915.

Isaac Husik, *A History of Mediaeval Jewish Philosophy*, pp. 388–405, New York, 1916.

I. I. Efros, *The Problem of Space in Jewish Mediaeval Philosophy*, pp. 78–86, 103–105. New York, 1917. Reprinted from the *Jewish Quarterly Review*, New Series, VI, VII (1916).

H. A. Wolfson, "Crescas on the Problem of Divine Attributes," in the *Jewish Quarterly Review*, New Series, VII (1916), pp. 1–44, 175–221.

Pierre Duhem, *Le Système du Monde*, V, pp. 229–232, Paris, 1917.

David Neumark, *Toledot ha-'Ikkarim be-Yisrael*, II, pp. 163–165, Odessa, 1919.

H. A. Wolfson, "Note on Crescas' Definition of Time," in the *Jewish Quarterly Review*, New Series, X (1919), pp. 1–17.

Meyer Waxman, *The Philosophy of Don Hasdai Crescas*, New York, 1920. Reprinted from the *Jewish Quarterly Review*, New Series, VIII, IX, X (1918–1920).

I. Ginzburg, *Idishe Denker un Poeten in Mitel-Alter*, pp. 250–256, New York, 1919.

S. Wininger, "Crescas," in *Jüdische National-Biographie*, I, pp. 607–608, Cernăufi, 1925.

H. A. Wolfson, "Spinoza on the Infinity of Corporeal Substance, in *Chronicon Spinozanum*, IV (1924–1926), pp. 79–103.

Jakob Klatzkin, *Anthologiyah shel ha-Pilosofiyha ha-'Iberit*, pp. 158–166, 331, Berlin, 1926.

S. Dubnow, *Die Geschichte des jüdischen Volkes in Europa*, V, pp. 268–270, Berlin, 1927.

Armand Kaminka, "Crescas, Chasdai," in *Jüdisches Lexikon*, I, pp. 1449–1450, Berlin, 1927.

Julius Höxter, *Quellenbuch zur jüdischen Geschichte und Literatur*, II, pp. 99–103, Frankfurt a. M., 1928.

INDEXES

I. INDEX OF SUBJECTS AND NAMES

A

II. INDEX OF PASSAGES

This Index is subdivided into four sections: A Greek Authors. B. Arabic Authors. C. Jewish Authors. D. European Authors. In each section, the authors are arranged in chronological order. For MSS. and editions used in connection with the works mentioned, see Bibliography II. MSS. are designated by asterisks.

A. GREEK AUTHORS

735

B. ARABIC AUTHORS

C. JEWISH AUTHORS

D. EUROPEAN AUTHORS

III. Index of Terms

Only those terms are recorded here which happen to be discussed in the Notes.
The Hebrew part, however, includes a few terms gathered from the text and
from the passages translated in the Notes.

A. HEBREW

B. ARABIC

اتصال 579	تقسيم 324, 332	ذبول 399
احاد 485	التقسيم ضرورة 324	
استحالة 500	تكاثف 400	رسم 388
استدلال 560	من تلقايه 673, 674	
اصطلاح 592		زنجار 697
اصل 319	ثمّ 465	
اصول موضوعة 466		لا سبيل 327
اضطرارا 324	جدل 397	سطح 420
اضمحلال 399	جسم مطلق 580	سمت الراس 387
اقدم 629	جمع ـ اجمع على 329	
الاقدمون 321	جملة 483	شرح 325-326
اقناع 397, 396	جهد 560	شغل 361
الى اقناع النفس 397	جيب 387	شكل 688, 687
والله شهيد 601		
امتداد 639, 640	حال 577, 517	الصورة الجسمية 583
امكان 693	حاوى 358	
انفعال 388	حدّ 388	ضرورة 324
انّما 534-535	حركة 421	ضرورية 324
اول 629	حل 544	
اولية 465	حلزونى 623	طبق ـ اطبقنا 345
		طبيعة 583
	خارجا 422	عبارة 639
برهان 326, 397, 526	خرج ـ اخرجا 465	عدّ ـ يعدّ 419
بعد 590	خاصّية 568	عدم 683, 422, 695
بعض 491	خبر 320	العدم المخصوص 699
بقاء 639, 640		علام 447
بلغ 491, 492	خصّ 365	علوما متعارفة 465
بيان 325-326	خطّ 420	عين 666
وذلك ما اردنا بيانه 339	خطابة 397	
	خلاء 357	غاية 358
تابع 497		
تحليل 399	دائرة معدل النهار 636	فساد 695
تخلخل 400	دلّ ـ يستدلّون 560, 574	فصل 351
تركيب 399	دمغ 570	فضاء 357

D. LATIN

In this reprinting many misprints have been corrected and fundamental changes have been made on page 151, last line, and page 662, lines 3–8.

Printed in the United States
By Bookmasters